Integrated Technologies for Environmental Monitoring and Information Production

Proceedings of the NATO Advanced Research Workshop on
Integrated Technologies for Environmental Monitoring and Information Production
Marmaris, Turkey
10–16 September 2001

A C.I.P. Catalogue record for this book is available from the Library of Congress.

ISBN 1-4020-1398-1 (HB)
ISBN 1-4020-1399-X (PB)

Published by Kluwer Academic Publishers,
P.O. Box 17, 3300 AA Dordrecht, The Netherlands.

Sold and distributed in North, Central and South America
by Kluwer Academic Publishers,
101 Philip Drive, Norwell, MA 02061, U.S.A.

In all other countries, sold and distributed
by Kluwer Academic Publishers,
P.O. Box 322, 3300 AH Dordrecht, The Netherlands.

Printed on acid-free paper

All Rights Reserved
© 2003 Kluwer Academic Publishers
No part of this work may be reproduced, stored in a retrieval system, or transmitted in any form or by any means, electronic, mechanical, photocopying, microfilming, recording or otherwise, without written permission from the Publisher, with the exception of any material supplied specifically for the purpose of being entered and executed on a computer system, for exclusive use by the purchaser of the work.

Printed in the Netherlands.

TABLE OF CONTENTS

Part VIII. Remote Sensing and GIS

Part IX. Transfer of Data into Information

Part X. Conclusions and Recommendations

PREFACE

This book presents the proceedings and the outcomes of the NATO Advanced Research Workshop (ARW) on *Integrated Technologies for Environmental Monitoring and Information Production,* which was held in Marmaris, Turkey, between September 10-14, 2001. With the contribution of 45 experts from 20 different countries, the ARW has provided the opportunity to resolve the basic conflicts that tend to arise between different disciplines associated with environmental data management and to promote understanding between experts on an international and multidisciplinary basis.

The prevailing universal problem in environmental data management (EDM) systems is the significant incoherence between data collection procedures and the retrieval of information required by the users. This indicates the presence of problems still encountered in the realization of; (1) delineation of objectives, constraints, institutional aspects of EDM; (2) design of data collection networks; (3) statistical sampling; (4) physical sampling and presentation of data; (5) data processing and environmental databases; (6) reliability of data; (7) data analysis and transfer of data into information; and (8) data accessibility and data exchange at local, regional and global scales. Further problems stem from the lack of coherence between different disciplines involved in EDM, lack of coordination between responsible agencies on a country basis, and lack of coordination on an international level regarding the basic problems and relevant solutions that should be sought. Such difficulties in EDM constituted the basis for the ARW, which, through its contents, was intended as a response to the current needs in data management. This essentially was the follow-up meeting of the previously organized ARW of 1996 (NATO ARW.951456), which focused on "Integrated Approach to Environmental Data Management Systems". The basic objectives of both meetings were: (a) integration of different aspects of EDM (i.e., the points (1) through (8) stated above); (b) state-of-the-art evaluation of problems and solutions relevant to each aspect; (c) integration of views and expertise of scientists and professionals from different disciplines and different countries; (d) integration of interdisciplinary approaches to data collection and information retrieval.

The keyword in the designated objectives of both workshops (1996 and 2001) is integration. The basic questions considered in the 1996 ARW were "why" we need to integrate EDM systems and "what" we need to integrate. The main focus of the recent ARW was on "how" integration should be accomplished to meet the prevailing demand for reliable information systems. Two main reasons were specified to explain the needs for integration. First, the multidisciplinary, global or regional character of various programmes requires strengthening of collaboration between data management activities of different organizations in order to ensure proper co-ordination of environmental data flow, collection and archiving and to avoid duplication of efforts both on national and international levels. Second, the requirements for a significant leap forward in the capacity to handle environmental data is occurring at a time when computer and communication technology has made significant advances in terms of technical capability and connectivity. As it was stated in Agenda 21 of the UNCED in

Rio de Janerio in 1992 and further in World Summit on Rio+10 in Johannesburg in 2002, the priority activities for environmental management should include establishment and integration of existing data on physical, biological, demographic and user conditions into a database; maintenance of these databases as part of the assessment and management databases; promotion of exchange of data and information with a view to the development of standard intercalibrated procedures, measuring techniques, data storage and management capabilities. The problems that must be addressed today require interdisciplinary approaches and much more sharing of data and information than in the past.

In addition to the above basic statements, the workshop has also resulted in a set of conclusions and recommendations regarding each step (i.e., steps (1) through (8) above) of environmental data management. An important outcome of the meeting was the need expressed by all participants for "concrete actions" towards integration of data management systems. Possible "actions" were suggested to include the development of: (a) large-team project proposals to be supported by regional or international programmes; (b) a forum or web site for exchange of views on various problems related to EDM; and (c) an international society or association on integrated EDM. The final recommendation stated in the ARW was that a follow-up meeting should be organized to complete the integration of data management systems by delineating possible means and procedures and by following up on the proposed "concrete actions". It is recommended that these results be communicated to national, regional and international programmes, users and data centers. In this respect, this book is intended to convey the conclusions and recommendations of the Workshop to the relevant communities.

The organization of this book follows the basic framework of the ARW. The introductory section addresses the current status of environmental data management systems in view of emerging new problems. Parts 2 through 9 delineate the prevailing conditions and the basic problems encountered in consecutive stages of data management, starting with objectives, constraints and institutional aspects (Part 2) and followed by design of data collection networks (Part 3), statistical sampling (Part 4), physical sampling and presentation of data (Part 5), environmental databases (Part 6), data processing, analysis and modeling (Part 7), remote sensing and GIS (Part 8), and finally by transfer of data into information for environmental decision making (Part 9). These sections basically focus on the problems stated in points (1) through (8) above. Part 10 presents the conclusions arrived at the workshop, together with recommendations on how integrated technologies for environmental monitoring and information production should be developed into concrete actions.

The editors wish to thank the authors for their participation in the ARW and for their contributions. Acknowledgment goes to the Scientific Affairs Division of NATO, to the NATO Science Committee and to the financial supporters complementing the major NATO grant, in particular Dokuz Eylul University, Water Resources Management Research Center (SUMER, Izmir, Turkey); and four private engineering and environmental companies, namely Ege Fren, NTF, Aydıner and Tugal Cevre of Turkey. Acknowledgment goes also to FIRATLI TOURISM, the authorized travel agent of the meeting, who played a significant role in success of the organization.

All of the participants owe a debt of gratitude to the staff at the Mares Hotel (Marmaris, Turkey), who provided the ideal environment for realization of such an international meeting.

Particular gratitude is expressed to Gulay Onusluel (M. Sc.), Mr. Ali Gul (M.Sc.), Mr. Cem P. Cetinkaya (M.Sc.) for their efforts devoted to preparing and organizing the workshop and to Dr. Filiz Barbaros for her endless support devoted to the production of this volume. Their untiring effort and dedication to the Advanced Research Workshop made possible a successful organization of the meeting as well as the publication of the proceedings.

March, 2003

Nilgun B. Harmancioglu
Sevinc D. Ozkul
Okan Fistikoglu
Paul Geerders
Editors

List of Contributors

MICHAEL ABBOTT
Knowledge Engineering BVBA
Avenue Françoise Folie 28B28
1180 Brussels, BELGIUM
Telephone : +31 15 2151803
Fax : +32 2 3720516
e-mail : mba@ihe.nl

DAFINA L. DALBOKOVA
EH Information Systems (STP)
WHO European Centre for Environment & Health Bonn Office
Herman-Ehlers Strasse 10, 6th Fl.
D-53113 Bonn, GERMANY
Telephone : +49 228 2094 408
Fax : +49 228 2094 201
e-mail : dda@who.nl

FRANCOIS DE TROCH
Ghent University
Department of Forest & Water Management
Laboratory of Hydrology &Water Manag.
Coupure Links 653
B-9000 Ghent, BELGIUM
Telephone : 32-9-264 6136
Fax : 32-9-264 6236
e-mail : francois.detroch@rug.ac.be

KURT FEDRA
Environmental Software & Services GmbH
P.O.Box 100 A-2352
Gumpoldskirchen, AUSTRIA
Telephone : 43-2252-633 050
Fax : 43-2252-633 059
e-mail : kurt@ess.co.at

OKAN FISTIKOĞLU
Dokuz Eylul University
Faculty of Engineering
Tinaztepe Campus
Buca 35160 Izmir, TURKEY
Telephone : +90 232 453 1008/1075
Fax : +90 232 4531191 (or 453 4279)
e-mail : okan.fistikoglu@deu.edu.tr

RAZIA D.GAINUTDINOVA
Institute of Physics
National Academy of Sciences
Chui Prosp.265-A
Bishkek 720071 KYRGYZ REPUBLIC
Telephone : +996 312 24 36 61
Fax : +996 312 24 36 07
e-mail : eco2001@rambler.ru

PAUL GEERDERS
P.Geerders Consultancy
Kobaltpad 16, 3402 JL IJsselstein
THE NETHERLANDS
Telephone : 31-30-688 4942
Fax : 31-30-688 4942
e-mail : paul@pgcons.nl

NILGUN B. HARMANCIOGLU
Dokuz Eylul University
Faculty of Engineering
Department of Civil Engineering
Tinaztepe Campus
Buca 35160, Izmir, TURKEY
Telephone : +90 232 453 1008/1016
Fax : +90 232 4531191 (or 453 4279)
e-mail : nilgun.harmancioglu@deu.edu.tr

KARSTEN HAVNØ
Director, Water Resources Division
DHI Water&Environment
Agern Alle 11
DK-2970 Horsholm, DENMARK
Telephone : +45 45 169200
Fax : +45 45 169292
e-mail : kah@dhi.dk

KAZIMIR A. KARIMOV
Institute of Physics
National Academy of Sciences
Chui Prosp.265-A
Bishkek 720071 KYRGYZ REPUBLIC
Tel/fax : +996 312 24 36 61
Fax : +996 312 24 36 07
e-mail : kazimir@academy.aknet.kg

M. LEVENT KAVVAS
Hydrologic Research Laboratory
Department of Civil and Environmental Engineering
University of California
Davis, CA 95616 USA

Telephone : 1 530 753 9584
Fax : 530 753 9584
e-mail : mlkavvas@ucdavis.edu

IAN LITTLEWOOD
CEH, Centre for Ecology and Hydrology
Crowmarsh Gifford, OXON
Wallingford, Oxfordshire OX10 8BB
UNITED KINGDOM

Telephone : +44 (0) 1491 692 385
Fax : +44 (0) 1491 692 424
e-mail : igl@ceh.ac.uk

ELENI CHAROU – MARMARINOU
N.R.C. "DEMOKRITOS"
Inst. Of Info. & Telecom
153 10 Ag. Paraskevi/GREECE

Telephone : +301 650 3149
Fax : +301 653 2175
e-mail : exarou@iit.demokritos.gr

ANDREY MARTYNOV
Institute of Computational Mathematics & Mathematical Geophysics
Siberian Division of RAS
Lavrentieva av., 6, Novosibirsk 630090
RUSSIAN FEDERATION

Telephone : 7 3832 181514
Fax : 7 3832 207410
e-mail : martynov-andrey@hotmail.com

THOMAS MAURER
Head, Global Runoff Data Centre (GRDC)
Federal Institute of Hydrology (BfG)
PO Box 20 02 53, D-56002 Koblenz
Am Mainzer Tor 1
D-56068 Koblenz, GERMANY

Telephone : 49-261 1306 5224
Fax : 49-261 1306 5280
e-mail : Thomas.Maurer@bafg.de

VLADIMIR MELENTYEV
Nansen International Environmental and Remote Sensing Center (NIERSC)
197101 St.Petersburg, B. Monetyana Str., 26/28
RUSSIA

Telephone : 7 812 234 3924
Fax : 7 812 234 3865
e-mail : vladimir.melentyev@niersc.spb.ru

GIUSEPPE MENDICINO
Dipartimento di Difesa Del Suolo
Università Della Calabria, Cont. S. Antonello
87040 Montalto Uff. (CS), ITALY

Telephone : 39-984-934316
Fax : 39-984-934245
e-mail : menjoe@dds.unical.it

NICKOLAY N. MIKHAILOV
Head of the Oceanographic Data Centre
RIHMI-WDC, Russian Federal Service for Hydrometeorological and Env. Monitoring
6, Korolev Str., Obninsk
Kaluga Region 249035 RUSSIA

Telephone : 7-095-255 2393
Fax : 7-095-255 2225
e-mail : nodc@meteo.ru

BEYHAN OGUZ
Istanbul Technical University
Faculty of Civil Engineering
Hydraulics Division
80626 Maslak, Istanbul, TURKEY

Telephone : +90 212-2853725
Fax : +90 212-2856587
e-mail : emineb@itu.edu.tr

BIHRAT ONOZ
Istanbul Technical University
Faculty of Civil Engineering
Hydraulics Division
80626 Maslak, Istanbul, TURKEY

Telephone : +90 212-2853723
Fax : +90 212-2856587
e-mail : onoz@itu.edu.tr

SEVİNC ÖZKUL
Dokuz Eylul University
Faculty of Engineering
Department of Civil Engineering
Tinaztepe Campus
Buca 35160 Izmir, TURKEY
Telephone : +90 232 453 1008/1025
Fax : +90 232 4531191 (or 453 4279)
e-mail : sevinc.ozkul@deu.edu.tr

JOHN PACKMAN
CEH, Centre for Ecology and Hydrology
Crowmarsh Gifford, OXON
Wallingford, Oxfordshire OX10 8BB
UNITED KINGDOM
Telephone : +44 1491 692 416
Fax : + 44 1491 692 424
e-mail : jcp@ceh.ac.uk

JOHN PAPADIMITRAKIS
National Technical University of Athens
Dept. of Civil Eng., Div. of Water Resources
Hydraulics & Maritime Engineering
5, Heroon Polytechniou Str.
15780 Athens, GREECE
Telephone : 30-1-7722876
Fax : 30-1-7722877
e-mail : ypapadim@central.ntua.gr

GREG REED
Intergovernmental Oceanographic Commission
Ocean Services Unit, UNESCO-IOC
1, rue Miollis
75732 Paris Cedex 15, FRANCE
Telephone : 33 1 - 45 68 39 60
Fax : 33 1 - 45 68 58 12
e-mail : g.reed@unesco.org

MARIA A. SANTOS
Laboratorio Nacional de Engenharia Civil
Av. do Basil, 101
Lisboa, PORTUGAL
Telephone : 351-21-8443607
Fax : 351-21-8443016
e-mail : masantos@lnec.pt

GERT A. SCHULTZ
Lehrstuhl für Hydrologie
Wasserwirtschaft und Umwelttechnik
Ruhr-Universität Bochum
D- 44780 Bochum 1, F. R. GERMANY
Telephone : 49-234-32-24693
Fax : 49-234-32-14153
e-mail : Gert.A.Schultz@ruhr-uni-bochum.de

SERGEI V. SEMOVSKY
Limnological Institute SB RAS
PO Box 4199 Irkutsk
664033 RUSSIA
Telephone : +7 3952 466933
Fax : +7 3952 460405
e-mail : semovky@lin.irk.ru

MARSEL SHAIMARDANOV
Oceanographic Data Centre NODC
RIHMI-WDC, Russian Federal Service for Hydrometeorological and Env. Monitoring
6, Korolev Str., Obninsk
Kaluga Region 249035 RUSSIA
Telephone : 7-095-255 2393
Fax : 7-095-255 2225
e-mail : marsel@meteo.ru

VIJAY P. SINGH
Louisiana State University
Department of Civil and Environmental Eng.
Baton Rouge. LA 70803 - 6405. U.S.A.
Telephone : 1-225-578-6697
1-225-578-6588 (personal)
Fax : 1-225-578-8652
e-mail : cesing@lsu.edu

SERGEY.V. STANICHNY
Marine Hydrophysical Institute
2 Kapitanskaya Str.
Sevastopol, UKRAINE
Telephone : +380 692 545065
Fax : -
e-mail : stas@alpha.mhi.iuf.net

WITOLD G. STRUPCZEWSKI
Water Resources Department
Institute of Geophysics
Polish Academy of Sciences
ks. Janusza 64, 01-452 Warsaw, POLAND
Telephone : 48-22-7877422
48-22-6915853
Fax : 48-22-6915915
e-mail : wgs@igf.edu.pl

UNAL SORMAN
Middle East Technical University
Department of Civil Engineering
Water Resources Branch
06531 Ankara, TURKEY
Telephone : +90 312 2105442 / 5446
Fax : +90 312 2101002
e-mail : sorman@metu.edu.tr

GOKMEN TAYFUR
Izmir Institute of Technology
Civil Engineering Department
Urla 35437 Izmir, TURKEY
Telephone : + 90 232 498 6280
Fax : + 90 232 498 6505
e-mail : tayfur@likya.iyte.edu.tr

VLADIMIR L. VLADYMYROV
Scientific Liaison and Information Manag.
Officer Programme Coordination Unit
Caspian Environment Programme
Room 108, 3-d Entrance, Government Build.
40 Uzeir Gadjibekov St.
Baku 370016 AZERBAIJAN
Telephone : 994 12 938003
Fax : 994 12 971786
e-mail: Vladymyrov@caspian.in-baku.com

ALEXANDER VORONTSOV
Oceanographic Data Centre NODC
RIHMI-WDC, Russian Federal Service for
Hydrometeorological and Env. Monitoring
6, Korolev Str., Obninsk
Kaluga Region 249035 RUSSIA
Telephone : 7-095-255 2393
Fax : 7-095-255 2225
e-mail : vorv@meteo.ru

EUGENY VYAZILOV
Oceanographic Data Centre NODC
RIHMI-WDC, Russian Federal Service for
Hydrometeorological and Env. Monitoring
6, Korolev Str., Obninsk
Kaluga Region 249035 RUSSIA
Telephone : 7-095-255 2393
Fax : 7-095-255 2225
e-mail : vjaz@meteo.ru

PAUL H. WHITFIELD
Manager, Science Division
Meteorological Service of Canada
Pacific and Yukon Region
700-1200 West 73rd Avenue
Vancouver B.C., CANADA V6P 6H9
Telephone : 1-604-664-9238
Fax : -
e-mail : paul.whitfield@ec.gc.ca

Part I

INTRODUCTION

INTEGRATED DATA MANAGEMENT: WHERE ARE WE HEADED?

N. B. HARMANCIOGLU
Dokuz Eylul University, Faculty of Engineering
Tinaztepe Campus, Buca 35160 Izmir, Turkey

Abstract. Current requirements in environmental data management constituted the basis for two NATO Advanced Research Workshops (ARWs) held in 1996 and 2001, which, through their contents, were intended as a response to the current needs in data management. The 2001 ARW was essentially the follow-up meeting of the previously organized ARW of 1996, the results of which were published by Kluwer Publishers under the title "Integrated Approach to Environmental Data Management Systems". Before elaborating further on the topic, it is considered useful to review the basic conclusions and recommendations derived at the 1996 Workshop and to decide where we are actually headed in integrated data management. This introductory chapter presents an overview of these conclusions/recommendations as the point of departure for the remainder of the volume.

1. Point of Departure

The prevailing universal problem in environmental data management (EDM) systems is the significant incoherence between data collection procedures and the retrieval of information required by the users. In this regard, an integrated approach to EDM has become a necessity in recent years. Two main reasons can be specified to explain the needs for integration. First, the multidisciplinary, global or regional character of various programmes requires strengthening of collaboration between data management activities of different organizations in order to ensure proper co-ordination of environmental data flow, collection and archiving and to avoid duplication of efforts both on national and international levels. Second, the requirements for a significant leap forward in the capacity to handle environmental data is occurring at a time when computer and communication technology has made significant advances in terms of technical capability and connectivity.

As it was stated in Agenda 21 of the UNCED in Rio de Janerio in 1992, the priority activities for environmental management should include establishment and integration of existing data on physical, biological, demographic and user conditions into a database; maintenance of these databases as part of the assessment and management databases; promotion of exchange of data and information with a view to the development of standard intercalibrated procedures, measuring techniques, data storage and management capabilities. The problems that must be addressed today require interdisciplinary approaches and much more sharing of data and information than in the past [1].

Such requirements in EDM constituted the basis for two NATO Advanced Research Workshops (ARWs) held in 1996 and 2001, which, through their contents, were intended as a response to the current needs in data management. The 2001 ARW was essentially the follow-up meeting of the previously organized ARW of 1996, the results of which were published by Kluwer Publishers under the title "Integrated Approach to Environmental Data Management Systems" [2].

The keyword in the designated objectives of both workshops is integration. The basic questions considered in the 1996 ARW were "why" we need to integrate EDM systems and "what" we need to integrate. The main focus of the recent ARW was on "how" integration should be accomplished to meet the prevailing demand for reliable information systems. Essentially, both workshops were intended to establish the basics of integrated approaches and cover EDM systems in a comprehensive framework, i.e., covering all aspects of data management to include objectives & constraints, design of data collection networks, statistical and physical sampling, remote sensing and GIS, databases, reliability of data, data analysis, and transfer of data into information.

Before proceeding further with new contributions on the topic, it is considered useful to review the basic conclusions and recommendations derived at the 1996 Workshop and to decide where we are actually headed in integrated data management. This introductory chapter presents an overview of these conclusions/recommendations as the point of departure for the remainder of the volume.

2. Integrated Environmental Data Management

Integrated Environmental Data Management is concerned with providing an opportunity to draw together relevant data on a transient or permanent basis both within the same or across disciplinary boundaries so as to address through analyses, modeling or other means, environmental issues of local, regional, national or international interest or concern [3].

Within the framework of the above definition for integrated EDM, the two basic questions were addressed at the 1996 Workshop:

a) do we need to integrate environmental data management systems? If yes, then *why*?
b) *what* needs to be integrated?

The reasons underlying the need for integration were described as the following:

a) There is a significant gap between information needs on the environment and information produced by current systems of data collection and management.
b) Various programmes on environmental management, e.g., World Climate (WRCP) and Geosphere-Biosphere (IGBP) Programmes, Cooperative Programme for Monitoring and Evaluation of the Long-Range Transmission of Air Pollutants in Europe (EMEP), Global Environmental Facility (GEF), United Nations Environment Programme (UNEP), World Weather Watch (WWW), and the similar, have a multidisciplinary regional or global character. They need strengthening of collaboration between data management activities of different organizations to ensure proper coordination of environmental data collection, data flow, and archiving and to avoid duplication of efforts on both national and international levels. Such collaboration can only be realized by integrated approaches to data management.

c) The solution to environmental problems often requires data exchange at local, national, and global (international) levels. Such an exchange may be needed for: (1) data of the same type, e.g., water quality data collected by different methods; (2) data of different types of one discipline, e.g., marine physical, chemical, biological, and other oceanographic data types; and (3) data of different disciplines, e.g., oceanographic, meteorological, geophysical, or demographic data.
d) Agenda 21 has emphasized that the priority activities for environmental management should include establishment and integration of existing data on physical, biological, demographic and user conditions into a database; maintenance of these databases as part of the assessment and management databases; promotion of exchange of data and information with a view to the development of standard intercalibrated procedures, measuring techniques, data storage, and management capabilities.

The reply to the second question comprised the integration of water quality data on the European scale; integration of hydrological and meteorological data to assess climate change; integration of data management activities for environmental management; integration of data for use in a series of models; institutional integration for data management, etc.

There are at least three levels of data integration: data of the same type (e.g., water quality data collected by different methods) into an integrated data set; data of different types of one discipline (e.g., marine physical, chemical, biological, and other oceanographic data types) into a comprehensive data bank; and data of different disciplines (e.g., oceanographic, meteorological, geophysical, or demographic data) for modeling and decision making purposes.

The new question to be addressed by contributions in this volume can now be formulated: What actions in data and information management should be taken to make integration possible and effective; stated another way, *how* should integration be realized?

The conclusions and recommendations of the 1996 Workshop, which are summarized in the following section [3], provide the basic inputs to the resolution of the above question.

3. Conclusions and Recommendations of the 1996 Workshop

3.1. OBJECTIVES, CONSTRAINTS AND INSTITUTIONAL ASPECTS OF ENVIRONMENTAL DATA MANAGEMENT

3.1.1. *Basic Conclusions*

With regard to objectives, constraints and institutional aspects, the following points were considered significant:

a) principles and value of the international agreements on free and open access to environmental data must be reconfirmed as being critical to the integrated approach to data management systems;
b) the principle that environmental data is a public good must be recognized;

c) the development of integrated environmental data management systems should be based on the experiences gained by international organizations and programmes and be coordinated with them as much as possible;
d) cooperation between different environmental agencies and institutions should be achieved at all levels of policy-making: national, regional and international. These cannot be achieved without scientifically sound, well-defined and robust agreements on common definitions, standards and protocols. This may imply institutional changes;
e) recognition that the institutional, political, and economic framework in which integration will take place will be characterized by:
   - devolution and increased regionalization,
   - possibility and potential danger of privatization of data collection, storage and management (forecasting),
   - reduced budget so that more information is expected from less data;
f) the principle of integration from national to international levels provides economic and environmental benefits at both levels and requires feedback in both directions;
g) success in integration of data management systems will not be achieved without:
   - associated development of technical infrastructure and expertise (hardware, software, and training),
   - possibilities of exchange of experience (visits, news groups, etc.);
h) the required integration cannot be achieved without close links between all data collectors, managers and researchers;
i) while environmental management traditionally included only considerations related to the natural environment, increasingly, the influences of economic and sociological developments need to be taken into account. This allows for a quantification in economic terms of the effects of these factors, which is an essential piece of data for environmental management. To facilitate the processing of socioeconomic data in relation with data from the natural environment, it is suggested to broaden the scope of the term "environment" to include socioeconomic variables and parameters;
j) as an essential element for achieving the required integration, international and interdisciplinary joint working bodies, meetings, and conferences should be implemented.

3.1.2. *Additional Comments*

In addition to the basic conclusions and recommendations stated above, additional comments made at the Workshop are worthy of note. These are summarized in the following:

a) The question of environmental management is a public welfare question. Therefore, the objectives, the institutional framework, and organizational structures in data management are public policy issues belonging to the realm of welfare economics but are not internal to the scientific community. It is on this ground that we need to address the integration of environmental data management systems.
b) Regarding constraints, the threats and possibilities of budget cuts for environmental monitoring are very real. Those involved in monitoring should always try to justify

the importance of their monitoring not only to policy makers (who control the budget) but also for themselves. For this purpose, a clear definition and documentation of the objectives of monitoring are necessary. The use of the collected data for meeting the objectives should also be defined, i.e., how the collected data meet the information needs or questions being asked.

c) Collection of data and establishing a database are expensive activities. If it is indeed useful or important to collect data, and if there is a desire in many different organizations to have access to data, it may be possible to think of data as a commercial product. If so, the organizations collecting and managing data could sell them to interested users at a nominal cost. The profits could be used to maintain data collection and storage activities. Furthermore, by thinking of data as a product that consumers would buy, such activities can be tuned to meet consumers' desires, i.e., there will be more emphasis on collecting what people want to use.
d) Integration of data management is not a static procedure; it has an adaptive nature since new environmental problems are emerging, which require new types of data to be collected.
e) It is often stated that objectives of a data collection system have to be specified and then, the design and/or operation of the system should be optimized in view of the objectives. This approach may be questioned. Since a data collection system designed today has to function for several decades and since the principle of sustainable development requires consideration of the needs of future generations, the objectives that the current systems have to meet are almost impossible to specify since the objectives of future generations are not known. The only possibility seems to be to define today's objectives, to try to anticipate potential objectives of generations to come, and to define objectives of a data collection system on the basis of both. The problem remains, however, that we cannot clearly anticipate those needs of future generations as our predecessors did not anticipate the high relevance water quality monitoring assumed today.
f) Although data collection networks are presently implemented in many countries of the world, we are faced with the constraint that there will always remain remote areas, e.g., Siberia, Sahara, Central Australia, etc., where data are not and will not be collected in the future. This is a problem since there is a growing interest in global data sets in order to understand and quantify global changes. Some approaches, e.g., the one adopted by the "Global Energy and Water Cycle Experiment GEWEX", may be applicable for generation of environmental data sets in remote regions.

3.2. DESIGN OF DATA COLLECTION NETWORKS

3.2.1. *Conclusions*

The basic conclusions stated on the design of data collection networks are the following:

a) *Environmental data networks can benefit from integrated approaches to their design.* There are both philosophical and pragmatic reasons for the integration of environmental data networks across various environmental phenomena.

The philosophical basis for this conclusion is that environmental processes are interdependent in nature. Thus, if one wants to understand any particular aspect of the environment, the data describing the web of processes whose interactions influence that aspect must be studied to attain adequate understanding.

From a pragmatic point of view, integration of environmental data networks makes sense because the interdependencies of the environmental processes permit information transfer among the processes. Thus, synergy and cost-effectiveness can result from integrated data networks.

b) *Design of data networks should be based on the purposes for which the data are to be collected.* There are many purposes for the collection of environmental data, and thus many network design tools are required. However, multipurpose networks are difficult to design rationally, so an approach that permits interactive designs of single purpose networks is the most feasible means of performing integrated design.

c) *A taxonomy of environmental data network purposes is useful in developing a strategy for integrated network design.* It is decided at the Workshop that the use of the following taxonomy for the classification of network design purposes could highlight commonalities among network design technologies that would facilitate their use under a more robust set of situations:

I. Decision support networks;
II. Academic-curiosity networks;
III. Contingency networks.

A decision support network has an explicit purpose that results in specific types of data being collected in specific locations, at specific times, with the use of specific data collection technologies to provide the most cost-effective information for decision making.

An academic-curiosity network ideally is designed to test a hypothesis about one or more environmental phenomena. The demands for specific locations of data collection, the frequencies of data collection, and the data collection technologies employed for such networks usually are not as rigid as for decision support networks.

Contingency networks are designed to ameliorate the impacts of not being able to forecast perfectly the future demands for environmental information. Such networks serve as insurance against unanticipated information needs, and they have designs that are less sensitive to location, frequency, and technology than do networks with other purposes.

d) *Basic understanding of environmental phenomena is the starting point for the design of environmental data networks.* Knowledge of the phenomena of interest is required to select an appropriate suite of network design tools. The choice of the actual tool or tools to be used for the design should be based on any existing data from the region of interest.

e) *Feedback from data collected in the initial network permits more complete descriptions of the environmental phenomena and the subsequent use of more complex approaches to redesign the network.* Knowledge and information gained from an environmental data network can be used for improvement of the network.

f) *Network design is but one link in an integrated environmental data management chain, and it must be harmonized with the constraints and opportunities provided by the complementary links.* The design of data networks should not be performed in isolation from the technologies that will be used to convert the data to environmental information.

g) *There currently is a paucity of robust technologies for the design of environmental data networks, and technology transfer for the existing technologies is not being carried out satisfactorily on an international scale.* Because of the great interest in the environment that exists today, there is a large investment internationally in the collection of environmental data. With the lack of adequate network design support, many of the data collection programmes probably are not being conducted in a cost-effective manner.

3.2.2. *Recommendations*
The following points are recommended for the design of monitoring networks:

a) Environmental data networks should be designed and operated in an integrated manner to take advantage of the informational synergies that exist among environmental phenomena.
b) Environmental data networks should be redesigned periodically to incorporate the new knowledge that is contained in the added data.
c) The development of more robust technologies for the design of environmental data networks should be supported by international environmental agencies.
d) New vehicles for the transfer of the technologies of data network analysis and design should be sought and implemented as they are demonstrated to be effective.

3.3. PHYSICAL SAMPLING AND PRESENTATION OF DATA

3.3.1. *Conclusions*
The conclusions derived under this topic are summarized separately for physical sampling, remote sensing, and GIS.

Physical Sampling

a) There should be standardized protocols for sampling procedures of water quality, air pollution, and solid waste. There seems to be a discrepancy between water quantity sampling, where standards are available (e.g., ISO regulations), and measurements of environmental parameters.
b) There should be standardized protocols for the preservation of samples for the time between sampling itself and the analysis of samples.
c) There should be a clear verification of sample analysis results, possibly including intercomparison of results.
d) Methods of presentation of results should be such that not only the values themselves are given but also the relevant statistics, e.g. the number of samples, mean, variance, confidence limits, etc., should be presented.

Remote Sensing

a) Remote sensing allows for retrieval of area information instead of point measurements of some parameters relevant to environmental and hydrologic monitoring and management.

b) Remote sensing has the potential of global coverage, i.e., it provides information also on remote areas, where no measurement can be acquired.
c) Remote sensing allows to acquire information on state variables relevant to environmental modeling, e.g., soil moisture status, snow cover, etc. in hydrological modeling.
d) New models should be developed such that they can use remote sensing data as state variables (initial and intermediate) as input and for model parameter estimation.
e) Deficiencies of remote sensing applications in relation to environmental data collection are:
   - problems with the resolution in space and time;
   - available spectral bands;
   - availability and costs of remote sensing data;
   - conversion of electromagnetic signals into parameters relevant to environmental monitoring.

Geographic Information Systems (GIS)
GIS is nowadays well developed and available for presentation, visualization and processing of spatially distributed data sets. There are also some deficiencies of GIS, and these are particularly relevant to real-time applications, i.e., when processing rapidly varying dynamic variables.

3.3.2. *Recommendations*

The following recommendations are made regarding physical sampling and presentation of data:

a) It is suggested that different spatial data sets should be integrated within GIS in order to make them compatible.
b) Cooperation is required between institutions performing physical sampling and analysis on the one hand and those dealing with remote sensing data on the other hand in order to verify both approaches.
c) Space agencies should be aware of the needs of the environmental community when setting up new programs.
d) The concept and practical use of remote sensing and GIS should be incorporated into educational curricula. International organizations should support such educational programmes.

3.4. STATISTICAL SAMPLING

The following basic points are made with respect to statistical sampling:

a) Statistical sampling is connected to the objectives of sampling as well as to those of network design. Basic considerations are sampling in time and space. Accordingly, locations of sampling sites, time interval between observations, cost-effectiveness in sampling and flexibility of the network are factors that have direct effects on statistical sampling.
b) The hydrological environment can be partitioned into different zones, i.e., surface zone (terrestrial, lakes, reservoirs, etc.), subsurface zone (vadose zone and

groundwater), wetlands, marine and coastal environments. The sampling schemes and design of sampling programs depend on where we sample the data in this environment.

c) Sampling is carried out to delineate both temporal and spatial variability. Selection of sampling frequencies in space and time depends on cost-effectiveness.
d) Emphasis should be put on errors of sampling in the form of instrumental failures, human errors, or laboratory analysis errors. A statement has to be made on the relative size of the errors that we can expect in sampling as well as on the type of the error, e.g. bias and randomness.
e) The effect of this error on the ultimate decision making process has to be quantified. Thus, the risk and reliability of not only the data but also those of the statistical analysis should be identified.

3.5. ENVIRONMENTAL DATABASES, DATA PROCESSING, RELIABILITY OF DATA

The following conclusions and recommendations are made regarding databases, data processing and reliability of data:

a) Access to hardware over the next 5 to 10 years will increase, and there should be no serious problems for most organizations.
b) There are good software systems for the prime archives, but there is a problem with the intermediate software connecting the sensor, data logger, and chemical analyses in the lab computer to the prime archive.
c) Documentation is essential to enable the data to be quality controlled.
d) While there are good hardware and software systems available, they still tend to be expert focused, and there is either the need for a major effort in training or the development of more user-friendly systems.
e) Regarding data validation:
   - expert validation (including lab/site comparison studies) is critically important;
   - there may be the need for standard validation procedures for each discipline;
   - integrating databases within the same environmental field may be an additional validation check.
f) Knowledge of existing environmental data is important at regional, national and international levels. More effort is needed to collect metadata. This initiative should be at a national level.
g) Data archives will continue to exist in a wide range of formats. The stress should be on the more widespread use of interface software to merge them into shell/application systems. New software developments at interface are needed.
h) Ideally all environmental data should be available (not necessarily free) as soon as they are validated and archived; but it has to be recognized that there may be local, regional or national sensitivities.
i) Requirements will vary between nations but, in general, there is no need for massive environmental databases to be held centrally. Systems are however needed that can call on a wide variety of dispersed databases when required.

j) Integration of databases is difficult without an appropriate institutional structure in place.
k) It is recognized that a number of international organizations have taken the lead in developing and publishing protocols (e.g., WMO, UNESCO, EA, etc.). However, these protocols are generally not given sufficient publicity or widespread dissemination.
l) Much environmental data collected at a national level is now of international importance in addressing global issues such as climate change, loss of biodiversity, atmospheric or marine pollution, etc. Nations should be prepared to make their key data sets more widely available possibly through the international agencies.
m) Because of the different spatial and temporal scales, it may be necessary to take different approaches to data integration in the different environmental sectors, e.g., the atmosphere, terrestrial, ecology, freshwater science, marine sciences, etc.
n) Validation procedures in relation to sensor location and operation, data transmission and storage, and data presentation are essential if good quality information is required.
o) The growth of computer technology and GIS systems is enabling comparisons of environmental data sets to be made in a way that was impossible five years ago. This is leading to new insights into many areas of environmental science. There is, however, a danger in that the images from a GIS are so powerful and seductive that they may mask an inadequate database(s).
p) There is great value in collecting a wide range of environmental variables at a number of carefully chosen sites so that comprehensive Environmental Change Networks can be established. The use of such networks is spreading worldwide.
q) It is invaluable to have close links between personnel who are collecting and managing the environmental data resources and the communities who are using such data for operational or research purposes.

3.6. DATA ANALYSIS AND TRANSFER OF DATA INTO INFORMATION

The major points made, regarding data analysis and transfer of data into information, are the following:

a) The difference between data and information can be explained as: data only need a collector whereas information is defined by the content of data, which is meaningful to the user. Information exists if it is useful to some audience or to decision makers in the most general and inclusive sense.
b) This information must have a number of properties. First, it has to be *timely*; that is information must be there when it is needed. Further, it has to be *accurate* and *precise*; otherwise, it is not useful information. It has to be *easy to understand* and must come in a *format* which meets the expectations and the capability of the specific audience who uses it. Context or context-rich information to allow or to ease interpretation is another important aspect. Finally, information has to be easily accessible. This last point brings about the question whether data access should be free or not.

c) There are a good number of well-established information products. However, there may be cases where more demand or requirement is put on expected information; in this case, planning an analysis scheme may be difficult. The requirement is that information products have to be well documented and well structured.
d) Data management systems have to support flexible exploitation and flexible transfer of information, which are subject to both technical and institutional aspects. This is another issue to be considered in future meetings.

3.7. ADDITIONAL REMARKS ON ENVIRONMENTAL DATA MANAGEMENT SYSTEMS

The following additional comments were made at the Workshop, regarding different levels of data management:

a) With respect to data processing and data exchange, another significant issue is monitoring of data flow, which means to know, throughout the flow process, where the data are at each time point, in what state they are, whether they are accessible or not, etc. In this respect, it is important that data are moved from the field where they are collected into a digitized form as quickly as possible. The basic principle here should be to transfer data from sensors, field and laboratories, put them through the validation procedures as quickly as possible, and get them firmly secured within databases.
b) Most often, data users' complaints are related to *lack of information*. Reversing the picture, we should also ask if we are using the available data or information in the most proper way. An important consequence of this question is the need to integrate available information with the decision making process. In this respect, it is not only the data managers but also the decision makers who should be involved in the process of information production and utilization.
c) With respect to availability and accessibility of current data sets, it appears that we are often unaware of the existing databases in different parts of the world. Thus, it may be recommended to try and compile an inventory of relevant environmental databases.
d) Regarding accessibility of data, another major question is whether environmental data should be made available free of charge. Two points may be made on this issue. First, a decision has to be made on what kinds of data could be or should be free of charge since some environmental data may be restricted and cannot be accessed free of charge. Second, open access should apply only to validated data.
e) Standards exist for different levels of the data management process. However, the important issue in this respect is to develop an agreement on which standards to apply, e.g., in the case of transboundary monitoring programs. Regarding standards, organizations like WMO have developed standards for hydrological measurements including water quality. However, one of the problems with international organizations is that they do not advertise the fact that such information is available. There is a need to encourage these organizations to disseminate information about the availability of standards to a much wider audience. Another point is that the only standards which are dependable and acceptable are those related to water,

particularly to water quantity. In case of air or solid waste monitoring, there is a lack of universally accepted standards.

It was agreed at the 1996 Workshop that integrated management of environmental data is definitely needed and that international and national authorities should spend every effort to realize such integration. Cooperation of scientists from different parts of the world is considered essential for achieving this goal. In this respect, two activities were recommended:

a) integration has to be attained via close links between data collectors, managers and researchers;
b) as an essential element for realizing the required integration, international, interdisciplinary joint working bodies, meetings, and conferences should be implemented.

4. Current Requirements on Integrated Data Management Systems

Agenda 21 of UNCED has officially stated the new outlook towards environmental management, namely that the environment should be managed by an integrated approach in respect of sustainability. It was further emphasized in Agenda 21 that effective management relies essentially on reliable and adequate information on how the environment behaves under natural and man-made impacts. Yet, Agenda 21 and several other similar reports have also recognized that current systems of information production, i.e., data management systems, do not fulfill the requirements of environmental management and decision making. This is a highly unfortunate situation in view of the rapidly growing environmental problems. At a time when we need informational support the most, we find that our data management systems experience a declining trend. Recognition of this trend has brought focus to current monitoring systems, databases, and data use. Accordingly, major efforts have been initiated at regional and international levels to improve the status of existing information systems [1].

The significance of informed decision making was readdressed at the World Summit on Sustainable Development, held in Johannesburg in September 2002. In essence, it was emphasized at the recent Summit that " The United Nations Conference on Environment and Development (UNCED), held in Rio de Janeiro in 1992, provided the fundamental principles and the programme of action for achieving sustainable development" [4]. Commitment was reaffirmed to the Rio principles, the full implementation of Agenda 21, and the Programme for the Further Implementation of Agenda 21.

With respect to environmental information, the Johannesburg Summit foresaw "urgent actions at all levels to ... provide information more effectively" (paragraph 99(a)) and further emphasized the following [4]:

- Use information and communication technologies, where appropriate, as tools to increase the frequency of communication and the sharing of experience and knowledge, and to improve the quality of and access to information and communications in all countries, ... (paragraph 106);

- Ensure access, at the national level, to environmental information and judicial and administrative proceedings in environmental matters, as well as public participation in decision making … (paragraph 119.ter);
- Promote the development and wider use of earth observation technologies, including satellite remote sensing, global mapping and geographic information systems, to collect quality data on environmental impacts, land use and land-use changes, including through urgent actions at all levels to (paragraph 119. septies):
  (a) Strengthen cooperation and coordination among global observing systems and research programmes for integrated global observations, taking into account the need for building capacity and sharing of data from ground-based observations, satellite remote sensing and other sources among all countries;
  (b) Develop information systems that make the sharing of valuable data possible, including the active exchange of Earth observation data;
  (c) Encourage initiatives and partnerships for global mapping.
- Support countries, particularly developing countries, in their national efforts to (paragraph 119.octies):
  (a) Collect data that are accurate, long-term, consistent and reliable;
  (b) Use satellite and remote-sensing technologies for data collection and further improvement of ground-based observations;
  (c) Access, explore and use geographic information by utilizing the technologies of satellite remote sensing, satellite global positioning, mapping and geographic information systems.
- Support efforts to prevent and mitigate the impacts of natural disasters, including through urgent actions at all levels to (paragraph 119.noviens):
  (a) Provide affordable access to disaster-related information for early warning purposes;
  (b) Translate available data, particularly from global meteorological observation systems, into timely and useful products.

In view of the current policies and perspectives delineated at the Johannesburg Summit, the present volume is deemed quite timely in addressing problems and recent solutions in EDM. In the following chapters, selected papers are presented on various steps of EDM (i.e., delineation of objectives, constraints, and institutional aspects of EDM; design of data collection networks; statistical sampling; physical sampling and presentation of data; data processing and environmental databases; reliability of data; data analysis and transfer of data into information; and data accessibility and data exchange at local, regional and global scales). Conclusions and recommendations regarding each step are covered in the last chapter.

An important outcome of the meeting leading to this volume was the need expressed by all participants for "concrete actions" towards integration of data management systems. Possible "actions" were suggested to include the development of: (a) large-team project proposals to be supported by regional or international programmes; (b) a forum or web site for exchange of views on various problems related to EDM; and (c) an international society or association on integrated EDM.

5. References

1. Harmancioglu, N.B., Singh, V.P., and Alpaslan, N. (eds.) (1998) *Environmental Data Management*, Kluwer Academic Publishers, Water Science and Technology Library, Volume 27, 298 pp.
2. Harmancioglu, N.B., Alpaslan, M.N., Ozkul, S.D., and Singh, V.P. (eds.) (1997) *Integrated Approach to Environmental Data Management Systems*, Proceedings of the NATO ARW on "Integrated Approach to Environmental Data Management Systems", Sept.16-20, 1996, Kluwer Academic Publishers, NATO ASI Series, 2. Environment, Volume 31, 546 pp.
3. Harmancioglu, N.B., Alpaslan, M.N., and Ozkul, S.D. (1997) Conclusions and recommendations in N.B. Harmancioglu, M.N. Alpaslan, S.D. Ozkul, and V.P. Singh (eds.), *Integrated Approach to Environmental Data Management Systems*, Proceedings of the NATO ARW on "Integrated Approach to Environmental Data Management Systems", Sept.16-20, 1996, Kluwer Academic Publishers, NATO ASI Series, 2. Environment, Volume 31, Part 9, pp.423-436.
4. World Summit on Sustainable Development: Plan of Implementation, Advance unedited text, 4 September 2002.

THE CONVERSION OF DATA INTO INFORMATION FOR PUBLIC PARTICIPATION IN DECISION MAKING PROCESSES

M. B. ABBOTT
Knowledge Engineering
Avenue Francois Folie, 28, B28
1180 Brussels, Belgium

Abstract. It is now becoming impossible in many countries to proceed with new constructions and management practices in the water sector of the economy without involving the general public in the associated decision making processes. This trend towards a greater public participation is essentially a *socio-technical* development that is itself dependent upon, and indeed inseparable from, sequences of technical innovations, as exemplified by the Internet and second and third generation telephony. It is a consequence of this situation that field data and model results, that were previously employed only by engineers and scientists, and employed for the most part technically, have now to be processed into forms which can be assimilated by the general population and employed politically, legally and altogether less technocratically. This process begins with the transformation of data into information, frequently through the intercession of modeling activities. The management of this activity falls within the purview of *Information Management*. The process may be taken further through the transformation of information into knowledge. This activity then falls, at least in principle, under the aegis of *Knowledge Management*. Beyond this again, it is essential that knowledge should transform into understanding if public participation is to proceed at all equitably and effectively. Within the sector of water and the aquatic environment, all of the above developments, as enabled by advanced information and communication technologies, fall within the ambit of *Hydroinformatics*. This kind of development that is occurring in hydroinformatics clearly necessitates a rather complete rethinking of data collection and processing strategies and associated network designs.

1. Introduction

It is now becoming almost impossible in many countries to proceed with new constructions and management practices in the water sector of the economy without involving the general public in the associated decision making processes. Whether, for example, the project involves the construction or the removal of a dam, or the raising or lowering of dikes, or the increase or decrease in pressure in a water distribution network, an ever increasing part of the general public insists upon participating in the associated judgmental and consequent decision making processes. This trend towards a greater public participation is a social development that is itself dependent upon, and indeed inseparable from, sequences of technical innovations, as exemplified by the Internet and second generation (or WAP) telephony. It is thus essentially a *sociotechnical* development and must be studied as such. This is to say that it cannot be studied as a social development any more than it can be studied as a technological development: its study is qualitatively different from the studies of its two constituting and enabling elements.

The sociotechnical study of technologically enabled public participation within the water sector can be viewed within a broader framework of a transition from predominantly representative democratic processes and institutions to increasingly participatory democratic processes. This transition, in turn, is necessitating the establishment of new institutions and institutional arrangements, including some new kinds of business enterprises.

In the situation that prevailed until quite recently, field data and model results were used only by engineers, scientists and other professionals and were deployed for the most part technocratically. Now, however, they have increasingly to be processed into forms, which can be assimilated by the general population and employed politically, legally, and altogether less technocratically. One immediate consequence of this change in the uses where data and model outputs are employed is that the transformations of data into information, and thence from information into knowledge, have to be largely rethought. In the current vernacular, we say that there now occurs a restructuring of the *knowledge chains* that are brought into play, and this restructuring has a major feedback effect on the type, the frequency and, most important of all, the speed with which data are collected and transmitted. All parties in a construction or management project, including the general public, must be provided with overviews of the instantaneous situation and its forecasted development if they are to be able to intervene meaningfully and responsibly in the corresponding processes. Only in this way can they be empowered as genuine stakeholders in the project and its environment.

2. Essential Definitions

In hydroinformatics, as in all other processes of representation and communication, we work with *tokens,* where a token is understood as any thing that can be offered up in place of another thing. In everyday language, tokens are often referred to quite generally as 'signs', but, as we shall shortly see, this is technically unsatisfactory within the context of hydroinformatics, and indeed also in other disciplines. There are then four basic types of tokens, each of which has another way of functioning, namely *indices, icons, symbols* and *signs in the strict sense* [1]. It suffices here only to define the last of these as *any thing that directs our attention to another thing.* There are several ways of partitioning sets of signs, but for the present purpose, we need to have recourse to only one such partition. We then distinguish between signs that indicate something to us but which do not express anything to us as, which we call *indicative signs,* and *expressive signs* that do predicate some degree of truth (including, possibly untruth) to us. Thus, '2' or 'fast' or 'up' or 'right' are all indicative signs, but `they do not tell us anything' as they stand. On the other hand, signs like 'that is a fast car' or 'the road to the right leads us home' do actually express something.

We then define a *model* as *a collection of indicative signs, which, in their collectivity, serve as an expressive sign.* A collection of indicative signs, which, as they stand, do not serve to provide an expressive sign, is then said to constitute *data* within this context. We see that the first definition applies entirely generally, whether to 'a fashion model', 'a 2001 model of a Mercedes SLK', 'a hydraulic scale model', 'a numerical model' or any other such connotation of 'a model'. The definition of data can be expanded similarly, while the two definitions taken together may serve to cast some doubt on the appropriateness of the expression 'data model'. Obviously, signs which are expressive at one level or order may, in their turn, be collected together to provide signs, each of which has only an indicative function at a higher level or order.

Hydroinformatics is obliged to define its terms very carefully, and, for this purpose, it is driven to scour the literature in such areas as epistemology, phenomenology, object theory, various branches of philosophy and theology in search of adequate formulations. Such terms as 'reality', 'truth', 'information', 'knowledge' and 'understanding', which the engineer could previously employ in a loose and informal way, now have to be defined as rigorously as possible within the contexts in which they are used [2, 3].

3. The Changing Structures of Knowledge Chains

These structures originate with the more systematic study of the transformation of data into information, frequently through the intercession of modeling activities. Within the context of modeling, this foundational informational structure has the basic form of:

$$\text{outgoing information} = \text{outgoing information (incoming information, data)} \quad (1)$$

We observe that the outgoing information is of a very different kind to the incoming information, as corresponds to the very different kinds of 'knowledges' (yes, in the plural!) to which the one and the other pertain. Thus, for example, the incoming knowledge may be concerned with the hydrodynamics of short-period waves and wave fields, together with the influences of harbor structures on these. The outgoing knowledge, on the other hand, may be much more concerned with the behaviors of various kinds of ships loading various kinds of cargos under the influences of various kinds of mooring and fendering systems when subject to a variety of wave fields. The information processes involved, the relation of these to their knowledge*s* and the resulting social/money-value relations that arise have been analyzed in Abbott [4]. The management of an activity of this kind nominally falls within the purview of *Information Management*. However, this process may be, and is in fact increasingly often, taken further through the transformation of information into knowledge, while, at the same time, it is the differences in the knowledge*s* that define the different ways in which the information is processed. This activity then falls, at least in principle, under the aegis of *Knowledge Management*. Beyond this again, it is essential that knowledge should transform into understanding, and this is by no means simple if public participation is to proceed at all equitably and effectively. In any event, the simplest knowledge production chain takes the form:

$$\text{outgoing knowledge} = \text{outgoing knowledge (outgoing information (incoming information (incoming knowledge)), data)} \quad (2)$$

With the last reference in mind, we may observe that in the case of modeling studies, (2) usually takes the form:

outgoing site-specific knowledge = outgoing site-specific knowledge
(outgoing site-specific information
(incoming generic information
(incoming generic knowledge)),
site specific chart and field data) (3)

It is a further consequence of the broader public participation that, in many activities such as those associated with environmentally sensitive construction operations and real time management of water resources or 'assets', all of the above actions and processes have to be accommodated within very short lead times. The *speed* with which data and information can be provided, and knowledge and understanding inculcated and induced, is often as important as are these products in themselves. Within the sector of water and the aquatic environment, all of the above developments, as enabled by advanced information and communication technologies, fall within the ambit of *Hydroinformatics*. This kind of development, which is now proceeding rapidly in hydroinformatics, clearly necessitates a rather complete rethinking of data collection and processing strategies and networks.

4. Practical Implementation

The above features may be illustrated by the examples of the design and construction of the (8 billion US dollar) road and rail links, between the two largest Danish islands (completed between 1997 and 1998) and between one of these islands and Sweden (completed in 2000). In the case of the second project, there was a particularly widespread and vocal opposition to the project, and this had to be accommodated. The water across which this link was constructed (called *Øresund* in Danish) is particularly clear and clean even while surrounded by conurbations housing some 3 million people. These projects could therefore only be built at all to the extent that they took a proper account of the interests and concerns of the general public. In the case of the particularly sensitive Denmark-Sweden connection, it was essential to keep this public informed in real time throughout the construction phase. This real-time requirement was essential, because a number of organizations had the legal right to modify or even to stop the project if they considered that certain agreements were not being respected. For this purpose, two new organizations had to be established, equipped and staffed in order to provide data and information in such a way that this could inculcate knowledge and, at least in some cases, induce understanding in the minds of a wide variety of stakeholders, commonly with widely different interests and intentions. The role of hydroinformatics in this project is described in some detail in a paper by two of the leading figures in the Consortium responsible for the design, construction and operation of the project [5]. Only the barest outline can be given here.

5. The Fixed Link between Denmark and Sweden

Early in the process of preparing the environmental impact assessment, or EIA, for this project, the potential major impact in the Øresund area was identified as originating in the dispersal of spilled sediments from dredging and reclamation activities. Consequently, a great effort was put into organizing the environmental monitoring in order to ensure that the criteria for the emissions would be met. The monitoring program was itself determined by the overall requirements set out by the Danish and Swedish authorities, and these were soon found to necessitate the development of a completely new environmental strategy. It became clear that this strategy should build on feedback principles of combining detailed planning of the dredging operations on-line with monitoring and modeling activities. The combination of monitoring and models could thereby assist in keeping track of the process of impact development, thus enabling adjustments to be implemented in the construction work in due time to ensure fulfillment of the objectives. The integration of the technical and environmental project requirements and the contractual commitments of the contractor in the fulfillment of the environmental requirements then also became a realistic option. The owner used several different tools in order to make sure that the distribution of responsibilities between the Consortium and the contractor functioned adequately in practice. The first of these was incorporated in a document on *'Dredging Instructions'*, which served as a quality document for administering the requirements set up by the Consortium to design, construct and operate the project. These were to reduce the spill to 5% of the original dredged material on average and to reduce the instantaneous spill in time, extreme intensity, and space. The second tool was a *Feedback Monitoring Programme* (FBMP), which provided both a planning tool and a tool for the timely control of the environmental consequences of the ongoing works. The authorities were, on this basis and on the basis of the owner's reports, responsible for reporting to the parliaments and governments of the two countries. In order to keep the public actively informed about the plans and progress relating to the environment, as well as to the other aspects of the project, the Consortium already, in its early days, established a *Public Affairs Department*. This department also functioned as the project's own instrument for monitoring public opinion.

6. The Hydrographic Monitoring and Modeling Employed at the Design and Construction Stage

The hydrographic conditions in Øresund were investigated during a comprehensive survey program that continued over the period from February 1992 to June 2000. The scope of this program was to contribute to the data and knowledge base used for the design of the link. In order to provide data for the environmental situation in Øresund, to prepare the general environmental impact assessment, and to target the control and monitoring program, a number of environmental studies were undertaken over the period 1992-1995. Detailed baseline studies were conducted for establishing the natural turbidity and for mapping the distribution of the dominant communities in Øresund.

These communities included eelgrass, shallow water vegetation, macro algae communities, mussels, and shallow and deepwater fauna. The aim of the baseline studies was that of describing the spatial and seasonal variation of selected key variables. Alongside the surveys, investigations were carried out and data collected aimed at determining the individual plant and animal communities' vulnerability to the products of the construction work.

The basic types of data recorded in the hydrographic survey, that covered the entire period from February 1992 to June 2000, were the speeds and directions of currents, the water temperatures, the salinities and turbidities, the water levels, the wind speeds and directions, and the significant wave heights and dominant directions of short-period waves. The permanent monitoring network utilized equipment with on-line data transmission facilities.

Great attention was given in this project to the accuracy of positioning of all components, whether these were anchored or towed instruments, model grid points or constructional elements. GPS equipment was employed extensively and played a major part in ensuring the success of the entire project. Within this context, it was shown again in this project that a much greater attention than hitherto had to be given to the chart and map projections that were employed for positioning and recording the positions of instruments and for setting up the models, as well as for construction purposes [6]. In particular, systems had to be set up to facilitate the transformation from one projection and one instrument survey model grid to another.

The overall objective of the FBMP was to ensure that the environmental objectives and related design criteria set up by the authorities in the two countries were fulfilled both during and after the construction work. The primary means to ensure compliance with the environmental requirements were provided by the planning of the dredging and reclamation works and the subsequent preparation of environmental assessments based on numerical model calculations and the introduction of data assimilation principles. The feedback made it possible to plan the execution of the construction work and implement adjustments of the ongoing activities even while the work was progressing, thus reducing the risk of violating the criteria, while at the same time minimizing the risk of stopping the ongoing activities. A comprehensive numerical model complex was used to simulate the environmental impact from the dredging and reclamation operations as part of the FBMP. The bases for the model complex were then frequently updated with the actual measured spill and the hydrodynamic boundary conditions. The integrity of the calibration and verification procedures used in the individual models was also checked regularly by comparing model results with hydrographic field data. The Feedback Monitoring Programme was operated by a *Feedback Monitoring Centre* (FBMC), which was staffed and equipped by the four major partners participating in the monitoring program. The FBMC provided content for both formal stakeholders (owner, contractor and authorities) and informal stakeholders (public, media, etc.), as well as for technical and environmental design and construction purposes.

7. The Transmission of Deliverables Using an Internet System

From the sociotechnical point of view of hydroinformatics, the key to the success of the Denmark-Sweden road and rail link was the introduction of new means to ensure the active participation of stakeholders in all environmentally sensitive decision making processes. These means were provided by an Internet-enabled system for the real-time provision of information about all environmentally critical processes, as these were actually occurring during the construction process. This system, which provided an overview of all significant events in text and graphical form, was called EAGLE. It was maintained by the FBMC. The FBMC was itself one of the major suppliers of environmental data and model results to EAGLE even as it functioned at the same time as the Consortium's feedback advisor and information manager. All feedback management and advice came in time to be based upon the EAGLE system. In the words of Jensen & Lyngby [7]: *"The backbone of the environmental control and management during the construction phase [was] the environmental information system EAGLE"*. The EAGLE system drew upon three databases. The first was provided, as already introduced, by the FBMC itself on the basis of its own feedback-monitoring program. This contained data collected during the monitoring surveys as well as data from the baseline studies, satellite images, coastal morphology photographs and other items. A second database was that run independently of the FBMC by two of its other partners, which contained immense amounts of material from surveys, simulations and physical model tests. The third source of data was the Consortium's database of the earthworks, tunnel and bridge constructions, positioning arrangements, surveys, construction qualities, soil samples and other such sources of information of an engineering nature. These three databases were linked together to provide a unified EAGLE database. The expert staff of the FBMC was responsible for assembling and formatting the material provided so as to make it 'digestible' to its 'end-users', or 'consumers'. The material was most commonly presented within a standard geographical information system (GIS) environment. This type of content was accompanied by other types, such as graphs, figures, images and reports. The latter included incident reports describing conclusions of investigations and corrective actions taken as a result of the feedback procedure. During the course of the project, EAGLE evolved into a powerful tool for decision making. The Consortium's management depended upon it for the latest environmental information, while the Danish and Swedish authorities and other stakeholders were provided with the same level of information at the same time. It was primarily the provision of information through the EAGLE Internet system that the general population and its own organizations could be empowered as genuine stakeholders in their own water resources.

8. References

1. Klinkenberg, J-M. (1996) *Précis de Sémiotique Générale*, De Boek et Lancier, Paris.

2. Abbott, M.B. (1994 a) Hydroinformatics: a Copernican revolution in hydraulics, *J. Hyd. Res.* **32**, special edition on Hydroinformatics, 1-14.

3. Abbott, M.B., Solomatine, D., Minns, A.W., Verwey, A., and Van Nievelt, W. (1994 b) Education and training in hydroinformatics, *J. Hyd. Res.* **32**, special edition on Hydroinformatics, 203-214.

4. Abbott, M.B. (1993) The electronic encapsulation of knowledge in hydraulics, hydrology and water resources, *Adv. Wat. Resour.* **16**, 21-39.

5. Thorkilsen, M. and Dynesen, C. (2001) An owner's view of hydroinformatics: its role in realising the bridge and tunnel connection between Denmark and Sweden, *J. of Hydroinformatics*, **3**, 2, 105-135.

6. Abbott, M.B. (1997) Range of tidal hydraulics, *J. Hydraulic Eng'g*, ASCE, **123**, 4, 257-275.

7. Jensen, A. and Lyngby, J.E. (1999) Environmental management and monitoring at the Oeresund Fixed Link, *Terre and Aqua*, **74**.

CHALLENGES IN TRANSBOUNDARY AND TRANSDISCIPLINARY ENVIRONMENTAL DATA INTEGRATION IN A HIGHLY HETEROGENEOUS AND RAPIDLY CHANGING WORLD

A View from the Perspective of the Global Runoff Data Centre

T. MAURER
Global Runoff Data Centre (GRDC)
c/o Bundesanstalt für Gewässerkunde, P.O. Box 20 02 53
56002 Koblenz, Germany

Abstract. A prerequisite for the sustainable management of the complex earth system or parts of it is a well-organized environmental database. However, authority over data and information is scattered regionally and sectorally, resulting in highly fragmented approaches to their management. Researchers and managers are caught in the deadlock of either spending too much time retrieving data or omitting relevant information, both ultimately leading to stagnation. The question as to how one may tackle this situation stands high on the agenda of international organizations, as outlined in this paper. Finally, upcoming integrating technologies are described and promoted. A point is made that societies have to accept to spend increasing overheads for the integration of information to foster the improvement of environmental management practices.

1. Why Do We Need Environmental Data Integration?

1.1. FRAMEWORK

The UN Conference on Environment and Development (UNCED) in Rio de Janeiro, 1992, stated in its Agenda 21 the need to adopt an integrated approach to the management of the environment. This need was stressed in recognition of the finite availability of essential resources such as clean water and air on our planet and in the context of an anticipated 1.5-fold increase of its population between the years 2000 and 2050. Since then, these ideas have been reflected in countless international meetings and programs on environmental issues, featuring titles and slogans involving such catchy adjectives as “integrated”, “multi- or transdisciplinary”, “transboundary” and “sustainable” (cf. section 3.1).

1.2. COMPLEX SYSTEM EARTH

Research aiming at understanding the earth system as a whole has revealed the picture of a complex system of highly interacting processes, the dynamics of which are largely determined by feedback cycles at various temporal and spatial scales and are featured by a large number of parameters and variables which all need to be accurately observed or estimated, stored and provided. In order to ensure optimal benefit, the access to stored data should ideally be free and unrestricted, and dissemination methods should be as comprehensive and comfortable as possible.

However, as opposed to comparatively simple, technical, man-made systems, the geometry of the earth system and its processes are still known only very roughly. This makes it difficult to thoroughly describe and model the transient state variables of the vital fluids of water and air within this geometry with sufficient accuracy, as well as their respective fluxes, which are driven by solar energy and gravitation. In fact, in the past and also at present, due to their complexity, these processes have by and large been studied separately by individual disciplines or small groups of collaborating disciplines, each concentrating on its respective "sphere" of concern (such as pedosphere, hydrosphere, limnosphere, atmosphere, oceanosphere, biosphere, and lithosphere, to name a few), thereby taking its viewpoint and its temporal and spatial scale of interest. Regularly, the interactions with adjacent spheres, their processes, and impacts from larger scales are being neglected or accounted for in an oversimplified manner. For instance, boundary conditions are set to constant, or significant feedback mechanisms are not incorporated at all (e.g., consideration of hydrological processes such as terrestrial evapotranspiration in some coupled global circulation models CGCMs).

1.3. "INFORMATION-INFARCT"

We have collected and are still collecting considerable amounts of data creating sectoral and/or regional information and knowledge documented in books, journals, local databases, and also in models. However, these materials have to be considered as being scattered and fragmented; and, due to their mere quantity, it has become impossible to make optimal use of them in new research efforts, i.e., to ensure the prior integration of all information from existing significant sources. Rather, due to poor information flow or low capacity data, information and knowledge are often redundantly derived by different individuals and organizations, thus binding resources and hindering real progress. Even worse, we are often spending significant portions of our time solving problems we are not primarily educated for (e.g., natural scientists developing information technologies).

It seems that societies are currently caught in a deadlock, suffering from a phenomenon which one might call the "information-infarct". Although modern communication technology has contributed immense improvements regarding the speed-up and automation of information flow, information still has to enter and leave individual human brains, in the end either by listening/talking or by reading/writing. Neither can the transfer rates across these interfaces be accelerated ad infinitum, nor the human storage capacity behind it is unlimited.

1.4. COUNTER MEASURES

Information flow can be characterized by its transfer speed, its amount, and the quality of its content (density, concentration or aggregation level). Today, we are possibly encountering for the first time the situation that men and their capacities are the bottleneck in the chain of information processing and, thus, the limiting factor for the further understanding of the system.

In the course of the centuries, solutions for improved information flow were always closely related to major advancements of human societies and their level of organization, e.g., mail services on Roman road networks and the invention of book printing techniques, to name only two.

In order to make significant progress towards a more complex understanding of the earth system and, based on that, to finally develop better integrated environmental management strategies, it is crucial to consider and integrate the ever growing amount of data and information. However, this may finally lead to the undesirable situation of using up the entire available capacity for communication without actually working on the problems. As the alternative cannot be to remain ignorant, information access and information processing urgently need to be facilitated by several measures, including:

- speeding up information transfer rates;
- homogenization and standardization of information representation;
- improvement of selection mechanisms for targeted information retrieval;
- improvement of aggregation schemes to summarize information;
- improvement of disaggregation schemes and interface definitions to facilitate sharing the work-load on a growing number of shoulders, i.e., decreasing the amount of information required by a single involved individual.

Achieving these goals calls for the organization of better-coordinated structures featuring higher degrees of complexity, both in a technical and an administrational sense. Examples in the technical domain are libraries, (meta)-databases, standardization efforts, generic concepts, expert and decision support systems, to name a few. Similarly, in the human domain, this relates to improved coordination of organizations and programs at international, national, regional and local levels.

Often, the development of new technical approaches simultaneously calls for an appropriate adjustment of organizational structures within the societies. Today, this may mean even more systematic approaches to work sharing on a global scale to cope with the more complex tasks to be solved. Tracing back the prerequisites for such goals to a more general level, it is evident that, in a knowledge society, this ultimately requires the establishment of very good education systems throughout the world.

2. A Heterogeneous World in Many Respects

The problem of environmental data integration is still often viewed as a minor organizational problem on the technical level. In the view of the author, in many projects and programs, the necessary efforts are underestimated, and data integration is regularly given a too low priority. This holds true already for small and local projects. In fact, however, the number of obstacles towards integration is nonlinearly increasing with the spatial scale and the number of disciplines to be covered by a project. At the upper end, i.e., where solutions have to be found on a global scale, factors like national policies and international relations will inevitably play their strong role and often complicate matters to an extent that they end up in stagnation or even regression.

However today, as no single country or national institution is able to bear alone the burden of installation, operation and maintenance of global observation and information systems, the challenge remains to join forces across boundaries, both of states and disciplines, to collect and organize the necessary data, thereby avoiding redundant efforts.

Building of integrating systems requires not only "hard" resources such as technology, infrastructure, and finances, but also "softer" resources such as well-functioning administration and organization, education and science, collaboration and standardization, and finally time.

However, if all these were available in abundance, it would not help anything if there were no political will for action. Therefore, the political will has to be seen as the primary prerequisite towards the goal of building integrated systems and, thus, is the primary starting point. The development of the political will for integration requires considerable insight in and understanding of the complex mechanisms in societies and, in this respect, it is closely related to education and, consequently, to freedom which is granted only by prosperity.

In fact, the UN-System is working towards this higher-level goal through its specialized agencies and programs. Its Administrative Committee on Coordination (ACC), together with its commissions and sub-committees, ensures further the integration of all sectoral aspects related to one crosscutting topic such as water resources. The following section elaborates in further detail on what is being undertaken on a global level.

3. Response of the International Community

3.1. FRAMEWORK

This section will be biased towards the freshwater aspects of the environment, for the mere reason of the author's and GRDC's affiliation to the water resources sector. However, many aspects mentioned here are likely to be transferable to other environmental issues, especially those bound to the terrestrial domain.

The authority over the management of the world's freshwater is fragmented among the nations of the world, hundreds of thousands of local governments, and countless non-governmental and private organizations as well as a large number of international bodies. The management issues have been the subject of numerous studies and debates in the international arena [1].

Though the water issue had been a topic of major concern already in 1972 at the UN Conference on the Human Environment in Stockholm, Sweden, the starting point of the debate on water resources is commonly attributed to the UN Water Conference in Mar del Plata, Argentina, in 1977, which was the first and the only intergovernmental conference exclusively devoted to water: a milestone in the history of water development. Since then, especially the 1992 International Conference on Water and Environment in Dublin, Ireland, and in the same year, the subsequent UN Conference on Environment and Development (UNCED) in Rio de Janeiro, Brazil, have helped to foster and guide global efforts in the sustainable use of water resources [2]. Activities will be reviewed and projected in the future at the upcoming International Freshwater Conference in Bonn, Germany, in December 2001, and the Rio+10 Summit in Johannesburg, South Africa, in 2002.

It was also in Dublin and at the Rio de Janeiro Earth Summit in 1992, where the idea of forming the World Water Council to take leadership in water affairs was first proposed. In a special session at the Eighth World Water Congress of the International Water Resources Association (IWRA) in 1994, a consensus was established for the creation of a common umbrella to unite the disparate, fragmented, and ineffectual efforts on global water management, which led to the legal incorporation of the World Water Council in 1996, with its headquarter established in Marseille, France.

At the First World Water Forum in Marrakech, Morocco in 1997, the World Water Council received the mandate to develop the World Water Vision for Life and Environment for the 21st Century, which was subsequently presented at the Second World Water Forum in The Hague, The Netherlands in 2000 to some 5,700 participants from all parts of the world. The Third World Water Forum will be held in Kyoto, Japan in 2003 to develop the vision further.

The Water Vision Project attempts to sketch the transition between today's practices and those we will need if we are to meet our water needs in the future. As a part of this project, three scenarios were developed, based on different water use strategies and different outcomes. Moreover, thematic panels were convened to discuss the effect of future developments in institutions, biotechnology, energy technology, and information technology. The World Water Vision describes the goals of sound water management and the actions required to sustain water resources.

As can be seen from this short and incomplete summary, there are plenty of global activities going on in the water sector, which, in essence, are all aiming at the development of integrated management practices for water resources, i.e., improving coordination and thus joining forces for a better future. In the following subsections, some exemplary international activities with integrating character are outlined for illustration. Here, extensive use is made of fragments of original text material presented at the (cited) respective homepages of the programs, as many facts can hardly be summarized more precisely.

3.2. HYDROLOGY AND WATER RESOURCES PROGRAMME (HWRP) OF THE WORLD METEOROLOGICAL ORGANIZATION (WMO)

The mission of the Hydrology and Water Resources Programme (HWRP) [3] is to promote activities in operational hydrology and to further establish close co-operation between Meteorological and Hydrological Services. The activities under the HWRP concentrate on:

- the measurement of basic hydrological elements from networks of hydrological and meteorological stations;
- the collection, processing, storage, retrieval and publication of hydrological data, including data on the quantity and quality of both surface water and groundwater;
- the provision of such data and related information for use in planning and operating water resources projects; and
- the installation and operation of hydrological forecasting systems.

The HWRP also promotes improvements in capabilities of the developing countries through technology transfer and technical co-operation, so as to enable them, on their own, to assess their water resources on a continuous basis, to respond to threats of floods and droughts, and, thus, to meet the requirements for water, its use, and management for a range of purposes. The Programme takes into consideration the existence of global change, its hydrological impacts, and the need to provide more information to the general public and to Governments so that they can better understand the importance of hydrology and the role of national Hydrological Services (NHSs) in their activities. The Programme also promotes increased collaboration between NHSs and NMSs, particularly in the provision of timely and accurate hydrological forecasts.

The overall objective of the Hydrology and Water Resources Programme can be summarized as: *"to apply hydrology to meet the needs for sustainable development and use of water and related resources; to the mitigation of water-related disasters; and to effective environmental management at national and international levels"*. The Programme is implemented through five mutually supporting components, namely the Programmes on:

Basic Systems in Hydrology: This programme's objective is "to provide guidance to, and support for national Hydrological Services in the development and maintenance of their activities for the provision of data and services, with an emphasis on quality assurance".

Forecasting and Applications in Hydrology: This programme's objective is "to apply hydrological modeling and forecasting techniques and related technology for the mitigation of water-related disasters of both natural and anthropogenic origin and for studies of global change."

Sustainable Development of Water Resources: This programme's objective is "to ensure the effective use of hydrology in support of sustainable development, including the protection and enhancement of the environment."

Capacity Building in Hydrology and Water Resources: This programme's objective is "to contribute to the rational development and operation of national Hydrological Services, including the education and training of their staff, increasing public awareness of the importance of hydrological work, and support to technical co-operation activities."

Water-related Issues: This programme's objective is "to increase the effectiveness of WMO's activities in hydrology and water resources through inter-organizational collaboration in the water field."

Though all programmes mentioned have implications with integration issues, in the context of environmental data integration, the WHYCOS programme [4] within the programme on Basic Systems in Hydrology is of particular interest and should be highlighted as an example. WHYCOS stands for *World Hydrological Cycle Observing System* and was launched in 1993 by WMO in association with the World Bank. Its objectives are to:

- strengthen the technical and institutional capacities of national hydrological services (NHS) to capture and process hydrological data, and meet the needs of their end users for information on the status and trend of water resources;

- establish a global network of national hydrological observatories which provide information of a consistent quality, transmitted in real time to national and regional databases, via the Global Telecommunication System (GTS) of WMO; and
- promote and facilitate the dissemination and use of water-related information, using modern information technology such as the World Wide Web.

WHYCOS is a global programme, modeled on the WMO's World Weather Watch. It has two components:

- a support component, which strengthens cooperative links among participating countries; and
- an operational component, which achieves "on the ground" implementation at regional and international river basin levels.

WHYCOS is based on a global network of reference stations, which transmit hydrological and meteorological data in near real-time, via satellites, to NHSs and regional centers. These data allow the provision of constantly updated national and regionally distributed databases of high quality. WHYCOS aims to support, in all parts of the world, the establishment and enhancement of information systems, which can supply reliable water-related data to resource planners, decision makers, scientists, and the general public.

WHYCOS does not replace existing hydrological observing programs but supplements them. An important product of WHYCOS is regional data sets that are of consistent quality and that can be used in preparing products for water resources assessment and management. However, WHYCOS has been conceived, perhaps more importantly, as a vehicle for technology transfer, training, and capacity building. WHYCOS is being developed in the form of regional components, HYCOSs, which meet the priorities expressed by the participating countries. In its initial phase, WHYCOS has focused on establishing components in international river basins, in the catchment areas of enclosed seas, and in regions of Africa, which are poorly served by hydrological information.

By providing a framework of common guidelines and standards, WHYCOS is intended to enable the use of information from the regional HYCOSs for larger scale applications, such as research into the global hydrological cycle. Accordingly, WHYCOS makes an important contribution to the work of the other WMO and international scientific programs which require water-related information.

HYCOSs are being implemented for the following regions: Mediterranean Sea, Southern Africa and West- and Central Africa. For others such as Caribbean Sea, Baltic Sea, Aral Lake, Black Sea, Danube basin, Amazon basin, and Nile basin, project documents or outlines are prepared. Others have been merely considered so far, such as Caspian Sea, Himalayan region, and the Arctic basin.

3.3. WORLD WATER ASSESSMENT PROGRAMME (WWAP)

The developments outlined in section 3.1 have successively reinforced the need for a comprehensive assessment of the world's freshwater as the basis for a more integrated water management. At the urging of the Commission on Sustainable Development and with the strong endorsement by the Ministerial Conference at the time of the Second World Water Forum in The Hague in March 2000, the UN Administrative Committee on Coordination Subcommittee on Water Resources (UN-ACC/SCWR) has undertaken a collective UN system-wide continuing assessment process, the World Water Assessment Programme (WWAP).

The WWAP [5], building on the achievements of the many previous endeavors, focuses on assessing the developing situation as regards freshwater throughout the world. The primary output of the WWAP is the biennial World Water Development Report (WWDR). The Programme will evolve with the WWDR at its core. There will be a strong need to include a Water Information Network comprising something like a global-scale geo-referenced metadatabase and knowledge management systems to facilitate the assessment and dissemination of information.

The recommendations from the WWDR will include capacity building to improve country-level assessment, with emphasis put on developing countries. This will include the building of capacity in education and training, in monitoring and database science and technology, and in assessment-related institutional management. The Programme will identify situations of water crisis and will thus provide guidance for donor agencies. It will provide the knowledge and understanding necessary as the basis for further capacity building.

The Programme, including the new WWDR, is undertaken by the UN agencies concerned and aided by a Trust Fund, with donors providing support in cash and in kind either through specific agencies or through the Trust Fund. UNESCO currently hosts the WWAP Secretariat and manages the FUND at its Headquarters in Paris.

The Programme is intended to serve as an "umbrella" for the coordination of existing UN initiatives within the freshwater assessment sphere. In this regard, it will link strongly with the data and information systems of the UN agencies, like GRID, GEMS-Water and the Global International Waters Assessment (GIWA) of UNEP, the Global Precipitation Climatology Centre and the Global Runoff Data Centre (GRDC) of WMO, AQUASTAT of FAO, the International Groundwater Resources Assessment Centre (IGRAC) being established by WMO and UNESCO, the water supply and sanitation databases of WHO and UNICEF, and the databases of the World Bank system.

4. Standards and Upcoming Integrating Technologies

The previous sections have summarized aspects of data, information and organization with respect to heterogeneous conditions and manifold environments, including not only the obstacles to reach integration at a global level, but also the measures taken so far to overcome these obstacles. Recapitulating the exemplary activities summarized in section 3, one might question whether there is already sufficient coordination within programs like WHYCOS or WWAP, i.e., whether data integration is significantly promoted and facilitated through such initiatives and meets the criteria outlined as one of the challenges of the WWAP entitled “Ensuring the Knowledge Base” [6]. It will never be possible to give a satisfying answer to this question in the sense of a clear “yes” or “no”. However, a point is made here that efforts need to be increased in order to significantly advance the integration.

In the opinion of the author, the management of international programs and good software engineering practices [7] must be evaluated simultaneously in the sense that it is a fundamental prerequisite for any good programs (for computers as well as for organizations) to tackle the problems in iterative cycles, where the problems are solved in each cycle, following an appropriate order:

1. find out the information needs and the corresponding sources;
2. find out how to store the required information;
3. find out how to access the information and prepare access routines;
4. develop the functional routines after having designed the data structure as the core element of a sound and extendable system.

The crucial point here is the order. If one fails to follow this sequence and starts off with step 4, as it often happens in over haste ("quick and dirty"-programming), the results are likely to become intricate and, after a while, cannot be extended any further due to structural shortcomings. In computer programming, such approaches are frequently called "spaghetti-code", using the metaphorical expression as a synonym for a chaotic and intricate structure.

However, it is neither an easy task to define all information needs prior to the development of complex models or management strategies, nor to locate the relevant sources or to ensure that all sources have been considered. This is the point where the call for Environmental Information Systems (EIS) or metadatabase systems is often raised. Even more difficult on a larger scale is the design of generalized storage schemes and access routines to the information, designed for shared application by many users.

Here, again a dilemma is encountered. On the one hand, one may work together from the start in an expensive coordinated effort to develop non-proprietary open standards. This is an enterprise, which always bears the potential of collapsing beyond a certain scale in case it runs out of resources prior to proving its usefulness by relieving enough potential supporters from significant problems. Alternatively, one may start developing relatively small proprietary stand-alone-solutions, which are easy to control at first, but which usually lead to insuperable problems when an attempt is made to combine too many of them in a multidisciplinary effort.

There have been considerable recent achievements in the field of standardization of metadata representation and transfer, especially in the field of geomatics, such as the:

- FGDC Metadata Content Standard of the Federal Geographic Data Committee [8] for geospatial scientific data developed in the US;
- Dublin Core metadata standard for biographical information, developed by the Dublin Core Metadata Initiative (DCMI) [9], which is an organization dedicated to promoting the widespread adoption of interoperable metadata standards and developing specialized metadata vocabularies for describing resources that enable more intelligent information discovery systems;
- Global Information Locator Service [10], a metadata standard used by Federal and other governmental agencies in the US.

Some of these standards are built on XML, the Extensible Markup Language [11], which is an Internet standard approved by the World Wide Web Consortium (W3C), allowing the separate definition of the logical and physical structure of a document.

Again, in the field of geomatics, there is also work underway to standardize not only the metadata but also the data exchange itself, conducted by two initiatives, which are collaborating:

- ISO TC211 [12] standard in the field of digital geographic information, aiming to establish a structured set of standards for information concerning objects or phenomena that are directly or indirectly associated with a location relative to the Earth;
- OpenGIS Project of the OpenGIS Consortium [13], which aims to provide a comprehensive suite of open interface specifications to enable transparent access to heterogeneous geo-data and geo-processing resources in a networked environment.

However, standards usually cover a sector (such as the sector of geomatics) and still are not accepted or implemented everywhere on a global scale. The reasons for this are manifold and are related to the heterogeneity discussed in earlier sections.

Denzer *et al.* [14, 15] have pointed out clearly in the context of developing Environmental Information Systems (EIS) that no matter how much effort will be invested to integrate distributed and heterogeneously spread information of different meaning, syntax and structure, there is no realistic way to combine them in a *single* unified system in a reasonable time frame for mainly three reasons, namely heterogeneity, autonomy and dynamics, which are discussed below:

Heterogeneity: In practice, different systems to be integrated are heterogeneous in different respects:

1. Syntactical heterogeneity means that systems differ with respect to hardware, operating system, storage technology, etc. Syntactical heterogeneity is a pure computer problem and should be hidden from the user.
2. Semantical heterogeneity means that there are different notions about the semantics of a single piece of information. This includes the development of different terminologies in parallel projects in different regions.
3. Structural heterogeneity evolves due to the fact that different parties combine different sets of simple information to different structures (or objects) denoting the same type of information, yet resulting in aggregates of different syntax and different meaning (although some part of these objects may have the same meaning).

Autonomy: Many EIS which have been built are information systems for public authorities, supporting public services in their every day work. Due to the legal authority of these institutions, they are completely autonomous in their decisions concerning information technology. As a result of the scattered sectoral and regional competences with regard to environmental management, one is confronted with a fragmented situation of approaches in different regions and subregions. This holds especially true for water as a traditionally locally managed resource in many regions of the world. The task of building a data network in such a situation means that it is not possible to apply a unique data model for such a network because one can never force anyone to use this data model or stick to it and its enhancements.

Software developers have to accept the fact that fragmented autonomy is something that will not vanish quickly. This makes integration more difficult but not impossible, i.e., this boundary condition has severe implications on the types of software architectures which can apply to autonomous systems.

Dynamics: It is usually impossible to thoroughly describe the tasks ahead in environmental management in a single step from scratch, and this consequently holds true for the definition of a final data model, too. Even if a perfect data model could be defined a priori, it is unlikely that this model remains valid for more than one year, given the rapidly changing demands. Considering the integration of hundreds of environmental data sources and linking them with hundreds of thousands of potential clients, it becomes clear what it means to keep such a system up-to-date when data sources change their features all the time.

To overcome these problems, Denzer *et al.* [15, 16, 17, 18] developed an Environmental Information System (EIS) that serves as a metadatabase system and as a data retrieval system, capable of integrating existing distributed database systems of different structures and levels of abstraction without interfering with their grown internal structure or their ownership, i.e., the control of local administrators. This system features a flexible internet-based client server architecture that ensures applicability across heterogeneous environments. The system is designed in a completely generic way by means of a communication server (termed SIRIUS [16]) between local service programs and distributed clients, featuring two interfaces. The system is furthermore prepared to automatically translate all features client-dependently.

Different local systems feature different levels of abstraction. Few of them give access to their catalogues; almost none of them is able to describe itself, e.g., by object-classes and structures they provide. To enable the outer world see what local systems have to offer in a unified way, each participating local system has to be equipped with a small local interfacing database and a number of service applications running on the local system. *These are the only parts of the system, which have to be adjusted and have to restrict to some standards of the system.* The local interfacing database can be regarded as the "table of contents" of the local database designated for integration. It remains completely under the control and responsibility of the local administration. It defines who is permitted to view or retrieve what information and also contains the methods of accessing the local system for retrieval. Once a local candidate system is set up as described, its data are readily available to the outer world.

On the client side of the system, the only prerequisite is a WWW-browser. A JAVA-application collects all meta-information, which a user is authorized, to view from the communication server. This application displays meta-information as a multi-hierarchical tree from which single data sets can be selected. Alternatively, the interface also allows a selective query of the metadata including location, which can be both selected and displayed in an integrated Internet map server window.

A system like the one described by Denzer *et al.* [15, 16, 17, 18] provides the slimmest possible approach to integration by ideally combining centralized and decentralized features, thus being flexible enough to be adjusted with minimal effort to the ever changing boundary conditions discussed above. The introduction of a new data source in the system is thus achieved:

- without making changes or enhancements to the communication server;
- without making changes or enhancements to the clients, i.e., all end-user applications;
- without having to write too many new codes for each new data source.

Even though Denzer *et al.* [14, 15] were among the first to promote the ideas outlined above, there are other initiatives with developments along similar lines, e.g., the Mercury approach for scientific data management [19] launched at the Oak Ridge National Laboratory, a federal research facility operated by Lockheed Martin Energy Research Corporation for the US Department of Energy. This metadata management scheme builds on existing WWW technology and commercial-of-the-shelf (COTS) products as well as on agreed metadata standards listed above. The basic idea here is to keep metadata sets in XML format on the servers of providers and to leave their maintenance to the hands of the providers who also maintain a "locator file" (the table of contents!) at their system. This "locator file" has to be registered with the Mercury staff. A specialized Mercury web-crawler extracts the latest versions of metadata sets in nightly "harvesting"-runs and stores them in a central database, which again is made available to the public by a web-browser application. Several US organizations already joined the system.

Finally, it should be mentioned that there are also endeavors not only to integrate environmental data but also to integrate model components developed by various specialists in a homogeneous environment, a prominent example being the Modular Modeling System MMS [20] developed by Leavesley *et al.* [21, 22, 23]. MMS is an integrated system of computer software that has been developed to provide the research and operational framework required for supporting the development, testing and evaluation of physical-process algorithms and for facilitating the integration of user-selected sets of algorithms into an operational model. MMS provides a common framework intended to focus on multidisciplinary research and operational efforts. Scientists in a variety of disciplines can develop and test model components to investigate questions in their own areas of expertise and can work cooperatively on multidisciplinary problems without each scientist having to develop the complete system model.

Continued advances in physical and biological sciences, GIS technology, computer technology, and data resources will expand the need for a dynamic set of tools to incorporate these advances in a wide range of interdisciplinary research and operational applications. MMS is being developed as a flexible framework to integrate these activities.

5. Conclusion

Advances in global water resources management are nowadays not primarily limited by a lack of data or information, but rather by the bottleneck of insufficient information management. The same most likely holds true for the broader field of environmental management. We are apparently lacking the ability to exchange our knowledge effectively, i.e., we are no longer able to avoid significant redundant work on the one hand and to assemble most of the relevant information available for the solution of a given problem on the other hand.

This leads to the conclusion that a much more intensively coordinated effort has to be launched to resolve the task of data and information integration and thus to raise environmental management to integrated and sustainable levels. The techniques for integration seem to be available, but they need to be installed and maintained, which is by no means a small task.

It has to be accepted by societies that the increased complexity of the tasks ahead will lead to a larger overhead with respect to data organization, i.e., these tasks will require an over-proportionally increasing share of our project's and program's budgets. Societies have to agree on this. In democracies where the idea of consensus and participative decision making is eminent, this requires a higher level of awareness of these relatively complex contexts. This issue is inevitably connected to a higher education of the larger part of the societies.

In fact, mankind has been stepping forward towards integration from its very beginning, and its advances have always been closely related to an improved management of information and its storage techniques. The amount of resources put into this in terms of manpower, technology, and money has been constantly growing and certainly will continue to do so. However it is in our hands to influence the rate of progress.

Nowadays, we dispose of networked libraries providing us with a wealth of information at our fingertips. We may have to think of the future developments towards an even better situation as an extension of the concept of libraries, including all kinds of electronic information and improved catalogues or metadatabases. We should step forward in this direction with realism, but should also be orientated by having in mind a vision of the ultimate goal. This goal should have high priority not only in research programs but also in the respective framework programs or long-term plans, such as the 6th European Framework Programme 2002-2006 for research, which already plans to allocate approximately 20% of its budget of 17,5 billion Euros to one of the priority areas on "Information Society Technologies".

The value of water is occasionally expressed by the metaphor "Water - the oil of the 21st century", thus reflecting the threatening scarcity of the once abundant resource. In fact, it is the second most vital resource for life after the air we breathe and much more essential than other important resources such as oil. It is unbearable to imagine that our efforts to manage the environment could be not sufficiently far-reaching and possibly not as professional as if it were an important business activity, promising substantial revenues such as prominent industries like banking or mobile telecommunication.

6. References

1. WWC, World Water Council (2001) *Homepage*, http://www.worldwatercouncil.org
2. WMO, World Meterological Organization (2001) *Exchanging hydrological data and information, WMO policy and practice*, WMO brochure No. 925, Geneva, Switzerland.
3. HWRP, Hydrology and Water Resources Programme of the World Meteorological Organization (2001) *Homepage*, http://www.wmo.ch/web/homs/hwrphome.html
4. WHYCOS, World Hydrologic Cycle Observation Programme (2001) *Homepage*, http://www.wmo.ch/web/homs/ whycos.html

5. WWAP, World Water Assessment Programme (2001) *Homepage*, http://www.unesco.org/water/wwap/index.shtml
6. WWAP, World Water Assessment Programme (2001) *Challenges*, http://www.unesco.org/water/wwap/targets/index.shtml
7. Sommerville, I. (1995) *Software Engineering*, Addison-Wesley, Harlow, England.
8. FGDC, Federal Geographic Data Committee (2001) *Homepage*, http://www.fgdc.gov/fgdc/fgdc.html
9. DCMI, Dublin Core Metadata Initiative (2001) *Homepage*, http://dublincore.org/
10. GILS, Global Information Locator Service (2001) *Homepage*, http://www.gils.net
11. XML, Extensible markup language (2001) *Homepage*, http://www.xml.org
12. ISO/TC 211 Geomatics (2001) *Homepage*, http://www.statkart.no/isotc211/
13. OGC, Open GIS Consortium (2001) *Homepage*, http://www.opengis.org/
14. Denzer, R., Schimak, G. and Humer, H. (1993) Integration in environmental information systems, in D. Russell (ed.), *International Symposium on Engineered Software Systems*, Malvern, USA, May 1993, World Scientific, Singapore.
15. Denzer, R. and Güttler, R. (1995) An overview of integration problems in environmental information systems, *J. Computing and Information* **1**(2), 1112-1120. (available at http://eig.htw-saarland.de/publications/english_pub/jci_95.pdf)
16. Denzer, R. and Güttler, R. (1995) SIRIUS - Saarbrücken Information Retrieval and Interchange Utility Set. in *International Symposium on Environmental Software Systems 1995 (ISESS 1995)*, Malvern, PA, USA, June 1995, Chapman & Hall.
17. Denzer, R., Schimak, G., Güttler, R., Houy, P. and TEMSIS-Consortium (1998) *TEMSIS - a Transnational System for Public Information and Environmental Decision Support*, HICSS-31, January 1998.
18. Denzer, R., Güttler, R., Houy, P. and TEMSIS-Consortium (2000) TEMSIS - a transnational system for public information and environmental decision support, *J. Environmental Modeling and Software* **15**(3), 235-243.
19. Mercury (2001) *Homepage*, http://mercury.ornl.gov
20. MMS, Modular Modeling System (2001) *Homepage*, http://wwwbrr.cr.usgs.gov/projects/SW_precip_runoff/mms/
21. Leavesley, G.H., Restrepo, P.J., Stannard, L.G., Frankoski, L.A., and Sautins, A.M (1996) The modular modeling system (MMS) - A modeling framework for multidisciplinary research and operational applications, in M. Goodchild, L. Steyaert, B. Parks, M. Crane, M. Johnston, D. Maidment, and S. Glendinning (eds.), *GIS and Environmental Modeling: Progress and Research Issues*, GIS World Books, Ft. Collins, Colorado, 155-158.
22. Leavesley, G.H., Restrepo, P.J., Markstrom, S.L., Dixon, M., and Stannard, L.G. (1996) *The Modular Modeling System - MMS: User's Manual*, U.S. Geological Survey Open File Report 96-151, 200 p.
23. Leavesley, G.H., Markstrom, S.L., Brewer, M.S., and Viger, R.J. (1998) The Modular Modeling System (MMS) - The physical process modeling component of the Watershed and River System Management Program, in *Proceedings First Federal Interagency Hydrologic Modeling Conference*, Las Vegas, Nevada, 19-23 April, **5**, 93-100.

INFORMATION TECHNOLOGY AND ENVIRONMENTAL DATA MANAGEMENT

M. A. SANTOS and A. RODRIGUES
Civil Engineering National Laboratory
Av. do Brasil, 101
1700-066 Lisboa, Portugal

Abstract. Over the last two decades, environmental data have increasingly received more attention, as they play a key role in planning and operational management. At the same time, the way data-related issues are being approached has been changing with the years. Firstly, the rising awareness on environmental problems presses governmental and non-governmental agencies to increase monitoring. At the same time, the growing use of simulation and optimization models to better understand environmental phenomena requires higher data volumes to calibrate and validate these models. Secondly, although modern sensor technology has made data collection far easier and bulkier than in the past, currently used digital forms of data capture, such as remote sensing, image processing, digital photography and GPS (global positioning systems), have brought in new types of data with different requirements for archiving and analysis. Thirdly, the recent trend for a multidisciplinary approach to environmental problems has made data integration an important issue. Finally, institutional and non-institutional data producers increasingly feel the need or are obliged to make their data available to students, researchers, or the general public. These aspects call for better and widely spread data collection and archiving, and for easier access to data and information. Data classification and analysis are two further aspects that require some consideration.

All these tasks have much to gain from using the most recent advances of information technology as well as the artificial intelligence based techniques. Among these new developments, geographic information systems and data warehouses are important tools in data archiving and retrieval. Data mining is a promising technique to improve data analysis and classification and to generate information from existing data in the process of decision making. Internet technology has become an excellent instrument for data dissemination.

After a brief review of these techniques and the ways by which they address the above-mentioned environmental issues, the paper presents a case study that illustrates the use of information and Internet technologies to support estuarine planning and management.

1. Introduction

Water resources data have always played an important role in water resources modeling and in the design of water works. The latter activity was even responsible for the installation of the first hydrologic stations, mainly water level gauges, in the early forties and fifties.

In the last two decades, data and the information they can generate have increasingly become key tools in the decision making process along with the simulation and forecasting processes. This new perspective led to a shift in the way data must be organized and presented to the user.

At the bottom of the management hierarchy*, operations control is concerned with operational activities and requires resources with a low degree of uncertainty. At this

* A classical model of the managerial activities [1] considers that these activities fall into three categories: strategic planning, management control and operational control.

stage, data must be accurate, detailed, and largely internal. At the top, strategic planning information is largely dependent on external sources, namely social, economical and political data. It must be presented in an aggregated form and does not have to be highly accurate, since strategic plans are broad rather than detailed in nature, and they require only approximate indications of future trends, rather than exact statements about the past or present. Actually, current strategic planning still relies much on judgment and past experience. In between, management control requires data with a reduced level of uncertainty, which must include the knowledge of surrounding environmental factors as well as internal data, and which must be predictive in the short-range [2].

More recently, a new paradigm has appeared. It is known as *Indicator*, and it is supposed to help policy makers to monitor the environment or to integrate environmental and sector-based policies. Environmental Indicators are variables or an aggregation of variables that summarize a large volume of data into information readily usable by decision makers and the general public.

In relation to policy-making, the main goals of environmental indicators are three fold [3]:

- *"To supply information on environmental problems in order to enable policy-makers to value their seriousness;*
- *To support policy development and priority setting, by identifying key factors that cause pressure on the environment;*
- *To monitor the effects of policy response".*

In addition to helping policy makers, indicators are also important tools to raise public awareness and to promote information exchange.

Still more recently, the European Union is following a new direction in policy development, taking initiatives in integrating environmental issues into sector-based policies [4], which means that environmental indicators are to be integrated with sector-based and structural indicators that deal with socio-economical aspects (Fig. 1). This calls for data integration, i.e., integration of large volumes of data of different types, detailed or summarized (depending on the target user), which should be made easily available in a convenient format. These data go beyond water- or environmental-related data, and most of them are geo-referenced. Furthermore, data integration is as much a technical as a political and organizational issue in the sense that integration facilitates sharing, and highly shared data are more used, becoming more cost-effective.

The storage of and an easy access to these data involve adequate information systems or data warehouses and such powerful tools as data mining to classify, select and analyze these data in order to produce the required information. Finally, the increasing public awareness on environmental problems creates the need for an easy disclosure of these data. Again, Information and Multimedia Technologies help to achieve this desideratum through visualization techniques or through the Internet.

A summary of the information technologies useful for the management of environmental data is presented in this paper. The second section gives a brief overview of different environmental data types (alphanumeric, geo-referenced, image, remote sensing data). Special emphasis is given to spatial data and to the specific issues of conceiving an integrated data store that must be useful to an interdisciplinary team.

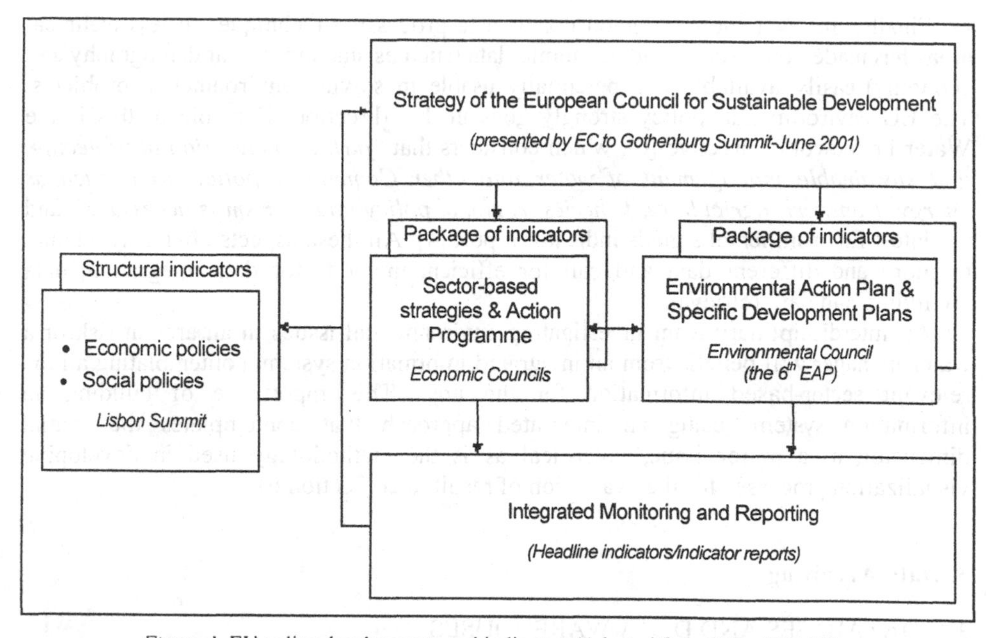

Figure 1. EU policy developments and indicators (Adapted from EEA, 2001 [4]).

The third section covers data archiving, focusing on how databases and geographic information systems respond successfully to the user's requests. The fourth section addresses the need for data analysis and processing and how data mining may be a promising tool on spatial data classification and pattern recognition. The fifth section covers the importance of Internet technology to convey environmental data to its users in a user-friendly way. Finally, in the sixth section, a case study, the Guadiana estuary information system, is presented. This case study shows a practical example of environmental data integration through a geographic information system, which also enables decision making and data dissemination. The paper ends by drawing some conclusions.

2. Environmental Data Explosion

Increased public and political awareness of environmental problems caused an increased need for new data. On the other hand, the emergence of new measuring sensors which provide data in a digital format, the widespread use of global positioning systems to collect location data, the use of video cameras for routine surveillance purposes, and the employment of airborne and satellite remote sensing technologies to monitor the Earth's surface have given birth to an explosion of environmental-related data. From a modeling perspective, it is recognized that hydrologic modeling can highly benefit from the combination of distributed hydrologic models with remote sensing or radar data, digital elevation models, digital maps as well as tracer techniques to better calibrate and validate these models [5]. The same rationale can be applied to other types of models such as water quality, solute transport, and sediment models.

Finally, the ever more common use of data processing techniques in less technical areas has made some social and economic data (such as statistics about demography and economy) easily available and potentially usable in solving environmental problems. The EU environmental policy strongly goes in this direction. Example of this is the Water Framework Directive [6], which considers that "*further integration of protection and sustainable management of water into other Community policy areas such as energy, transport, agriculture, fisheries, regional policy and tourism is necessary*", and the latest Environmental signals indicator report [4]. All these aspects obviously amount to more and different data and call for efficient methods for data integration, data archiving, and data retrieval.

An interdisciplinary team investigating environmental issues in an area at risk or a water manager will benefit from an integrated information system, contemplating all the relevant sector-based information for the area. The importance of building an information system, using an integrated approach that contemplates the spatial dimension as a major issue, is critical as is the methodology used in developing visualization processes for the evaluation of results (see Section 6).

3. Data Archiving

3.1. DATABASES AND DATA WAREHOUSES

A collection of data organized in such a way that a computer program (database management system) can quickly select, manipulate and retrieve desired pieces of data is a database. Although the value of environmental databases is well acknowledged and may be sufficient for operational management, a set of different and disperse databases (one for each type of data, e.g., surface water database, groundwater database, legislation database, ...) may not constitute an efficient tool at the strategic level. Information technology has responded to this need by introducing the concepts of data warehousing and database clusters. A Data Warehouse is a collection of subject-oriented, integrated, time-variant, non-volatile data (known as data marts) stored and maintained in an organization in support of the decision making process [7].

By subject-orientation, in contrast to process-orientation applications, it is meant that data warehouses are organized around major subjects such as rainfall, water levels, climate, water table levels, etc., as opposed to surface water (that includes the first three subjects), ground water databases, etc.

Integration in a warehouse is supported by data consistency: naming conventions are applied, attribute measurements are unique, and encoding structures are previously defined. For example, if "millimeter" is the unit chosen to represent rainfall, all rainfall depths stored in the warehouse must be measured in mm or converted to mm beforehand. Time variant means that data in the warehouse are correct "as of some moment in time", not necessarily "at the moment of access". Data in a warehouse are like a series of snapshots over a (variable) medium- to long-time horizon. Once stored, these snapshots are not supposed to be altered unless they are incorrect. Non-volatile data means that, once data are loaded in the warehouse, they do not change. This implies that the only operations allowed are loading and retrieval. There is no data update or change. New versions of the data may be added, but they will not replace what is already there.

These four features make the data warehouse environment quite different from the classical operational database environment. When they co-exist in the same organization, they must complement themselves and avoid redundancy.

In what follows, the term "database" is used in a general way and may include a simple database, a distributed database, a warehouse, or a database cluster. The first phase in building a database is the *conceptual analysis* that includes a study of the available data and of how they relate among themselves. At this stage, the database builder must meet with the domain experts in order to create the database structure.

Then follows the *logical analysis* when the most convenient database model must be selected. There are four types of models: hierarchical, network, relational, and object oriented (OO). The relational model introduced around 1990 is by far the mostly used due to its simplicity. Several attempts have been made to use the OO model, without much success, as the database management systems based on this model are not standardized yet. One example of an application of this model to the water domain is the AMHY database, an international database containing rainfall and discharge data from a set of countries in Southern and Eastern Europe [8, 9, 10]. The AMHY database has been set up since 1992 in the framework of the UNESCO International Hydrological Programme.

The third phase of this process is the *physical analysis* that corresponds to the computer implementation of the structure previously defined. System design and implementation, which involve the definition and development of processes to be implemented in association with the database structure, are of major importance at this stage. The result of this will be the availability of the required user interfaces for inserting, manipulating, and updating the database. Current Database Management Systems (DBMS) include tools to develop these interfaces. However, major operating system developers also contemplate these tools, which can normally be applied to several of the major DBMS in the market.

A very important issue in data archiving refers to data quality. Information technology addresses this issue by introducing the concept of Metadata or Data about data. Metadata describe how and when and by whom a particular set of data was collected and how the data are formatted. Metadata are essential for understanding and documenting information stored in databases, facilitating data dissemination and data sharing. In order to harmonize metadata specifications, several attempts are being made to establish metadata standards, which will help to catalogue metadata and to keep track of them. Metadata are mainly used in association with geographic information systems [11].

3.2. GEOGRAPHIC INFORMATION SYSTEMS

A Geographical Information System (GIS) is a set of tools for collecting, storing, retrieving and displaying spatial data in the form of maps. Due to the high cost of obtaining and analyzing spatial data, GIS applications have used to be associated with the public sector; but, currently, private organizations such as environmental consulting companies are starting to use and produce geo-referenced data. Such environmental data as land use, infrastructures inventory, monitoring data, river network, coastal zones, and other sensitive areas are examples of data that should be stored and managed through a GIS.

Nowadays, most GIS commercial packages can combine alphanumeric data, spatial data and environmental models, becoming a powerful tool in water resources planning and management. Fed with real data, a GIS must endow the managers and decision makers with the required information (Fig. 2).

Various authors stress that GIS is currently shifting from an operational tool to a strategic decision support system, requiring the incorporation of more powerful analysis techniques [13]. This can be due to the following developments:

1. visualization techniques have evolved considerably, and the introduction of multimedia and virtual reality functionalities in GIS, as well as 3D-GIS, have greatly improved and enlarged the range of visualization opportunities in a GIS;
2. the progress in GIS technology has made it possible for spatial data to be accessible through the Internet, transforming closed, proprietary geographic information management systems into open GIS (see Section 5);
3. innovative spatial data analysis methods, some of which are based on neurocomputing tools, have provided the solution to specific GIS problems, namely those problems which occur due to the existence of very large volumes of data as well as missing and fuzzy data.

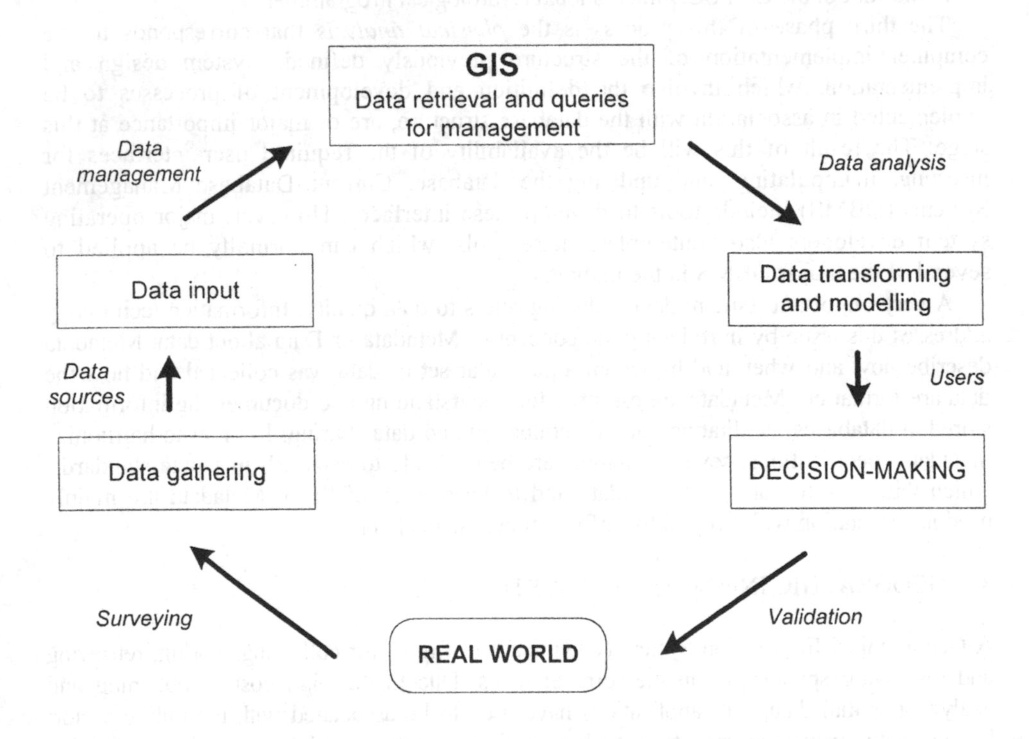

Figure 2. GIS as a management tool (adapted from Valenzuela [12]).

Therefore, GIS and other emerging technologies influence GIS-based applications in a positive way. Along with software, hardware progresses also contribute to better GIS applications. For example, parallelism enables the use of more efficient algorithms for computational geometry. Moreover, GIS technological developments are also application-driven [14], since these applications frequently require new hardware (e.g., new computer architectures and mobile internet devices) and new software.

At this point, the trend (still at the research level) is to make GIS available to mobile users. Hypergeo (http://www.hypergeo.org/hypergeo/home/index_new.htm) and WebPark (http://www.soi.city.ac.uk/is/research/gisg/research/) are two EU-funded projects that investigate how to supply geographic information to mobile users. The research results are expected to be applied to the tourism industry.

4. Processing and Retrieval Methods

The data glut referred to in Section 2 makes it impossible for a human being to deal directly with the data. Traditional methods of analysis cannot handle large data sets in a big database or in a warehouse and may not be able to process such types of data as spatial data. Dealing with such a data explosion requires a new generation of analytical tools, which should involve developments in automation and intelligent reasoning [15].

Data mining may be the answer to this problem. From a database perspective, "data mining" means extracting knowledge from large amounts of data. Some authors use data mining and knowledge discovery in databases (KDD) synonymously; others consider data mining as a step in the process of knowledge discovery in databases [16]. According to these authors, the knowledge discovery process is a sequence of seven steps (Fig. 3):

i) *Data cleaning*, to remove noise and inconsistent data;
ii) *Data integration*, to combine different sources of data;
iii) *Data selection*, to retrieve relevant data for the analysis task;
iv) *Data transformation*, to transform the selected data, through summary or aggregation operations, into forms suitable for data mining;
v) *Pattern evaluation*, to identify the interesting patterns representing knowledge;
vi) *Knowledge representation*, to present the mined knowledge to the user [16].

Data mining has an interdisciplinary nature, drawing from disciplines such as database technology, statistics, machine learning, information science, and visualization.

Data mining tasks can be classified as either *descriptive tasks*, which provide a general data characterization, or *predictive tasks*, which perform inference on data in order to make predictions.

Data mining functionalities include: data characterization and data discrimination, association analysis, classification and prediction, cluster and outlier analyses. These tasks are performed through different methods, which comprise neural networks, fuzzy theory, knowledge representation, inductive logic programming and high performance computing; or through such techniques as spatial data analysis, pattern recognition, image analysis, or computer graphics.

Up to now, data mining has been primarily applied in marketing and sales telecommunications with limited applications in environmental studies. However, its potential to analyze environmental data is already being explored, and some authors are already addressing the issue of spatial data mining, specially on what concerns data clustering and pattern recognition (e.g., [17, 18, 19, 20, 21]). Spatial data mining and its association with GIS, allowing efficient processing and retrieval methods to classify geographic data and to uncover patterns, is particularly important in dealing with environmental data.

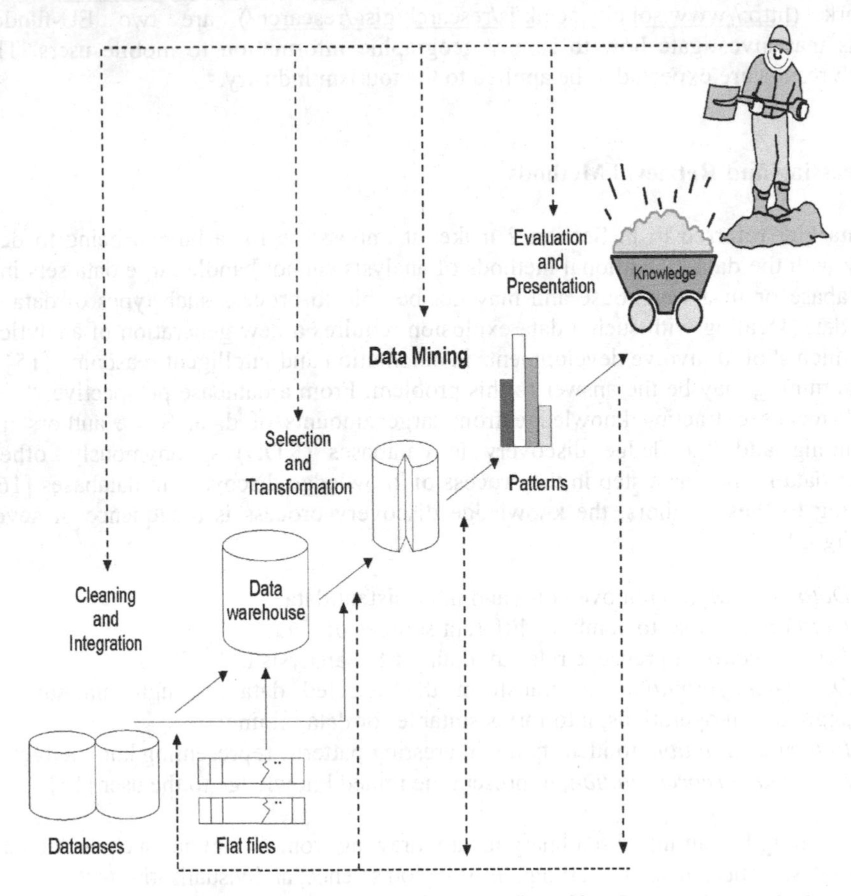

Figure 3. The process of knowledge discovery (adapted from Han and Kamber [16]).

According to Scholten and LoCascio [13], "*there are four major areas where statistical spatial data analysis techniques can strengthen current GIS practice*:

- *Sampling objects from the database and the choice of an adequate spatial scale of analysis;*
- *Data rectification to compare variables, which are defined for the same study area, but for different and incompatible set of zones;*
- *Exploratory spatial data analysis (ESDA) aiming to explore and exploit the GIS database to arrive at new insights, including the search for data characteristics such as trends, spatial outliers, spatial patterns and associations;*

- *Confirmatory or exploratory spatial data analysis (CSDA) concerned with systematic analysis of data and hypothesis testing based upon specific assumptions."*

Although this is a young field still under research, some data mining techniques are already available in commercial software, e.g., SPSS, SAS and STATISTICA, just to name three statistical software products used worldwide. Spatial data mining and specially multimedia data mining are still at an experimental stage [22, 23].

5. The Importance of the Internet

The Internet and mainly the World Wide Web (WWW) are excellent means to convey updated information to users in a friendly manner, using existing facilities to view either alphanumeric or spatial data. Fortunately, in recent years, geographic information (GI) technology has evolved in such a way that it is now possible to publish and access GI on the Internet, taking advantage of analysis capabilities and of the potential to develop fully customized interfaces to online mapping applications. On the other hand, several factors associated with current Internet technology have made it possible to retrieve, in an efficient way, raster images such as digital orthophotomaps or satellite images, digital terrain models and other types of 3-D models. Among these factors, a special reference can be made to:

- the extended use of the TCP/IP protocol with increasing bandwidths;
- the generalization of standards for different data types: graphics, multimedia, metadata, geographic data formats ...;
- the appearance of HTML (HyperText Markup Language) and VRML (Virtual Reality Markup Language) and more recently DHTML and XML;
- the production of plug-ins that enable most WWW browsers to present non-standard image formats.

To access these data, the user (client) requires only a network connection, an intelligent browser (and the installation of the most popular plug-ins), or client versions of proprietary freeware. From the owner/server side, in addition to the database management systems, the main requisite is high volumes of disk space, which is currently quite inexpensive.

6. The Guadiana Case Study

6.1. BRIEF PRESENTATION

A case study for this paper has been achieved by development of the Guadiana Estuary Information System, which was a part of the Study on the Environmental Conditions in the Guadiana Estuary and Adjacent Areas, developed by The National Laboratory of Civil Engineering in association with the Universities of Évora and Algarve (Fig. 4). This is a study under contract to the Portuguese Water Institute (INAG), which aimed to environmentally characterize the area and to identify management strategies for the

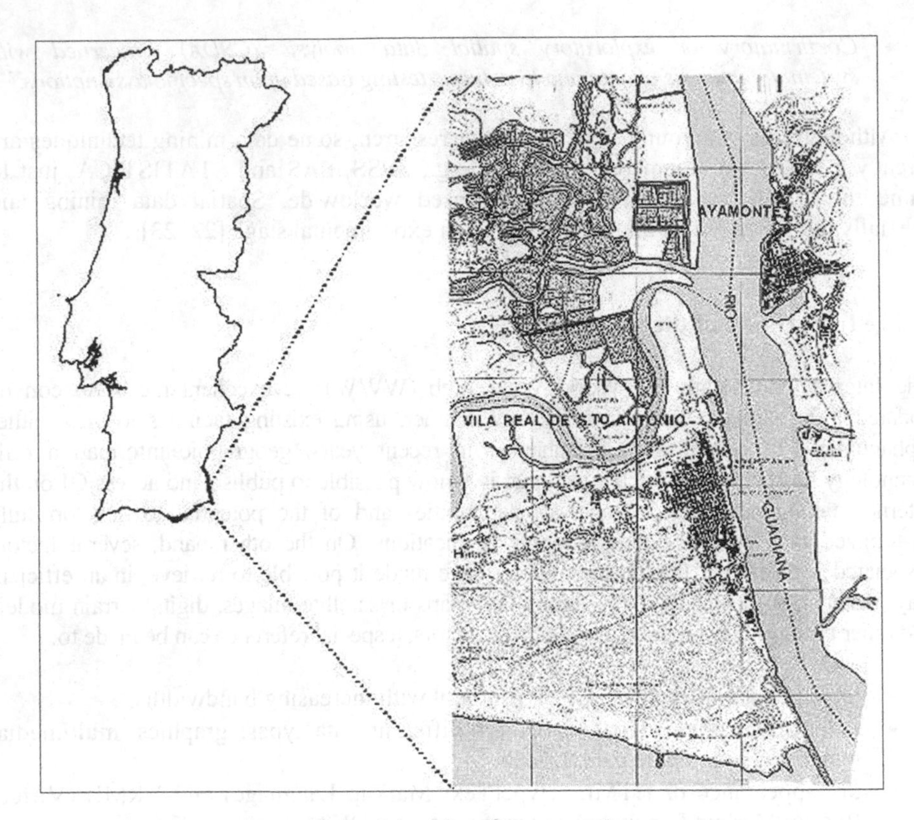

Figure 4. Location of the case study.

situation, given the construction of the Alqueva dam in South-eastern Portugal. Eventually, the system will help INAG to manage the estuary, taking into account the exploitation conditions of the Alqueva dam.

The study involves several components and, therefore, several interdisciplinary teams:

- River component;
- Estuary component;
- Groundwater component;
- Ecosystems component;
- Socio-economy component;
- Information System component.

The information system component encompasses the development of a database structure that will store all the data required by several teams involved in the study to facilitate their assessment. It also comprises all the relevant information resulting from the work. These data should all be geo-referenced. Moreover, the system should include the possibility of mapping the data spatially by using base cartography and thematic spatial analysis.

6.2. DATABASE STRUCTURE DESIGN

In order to design a database structure for the system, the information system team (IST) started the work by meeting with all the remaining teams. The issues encountered at these meetings were the following:

- there was a preference of the researchers towards the maintenance of their specific data by someone inside the component team instead of leaving that task to IST, developing what could be called a component database;
- each team felt that the existence of one global database would be useful, but they were not sure of how they would personally benefit from that while doing their research.

As the database structure gained form, and as it was made available to all the teams, the researchers became aware of how the integration of their data with that of the other components could be used to provide input to their evaluations. Eventually, their interest in building one common database grew.

6.3. SYSTEM DEVELOPMENT – TECHNOLOGICAL ISSUES

The system is currently under development, and decisions on technological issues have already been made. Since this is an interdisciplinary study, composed of information from different sources, several issues are considered:

- users may present several levels of capacity in manipulating computational tools, specifically GIS;
- data may be provided using different units, depending on data sources;
- spatial data may also have different origins, possibly using different projection systems;
- two research teams are external to LNEC and must be given access to the data, thereby raising security issues;
- the results of the project may include public material; making part of the information available to the public will be an issue.

Since development is underway, some decisions have been made on the system architecture, while other decisions are still dependent on the volume and types of data resulting from the study. The evaluation of the issues described above has led to some key decisions in terms of the system architecture.

The developments in GIS software packages have facilitated some decisions in the manipulation capacity of the database. Improvements in hardware have now led manufacturers to incorporate both parts of the database (alphanumeric and graphical) into one structure. In this way, entities in one database can be represented through their specific parameters, while their visual mapping becomes a representation of the information store in the geodatabase. Along these lines, the potential for data handling is much greater, specifically in the links that may later be made with data mining tools.

Since the complete data sets are still not available, it is not clear whether all of the knowledge discovery process (see Section 4) can be integrated in this system. However, it is already obvious that the data to be handled can benefit from the steps of:

- Data cleaning: to evaluate inconsistencies in the data received from the different teams. It would also be interesting to evaluate the possibility of checking for global inconsistencies in the data;
- Data integration: to define strategies to combine data coming from the different data sources of the study;
- Data transformation: to generate aggregate data for global evaluation in the study;
- Pattern evaluation: to search for patterns in the data, specifically spatially mapped patterns.

Figure 5 represents the architecture of the system under development. The resulting database, which includes the complete data sets for the study, has been defined as one coherent whole, not only for the project research teams but also in relation with the client of the study, the National Water Institute. This means that the resulting data structures could be integrated with their current information system, the National Water Resources System (SNIRH). The database system will be available on the project internal network to the several interdisciplinary teams. Public domain information will be made available through the Internet, using online mapping technology, which will be based on a commonly used Internet browser.

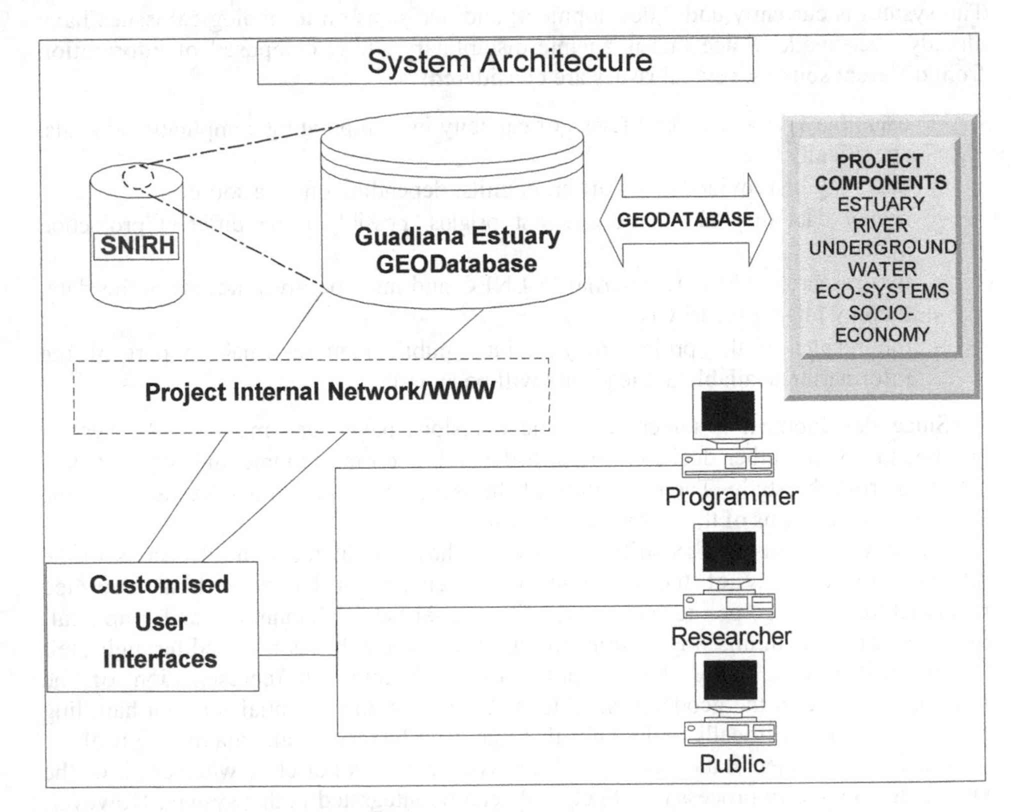

Figure 5. System Architecture for the Guadiana estuary study.

The different profiles of the team researchers have led to the definition of two internal profiles for access, Programmer (given by the system developer profile) and researcher (user of the complete GIS environment, including privileges for data update). These two profiles will be implemented, using current GIS technology. Access to public domain data will be provided through the Public profile, which will be made available through an online mapping commercial package, as described above.

7. Conclusions

This paper attempts to highlight how water resources data management can benefit from information technologies. This is emphasized with the Guadiana estuary Information System. Since the system is currently under development, it is only possible to draw preliminary conclusions as a result of the evaluation done internally in the information system team.

It is the authors' belief that the development of an integrated system for this study, including the different interdisciplinary needs and specifically the ones of the client, will enable the optimized management of the produced information as it becomes available. This will also facilitate future developments for this system and extensions to the database structure, perhaps in the context of other types of study. Moreover, the potential for an interdisciplinary work is enhanced as new opportunities and possibilities become clear from the available information. The system also promotes sharing of experiences as well as information between the different research teams.

Finally, the creation of customized user profiles, with different security definitions, will not only control access to the data, but will also facilitate the use of the system by people from different backgrounds.

8. Acknowledgements

This paper was prepared under the project on the Environmental Conditions in the Guadiana Estuary and Adjacent Areas, financed by the National Water Institute. Acknowledgements are due to the project team who created the opportunity for such an interesting interdisciplinary study.

9. References

1. Anthony, R. (1965) *Planning and Control Systems: A Framework for Analysis*, Boston USA, Harvard University.
2. Santos, M.A. (1991) *Decision Support Systems in Water Resources*, Research program prepared as partial fulfillment of the requirements for obtaining the rank of Principal Research Officer, Lisbon, Portugal, LNEC, 152 p.
3. Smeets, E. and Weterings, R. (1999) *Environmental Indicators: Typology and Overview*, Copenhagen, Denmark, European Environment Agency, 19p.
4. EEA (2001) *Environmental Signals 2001*, Denmark, European Environmental Agency, 113p.

5. Leibundgut, C., McDonnell, J., and Schultz, G. (eds.) (1999) *Integrated Methods in Catchment Hydrology-Tracer, Remote Sensing, and new Hydrometric Techniques*, IAHS Publication no. 258, IAHS Press, Wallingford (UK), 284p.
6. EC (2000) *Directive 2000/60/EC*, Official Journal L 327, 22/12/2000 pp 0001-72.
7. Inmon, W.H. (1996) *Building the Data Warehouse*, New York, John Wiley & Sons.
8. Lang, A. and Manea, A. (1994) BRECHE project of hydrological database for ERB: Modeling of data series, in P. Seuna, A. Gustard, N.W. Arnell, and G.A. Cole (eds.), *FRIEND: Flow Regimes from International Experimental and Network Data*, Proceedings of the Braunschweig Conference, October 1993, IAHS Publication no. 221, Wallingford (UK), IAHS Press, pp 21-31.
9. Manea, A. (1997) Adequate DBMS tools, in *FRIEND: Flow Regimes from International Experimental and Network Data, Third report: 1994-1997*, Paris, France, Cemagref, pp 38-42.
10. Breil, P. (1997) Data selected for the AMHY database, in *FRIEND: Flow Regimes from International Experimental and Network Data, Third report: 1994-1997*, Paris, France, Cemagref, pp 43-44.
11. Rodrigues, A. (1998) The importance of metadata within GI Infrastructures, *Proceedings of the Workshop on Challenges and Future Developments of GI Infrastructures: The Portuguese Experience*, First International Conference and Exhibition on Geographic Information (GIS Planet'98), Lisbon, Portugal, 8 September.
12. Valenzuela, C.R. (1991) Basic principles of GIS, in A.S. Belward and C.R. Valenzuela (eds.), *Remote Sensing and GIS Resource Management in Developing Countries*, Kluwer Academic Publishers, London, pp. 279-295.
13. Scholten, H.J. and LoCascio A. (1997) GIS application research: History, trends and developments, Paper presented at the *ESF GISDATA Final Conference on Geographical Information Research at the Millennium*, Le Bischenberg, France 13-17 September, (http://shef.ac.uk/uni/academic/D-H/gis/key3.html).
14. Laurini, R. (1997) Novel technological for the management of located and mobile information and intelligence, Paper presented at the *ESF GISDATA Final Conference on Geographical Information Research at the Millennium*, Le Bischenberg, France 13-17 September, (http://shef.ac.uk/uni/academic/D-H/gis/laurini.html).
15. Hooper, R.P. and Aulenbach, B.T. (1993) Managing the data explosion, *Civil Engineering* **63(5)**, 74-76.
16. Han, J. and Kamber, M. (2001) *Data Mining: Concepts and Techniques*, The Morgan Kaufmann Series in Data Management Systems, Academic Press, Morgan Kaufmann Publishers, SanDiego, CA, USA.
17. Koperski, K. and Han, J. (1995) Discovery of spatial association rules in geographic information databases, *Proceedings of the International Symposium on Large Spatial Databases*, Portland, Maine (USA), Aug. 6-9, 1995, pp. 47-66.
18. Knorr, E.M. and Ng, R.T. (1996) Finding aggregate proximity relationships and commonalities is spatial data mining, *IEEE Transactions on Knowledge and Data Engineering* **8(6)**, 884-897.
19. Koperski, K., Han, J., and Stefanovic, N. (1998) An efficient two-step method for classification of spatial data, *Proceedings of the International Symposium on Spatial Data Handling SDH'98*, Vancouver, BC, Canada, July 1998, pp. 45-54.
20. Wachowicz, M. (2000) How can knowledge discovery methods uncover spatial-temporal patterns in environmental data, in B.V. Dasarathy (ed.), *Data Mining and Knowledge Discovery: Theory, Tools, and Technology II*, Proceedings of the SPIE 2000 Conference, pp 221-229.
21. Wachowicz, M. (to appear) A knowledge construction process for uncovering spatial-temporal patterns, Paper submitted to the Water Resources Management Journal Spatial Issue on Geoinformatics.
22. Zaïane, O.R., Han, J., Li, Z-N., and Hou, J. (1998) Mining multimedia data, *Proceedings of CASCON'98: Meeting of Minds*, Toronto, Canada, November 1998.
23. Koperski, K., Han, J., and Adhikary, J. (1999) *Mining Knowledge in Geographic Data*, Comm. ACM.

Part II

OBJECTIVES AND INSTITUTIONAL ASPECTS OF ENVIRONMENTAL DATA MANAGEMENT

INFORMATION – INTEGRATION - INSPIRATION

Aspects of Integrated Management of Environmental Data

P. GEERDERS
P. Geerders Consultancy
Kobaltpad 16
3402 JL Ijsselstein, the Netherlands

Abstract. Management of the environment and decision making on the use of its resources require in-depth knowledge and analysis of complex processes and their interactions. Numerical data form an essential basis for the development of models that, in turn, allow for forecasting and simulation. Technology permits ever-increasing amounts of a wide spectrum of data on the environment to be acquired, often in real-time. However, the user in environmental management and decision making is seldom interested in a direct access to these vast amounts of diverse data. Reduction, compression and above all, integration need to be applied, generating user oriented and dedicated information products. Examples will be given on how this approach can be implemented, and the requirements to make such integration successful will be investigated. Only in this way, information extracted from raw data and integrated into a new product can become a valuable source of inspiration for management and decision making on the environment and its resources.

1. General

Mankind is increasingly discovering that its activity, even its mere presence, has a notable and not always a positive influence on the environment and its resources. In many parts of the world, unique ecosystems are disappearing at a frightening rate, leaving behind a barren wilderness of depleted resources and annihilated biodiversity. The impact of these developments on the long term can only be remotely assessed, considering our still very limited understanding of the relevant processes and their interactions on local, regional and global scales.

A call for a more responsible and integrated management of the environment and its resources is being heard. Unfortunately, this call is frequently masked by the noise of the battlefield of seemingly conflicting interests, strongly controlled by financial, economic and political powers, rather than by objective, scientific reasoning. It is the aim of integrated environmental management to take into account the full width of interests and to assess the possible consequences before any actions are implemented. Moreover, this process requires convincing all stakeholders of their common interests and of the need to collaborate, rather than to emphasize and maintain their different viewpoints.

Integrated management of the environment, combined with decision making on the use of resources, requires an in-depth understanding of complex processes and their interactions. It is necessary to understand the processes to be managed before taking any actions that could have irreparably bad consequences for the ecosystem, for the environment, and for us.

2. Data and Information

An essential basis for understanding the environment is formed by data and information on the environment. Besides providing an insight into the actual situation, measurement data form an important contribution to the development of numerical models. Models are a way to visualize scientific understanding of a process or a phenomenon in a mathematical and graphical form. Complexes of models, or meta models, describe the relationships between many processes, each represented by a specific model. Models and meta models allow for forecasting – an expectation of future trends and developments, and for simulation – the assessment of the possible consequences of human actions and interventions through the simulated application of different scenarios, as a basis for selection from available alternatives and preparation for the consequences. Models and meta models form an essential element in integrated management of the environment; however, models always have inherent limitations which can strongly impact upon the results.

While,. a number of years ago, only the availability of proper measurement technology and human resources determined the amount of data that could be acquired on the environment, nowadays technology permits ever-increasing amounts of a wide spectrum of data to be acquired in many ways, often in real-time. In many cases, such data are stored almost routinely, even without a proper view on the specific necessities of data (base) management, and without a proper assessment of the costs and resources involved for long-term storage.

Today, the major bottleneck has become our capacity to process and analyze the huge amounts of different types of data that become available at a frightening pace. It is even more challenging to do so within a reasonable time in order to permit adequate action to be taken, e.g., in cases of emergencies and disasters. The end-user in environmental management, or the decision maker, is seldom interested in a direct access to these vast amounts of diverse data.

Optimizing the monitoring activity and taking into account the previous knowledge of the processes to be monitored can help to significantly reduce the flow of data. In this case, models are used to generate the process data, while a limited amount of monitoring data is used to calibrate and validate the model performance. Of course, a model is an approximation and is never perfect; it only covers "half of the reality". Care should be taken to ensure that the frequency of the relevant changes in the process and the sampling frequency used in monitoring are compatible. This implies the application of the Shannon Theorem, prescribing a sampling frequency twice the frequency of the variations still to be detected reliably.

Decision making requires data to be presented in a dedicated form in order to support and facilitate the process of assessing the possible consequences of measures and actions, before making a proper selection from the available alternatives. This dedicated form needs to be the result of an integrative process, including data as well as models, reducing and compressing the flow of data into a condensed form that provides an "intelligent answer to intelligent questions".

3. Implications

We cannot limit ourselves to integration in the form of simply merging data from different data archives. Experience shows that this is a tedious and mostly unrewarding process, mainly because of the largely different criteria, standards, and methods used for the acquisition and processing of the data. These differences are the result of the largely individualistic and specialized character of environmental sciences, which has maintained itself for a (too) long time. In the last few decades, it has been recognized that only a truly multidisciplinary, integrated approach can achieve the answers to our questions on the environment, its behavior, and its management. We need to aim for a full and consistent integration along the full chain: from the integrated acquisition of data in the field, through various integrated models, to the generation of dedicated, integrated products for decision making.

This results in a view of a generic, end-to-end integrated environmental monitoring system with the following characteristics:

- a wide range of parameters and variables covered;
- well established, standardized methods for data acquisition, quality control, and processing;
- systematic, long-term archival of data in order to allow for assessment of long term changes;
- standardized methods for generating aggregated information in the form of indicators, including status and trends, based upon the basic data and data from appropriate, recognized models;
- standardized criteria for the evaluation and assessment of indicators ("what is good and what is bad?");
- common, interactive, dedicated forms of presentation of indicators, taking into account the needs and requirements of the decision making process.

4. Present Status

Today, operational meteorology provides an excellent example of an integrated observing system: the World Weather Watch. This system includes standardized measurements and observations with standardized instruments and methods at a global scale and at fixed daily intervals. These data form the basis for the operation of complex global, regional and local models and meta models that, in turn, form the basis for weather forecasts in all countries.

In oceanography and marine sciences, be it for a specific set of parameters and variables, certain standards have been developed and are being promoted, e.g., by Intergovernmental Oceanographic Commission of UNESCO (IOC) and International Council for the Exploration of the Seas (ICES). Similar trends are underway in the form of Global Ocean Observing System (GOOS). The GOOS system was initiated by IOC and aims at the establishment of a monitoring and forecast system, similar to that existing in meteorology, for the world's oceans and coastal waters. However, marine and coastal scientists in various countries and centers use largely different instruments and methods for data acquisition, processing, presentation, and archival. It will be one of the challenges for GOOS to overcome these differences and achieve a certain level of comparability and compatibility of its data.

For terrestrial sciences and the monitoring of the terrestrial environment, still much needs to be done to achieve the desired integration and standardization. The many different scientific disciplines involved have only recently started to consider coordinating, homogenizing and standardizing their methods and tools for data acquisition and processing, partially under the influence of large-scale programs and projects such as the International Geosphere Biosphere Project (IGBP).

The situation is even more complicated in the case of socio-economic data of which there are many different types, often not geo-referenced and collected through largely different procedures. Developments are underway to achieve a certain level of standardization for the main variables.

The ultimate challenge is formed by the full integration of data from atmosphere, water, terrestrial environment, as well as from socio-economy, to form a complete and solid basis for integrated management of the environment and its resources.

5. Constraints

A primary constraint for effective integration is the lack of compatibility between environmental data and information from different sources. This relates to a range of aspects including definitions, procedures and methodologies for observations and measurements, as well as the calibration and intercalibration of the measurement systems themselves.

An example of an effort to improve this situation is the recent feasibility study on a Regional Centre for Marine Instrumentation and Related Issues in South America. Such a centre would help to improve comparability, compatibility and traceability of marine and coastal environmental data in the region. The Centre aims to provide services related to calibration, intercalibration, reference materials and information, as well as to provide training in the use of specific instrumentation. However, still too little has been achieved in this respect in most countries: large differences and discrepancies remain, complicating the integration of environmental data.

Another constraint is the lack of a clear description of objectives and criteria for environmental management in specific situations. While some general aims are often mentioned (e.g., "protection of the biodiversity"), these are seldom translated into specific, concrete and quantified requirements, needs and priorities. This implies that the integration process lacks a definition for its desired result and, consequently, the internal structure of the corresponding information system cannot sufficiently be optimized for the generation of a satisfactory output.

6. Institutional Aspects

At the institutional level and in most environmental disciplines, several aspects are of importance. Institutions do not have a ready and consistent policy with regard to the management of their environmental data: how to manage and process them specifically, who is responsible, etc. This negatively affects the availability and accessibility of environmental data from the past; especially on the long run, this may even result in the loss of precious historical data. Projects such as the Global Ocean Data Archaeology and Rescue (GODAR) project initiated by IOC have clearly demonstrated the urgent need for and the great usefulness of well-preserved historical oceanographic data sets at the local, the regional and the global levels.

Meanwhile, many scientific institutions and individual scientists consider environmental data as a potential source of income, as a possible means to recuperate part of the cost of acquisition and processing of the data and maintenance of the archive. Therefore, they actively promote sales of their data and data products, rather than allowing for a free and uninhibited exchange, as promoted by international organizations such as IOC and UNESCO. This leads to undesirable conflicts about property and proprietary rights on the data.

However, while these institutions may be strong in environmental sciences, their capabilities for (commercial) promotion and sales of data often stay largely behind, and, for this reason, the expected profits almost never really emerge. At a global scale, this situation strongly hampers the free, open and integrated access to environmental databases, and consequently forms a hindrance for the effective data exchange necessary to resolve the current problems of global climate and environment. A consistent data policy needs to be developed and implemented to regulate these aspects at institutional, national and international levels.

With regard to integrated data management, the following aspects of data policy are of importance at the institutional level:

- amounts and types of data to be acquired and consequently archived;
- procedures for quality control and assessment of the data in the archive;
- protection of long term consistency of the data archive;
- protection of integrity of the data in the archive;
- procedures for the updating of data archives;
- availability of data documentation and metadata;
- procedures for access to the data archive (options, rights, liability);
- procedures for backup;

- procedures for transcription of archives to new media due to older media becoming obsolete;
- responsibility for long term archival (more than 100 years? Always?).

Clear and unambiguous decisions on these issues need to be made by environmental institutions, including the setting of priorities and the allocation of funding. Maintaining environmental data archives is not a simple task and requires thorough planning and firm long-term commitment! In this context, a certification comparable to ISO 9000 might be considered for institutions that have implemented proper management procedures for their data.

7. View of the Future

The Internet, and its next generation versions, will undoubtedly play a main role in a future integrated system for distribution of and access to environmental data and information. Such a system will include today's functions of telephone, television, radio, as well as the current functions of Internet such as e-mail and access to WebPages. This system will be enabled by a worldwide communication infrastructure consisting of land-based, radio-based and satellite-based elements, forming a global coverage, high-speed network.

As most of the data and information on the environment is georeferenced, it is obvious that Geographic Information Systems play a major role in the presentation and analysis of these data and information. However, these systems still represent reality in a largely reduced and simplified form and, moreover, they usually are static and therefore lack the necessary dynamics to depict the temporal changes occurring in the environment.

New trends can be observed that make use of virtual reality (VR) technology to provide almost lifelike images of complicated environments. The sceneries of Flight Simulator games clearly demonstrate this advanced capability. Such technology could effectively be used to represent digital environmental data and information, as well as model outputs, in a virtual "natural" environment. It has been demonstrated that georeferenced environmental data can be mapped upon a true 3-dimensional, highly detailed landscape. Such a landscape could include, as a realistic representation, all necessary elements, now usually depicted in the form of symbols on maps: buildings, roads, infrastructure, morphology, land use and vegetation cover. It has been demonstrated that the necessary regular updating of the landscape information can be realized by deriving the relevant landscape elements almost in real time from satellite and aerial imagery through a process of intelligent recognition.

As another example, a system was recently presented for the location of personnel in a large company building during office hours. All were provided with a small unit containing a GPS, recording in real time their position in the building and their direction of movement. The unit continuously transmitted this information to the location system. The system then presented this information in a 3D/4D dynamic virtual world, using a schematic presentation of the building and of each personnel member. In the future, using more sensors on the person and a more detailed presentation of the office, the system could present in more detail what the people are doing and make the virtual presentation much more life-like.

The environment is being monitored more and more frequently and, in many ways, both remotely (from space and air) and in the field. Technology is advancing at a rapid pace, and capabilities to represent the environment in a realistic manner are readily available. Integrated environmental management requires a high degree of integration of data of many different types and from different sources, including those from models and meta models. Current technology offers many sophisticated tools and options for realistic "almost life-like" representation of the environment, that easily impress the users. However, in order to ensure that responsible decisions are taken on the basis of these attractive presentations, proper consideration should be given to the inherent limitations and inaccuracies of the original data and information that form the basis of such representations.

8. Conclusions

Although current technology offers a wide range of facilities to achieve integration of environmental data, a successful implementation of this integration requires a set of preparatory steps to be taken, that are largely outside the realm of technology and that precede the phase of integration. These steps are of an organizational character and define the essential consistent context required for the integrative process to lead to useful results and successful applications. They include:

- strong, explicit and long-term institutional commitment to the principles of integrated data management throughout the full chain from acquisition to final application;
- unambiguous and committing definitions of terminology and methodology for data acquisition, processing and presentation, and for the generation of environmental data products;
- quality assurance through regular calibration and intercalibration of measurement instruments and systems;
- development and consistent application of criteria for the evaluation of environmental indicators, in accordance with the established objectives of environmental management in each specific case;
- development of a data policy that satisfies all stakeholders.

Only in this way, information obtained from measurements and observations of the environment, through a process of integration, can develop into a valuable source of inspiration for management and decision making on the environment and its resources.

9. Web References

1. World Meteorological Organization data management: http://www.wmo.ch/indexflash.html (click on: data management)
2. Intergovernmental Oceanographic Commission: http://ioc.unesco.org/iocweb/
3. ICES data aspects: http://www.ices.dk/ocean/Global Ocean Observing System: http://ioc.unesco.org/goos/
4. Example of software for feature extraction: http://www.definiens-imaging.com/ecognition2/
5. Data policy aspects: http://www.geog.ucl.ac.uk/eopole

OCEAN TEACHER: A CAPACITY BUILDING TOOL FOR OCEANOGRAPHIC DATA AND INFORMATION MANAGEMENT

G. REED
Consultant, Ocean Services Section
Intergovernmental Oceanographic Commission (IOC) of UNESCO
1 rue Miollis, 75732 Paris Cedex 15, France

Abstract. Within the framework of the International Oceanographic Data and Information Exchange (IODE) program, the IOC is assisting developing countries with the establishment of National Oceanographic Data and Information Centres. Capacity building comprises an important component of the IODE program and includes regional group training courses and workshops to familiarize data center managers in the different aspects of oceanographic data and information management.

Ocean Teacher is an internet-based system designed to provide training tools for oceanographic data and information management. These tools are used during IODE training courses but can also be used for self-training and continuous professional development. The Ocean Teacher site comprises two components: the IODE Resource Kit and the Resource Kit Training Manual. The Resource Kit contains a range of marine data-management and information-management material, including software, quality control and analysis strategies, training manuals, and relevant IOC documents. The Resource Kit Manual is a collection of outlines, notes, examples, and miscellaneous class work documents used in conjunction with the Resource Kit to organize a training program in marine data and information management.

This paper describes the approach that is being taken by IODE in developing capacity building programs and details the development of the Ocean Teacher system and its content.

1. Introduction

The International Oceanographic Data and Information Exchange (IODE) program of the IOC has developed the Ocean Teacher system, a comprehensive internet-based training and self-study tool for ocean data and information management to support its capacity building strategy.

Ocean Teacher contains a range of marine data management and information management material, including software, quality control and analysis strategies, training manuals, and relevant IOC documents. It is a comprehensive self-training and resource tool for newly established Oceanographic Data Centres, designed to assist managers and staff members to acquire the skills to set up and run new IODE centers. It provides a broad spectrum of background information on global data and information archiving activities, specifications for data storage in standard formats, and the software tools needed to perform many quality control, sub-setting, and analysis techniques. In addition, datasets and information relevant to specific geographical regions are provided as a plug-in "custom pack". While aimed at developing countries, Ocean Teacher will be of considerable value to developed countries and their marine science agencies.

2. IODE Data and Information Management Capacity Building

Capacity building has always been an important component of the IODE program since its inception in 1960. Assisting member states to acquire and build the necessary resources for ocean data and information management is an essential and critical part of all IODE activities. Training courses, mostly hosted by National Oceanographic Data Centres, have been held annually with resource persons from the IODE community providing instruction.

Although these training courses have been successful in training data center staff, the IODE program had never reached the stage where an agreement was made on a 'standard curriculum'. This meant that courses and their content could vary from course to course and that the course programs were not necessarily in line with the requirements of the participants. As these training courses were mostly one-off activities, the long-term impact was often disappointing, as the participating countries did not have the necessary infrastructure to implement the acquired knowledge. There was little or no follow-up to the courses, so the impact of the course on the participants' day-to-day work could not be monitored.

2.1. REGIONAL NETWORKS AS A STRUCTURE FOR IODE CAPACITY BUILDING

IOC activities in developing countries are mostly organized using a regional approach. The IOC regions include IOCINCWIO (North and Central Western Indian Ocean - 'Eastern Africa'), IOCEA (Central Eastern Atlantic - 'Western Africa'), IOCARIBE (Caribbean and adjacent region), IOCINDIO (Central Indian Ocean), WESTPAC (Western Pacific region), Black Sea and Southern Oceans regions. The IODE program has recognized the importance of building networks based on this regional structure. This approach has multiple advantages such as:

- promotion of communication between members of the network;
- promotion of south-south cooperation in training and institutional capacity building;
- facilitating collaboration with other programs and projects;
- facilitating follow-up for training activities;
- facilitating exchange of data and information using compatible technologies and formats.

2.2. OCEAN DATA AND INFORMATION NETWORKS

Since 1998, the IOC has been developing and enhancing the new ODIN (Ocean Data and Information Network) capacity building strategy that links training to infrastructure and operations. These networks promote regional cooperation, data exchange and creation of regional data products. The first project to be implemented under this new strategy was the ODINEA project for the seven East African countries in the IOCINCWIO region. In order to support this new capacity building strategy, IODE started the development of a CD-ROM based "*NODC- In-A-Box*" product. This product proved to be a useful capacity building tool and provided both the tools and the instructions on how to manage and manipulate oceanographic data as well as providing basic regional datasets. In 1998, it was decided to implement a Pilot Project for the development of a computer-based training tool based on the ODINEA CD-ROM, to be called the 'IODE Resource Kit'.

The success of the ODINEA project led to the development, and approval by a donor, of the ODINAFRICA II project involving 19 African member states. This project has duration of three years and a budget of US$4 million. The objectives of the ODINAFRICA-II project are:

- providing assistance in the development and operation of National Oceanographic Data (and Information) Centres and establishing their networking in Africa;
- providing training opportunities in marine data and information management, applying standard formats and methodologies as defined by the IODE;
- assisting in the development and maintenance of national, regional and Pan-African marine metadata, information and data holding databases;
- assisting in the development and dissemination of marine and coastal data and information products responding to the needs of a wide variety of user groups, using national and regional networks.

The first ODINAFRICA II capacity building workshop on ocean data management was held in April 2001, and the existing IODE Resource Kit was expanded to include a Training Manual and was renamed Ocean Teacher.

3. Ocean Teacher

Ocean Teacher is an internet-based training and self-study tool for ocean data and information management. It is used in IODE capacity building programs and also for self-training and continuous professional development. Ocean Teacher comprises two components: (i) the IODE Resource Kit, and (ii) the Resource Kit Training Manual. The Resource Kit contains a range of marine data-management and information-management material, including software, quality control and analysis strategies, training manuals, and relevant IOC documents. The Resource Kit Training Manual is a collection of outlines, notes, examples, and miscellaneous class work documents used in conjunction with the Resource Kit to organize a training program in marine data and information management. As Ocean Teacher is internet-based, it can be viewed with a web browser either on-line or off-line. A CD-ROM version can be prepared at any time for use in capacity building workshops or for distribution to interested organizations or individuals. It is not essential that users need Internet connection to use the system.

3.1. IODE RESOURCE KIT

The IODE Resource Kit is a comprehensive self-training and resource tool for newly established Oceanographic Data Centres, designed to assist managers and staff members to acquire the skills to set up and run IODE data centers. It contains a range of marine data-management and information-management materials, including software, quality control and analysis strategies, training manuals, and relevant IOC documents. The Resource Kit provides a broad spectrum of background information on global data and information archiving activities, specifications for data storage in standard formats, and the software tools to perform many quality control, subsetting, and analysis techniques.

The Resource Kit is modular in design and contains three basic modules:

- Module 1. IODE Data Centre System
- Module 2. Data Management Systems
- Module 3: Data Analysis and Products

Module 1 discusses the roles and responsibilities of an oceanographic data center and describes the IODE global network system of data centers. It further describes data and information management within a science program and shows how the data manager can provide valuable data and information sources to managers and project scientists during a science program. A comprehensive reference library containing relevant IOC manuals and guides, online tutorials, and standard reference material is also included.

Module 2 describes some of the skills essential for an ocean data manager including computer systems, database technology, metadata and information management, data observation and collection instructions, data quality control, the use of the internet for data and information exchange, and an introduction to geographical information systems.

Module 3 describes in detail a number of data formats and the source of collateral data. It also includes a data classroom and software toolbox. The data classroom provides a training curriculum in the use of selected software to quality control and analyzes ocean station data, using software tools such as the Ocean Data View and Java Ocean Atlas programs, and standard spreadsheet and relational database programs. The data classroom emphasizes the connections between available software and global databases, based on the use of common formats. The software toolbox provides a number of useful software tools that can be immediately installed and run. Manuals and test datasets are included. These software packages are freeware and shareware applications.

Full details of the content of the Resource Kit are listed in Annex.

In addition, a regional data and information module that includes environmental datasets, ASFA extracts, and information and data products for a specific region is included as part of the Resource Kit. Currently, the regional data and information module is available for the Western Indian Ocean (IOCINCWIO) region, and a further data and information module is under preparation for the Eastern Atlantic Ocean (IOCEA) region.

The modular approach taken with the development of the Resource Kit enables: (i) selected experts to contribute and regularly update the content; and (ii) course programs to be designed, based on individual, national or regional priorities, using material from each module.

3.2. RESOURCE KIT TRAINING MANUAL

The Resource Kit Training Manual is a collection of outlines, notes, examples, and miscellaneous class work documents that can be used in combination with the IODE

Resource Kit. The aim of the Training Manual has been to organize the original source documents and reference materials into the Resource Kit itself, while saving the instructional materials that point to these documents for the Training Manual. The Manual and its inherent course outline have been developed over a number of years, principally during IODE training workshops for ODINEA project. The long-term exercises, referred to as "Intersessional Goals," are those pioneered by the ODINEA participants. It was felt that the best way to learn the material in the Resource Kit was to undertake real-world projects to find, quality control, analyze, synthesize, and publish marine data and information.

A typical ocean data management workshop would cover:

- basic computer skills;
- the importance of marine data in general, and within the national and regional environment in particular;
- how to set up an oceanographic data center within the IODE System;
- the infrastructure requirements, including hardware and software tools;
- how to manipulate and analyze the principal types and formats of marine data;
- how to produce ocean data products and to disseminate these products, both over the internet and by traditional methods.

The topics covered are based on material from the Resource Kit, although not necessarily all during the same workshop session. Some topics are covered only once, while others are taught more than once, but at increasing levels of difficulty during the 3-year training cycle.

The aim of Ocean Teacher is not just to train individual scientists, but also to build new training capacity in participating institutions. Toward this end, the entire Resource Kit and Training Manual are designed to let former students become new local trainers. This means that the Training Manual is designed not just for the IODE training staff, but also for use by the course participants themselves.

3.2.1. *Information Management Component of the Training Manual*

In the ODINEA project, a clear distinction was made between the data management component and the information management component of the capacity building program. Ocean Teacher now brings these two components together into a single system.

The marine information management component provides a foundation for the professional education essential for the modern information worker. It consists of three parts:

Part 1. Information concepts, information technology and software, information seeking, and the organization of the knowledge base using a defined Integrated Management System;

Part 2. Creation of research support services, information management and retrieval, information resources for marine science particularly in the electronic environment;

Part 3. Document and e-publishing, internal knowledge management, archives, professional networking, continuing professional development, information skills training techniques.

As well as specific information handling techniques, particularly in the electronic environment, students examine the political, economic and social context of the formation of an information center. Teaching is underpinned by relevant research and utilizes web-based learning modules as well as practical exercises. Topics covered include:

- evaluation and assessment of the information requirements of the organization;
- project management on the creation of an information center in the organization;
- understanding the software and hardware required to underpin information management;
- application of efficient techniques to information seeking exercises;
- building, organizing and documenting a library collection, both paper and electronic;
- use of the defined integrated library management system to support all management activities;
- setting up and maintaining research support services;
- identification of the major electronic resources in marine science and organization of access;
- overseeing the production of internal publications and advising on e-publishing;
- introduction of the concept of knowledge management to the organization;
- networking within the activities of marine science information associations;
- identification of opportunities for continued professional education;
- provision of information skills training on marine science resources to Information Centre users.

Instructional material for the information management component of the Training Manual is currently under preparation and will be completed in time for the first ODINAFRICA II Marine Information Management Training workshop to be held in November 2001.

4. Conclusion

The Ocean Data and Information Network (ODIN) capacity building strategy, developed by IOC, links training to infrastructure and data center operations. This model has proved to be successful within the African framework and will be expanded to include other regions. The Ocean Teacher system is an important component of this strategy, providing an internet-based training and self-study tool for ocean data and information management for use in IODE capacity building programs and also for self-training and continuous professional development.

The IODE program of IOC will continue to develop and maintain the Ocean Teacher system as both an on-line and off-line resource to support its capacity building strategy. The on-line version is available at www.OceanTeacher.org. A CD-ROM version, for off-line viewing, can be ordered from IOC.

ANNEX. DETAILED CONTENTS OF THE IODE RESOURCE KIT

Module 1. The IODE Data Centre System

- ❑ Ocean Data Centres
  - ➢ What is an Oceanographic Data Centre?
  - ➢ The Roles and Duties of an NODC
  - ➢ The IODE System Description
- ❑ Global Programs
  - ➢ Science Programs
  - ➢ Intergovernmental Coordination Programs
- ❑ Science Plans
  - ➢ Examples of Science and Implementation Plans
- ❑ Data Policies & Guidelines
  - ➢ Examples of Data Management Policies
- ❑ Reference Library
  - ➢ IOC Manuals & Guides
  - ➢ Online Tutorials
  - ➢ Standard Reference Material

Module 2. Data Management Systems

- ❑ Computer Systems
  - ➢ Hardware
  - ➢ Operating System
  - ➢ Applications Software
  - ➢ Networks
  - ➢ Computer Maintenance
  - ➢ Computer Viruses
- ❑ Databases
  - ➢ Database Management Systems
  - ➢ MS Access 2000 Tutorial.
  - ➢ Cruise Report Database Tutorial.
  - ➢ FileMaker Getting Started Guide
  - ➢ FileMaker User Guide.
  - ➢ Other Database Management Systems.
  - ➢ Oracle 8
  - ➢ Informix Universal Server
  - ➢ PostgreSQL
- ❑ Metadata
  - ➢ Overview of Metadata
    - • What are Metadata?
    - • Why Use Metadata?
    - • The Role of a Data Directory
  - ➢ The MEDI Metadata System
    - • Background to MEDI
    - • The MEDI Catalogue
    - • Install the MEDI software

- Cruise Summary Report
    - Introduction to the Cruise Summary Report
    - Paper version of the Cruise Summary Report
    - Install the ROSWIN software to fill a Cruise Summary Report form
    - Install the Cruise Summary Report data files
    - Install the ROSEARCH software to search for cruise summary reports
- Global Change Master Directory
- Distributed Oceanographic Data System (DODS)
- Metadata Standards

- Data Collection
    - Instrumentation
        - WOCE Operations Manual
        - Oceanographic Instrumentation
        - Protocols for JGOFS core measurements (JGOFS Report No. 19)
    - Data Collection Forms
        - Hardcopy Logsheets
        - Spreadsheets
    - Instruction Manual for Data Collection
- Quality Control
    - Overview of Quality Control
        - Objectives of quality control
        - Quality control procedures
        - Quality Control & Processing of Historical Oceanographic Nutrient Data
        - NODC Data Quality Control and Quality Insurance
    - Quality Control of Data from Global Programs
        - MEDS Quality Control Procedures for Ocean Profile Data
    - TOGA/COARE Handbook of Quality Control Procedures for Surface Meteorology Data
        - BODC-WOCE Sea Level Data Assembly Centre Quality Assessment
        - TOGA Sea Level Centre Quality Assessment Policy
        - Quality Control of data received by Ocean Climate Laboratory
    - Quality Control References
        - GTSPP Real-time Quality Control Manual (Manuals & Guides 22)
        - Manual of Quality Control Procedures for Validation of Oceanographic Data (Manuals & Guides 26)
        - Quality Control Cookbook for XBT Data
- The Internet
    - Introduction to the Internet
    - History of the Internet
    - Electronic Mail (email)
    - File Transfer Protocol (FTP)

- Telnet
- Discussion Groups
- Mailing Lists
- World Wide Web (WWW)
    - Web Browsers
    - Netscape Interface
    - Internet Explorer Interface
    - Browser Errors
- The URL
- Search Engines
    - Search Syntax
- Netiquette
- Glossary of Internet Terms
- Beginners Guide to HTML
    - HTML Tags - Quick Reference
- eXtensible Markup Language (XML)

❑ Geographic Information Systems

- Overview of Geographic Information Systems
    - Benefits of a Marine GIS
    - GIS Glossary
    - The Emergence of Marine GIS
- GIS Tutorial
    - Training Module on the Applications of Geographic Information Systems (GIS)
- GIS Resources on the Internet

Module 3. Data Analysis & Products

❑ Data

- An Introduction to Atmospheric and Oceanographic Datasets
- Oceanography Primer
- Datasets
    - Major Publishers
    - Major Publications
    - Data Directories/Indexes
    - WWW Data Sources Catalog
    - Data CD-ROM Catalog
- Quality Control
    - Program Planning
    - Manuals, Methods & Protocols
    - Standards and Reference Materials
    - Intercalibration
    - Managed Data Flow
    - Statistics & Graphics
    - Analysis
    - Bad Data?
    - Final Data

- Formats
  - Format ABC's
  - Format Types
  - Integrated Data Formats
  - Formats Catalog
  - Examples
- Software
  - The Software Toolbox
  - IOC/IODE Catalog of Marine Software
- Data Products
  - Metrics
  - Centre Documents
  - Maps
  - Principal Formats
  - Dataset Products
  - Analysis Products
  - Data Atlases
  - Web Options
- Data Classroom
  - Resource Integration
  - Format Conversion
  - Tutorials

Part III

DESIGN OF DATA COLLECTION NETWORKS

ENVIRONMENTAL MONITORING TIME SCALES: FROM TRANSIENT EVENTS TO LONG-TERM TRENDS

P. H. WHITFIELD
Meteorological Service of Canada – Pacific and Yukon Region
Environment Canada
700 - 1200 West 73rd Avenue
Vancouver, B.C., V6P 6H9, Canada

Abstract. Environmental studies and monitoring programs provide information that is required to make wise decisions regarding the present and the future state of resources. Too restricted a perspective on either temporal or spatial scales curtails the adequacy of the associated decisions. As environmental scientists, we have often over-sampled the one-week to five-year time scale and the 1 to 1000 km spatial scale, neglecting other time and spatial scales. Environmental studies seldom collect samples more frequently than once per week, and studies seldom exceed five years and very rarely exceed 25 years. Correspondingly, environmental studies usually focus upon repeated sampling at specific locations, usually many tens or hundreds of kilometers apart. However, many important environmental events take place on temporal and spatial scales outside this narrow focus. Understanding processes that take place outside this 'time and space window' is critical to making wise decisions. The scale at which processes operate may change; such changes may be more significant than minor changes in the mean level. In the environmental sense, all time scales and spatial scales that exist in ecosystems are relevant. Additionally, time scales at which we sample must be appropriate for the spatial scales of the phenomena with which we are concerned. This paper provides some examples of results from data gathered outside the usual temporal 'window' and highlights the relevance and complementary nature of such results to environmental issues. Inclusion of additional time scales is crucial to understanding environmental processes and fundamental to identification of the functioning of ecosystems.

1. Introduction

Environmental monitoring is necessary to provide the data and information necessary for making informed decisions regarding the current and the future state of environmental systems. Outputs from these systems take place at various time and space scales, ranging respectively from fractions of seconds to millions of years, and from millimeters to the global extent [1, 2]. Study of responses on various time and space scales is crucial to understanding environmental processes and fundamental to identification of the functioning of ecosystems [2, 3]. Extrapolating in space and time from network data is difficult [4]. Also, the scale of concern is not often explicitly considered neither in designing monitoring programs [5] nor in the redesign, curtailment or cancellation of existing long-term programs [6]. It is essential that all data collection and evaluation programs consider how ecological processes, including those of man, might be affected by changes at a variety of time scales. Therefore, from an environmental perspective, all time scales that exist in an ecosystem are relevant.

Environmental monitoring or data collection programs consist of five parts as identified by Whitfield [7]: (1) establishing a goal for monitoring, (2) selecting a monitoring strategy to attain that goal, (3) reviewing the adequacy of the sampling to meet the goal, (4) optimizing the sampling, and (5) reviewing the adequacy of the monitoring goal. The many specific needs for data can similarly be categorized into five goals: (1) assessing for trends, (2) assessing compliance with objectives, (3) estimating mass transport of materials, (4) assessing environmental impact, and, (5) gathering surveillance data. Most existing programs fit within one, or more, of these monitoring goals.

While the desire of most resource management organizations is 'perfect' decisions, most recent programs intended to generate the information that serves as the basis of such decisions have restricted perspectives on both temporal and spatial scales. On the temporal scale, most studies consider seasonal and year-to-year variations. On the spatial scale, most programs consider stations that are tens to hundreds of kilometers apart. It is important to understand that these scales are convenient and expedient, from both the operational and the political perspectives. They are not necessarily chosen because they are environmentally significant. During such studies, samples are only collected infrequently and sparsely. It seems very likely that such under-sampling cannot provide the information needed about the system studied, failing to give any sense of either the short-term features or the long-term features of the environment. This perspective often results in lack of full appreciation of much of the structure of systems.

There is another limiting feature to data collection and analysis programs. The nature of the training of most professionals has been concentrated on a time scale from several months to several years. In this training, there is a distinct focus on seasonal and diurnal variations in the environment. This "culture" is further promoted by many agency mandates and, for government organizations, by the nature of the operating systems that are in place. The result of this "cultural bounding" is that we are often faced with trying to address environmental issues and problems from an inappropriate time focus. The spatial and temporal scales of the environment are related in Fig. 1. The band across the scales suggests where sampling frequency and spatial extent may be effectively studied. The "cultural bounding" area shows the approximate limits of studies focused on seasonal and diurnal processes. The time scales at which sampling commonly occur are inappropriate for the spatial scale of the phenomena, and *vice versa* [8, 9]. This paper discusses the importance of gathering data that provide information about the environment in addition to those gathered inside this usual 'window' and emphasizes the importance of obtaining relevant data from all time scales present in the environment.

2. Time Scales

Ecological systems respond on three time scales. First, transient events are deviations from usual conditions. These deviations are of short duration relative to the time frame studied and are often of significant environmental consequences. It is important to note that studies of minutes may have transients of seconds, studies of days may have transients of hours, studies of centuries may have transients of years, and studies of

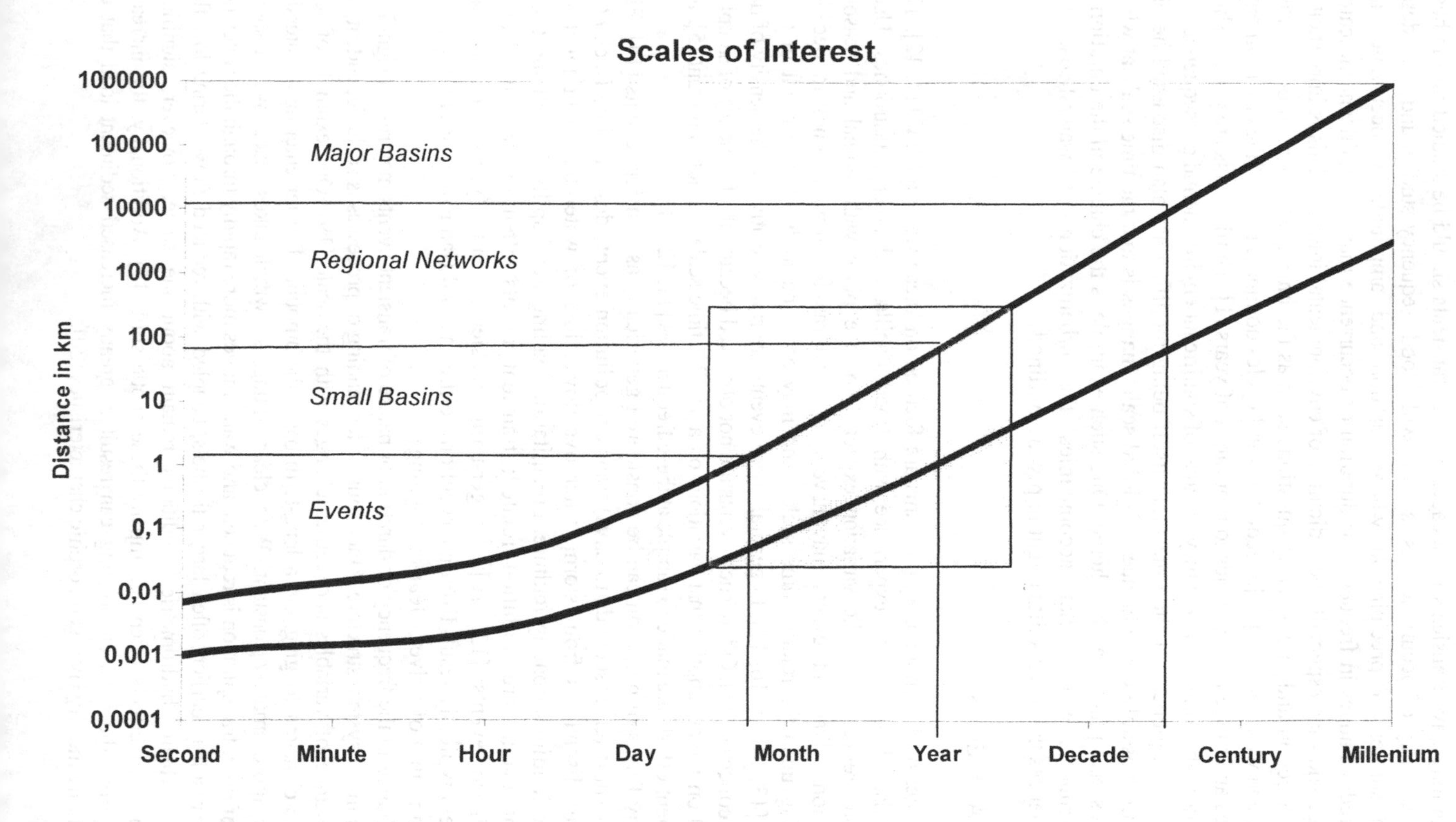

Figure 1. The relationship between the Spatial and Temporal Scales of environmental monitoring; effective sampling occurs in the area between the curves. The black shaded area represents the area of sampling corresponding to seasonal and diurnal processes.

millennia may have transients of decades. These transients should be studied on the basis of the frequency of occurrence, as is done with flood frequency studies and with depth duration studies of precipitation where duration and amplitude of occurrence are considered. Changes in frequency or duration of occurrence may be significant indicators of change and may especially be indicative of environmental impacts. This is important in change detection and in answering questions such as the impact of urbanization on water quantity and water quality [10]. Second, natural cycles dominate most of the environmental processes and range from seconds to thousands of years [1]. Third, trends are pattern shifts that occur in the long term, and they are also of significant environmental consequence.

The consequences of the time scale itself being modified are also discussed herein. One of the possible consequences of global scale changes is that the time scale at which processes take place may be altered. One such example is the change in the duration of rainfall-runoff events, that accompanies the urbanization of watersheds. The consequences that such shifts might imply are outlined.

2.1. TRANSIENTS

Many water quality monitoring programs are focused on 'mean water quality' [11, 12]. This focus exists despite man's experience with floods, spills, and other catastrophes. Usual conditions reflect the cyclic variation expected to take place with diurnal and seasonal progressions. Transient events may have a duration ranging from seconds to decades, depending upon the relative time scales, and may occur naturally or as a result of man's activity [13, 14, 15, 16, 17]. In general, these events are rarely relative to the density of data that are collected [16]. Often, such events are not observed because the frequency of sampling is based on the assumption that variations on a smaller time scale are not important [5], and insufficient efforts are made to sample at these frequencies [11, 12, 18].

Many forms of pollution can be considered generically as transient events [14]. Five types of short term (seconds to days) transient pollution events described by Beck *et al.* [14] are: discharges from storm water overflows, loss of water treatment functions, episodic loadings from agriculture, precipitation loadings, and spills. Not all short-term transient events are pollution related; transient events often occur coincident to atmospheric events [13, 16]. Longer-term transient events (years, decades, and centuries) typically result from interventions, either natural (landslide) or anthropogenic (e.g., damming of a river, clear-cut logging).

A change in the frequency or duration or nature of transient events may reflect significant changes in ecosystem structure. Often, our understanding of processes is scale dependent, and the relevance of variables and causes changes with the scale [19]. One example of scale effects is clear-cut logging. On a long-term basis, the amount of water entering a watershed remains approximately constant. With clear cutting, snowmelt takes place over a shorter period of time, transpiration is reduced, and water moves more rapidly through the watershed. The net result is earlier, often larger freshets, coupled with reduced flows during low flow periods. "Mean" hydrometric conditions remain about the same; however, significant environmental effects often accompany these subtle shifts [5]. Additionally, the increased erosion from the clear-cut logging can result in greatly increased sediment load that can substantially modify river geomorphic characteristics [20].

In natural systems, transient events provide some pressures that help ecosystems remain resilient [21]. Uncoupling these events from the natural systems may lead to ecosystem state changes. Transient events may be significant in a number of ways [16, 22]. First, the nature of a transient event may explain observed ecological impacts. Second, transient events may be the significant feature of an environment or of an environmental problem. Some transients may be characteristic of system functioning, and other transients may be symptomatic of human impacts [14].

Transient events are difficult and costly to monitor [16, 23]. At present, our understanding of such events is quite restricted, and often, we end up focusing on the aftermath of an event rather than the event itself. Spangberg and Niemczynowicz [15] report on the runoff events from a parking lot. They found that the significant portion of the pollution wash-off occurred during the initial stages of the runoff process before the peak runoff, and that the amount of wash-off was a function of the rainfall intensity. Harriman *et al.* [18], using two sampling regimes, found that a more frequent sampling enhanced the ability to interpret hydrochemical processes by improving the level of detail. This is essentially improving the resolution of the signal being measured. Weatherley *et al.* [12] suggested that the exclusion of transient events from models may cause problems in interpretation for individual catchments but are unlikely to cause major errors in regional scale models. Whitfield and Dalley [13] report rainfall driven pH depressions, events lasting from several hours to several days. Different types of transients have different driving mechanisms; Whitfield and Wade [16] observed three types of transient events: one driven by rainfall, one driven by the application of road salt, and one resulting from the wash-out of residential herbicides. Wade and Whitfield [22] describe variations in the impacts of rainfall events, and the impact of a sewage spill. These events all took place over a short time period, and less frequent water sampling would have likely failed to observe any trace of the events.

One example of a transient event is given in Fig. 2. This example was observed, using the methods described by Whitfield and Wade, where pH and water temperature were observed at one-minute intervals [16, 24]. The transient event observed is the result of a storm front crossing over the watershed. The changes in temperature are rapid; recovery is somewhat slower. In addition to the general diurnal nature of pH, there is a distinct depression of pH (~0.4 units) at the time of the temperature drop, shown on expanded scale in Fig. 3. At this resolution, events that last only a brief period of time exist here only for the five to ten minutes during when the storm front passed over the site.

Relatively, little is known about transient events in environmental systems, and even less about the processes by which they are generated. Whitfield [25] describes spectral decomposition as an approach to analysis of the nature of transient events. Spectral decomposition provides information about the signature of each variable; some variables have common signatures for different variables in different streams, while some variables have unique signatures [25]. In comparing events that affect the entire watershed (i.e., rainfall) to events that take place within the stream channel (i.e., sewage spills), Whitfield [26] found that the two types of events had distinctive responses. Such signatures might be useful in classifying types of transient events, and they also have applications in screening records for known and unknown transients [25, 26].

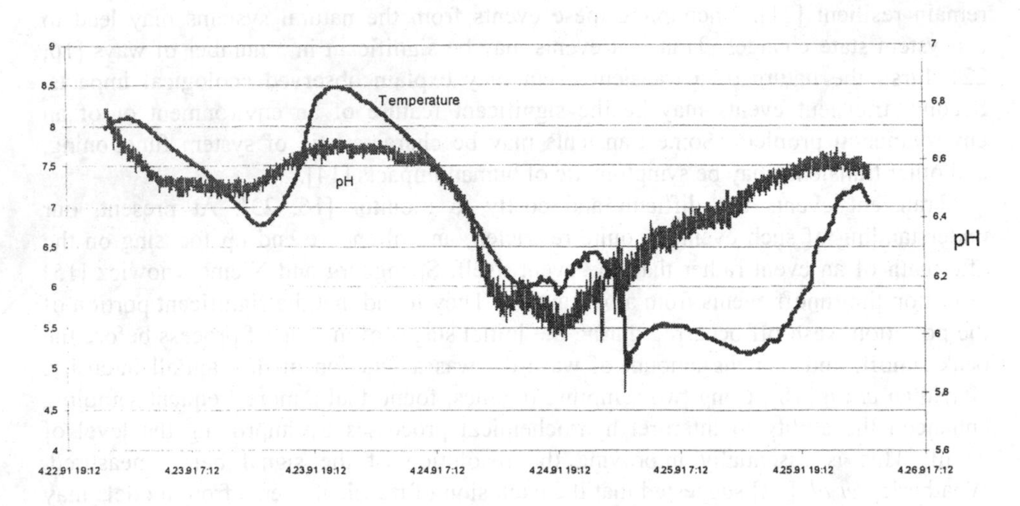

Figure 2. Transient variations in water temperature, and pH in Kanaka Creek, B.C during April 1991. Sampling interval is 1 minute.

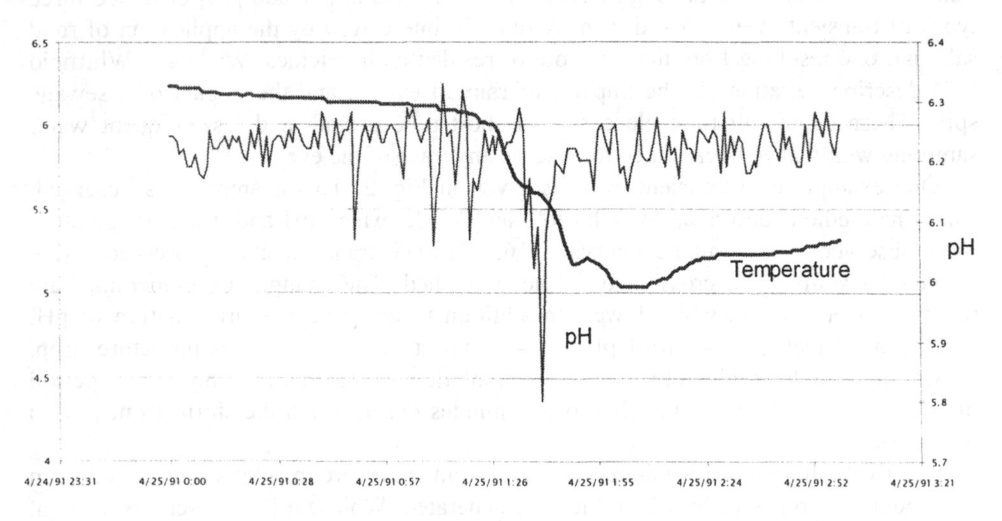

Figure 3. Expanded 3 hour portion of Figure 2.

Environmental managers also need to be concerned with transient events from a number of longer-term perspectives. These include but are not limited to:

Changes in Frequency - how often does a transient occur, and is the process susceptible to changes in recurrence? Examples of increasing frequency of fish kills and spills might be included in this list. We should also be aware of the consequences of reducing the frequency of a class of event. One example of this might be reducing the inundation of wetlands by floods; decreased flood frequency reduces water and nutrient supply and impacts on the productivity of wetlands.

Changes in duration - how long does a transient last? Examples in changing duration of an event include the urbanization of watersheds, reducing infiltration and reducing the duration of runoff events. Fish may be able to withstand a short period of low dissolved oxygen or higher temperatures; however, the consequences of increased duration of such events can be significant.

Changes in magnitude - how much does the event deviate from the normal condition? Examples include increased flood magnitudes; increased levels of toxic materials, and reductions in essential materials. The response of ecosystems to such changes has both immediate and long-term effects.

Trends in transients - are multiple changes in the nature of transients taking place in a monotonic fashion? Are such events as floods and spills increasing in a combination of frequency, duration and magnitude? How susceptible are ecosystems to such changes?

Transient events are ecologically significant; they occur in nature and can be significantly altered by human activities. Transient events may have consequences that are not a function of duration [27]. In other words, a short duration event can have a large, and/or a lasting impact on an ecosystem. A forest fire, or a flood, is an example of such an event where the consequences are independent of the duration of the event. How an ecosystem responds to such events may be a characteristic of the ecosystem [6, 26]; changes in the nature of the event or of the ecosystem might not result in the characteristic response.

Some transients are symptomatic of human impacts [14, 16, 22, 23, 25]. Such events may be outside the natural range of the ecosystem, and hence, the impacted ecosystem may have no resilience for such events. One example would include the consequences of an oil spill. Table 1 provides some examples of the time scales of ecologically significant events in relation to the spatial scale. The table indicates that different events have effects on different spatial scales and suggests that relative scale is determined by the nature of the driving force. Weins [3] suggests that many ecologists have behaved as if patterns and the processes that produce them are insensitive to differences in scale and have designed their studies with little attention to scale. Similarly, investigators have often addressed the same question of different scales; not surprisingly, the findings do not always match [3].

2.2. CYCLES

Natural ecosystems have temporal patterns. These cycles range from biologically driven diurnal patterns to multi-year population cycles. Physical and chemical processes also exhibit cycles; snowmelt-runoff cycles operate on an annual basis, and rainfall-runoff cycles operate over a few days. Some aquatic systems also have lunar driven cycles. Temporal patterns are characteristic of system functioning.

TABLE 1. Selected examples of ecological events and their most common temporal durations.

	Small Basin	Regional Network	Major Basin
Flood	Minutes-hours	Hours-days	Several days
Freshet	Days	Weeks	Months
Spill of waste	Minutes-hours	Days-weeks	Weeks-months
Plankton Bloom	Days-weeks	Weeks	Several weeks
Acid Rock Drainage	Hours-days	Months-years	Hundreds of years
Hydro-Dam Lifetime	Decades	Tens of decades	Centuries
Forest Fire	Decades	Weeks	Days
Clear-cut Logging	Centuries	Tens of Decades	Years
Drought	Weeks	Months	Years

These natural cycles also vary on a spatial scale [2, 3]. Spatial variations in these cycles may result in the existence of an ecosystem type. Similarly, the ecosystem type may generate a characteristic temporal cycle. In a sense, this is the dilemma of ecosystem science: separating what processes generate which ecosystem characteristics and what ecosystems generate which processes.

Sampling seasonal and diurnal cycles is a complex task. If we sample at one rate, for instance monthly, we cannot identify cycles shorter than seasonal. The resolution that the sampling provides needs to account for all the important time scales. Such sampling can be approached using the Nyquist sampling theorem; the sampling rate must be at least twice the frequency of the highest frequency component in the waveform being sampled [28]. However, cycles at shorter intervals, and transient events can significantly impact the data from lower sampling intensities.

2.3. TRENDS

A trend is the general tendency to change in a certain direction or manner. Therefore, when we discuss trends, we include such changes over time as step trends in the mean, or in precision, long term monotonic trends, pattern shifts, transient event frequency/duration/magnitude shifts, shifts from stationarity to non-stationarity (or *vice versa*). Trends are symptomatic of environmental changes. With many social and economic consequences to ecological change, concern for the potential existence and effects of change maintain a high profile. Unfortunately, many data collection programs that are identified as trend assessment programs are not operated in the appropriate time domain.

Intuitively, we view the universe from a local perspective. We assume that today's conditions are 'typical' or 'normal,' and collaterally; we also assume that changes from the present state are not desirable from an ecological perspective unless there is clear evidence of an anthropogenic modification or degradation of some location. Trends towards better conditions also have environmental significance. In the latter case, the concern is that controls, legislation, or other actions are having the desired effectiveness [29]. Present-day interest is primarily a function of the impact that human activities have on water quality [30], and secondarily, a function of long-term sustainability of aquatic or water related resources.

Statistical techniques for data collection and analysis make the assessment of some classes of trends possible on a routine basis [e.g., 31, 32, 33, 34, 35, 36, 37]. However, it is not presently possible to design data collections that permit analysis of all classes of trends. Data collection programs then need to be structured around maximizing the likelihood of finding only certain classes of trends. There are many definitions of a trend; if you are looking for a trend then you need to have an appropriate definition of what you are seeking [38]. It is not practical to look for all types of trends. Long-term data collections require the commitment of significant resources over a long period of time in order to achieve the desired results. Often, our concern for long-term change is compromised by our "cultural focus" on the short-term return. While all data collection programs need to be evaluated as to efficiency and effectiveness, long-term programs are often measured against short-term criteria. The net result of this type of evaluation is that short term "time scale bound" programs are made to look more effective, and funding for long-term monitoring is curtailed. As a consequence, we often see many long-term hydrometric sites being discontinued [39].

While there has been considerable recent interest in statistical detection of trends in time series of data [40, 41, 42, 43], little has been done to determine objectively the underlying causal factors generating the trends [30]. More studies of trends are needed, which relate the trend to the changes in process that generate the trend. Clair and Whitfield [44] assessed trends in acid related variables in Atlantic Canada and found trends that were in disagreement with those suggested by hydrochemical models. Uri [30] showed that changes in cropping practices have a significant impact on the detected trend in sediment concentrations. Development of a reservoir for flood control and hydroelectric generation changed the natural cycles in a river, and stochastic models relate these changes to environmental processes [45, 46]. Welsh and Stewart [47] studied the effect of drought on stream conductivity and bushfires on flow and turbidity of two Australian streams; such studies are effective in revealing the nature of such interventions.

Environmental trends may be significant in a number of ways. Of direct human concern are trends that indicate changes in ecosystem states. Changes in ecological states may restrict the options for use [21]. In some cases, such a restriction may result in option foreclosure or the elimination of particular choices. Trends may indicate an ecosystem change or response to changes. Where an ongoing change is indicated, the consequences can be evaluated. Finally, certain trends may validate the effectiveness of remedial measures [29].

Monitoring of ecosystems should use a continuum of time scales [2]. Loftis *et al.* [5] advocate a more explicit consideration of time scales in data collection and analysis. This range of time scales needs to be associated with the class of trend that is of interest.

2.4. TIME SCALE BOUNDING

What are the appropriate time scales to be considered? A data collection program should be viewed from the point of the design time scale in terms of efficiency and effectiveness. As an example, let us consider hydrometric data collection. Most hydrometric data collection programs are similarly structured. Generally, all hydrometric stations are operated under a single set of protocols. These protocols are based on common operating and reporting procedures. In effect, every hydrometric data collection station operates in approximately the same fashion. This is logistically and administratively simple for reporting and exchanging data. However, no thought is given to maintaining the same level of precision and accuracy between stations; rather, the same level of operating procedure is maintained. Large area watershed stations are more precise than are needed, while small watersheds are unlikely to ever achieve the same level of precision and accuracy [48].

Similarly, the scale of reporting is also a time/space scale phenomenon. Large watersheds are generally insensitive to short time scale events; for example, one hour of rain at some location in a large basin does not generate an event at the same time scale downstream in the basin. At smaller spatial scales, short-term events do result in short time scale events. Using a single set of reporting criteria, such as that typical of hydrometric programs, [annual runoff, monthly mean flow, daily mean flow] may compromise the usefulness of the data for specific analyses. The design and operation of each station should be an optimization of the measuring and reporting of data on a time scale appropriate for the specific size basin. In this way, the data collection can be focused on data needs and analyses rather than on operating conditions. Gupta [49] suggests that stations and networks of stations should be operated to optimize the information content of the records obtained.

In terms of environmental change, some effects of long-term change are better considered as changes in scale rather than changes in amount. Often, when we talk of climate change, we focus on such issues as the warmer and the wetter. Global climate models suggest that we will see changes in mean conditions. It is likely that these will not be the only changes that will take place. At the ecosystem level, there are potentially changes at several scales that we should be concerned about. One example is eutrophication. Eutrophication is a natural process that is greatly accelerated by the addition of nutrients from sewage or agricultural sources. The concern is for the changed rate of the process rather than for the process itself.

An example of this issue is demonstrated by the analysis of Leith and Whitfield [50] in seeking a hydrologic signature of climatic variations. They found that monthly series was too coarse to determine change, and daily series too noisy to determine changes in amounts and timing. They focused on five-day average streamflow that enhanced the signal to noise ratio and showed allowed shifts in streamflow patterns to have occurred.

Will climate change result in changes in extremes or rare events such as floods and droughts? Will these events change frequency, becoming more common or more rare? Will we be able to tell? Our existing data records do not suggest that stationarity exists for ecosystems. This will make it difficult to assess whether or not the change is linked to climate change or to the lack of stationarity. On a geological time scale, ecosystems may never achieve stationarity. Are the present patterns we observe truly characteristic of 'normal' or 'usual' conditions?

Will ecosystem changes accompanying climate change affect the domain of processes or the temporal patterns characteristic of the functioning of systems: seasonal, diurnal and event-driven? With climate change, ecosystems will change. These changes will likely range from quite distinct effects such as shifts in ecotypes to subtle changes in specific processes. The runoff chemistry of a watershed is characteristic of the relationships between the hydrology, geology and ecology of a watershed. These patterns might shift in response to climatic changes. Recognition of these more subtle changes may be important in identifying resources at risk. Can we anticipate what changes might take place, which are outside our current experience?

3. Optimum Frequency

Over the years, there have been evaluations of sampling frequency [48, 51, 52], but these have largely been retrospective. In a retrospective approach, a statistical analysis is performed to assess the 'optimum' sampling, either by reducing redundant data or by maximizing the signal or an existing data set. It has been suggested that a table of recommended sampling frequency could be used to summarize the knowledge in this area. One should be concerned about this approach since someone, somewhere, will take it quite literally. What is offered here is an extension of the concepts developed by Whitfield and Clark [6], who provide a framework for design monitoring using 'force analysis'. The method they proposed involves an assessment of the six aspects of the watershed of interest; the reader should recognize that the basis of these categories is experience with hydrologic systems in western Canada. The user would assess the hydrologic process, watershed size and slope, current land use and land cover, and the potential for change in the watershed. Each of these is ranked on a scale from one to four, and a score is generated from the product of these six values using the criteria listed in Table 2. An estimate of the sampling frequency (number of days between samples) is then obtained by dividing 30 by that score. Table 3 shows the range of sampling frequencies for selected scores in the range from 1 to 4096.

TABLE 2. Scoring system for assessing the Time Scales of Watershed Process.

Score	1	2	3	4
Process	Nival	Nival Hybrid	Pluvial Hybrid	Pluvial
Basin Area	>2000 km^2	200-2000 km^2	50-200 km^2	<50 km^2
Slope	<1%	1-5%	5-10%	>10%
Urban Development	<1%	1-10%	10-20%	>20%
Forest Cover	>95%	65-95%	35-65%	<35%
Potential for Change	none	low	medium	high

TABLE 3. Sampling frequency for assessment scores. Note that these should be used only for guidance.

Score	Estimated Days 30/Score	Proposed Sampling Frequency
1	30	Monthly
2	15	Two Weeks
4	7.5	Weekly
6	5	5 days
16	1.8	2 days
32	0.937	Daily
64	0.468	12 hours
144	0.208	5 hours
288	0.104	2.5 hours
324	0.093	2 hours
432	0.069	1.5 hours
648	0.046	Hourly
1728	0.017	25 minutes
3072	0.009	15 minutes
4096	0.007	10 minutes

This approach, while simplistic, leads the user to consider important aspects of the system of interest. Consider the difference in scores between a large (4000km^2) (1) nival (1) watershed on an upland plateau (2), where 5% of land is developed (2), and the forest cover is largely intact (1), and there is no anticipated potential for change (1) – (Score 4), and a small (4) pluvial (4) watershed in mountainous area (4) with substantive urban development (4), loss of more than 35% of the forest cover (4), and medium potential for change (3) – (Score 3072). The proposed sampling frequency in the first case would be once each week, whereas in the second case, it would be every 15 minutes. These sampling frequencies reflect recommended practice in Western Canada [48, 51]. Table 3 shows the range of sampling frequencies for selected scores in the range from 1 to 4096. This method should be refined to reflect local conditions and experience in other geographic areas.

4. Situational Analysis

Designing an effective monitoring program is a complex process. Of primary concern is determining what information the network must provide to be effective. The resources that an agency invests should provide a return that meets specific predefined needs. Data alone will not meet this need; rather the data must be converted into useful information. The network then must be designed to maximize the potential for acquiring information. This involves assuring that there is some consideration for important scales and for a level of redundancy in time and space. Such overlap allows for confirmation and extrapolation and reduces the risk of data loss. We can describe this problem as wishing to have perfect information - everywhere all the time - and yet only being able to obtain a small sample. Coupling selected data collection with scientific understanding [6] or with statistical models [53] provides a better framework for providing environmental information.

In a changing world, particularly in the face of issues such as changing climate, will the network provide the level of detail we will need to assess changes in conditions? How will our ability to provide valid interpretations be affected as the processes of interest become non-stationary? This is a matter of resolving several signals from noise [54]. In many cases, the signal that is necessary may depend on the application and the analysis methods, most of which assume the processes to be stationary. This suggests that there is a need for more types of data, or data collected to different standards, in addition to analysis techniques.

In addition, the information we seek is context sensitive. We need to develop a better understanding of the context of the network to allow for the augmentation of our data collection. Many networks are operated in a standard fashion that often fails to recognize the impacts of scale on the information content of the records generated [48]. An alternative would be to operate networks to produce data that have standard information content as proposed above. Applications of this approach will allow to extrapolate data to ungauged areas, to couple data collection and models, and to particularly focus data collection where uncertainty is the greatest.

5. Conclusions

The design of environmental monitoring programs needs to account for events at time scales outside those covered by a single standard approach. Some consideration must be given to transient events, seasonal and diurnal cycles, and long-term trends. In particular, events from a broader range of time scales need examination and study. Environmental monitoring needs to be operated in a manner that adequately covers a broad range of time scales. The technologies for observing short term and long-term phenomena are now available. Data collection programs need to be assessed and audited as to the efficiency and effectiveness on the time scale for which they were designed.

6. Acknowledgements

Drs. Rory Leith and Malcolm Clark provided useful suggestions to earlier drafts of this manuscript.

7. References

1. Steele, J.H. (1991) Marine ecosystem dynamics: comparison of scales, *Ecological Research* **6**, 175-183.
2. Smol, J.P. (1992) Paleolimnology: an important tool for effect ecosystem management, *J. Aquatic Ecosystem Health* **1**, 49-58.

3. Weins, J.A. (1989) Spatial scaling in ecology, *Functional Ecology* **3**, 385-397.
4. Ongley, E.D. (1987) Scale effects in fluvial sediment-associated chemical data, *Hydrological Processes* **1**, 171-179.
5. Loftis, J.C., McBride, G.B., and Ellis, J.C. (1991) Considerations of scale in water quality monitoring and data analysis, *Water Resources Bulletin* **27**, 255-264.
6. Whitfield, P.H. and Clark, M.J.R. (in press) Force Analysis as a Strategy to Target Key Environmental Information, *Environmental Management.*
7. Whitfield, P.H. (1988) Goals and data collection designs for water quality monitoring, *Water Resources Bulletin* **24**, 775-780.
8. Magnuson, J.J. (1990) Long-term ecological research and the invisible present, *BioScience* **40**, 495-501.
9. Swanson, F.J. and Sparks, R.E. (1990) Long-term ecological research and the invisible place, *BioScience* **40**, 502-507.
10. Leigh, R.M. and Whitfield, P.H. (2000) Some Effects on Streamflow records of Urbanization in a Small Watershed in the Lower Fraser Valley, *B.C. Northwest Science* **74**, 69-75.
11. Gocke, K., Kremling, K., Osterroht, C., and Wenck, A. (1987) Short-term fluctuations of microbial and chemical variables during different seasons in coastal Baltic waters, *Marine Ecology Progress Series* **40**, 137-144.
12. Weatherley, N.S. and Ormerod, S.J. (1991) The importance of acid episodes in determining faunal distributions in Welsh streams, *Freshwater Biology* **25**, 71-84.
13. Whitfield, P.H. and Dalley, N.E. (1987) Rainfall driven pH depressions in a British Columbia Coastal Stream, in Symposium on Monitoring, Modeling, and Mediating Water Quality, American Water Resources Association, pp. 285-294.
14. Beck, M.B., Adeloye, A.J., Finney, B.A., and Lessard, P. (1991) Operational water quality management: Transient events and seasonal variability, *Water Science and Technology* **24**, 257-265.
15. Spangberg, A. and Niemczynowwicz, J. (1992) High resolution measurements of pollution wash-off from an asphalt surface, *Nordic Hydrology* **23**, 245-256.
16. Whitfield, P.H. and Wade, N.L. (1992) Monitoring transient water quality events electronically, *Water Resources Bulletin* **28**, 703-711.
17. Whitfield, P.H., Rousseau, N., and Michnowsky, E. (1993) Rainfall induced changes in chemistry of a British Columbia Coastal Stream, *Northwest Science* **67**, 1-6.
18. Harriman, R., Gillespie, E., King, D., Watt, A.W., Christie, A.E.G., Cowan, A.A., and Edwards, T. (1990) Short-term ionic responses as indicators of hydrochemical processes in the Allt a' Mharcaidh catchment, Western Cairngorms, Scotland, *J. Hydrology* **116**, 267-285.
19. Clayton, K. (1991) Scaling environmental problems, *Geography* **76**, 2-15.
20. Miles, M. (2001) BC's Rivers are changing: What are the engineering and environmental consequences? Presentation at the CWRA BC Branch Conference "Changing Water Environment's Research and Practice", Whistler B.C.
21. Whitfield, P.H., Valiela, D., and Harding, L. (1992) Monitoring ecosystems for sustainability, in Managing Water Resources During Global Change, American Water Resources Association, pp. 339-348.
22. Wade, N.L. and Whitfield, P.H. (1994) Observing transient water quality events using electronic sensors, in National Symposium on Water Quality, American Water Resources Association, pp. 105-112.
23. Whitfield, P.H. and Wade, N.L. (1996) Transient Water Quality Events in British Columbia Coastal Streams, *Water Science and Technology* **33**, 151-161.
24. Whitfield, P.H. and Wade, N.L. (1993) Quality assurance techniques for electronic data acquisition, *Water Resources Bulletin* **29**, 301-308.
25. Whitfield, P.H. (1995) Identification and Characterization of Transient Water Quality Events using Fourier Analysis, *Environment International* **21**, 571-575.

26. Whitfield, P.H. (1994) Identification and analysis of transient water quality events, in W.R. Blain and K.L. Katsifarakis (eds.) *Hydraulic Engineering Software V, Volume 1- Water Resources and Distribution*, Computational Mechanics Publications, pp. 293-299.
27. Valiela, D. and Whitfield, P.H. (1989) Monitoring strategies to determine compliance with water quality objectives, *Water Resources Bulletin* **25**, 63-69.
28. Ramirez, R.W. (1985) *The FFT, fundamentals and concepts*, Prentice-Hall, 178 pp.
29. Erlebach, W.E. (1978) A systematic approach to monitoring trends in the quality of surface waters, in Establishment of Water Quality Monitoring Programs, AWRA, pp. 7-19.
30. Uri, N.D. (1991) Water quality, trend detection, and causality, *Water, Air, and Soil Pollution* **59**, 271-279.
31. El-Shaarawi, A.H., Esterby, S.R., and Kuntz, K.W. (1983) A statistical evaluation of trends in the water quality of the Niagara River, *J. Great Lakes Research* **9**, 234-240.
32. Hipel, K.W., Lennox, W.C., Unny, T.E., and McLeod, A.I. (1975) Intervention analysis in water resources, *Water Resources Research* **11**, 855-861.
33. Harcum, J.B., Loftis, J.C., and Ward, R.C. (1992) Selecting trend tests for water quality series with serial correlation and missing values, *Water Resources Bulletin* **28**, 469-478.
34. Hirsch, R.M., Slack, J.R., and Smith, R.A. (1982) Techniques of trend analysis for monthly water quality data, *Water Resources Research* **18**, 107-121.
35. McLeod, A.I., Hipel, K.W., and Comacho, F. (1983) Trend assessment of water quality time series, *Water Resources Bulletin* **19**, 537-547.
36. Montgomery, R.H., and Reckhow, K.H. (1984) Techniques for detecting trends in lake water quality, *Water Resources Bulletin* **20**, 43-52.
37. Zetterqvist, L. (1991) Statistical estimation and interpretation of trends in water quality time series, *Water Resources Research* **27**, 1637-1648.
38. Brillinger, D.R. (1994) Trend analysis: Time series and point process problems, *Envirometrics* **5**, 1-20.
39. Clark, M.J.R., and Whitfield, P.H. (2001) The Water Management Shuffle - Three steps forward, two steps back, Changing Water Environments: Research and Practice, CWRA, May 2001, pp. PL-1-9.
40. D'Astous, F., and Hipel, K.W. (1979) Analyzing environmental time series, *J. Environmental Engineering Division ASCE* **105**, 979-992.
41. Leith, R.M. (1991) Trends in snowcourse and streamflow data in British Columbia and the Yukon, in Proceeding of the Western Snow Conference 1991, pp. 49-56.
42. Smith, R.A., Alexander, R.B., and Wolman, M.G. (1987) Water quality trends in the nation's rivers, *Science* **235**, 1607-1615.
43. Whitfield, P.H. (1986) Spectral analysis of long-term water quality records, in A.H. El-Shaarawi and R. Kwiatkowski (eds.), *Statistical Analysis of Water Quality Monitoring Data, Advances in Water Research*, volume 27, pp. 388-403.
44. Clair, T.A., and Whitfield, P.H. (1983) Trends in pH, Calcium, and Sulfate of Rivers in Atlantic Canada, *Limnology and Oceanography* **28**, 160-165.
45. Hipel, K.W., McLeod, A.I., and McBean, E.A. (1977) Stochastic modeling of the effects of reservoir operation, *J. Hydrology* **32**, 97-113.
46. Whitfield, P.H., and Woods, P.F. (1984) Intervention Analysis Of Water Quality Records, *Water Resources Bulletin* **20**, 657-667.
47. Welsh, D.R., and Stewart, D.B. (1989) Applications of intervention analysis to model the impact of drought and bushfires on water quality, *Aust. J. Mar. Freshwater Res.* **40**, 241-257.
48. Whitfield, P.H. (1998) Reporting Scale and the Information content of hydrometric records, *Northwest Science* **72**, 42-51.

49. Gupta, V.L. (1973) Information content of time-variant data, *J. Hydraulics Div ASCE* **99**, 383-394.

50. Leith, R.M. and Whitfield, P.H. (1998) Evidence Of Climate Change Effects On the Hydrology Of Streams In South-Central B.C., *Canadian Water Resources Journal* **23**, 219-230.

51. Whitfield, P.H. (1983) Evaluation Of Water Quality Sampling Locations On The Yukon River, *Water Resources Bulletin* **19**, 115-121.

52. Whitfield, P.H. and Covic, A. (1998) Memory and the Statistical Independence of Events in Rainfall and Runoff, *Canadian Water Resources Journal* **23**, 21-29.

53. Whitfield, P.H. (1997) Designing And Redesigning Environmental Monitoring Programs From An Ecosystem Perspective, in Harmancioglu, N.B., *et al.* (eds.), *Integrated Approach to Environmental Data Management Systems, Proceedings of the NATO Advanced Research Workshop*, Izmir, Turkey, September 1996, pp. 107-116.

54. Whitfield, P.H., Clark, M.J.R., and Cannon, A. (1999) Signals and Noise in Environmental Data - characterization of non-random uncertainty in environmental monitoring, in V.P. Singh, I.W. Soo, and J.H. Sonu (eds.), *Environmental Modeling: Proceeding of the International Conference on Water, Environment. Ecology, Socio-Economics, and Health Engineering (WEESHE)*, pp 86-96.

REGIONAL STREAMFLOW NETWORK ANALYSIS USING THE GENERALIZED LEAST SQUARE METHOD: A CASE STUDY IN THE KIZILIRMAK RIVER BASIN

A. U. SORMAN
METU, Civil Engineering Department
Water Resources Division
06531 Ankara, Turkey

Abstract. Hydrologic information is used to estimate the parameters of regional models so that they can be transferred to other sites where information is needed. In this study, an analysis is performed to obtain the best representative model with the available parameters to minimize the model error variance. The number of runoff stations, their locations, and the length of records are used to determine the sampling error variance for each station. The prediction error variances are determined with different station numbers by considering the length of records and cross correlation values.

1. Introduction

The assessment of water resources is a prerequisite for the management of world's water resources. To that end, one of the fundamental issues is the availability of hydrologic data at spatial and temporal scales to meet the users' requirements. Sometimes, many years of data collection are required to analyze basic hydrologic networks.

Due to the variability of hydrologic data, long-term records at various sites of a homogeneous region are necessary to obtain surface water information. This requirement implies the importance of developing and upgrading of well-designed existing hydrologic data networks that are in balance with today's needs and future requirements for information.

Hydrologic analysis can never be precise. Errors always exist due to insufficient record lengths and numbers of stations for data collection (sampling errors). Errors also result from procedures used in the estimation of information conveyed by data. Regional network analysis can assist in obtaining the optimum information from the existing data.

There are several techniques for the regression of a hydrologic variable against basin characteristics, such as the Karasseff method [1] used in USSR, the square grid method used in Canada, and the World Meteorological Organization standards [2]. Moss *et al.* [3] proposed the use of regional regression models based on ordinary least squares (OLS) for the assessment of regional information. Stedinger and Tasker [4, 5] proposed a network design method known as the Generalized Least Squares and Network analysis method (GLSNET), which is known as the US Geological Survey (USGS) network analysis method.

The usefulness of each existing or proposed station in the gauging network in reducing the error in the regional regression model can be determined. In addition, coefficients of variation and cross correlation coefficient can be varied over the region. The outputs from the model enable objective decisions to be made on how many stations should be operated or should be discontinued in order to achieve network accuracy. The model error can be compared to the sampling error to infer whether future expenditure should go for improving the model with more parameters or for collecting more data.

In this paper, regional network analysis is carried out to derive the best regression model of gauged basins in Turkey. The model encompasses minimum number of parameters representing basin characteristics and uses the hydrologic data available at the downstream station located at the outlet of the representative basin in a region. Therefore, the basic aim is satisfied by presenting a set of GLS network analyses for different sets of homogeneous stations to develop a regional network system, where the available information at gauged locations can be transferred to ungauged sites.

Various computer programs developed by the USGS staff are used. The existing network information (instantaneous maximum discharges) is correlated with the basin characteristics such as drainage area, relief, channel length, basin shape and drainage density.

2. Background

Numerous studies on streamflow frequency and network have been done for various regions in the USA. The early studies, although differing in scope, have all used linear regression based on ordinary least square estimates. Some other studies using weighted least squares have also been realized for several locations in the USA. A more basic research oriented study using a generalized least square (GLS) is done by Tasker and Slade [6]. They demonstrated that the site-specific approach coupled with regression provides smaller root mean square errors than the traditional geographic approach.

The technique known as Network Analysis for Regional Information (NARI) is used for a network of stream gauging stations in Arizona by Tasker and Moss [7]. Stedinger and Tasker [5] proposed a method to adjust the GLS estimator to account for possible information about historical floods. In other studies by the same investigators, alternative methods are considered. GLS techniques are compared with the commonly used ordinary least squares method. The GLS technique is shown to be better than the others in terms of the average variance of prediction based on a split sample study, which is carried out by Tasker *et al.* [8] on 89 steamgauges in Pime Country.

A network analysis using GLS techniques is employed by Williams [9] to evaluate the peak flowgauging network for North Dakota, USA. He proposed to add new gauging stations to the current network on streams with small drainage areas and mild to steep main channel slopes in order to obtain improved regional peak flow information.

3. Regional Network Analysis

3.1. DESCRIPTION OF THE MODEL

Streamflow network analysis provides hydrologic information, which is used to derive relationships between physiographic variables of gauged basins and streamflow statistics. Traditionally, the regression model parameters are estimated using ordinary least (OLS) and weighted least squares (WLS). A method developed by Stedinger and Tasker [4, 5] proposes the use of generalized least squares (GLS) techniques.

For most cases in hydrology, where the model residuals do not have constant variance and are not independent, the OLS procedure does not identify the most efficient estimates of model parameters. In alternative methods, the covariance matrix of the residual errors must be estimated. The estimation of the matrix elements is not an easy task if the streamflow records are of different lengths and if flows at different gages are cross-correlated. The GLS procedure provides more accurate parameter estimates, which are unbiased of the variance of the model's residual errors. However, the method also requires estimates of the cross correlation function. Thus, it is clear that the GLS procedure is a better approach than the others for regional network analysis. The method is fully described in the GLS manual presented by Tasker *et al.* [8], who described how to get a better estimate of the covariance matrix and of the GLS estimator of model coefficients, b_i. The elements of the matrix are presented fully by the respective equations within the manual.

A regional regression model is given in logarithmic units and the cross correlation coefficients (r_{ij}) are estimated by developing an empirical relationship between sample r_{ij} and the distance between the stations d_{ij}. The parameters to be found in the model are the model error variance (γ^2) and the sampling error variance (SE). In order to estimate the required quantities, two different regression models are determined:

- the model of the regression of logarithms of standard deviations of flows on basin characteristics; and,
- the regional regression of the T-year quartile on basic parameters including the model coefficients.

3.2. COMPUTER PROGRAMS

In order to facilitate the type of activities listed, three main package programs (IOWDM, ANNIE and GLSNET) and a format file (WDM) are developed by the USGS. These programs are downloaded from the USGS web page through Internet and are used in this study.

The GLSNET program uses the watershed data management (WDM) file for managing the records of station peak discharges and the associated basin characteristics. This file is initially created for use by the interactive program called ANNIE for storage, manipulation, display, and statistical analysis of watershed data. ANNIE enables the user to perform various tasks related to data management and prepare the inputs for hydrologic models. The IOWDM program is used to enter data in selected file formats into a WDM file. Four steps are needed to arrive at the final model to compute regression coefficients, run error analysis, and configure the network for future data collection needs:

- identify data needs;
- acquire the data that have been identified;
- use IOWDM to build the data management file (WDM) and enter the data;
- use ANNIE to add additional attributes in order to run GLSNET.

The regression summary tables are printed out as the output of GLSNET, including station number, observed and predicted values of y_i, residuals, standardized residuals, model error variance (γ^2), estimated sampling error variance [$x_i (X^T \Lambda^{-i} X)^{-1} x_i^T$], equivalent years of record, L-average, and the statistic of Cook's D depending on both the leverage and residuals. The average values summarize the strength of the established model; the square root of the predicted model error variance is compared to the standardized error of estimate for the sampled data.

3.3. CLUSTER ANALYSIS AND OTHER STATISTICAL MEASURES FOR HOMOGENEITY

Regional regression analysis involves the formation of homogeneous sub-regions that are required for an effective transfer of information in space. The goal of this process is the identification of groupings of basins that are sufficiently similar. The clustering technique, which employs physiographic basin characteristics, is used to define similarity between them. The initially formed regions are subsequently modified, following a process to improve the regional characteristics.

Regionalization approaches require the selection of variables to define the similarity for the basins. The two most common variables used are basin physiographic characteristics and flood statistics. Sometimes, the clusters formed may not be hydrologically homogeneous unless flood statistics (L-moment ratios, Cook's statistics, etc.) are used as similarity measures. Therefore, the output resulting from the cluster analysis needs to be adjusted to improve the physical coherence and to reduce the heterogeneity of the regions, measured by some statistics like the L-coefficient of variation [10, 11]. The homogeneity measure defined by H is calculated for three different cases of L-coefficient of variation. The region is considered to be completely homogeneous if $H < 1.0$ and possibly heterogeneous if $1 \leq H < 2$ [12].

There are two steps in the process of determining if a basin should change regions. The first one is to identify a basin that is a candidate for moving, and the second one is to evaluate the move. The Cook's D statistic is calculated in the program package of GLSNET. Two criteria are used to determine if a basin should be moved or removed. The first is a discordancy measure that is calculated as a part of the homogeneity test, and the second criterion is to identify potential new homes for a catchment. This is the closeness of a basin to the centroid of each of the other cluster units.

4. Application of GLSNET Model to Kizilirmak Basin

4.1. GENERAL

The above-presented methodology on regional network analysis is applied to Kizilirmak Basin in Turkey. There are 185 runoff stations in the basin. Due to the availability of information on basin characteristics and discharge records at the basin's outlet, the number of stations was decreased to 18. Later on, it was increased to 26 with the addition of stations operated by governmental organizations. The model outputs of the enlarged network are compared with the initial regression results, using model and sampling error variances. The homogeneity tests for subsetting of two groups are performed, using clustering techniques and L-moment statistical measures.

4.2. REGRESSION MODEL AND GLSNET OUTPUTS

First, the analysis of the regression model is carried out for N = 18 stations, using all possible combinations of basin parameters (case A). The dependent variables are the instantaneous maximum discharges of various return periods, and the independent variables are the physiographic basin characteristics to represent topography (area, length, perimeter), morphology (stream density, frequency), shape (circularity (CR)), relief (Slope (S), max relief (MR)), elevation (E) and precipitation (PR). The GLSNET model outputs are examined in order to set up the best regression model, using the criteria of error variances.

A more realistic (rational) and better model is formed by using a combination of two basin parameters (case B). One runoff station is dropped out, for which the Cook's D value is found to be very high, and the model is re-run for N = 17. The homogeneity test is applied, using discordancy measure (D) and the homogeneity measure (H). All the discordancies were below the critical value of 3, and the H-measure for L-coefficient of variation was below 1, indicating that the basin is hydrologically homogeneous. Drastic drops in the error terms are noticed. They are decreased to more than half of the errors determined for the case of 18 stations.

A regression model is created for the first case (case A), using the basin parameters of circularity (CR) and maximum relief (MR), for which the results are obtained in logarithms and percentage units. Although the set up model seems to have the optimum number of parameters, it actually contains four basin characteristics, namely, $CR = 4\Pi A/P^2$ where A is the area and P is the basin parameter; and $MR = (E_M - E_o)$ where E_M is the maximum elevation within the basin and E_o is the station elevation at the outlet.

The second regression model for case B is set between the instantaneous discharge (Q_{Tr}) and two basin characteristics (CR and E). It is known that an increase in the number of model parameters increases the time and effort required for the prediction. Thus, from a practical point of view, it is preferable to select the second model because it is comparatively easy to find the model parameters for case B for the ungauged sites than for case A.

Next, the number of runoff stations is increased to 26 to run the model, using the same parameters CR and E. The heterogeneity measures based on L-coefficient of variation are calculated for three cases. H_1 and H_2 are found to be less than 1; but the standardized test value (H_3) for the ratio of L-C_s versus L-C_k is just over 2 (possibly heterogeneous). Hosking [10] recommends the use of H_1, since both H_2 and H_3 may give a false indication of homogeneity, especially in regions with small number of sites.

When the GLSNET output values for 17 and 26 stations are compared, the sampling error values are expected to decrease for the latter case because of the increase in the number of stations. Despite this expectation, a slight increase is noticed in the sampling error variances. This term has two components: temporal error and spatial error. In order to decrease the temporal term, stations with long historical records should be added. To decrease the spatial error, the number of stations should be increased, and their cross correlations should be reduced. It may be the case in this study that temporal errors, due to equivalent length of records, increased more than what has been gained in the spatial error term due to the increase in the number of stations.

On the other hand, the model error percentages are all increased significantly because the model coefficients do not seem to be very rational in sign. This may be attributed to the additional nine stations, which are not well represented with the same basin parameters (CR and E) as in the case for N = 17. Thus, some additional physiographic variables (slope, soil parameters, etc.) are needed to test the model error variance to reduce it to an acceptable level.

4.3. GROUPING OF RUNOFF STATIONS USING HOMOGENEITY TESTS

Cluster analysis and other statistical measures based on L-moment techniques are performed for the two sets of data formed for N = 26 and N = 17, respectively. It is expected that the newly formed sub-groups will provide more insight into the analysis.

The K-means algorithm in cluster analysis resulted in the largest subgroups of 12 and 8 stations for the total number of stations 26 and 17, respectively. The model coefficients of the GLS analysis and their error variance percentages are presented in Tables 1 and 2 for the representative subgroups of 12 and 8 stations, with the regression coefficients (b_0, b_1, b_2), model error variance (MEP), and sampling error variance (SEP) with respect to the return period (T_r), which varies between 2 ~ 100 years.

The MEP decreased for T_r greater than 5 years when the number of stations decreased to 12 from the total of 26, but the SEP values increased as expected. The model coefficients in Table 1a became more rational when they are compared to the earlier results discussed.

TABLE 1a. GLS regression model coefficients for independent parameters of circularity and elevation for station group D (N=12)

Return Period (Tr)	Constant (b_o)	Coefficient of log (Circularity)	Coefficient of log (Elevation)
2	2.27955	0.35398	-0.37307
5	4.08279	0.57497	-0.86650
10	5.14596	0.68201	-1.16619
25	6.37284	0.78844	-1.51833
50	7.20195	0.85501	-1.75829
100	7.96430	0.91522	-1.97934

TABLE 1b. GLS regression results for the model of circularity and elevation for station group D (N=12)

Tr	Sampling Error Variance (in logarithms)	Model Error Variance (in logarithms)	Prediction Error Variance (in logarithms)	Sampling Error (%)	Model Error (%)
2	0.0304	0.1075	0.1379	16.15	87.65
5	0.0240	0.0764	0.1004	13.17	70.67
10	0.0235	0.0680	0.0915	13.13	65.89
25	0.0258	0.0673	0.0931	14.41	65.48
50	0.0292	0.0730	0.1022	16.06	68.75
100	0.0340	0.0834	0.1174	18.35	74.57

TABLE 2a. GLS regression model coefficients for independent parameters of circularity and elevation for station group F (N=8)

Return Period (Tr)	Constant (b_o)	Coefficient of log (Circularity)	Coefficient of log (Elevation)
2	2.08741	-0.31145	-0.38270
5	2.99640	0.23646	-0.52492
10	3.32189	0.52301	-0.54634
25	3.55730	0.83405	-0.52862
50	3.61778	1.04502	-0.48325
100	3.59030	1.24550	-0.41194

TABLE 2b. GLS regression results for the model of circularity and elevation for station group F (N=8)

Tr	Sampling Error Variance (in logarithms)	Model Error Variance (in logarithms)	Prediction Error Variance (in logarithms)	Sampling Error (%)	Model Error (%)
2	0.0395	0.0954	0.1349	21.07	81.14
5	0.0363	0.0806	0.1169	19.64	73.02
10	0.0402	0.0830	0.1232	21.65	74.35
25	0.0503	0.0975	0.1478	26.79	82.27
50	0.0613	0.1160	0.1773	32.72	92.18
100	0.0751	0.1409	0.2160	41.00	105.39

On the other hand, the number of stations in the other subgroup decreased to 8 from 17 when clustering analysis is used. The discordancy and homogeneity measures are also determined, following Hosking's procedure of the L-moment method. All of the D_i's were less than the critical value of 2.14, and H measure is found to be less than 1. The MEP increased for T_r greater than 5 years; on the other hand, the SEP values are found to be higher than those obtained for 12 stations. Sampling error percentages are increased because of an increase in spatial error since the number of stations is less. When the Cook's D values are examined in model outputs, it is noted that two of the runoff stations, namely 15-25 and 15-43, have values of 1.13 and 1.17, which are greater in magnitude than the limit value of 4/N = 0.5. The results clearly indicate that these two stations have a great effect on the final regression model error.

5. Discussion of Results

GLS analysis is performed for a two-parameter combination of different groups of runoff stations. The model runs are performed to derive a regression model and its error variance to obtain the sampling error variances for individual stations, considering their record lengths, cross correlation coefficients, and concurrent records.

The traditional method of LSE techniques is based on the assumption that the variance over the data range is constant and the residuals are independent. These assumptions are violated in the regional analysis because of unequal variances of streamflow characteristics. In the GLS analysis, the standard deviations of peak discharges are estimated by a regional regression equation. The final regression estimation is realized to find the regression coefficients for a network of stations, where a covariance matrix is formed to compute standard errors of prediction for new observations. The square roots of the diagonal elements of the matrix are the standard errors of the regression coefficients. The program computes the Cook's D value, which shows the influence of each station on the final regression equation. The D values of the records at each station should be investigated carefully to infer on the necessity of each station.

In order to obtain homogeneous regions, regions with a large number of stations should be split into new homogeneous sub-regions. This will lead to a decrease in the number of stations with an increase in the sampling error terms. Sampling error is one of the most important elements in the analysis. It is concluded that, as the number of stations with long records increases, the sampling error of the model decreases significantly. This also leads to a decrease of the model error. Both decreases result in an overall decrease in the prediction error. The effect of the number of stations (N) and return period (T_r) on the SEP for the GLS model, using CR and E, is presented in Fig 1. The sampling error follows a tangential path when the optimum number of stations is reached. There is little advantage in using a very large number of runoff stations, and the accuracy gained in quantile estimates is limited when there are more than 20 runoff stations in the homogeneous region [12].

The effects of the number of stations and the return period on the MEP are presented in Fig. 2. The increase in the number of stations decreases the model error up to 20 stations; above 20, the model coefficients get irrational, and high error terms are produced.

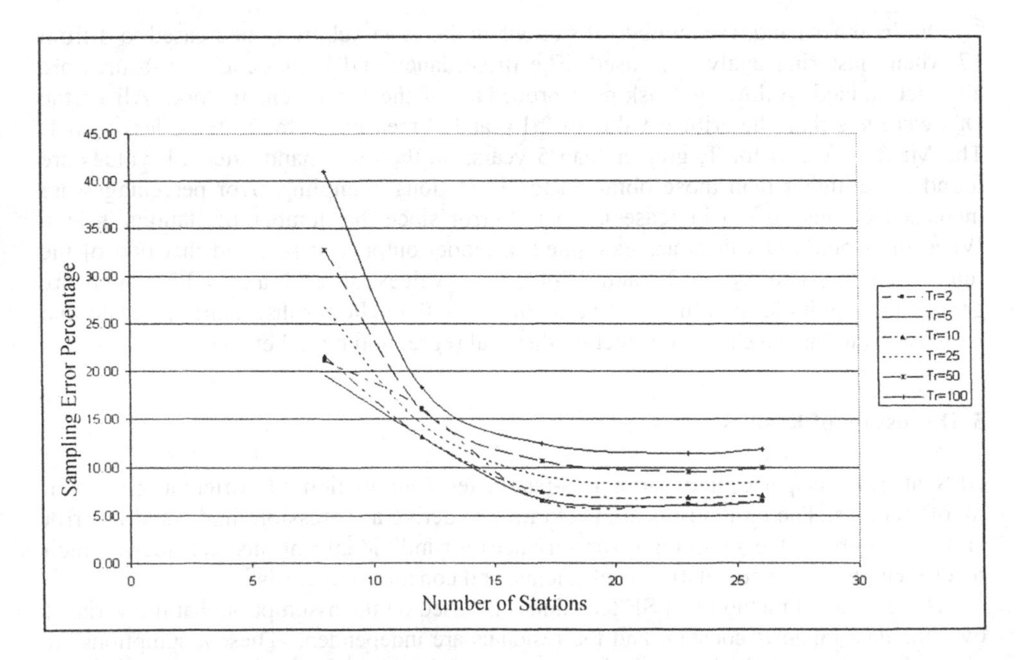

Figure 1. Relation of average sampling error percentage, number of stations, and return period.

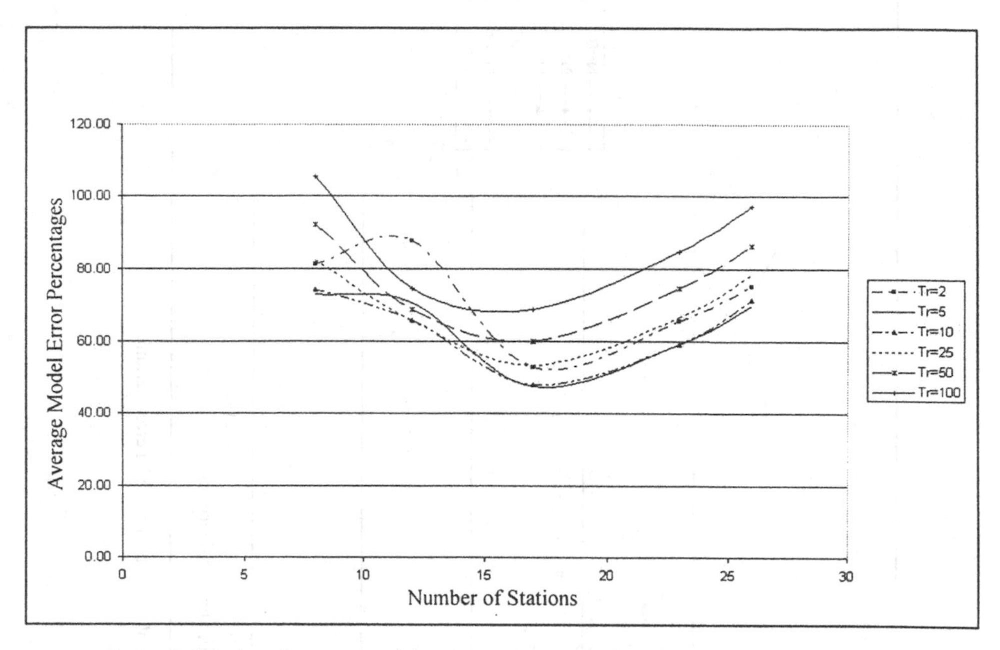

Figure 2. Relation of average model error percentage, number of stations, and return period.

One may convert the average sampling and model error variances given in Table 3a to the expected percent increases in equivalent years of record (Table 3b) for each return period T_r (yrs) and for the group of stations varying between 8 to 26 by the formula:

$$\Delta n_e = 100 \times \left[\frac{S_0^2 - S_f^2}{S_m^2 + S_f^2} \right] \tag{1}$$

where S_0^2 and S_f^2 are the average mean square sampling errors for the current and extended network, and S_m^2 is the mean square model error variance in logarithms.

The values in Table 3b express the accuracy of an increase or decrease of prediction in terms of two variables, namely, the return period and the number of stations, so that the latter variable and the required length of records can be determined to achieve an equal accuracy of results. The increase in accuracy is not much beyond 20 stations, and the increase in the number of stations becomes more effective for small groups of stations at higher return periods than the network of stations when it reaches its optimum size for the homogeneous area under consideration.

When the variance of the standard error of an estimate and model error variances are compared on the same graph as shown in Fig 3, it is seen that the model error values decrease as the number of stations increases up to 17. The increase to above 17 to 23 and 26 causes the model error to increase. The line drawn at 45° angle, which passes between the 2 – 5 years of return periods, indicates equal errors. This is the case where the mean annual maximum discharges occur at a return period of $T_r = 2.33$ years.

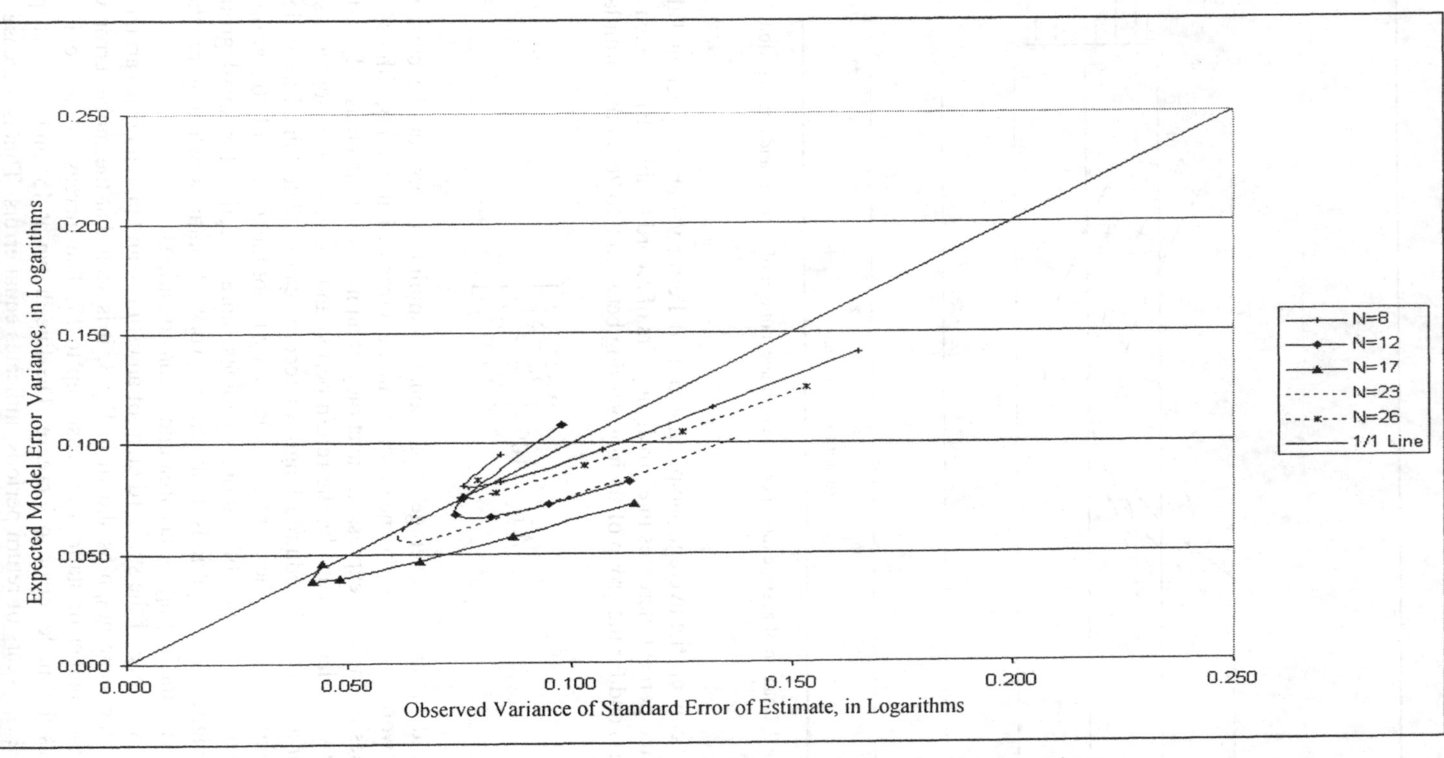

Figure 3. Relationship of variance of standard error of estimate and model error variance.

TABLE 3a. Sampling and model error variances, in logarithms.

		N=8	N=12	N=17	N=23	N=26
Tr=2	$S_{o,f}^2$	0.0395	0.0304	0.0104	0.0108	0.0118
	S_m^2	0.0954	0.1075	0.0464	0.0675	0.0842
Tr=5	$S_{o,f}^2$	0.0363	0.0240	0.0101	0.0103	0.0115
	S_m^2	0.0806	0.0764	0.0383	0.0568	0.0748
Tr=10	$S_{o,f}^2$	0.0402	0.0235	0.0115	0.0116	0.0128
	S_m^2	0.0830	0.0680	0.0392	0.0568	0.0777
Tr=25	$S_{o,f}^2$	0.0503	0.0258	0.0148	0.0145	0.0157
	S_m^2	0.0975	0.0673	0.0472	0.0694	0.0900
Tr=50	$S_{o,f}^2$	0.0613	0.0292	0.0183	0.0176	0.0187
	S_m^2	0.1160	0.0730	0.0581	0.0833	0.1052
Tr=100	$S_{o,f}^2$	0.0751	0.0340	0.0226	0.0215	0.0223
	S_m^2	0.1409	0.0834	0.0732	0.1020	0.1252

TABLE 3b. Percent increase in equivalent years of record.

Tr	N=8 to N=12	N=12 to N=17	N=17 to N=23	N=23 to N=26
2	7.23	16.96	-0.70	-1.26
5	11.76	16.07	-0.41	-1.76
10	15.68	15.09	-0.20	-1.72
25	19.87	13.40	0.49	-1.41
50	22.11	11.94	0.92	-1.08
100	23.50	10.75	1.16	-0.64

The proposed regional regression model becomes questionable when the number of stations increases to more than 20 because the physiographic parameters may not be representative any more so that additional basin variables must be searched to minimize the model error.

The sampling error approaches an asymptotic value of 0.01 – 0.02 when the number of stations reaches the optimum value. The stations must then be sub-setted into smaller homogeneous units, not only to keep the basin similarity but also to preserve the hydrologic similarity. Besides physiographic homogeneity, the hydrologic similarity should be checked with an additional statistic, using L-moment ratio techniques, which may be obtained after clustering analysis.

6. References

1. Karasseff, J.F. (1972) Physical and statistical methods for network, *Design Casebook on Hydrological Network*, Design Practice, WMO No. 325.

2. World Meteorological Organization (WMO) (1991) Technical reports in hydrology and water resources, *International Workshop on Network Design Practices*, Koblenz, Germany, WMO/TD No. 671.

3. Moss M.E., Gilroy E.J., Tasker G.D. and Karlinger M.R. (1982) Design of Surface Water Date Networks for Regional Information, USGS Water Supply Paper No. 2178.

4. Stedinger J.R. and Tasker G.D. (1985) Regional hydrologic analysis, ordinary, weighted and generalized least squares compared, *Water Resources Res.* **9**, 1421-1432.

5. Stedinger J.R. and Tasker G.D. (1986) Regional hydrologic analysis 2, Model error estimators, estimation of sigma and log Pearson Type III distributions, *Water Resources Res.* **22-10**, 1487-1499.

6. Tasker G.D. and Slade, R.M., Jr. (1994) An interactive regional regression approach to estimating flood quantiles in D.G. Fortrane and H.N. Tuvel (eds.), *Water Policy and Management Solving the Problems: Proceedings of the 21st Annual Conference of the Water Resources Planning and Management Division*, American Society of Civil Engineers, pp. 782-875.

7. Tasker G.D. and Moss M.E. (1979) Analysis of Arizona flood data network for regional information, *Water Resources Research* **15-6**, 1791-1796 (http://water.usgs.gov/index.html).

8. Tasker G.D., K.M. Flynn, Lumb A.M., and Thomas W.O. (1995) Hydrologic Regression and Network Analysis Using Program GLSNET, U.S. Geological Survey, Water Resources Investigations Report 95, Reston, Virginia.

9. Williams T.S. (1996) Analysis of the peak-flow gaging network in North Dakota, *Water-Resources Investigation,* **96-41**, 78 (http:www.usgs.gov./).

10. Hosking J.R.M. (1991) Fortran Routines for Use with the Method of L-Moments Version 2. IBM Research Report RC. 17097, IBM Research, York Town Heights, New York.

11. Rao A.R. and Hamed K.H. (1997) Regional frequency analysis of Wabash River flood data by L-Moments, *Journal of Hydrologic Engineering*, 169-179.

12. Hosking J.R.M. and Wallis J.R. (1997) Regional Frequency Analysis, *An Approach Based on* L-Moments, 54-61.

AUTOMATED WATER QUALITY MONITORING IN WATER DISTRIBUTION NETWORKS

Y. A. PAPADIMITRAKIS
National Technical University of Athens (NTUA)
Department of Civil Engineering
Hydraulics, Water Resources and Maritime Engineering Div.
5 Heroon Polytechniou St., Zografos
15780 Athens, Greece

Abstract. An integrated methodology is presented for monitoring, *in a continuous and automated fashion*, the quality of drinking water in pipe distribution networks. Such water quality monitoring may be accomplished by observing simultaneously several quality indices (from those included in the E.E. regulations) at various sites of a water distribution system. Criteria are described for selecting these indices and determining both the geographic distribution and the density of the monitoring sites. At these sites, proper hardware (several electronic sensors and a computer) is installed inside small terminal stations where sampling, analyzing, storing, and transmission to a central station, of the observed water quality characteristics are taking place. The monitoring system is complemented and supported by a small number of mobile laboratory units that perform, *in situ*, quick chemical analyses, on a periodic and/or need basis, for validating the performance of the continuously observing sensors at the various terminal stations. The sampling rates of the collected data may vary, depending on the fluctuation rates encountered. The data collected at each terminal station, in the form of analog or digital time series, and those from the mobile units, are transmitted (wirelessly or via leased lines or Internet Service Providers) to a central station. At the central station, data are *either* analyzed in a statistical fashion to produce index levels, say, on a daily basis *or* used, via an assimilation technique, in conjunction with a software that simulates the water quality in the entire distribution network in order to provide a realistic picture of the water quality in the entire network.

1. Introduction

Water is necessary for both life and the production of other goods, used either as a consumable good or as a raw material. Although most of our planet is covered by water, only a small percentage of the worldwide available quantities satisfies certain quality criteria (standards) and is appropriate for human consumption. Water management is, therefore, of prime concern in most human societies.

More specifically, drinking water is of great importance to human health, particularly in large cities. The existence of aged water distribution networks, extending over large geographic areas and having frequently corroded pipes with inappropriate and inadequate joints, mandates the conduction of systematic water quality checks in both the supply reservoirs and the water treatment plants, prior to release to the distribution system, as well as within the latter prior to consumption.

Recently, some of the authorities in charge of the operation of supply and distribution networks have started to add chemicals into the drinking water in order to reduce corrosion, as the latter is also a prime reason for lead and copper contamination.

The need for adding chemicals arises from plumbing practices, which usually cause lead problems in buildings, although conventional chemical analyses of samples taken from distribution networks have not detected lead there. Yet, it becomes apparent that, with stricter water quality standards applied nowadays in most cities, lead and other heavy metals may become a concern requiring systematic examination in distribution systems. In systems where corrosion control chemicals have been injected for some extended period of time, water pH has been found to increase dramatically, occasionally reaching values of up to 9.4. Water pH of 8.0 or higher may reduce chlorine disinfection effectiveness and, thereafter, the system's ability to combat pathogens. Again, such a situation can be assessed by accurate real-time monitoring of the parameters of interest and may require chlorine injection (by means of boosters) at strategic points of the distribution system, if it becomes necessary. References for these facts may be found in [1, 2] and the work by Koutra [3].

Catastrophic events, such as earthquakes, fires, flooding, etc., that usually lead to power outage, may also create serious problems in the drinking water supply system and lead to further disasters in the quality of water from pollution episodes, as contaminants and pathogens may enter the distribution system and cause public health problems. Another recent concern is terrorism, especially in cities where the water supply system is relatively simple and vulnerable to contamination. To handle such extraordinary events, particular information, regarding both the extent and the degree of water quality degradation imparted to the system, is required immediately after the episodes and/or the terrorist actions, for proper response of those in charge to minimize the expected health risks and other possible detrimental effects associated with the polluted water flowing into the distribution system. These events also stress the need to implement systematic monitoring in both the water supply and the distribution systems.

Some degree of automation may be found in distribution networks today, associated mainly with the monitoring of quantities such as the discharges through the system, water losses, pressure heads at junction sites and, perhaps, other hydraulic parameters important for maintaining the regular functioning of the network. Software that simulates the above quantities is also available for some time, and it may be used along with the observations collected at the various sites of the system, periodically and/or continuously, to improve the simulation performance. Such software related to available technologies for automated monitoring of hydraulic quantities in distribution networks is described in Theodoulidis [4]. It is true that information and automation technologies had a significant impact on Water Utilities over the past few years and are already considered an indispensable management tool for their entire organization and, in particular, for monitoring and control of the water distribution networks.

Yet, in most distribution networks, the current water quality monitoring program is manual and expensive. *In situ* sampling of drinking water from the distribution system by conventional means, shipment of the samples to a nearby-authorized chemical laboratory, and subsequent analysis of the samples is done routinely. However, these checks are done infrequently, perhaps once a month or a few times per year and, therefore, do not guarantee that other changes in the quality of drinking water do not take place in between these infrequent checks. Such monitoring capabilities are primarily the result of past philosophy prevailing in the design and operation of distribution systems without paying attention to automated monitoring of the quality of drinking water, perhaps due to the difficulties encountered in such complex dynamic systems. Thus, new ideas and approaches must be examined with respect to monitoring the quality of water in distribution systems.

The widely spread practice, today, for a continuous and remotely (or tele-) controlled monitoring of the quality of both the atmospheric and the oceanic environment appears sensible and is necessary to be extended to the water distribution systems. The theoretical concepts and scales (temporal and spatial) upon which the design and subsequent operation of automated and continuous monitoring systems of the water quality in coastal areas, deep seas and inland waters (rivers, lakes, etc.), have been presented by several authors [5, 6, 7, 8, and others]. Extended references for methodologies (or guidelines) of designing systems to monitor the quality of inland waters may be found in the National Handbook of Water Quality Monitoring [9].

There are many private firms today designing, producing, and installing such integrated water quality monitoring systems for operational use in either coastal (and deep sea) or inland waters.

It is worth noting that an automated monitoring network, the POSEIDON system, already operates in some of the Greek seas (in northeastern and central Aegean, in Saronikos Gulf, and in the Cretan Sea). Such a network monitors, in real time, the quality of seawater in these areas, along with a number of meteorological and oceanographic parameters, which provide useful information on local wave climate, marine currents, etc. The POSEIDON network uses a number of electronic sensors that monitor air pressure and temperature, wind speed, humidity and other meteorological parameters, sea surface elevation, subsurface currents, sea temperature, conductivity, and a few other oceanographic parameters, as well as various water quality parameters (e.g., pH, dissolved O_2, chlorophyll a, nutrients,...). The network operates in conjunction with a software that simulates the corresponding hydrodynamic and water quality processes in the above areas, utilizing the collected data through an assimilation scheme to maximize the accuracy of the simulated results. The POSEIDON network has been documented in various articles, as for example in [10], presented in a recent Conference.

In this work, an integrated methodology for designing and operating remotely an automated system to monitor the quality of drinking water in pipe distribution networks is presented. The methodology is currently developed at the National Technical University of Athens (Greece), through multidisciplinary research efforts among three Laboratories that belong to Civil, Chemical and Electrical Engineering Departments. The article begins with a short justification of the necessity to monitor the quality of drinking water in distribution systems. The basic groups of parameters, specified by EE regulations for monitoring drinking water quality, are then presented. A short description of the basic idea behind this remotely controlled monitoring system follows, and the necessity of an interdisciplinary cooperation for the design, production and operation of such a monitoring system is consequently stressed. The benefits from the use of an automated water quality monitoring system are subsequently presented. Available sensors, operating in automated and continuous mode, are briefly described next. Data transmission from the monitoring sites to a central control -and- command station and available communication alternatives among them, are also discussed. A short description of software that simulates the physicochemical and other processes, that govern water quality inside a distribution system, is given in a subsequent section.

Assimilation aspects of the collected data into the simulation software are also briefly mentioned. At the end of the article, the development stages of an automated monitoring system are presented, and comments are made for the system's future perspectives.

2. Ideas Behind a Remotely Controlled Monitoring System of Water Quality

There are several reasons for monitoring, in real time, the quality of drinking water in water distribution systems. The existence, for example, of corroded pipes in aged distribution systems may facilitate the presence of undesirable chemical substances in the network, either as a result of pipe wall oxidation-corrosion or through wall entrainment of various contaminants from the surrounding soil and/or the groundwater table. Differential ground settling, pipe joint relaxation, and accidental pipe breaking (during related construction or repair work) may also facilitate the entrainment of either contaminated water or contaminated soil or both into the distribution system. It is also possible that untreated water may enter accidentally the latter.

If, for any of the reasons cited above and in the previous section, contaminated or untreated water enters the distribution system, the contaminants contained in the water (and/or perhaps the products of pipe wall corrosion-oxidation) are transported along the network by the flow and transformed by various physical and biochemical processes that take place within the system.

There are other reasons as well, favoring the ideas for the establishment of an automated monitoring system. They are mostly related to treatment practices and their possible adverse consequences on the quality of water inside the distribution system. Nowadays, many large cities (metropolitan capitals, for example) exchange ideas, share the objectives to comply with stricter environmental standards, and adopt new (stricter) regulations in order to provide high quality drinking water. Such regulations deal with various issues related to: universal filtration, stricter standards for some substances (e.g., lead, copper,…), and the use of chlorine as disinfectant, etc., and they have been already authorized by local councils in various cities.

It is also a common practice, today, the required quantities of drinking water to be obtained from surface and groundwater sources. Yet, in many instances, due to agricultural practices, groundwater is contaminated by nitrates and pesticides. Then, the strict European water quality standards for pesticides are often violated, and groundwater is either discharged back (to rivers, etc.) without being used, or treated heavily and disinfected in water treatment plants by adding (excessive) chlorine. Such practice is frequently justified by the poor quality of water due also to fertilizers and the presence of algae, etc. Other chemical substances, such as caustic soda for pH control, copper sulphate for algae control, and alum for turbidity control, are also used to alleviate the respective problems.

Excessive chlorination is known to cause undesirable effects, as some of its by-products may be carcinogenic. Although in many treatment facilities and/or storage water tanks, dechlorination devices eliminate the remaining taste of chlorine and reduce the chlorine concentrations to acceptable levels, it becomes apparent, and often necessary, that the remaining chlorine in the distribution system be monitored carefully prior to consumption. Residual chlorine concentrations in distribution networks are often found at levels of 0.1 mg/l, and many residents complain about such unpleasant experiences. It is reminded, however, that the residual chlorine concentrations must also remain above some minimum level in order to maintain the system's disinfection capability or its ability to control pathogen organisms, etc.

There are many strict regulations that describe, nowadays, how monitoring and testing the quality of drinking water may be done. Such EE (European Union) regulations (e.g., 80/778 and its 10/1997 modification) determine: the characteristics of water quality (in terms of several indices) that may be monitored, their upper and lower bounds, the methods of testing and monitoring these characteristics, as well as the number and frequency of required tests. According to these regulations, water quality is checked by monitoring 61 parameters-indices which may be classified into five groups, such as: (1) physico-chemical parameters (e.g., temperature, conductivity,...); (2) organoleptic parameters (e.g., color, turbidity, odor, taste,....); (3) parameters characterizing undesirable substances (e.g., nitrates, nitrites, ammonia,...); (4) parameters characterizing toxic substances (e.g., arsenic, mercury,...); and (5) microbial parameters (e.g., total coliforms, etc.).

The question arising then is, certainly, not whether the quality of drinking water has to be monitored in distribution systems, but rather how this monitoring ought to be done. Any infrequent and sporadic control of the quality of drinking water in a distribution system does not necessarily guarantee that some of the water quality parameters (indices) remain always within the acceptable bounds specified by EE and/or other regulations.

Thus, the proper answer to the latter question is expected to be by means of a monitoring network whose sensors (and/or other devices) are installed at pre-selected (junction) sites of the distribution system, where information is collected in an automated and continuous fashion. This answer comes as a natural continuation of the contemporary trend for automatic and continuous remote monitoring of the quality of both the atmospheric environment (above land and water) and the water environment (oceanic and inland waters).

3. The Monitoring System

The idea behind a system for remote and continuous monitoring of the water quality in a water distribution network is based on the presence of a limited number of terminal stations and a central control -and- command station. Terminal stations are positioned at properly selected specific junction sites of the distribution network, where water is pumped from the network to a small tank. There is a real time -on line- connection among the terminal stations and the central control –and– command station, in order to maintain a continuous picture of the monitored water quality indices.

Each terminal station hosts the monitoring instruments, i.e., sensors plus connecting signal transducers. The sensors sample the water and perform chemical (and other) analyses with the aid of a computer that receives the incoming information at preset rates, stores it on its hard disk and may process the raw data before transmitting it to the central control -and- command station. At this station, a central processor receives and stores the data or transmits various commands to each or some of the terminal stations. The total capabilities of such a system are determined, to a large extent, by the software used in both the central station and the terminal stations.

It has also been suggested, as an alternative, that part of the terminal stations may be formed as mobile (or vehicle-portable) units such that, when needs arise (as, for example, from damage of network pipes, etc.), the system is served better by moving some of the mobile stations to places where extra needs exist. Furthermore, it should be clarified that it is also necessary to perform periodic crosschecks of water quality at a limited number of terminal stations by means of such mobile (or field-portable) units that will also serve as small chemical laboratories. These complementary tests and separate monitoring, however, aim at cross checking the data collected from the continuously monitoring instruments and assist in maintaining their calibration and proper performance. Such mobile units may be equipped with analytical (in a chemical sense) instruments that have been growing, nowadays, at unexpectedly fast rates. Recent references on the advances in field-portable, mobile instrumentation may be found in the Proceedings of the International Conference on ''Instrumental Methods of Analysis, Modern Trends and Applications'' [11] (see, for example, [12]). The capabilities of these mobile units, along with the requirements for remote transmission of the collected data, are also described in several other references in Greek.

Thus, the remotely controlled monitoring system will consist of: (1) a network of fixed terminal stations built at pre-selected junction sites of the distribution system for continuous observations of several drinking water quality indices and continuous transmission (wirelessly or otherwise) of the monitored information to a control center; (2) a control -and- command center where the received information is either analyzed statistically or embedded in a simulation software with the aid of assimilation techniques; and (3) a limited number of mobile (vehicle-portable) biochemical units, necessary for both cross checking the quality of monitored data from the continuously observing network instruments and the verification of their calibration maintenance, as well as for covering special situations where extra needs may arise from accidents or crisis events.

It, thus, becomes apparent that the realization of drinking water quality monitoring in distribution networks, in an automated and continuous fashion, requires an interdisciplinary cooperation among various scientific fields, as for example those of: (a) site chemical analysis; (b) portable analytical instruments; (c) informatics-telematics; (d) environmental fluid mechanics; (e) applied hydraulics; (f) mathematical modeling of water hydraulics and water quality in closed conduits; and (g) inverse (assimilation) techniques, etc.

It is also evident that, besides the many scientific and technological challenges associated with the realization of such a monitoring system, many practical benefits may arise from the daily use of the latter, such as: the reduction of human errors in the measurement chain, the continuous and simultaneous monitoring of several water quality indices in the network, the low cost of monitoring resulting in a cost-effective operation of the system, and the ease of overall network maintenance. The latter is possible in conjunction with monitoring of other quantitative aspects of water supply. It is also possible to eliminate stagnant water from the pipe network by adjusting various remotely controlled valves and to upgrade the water quality, at least locally, and by proper monitoring of the latter to adjust the chlorine levels in the entire network. Last, the benefit list must include the -on time- diagnosis of any undesirable water quality change in the water distribution system and the ability for a subsequent immediate intervention to limit such undesirable water quality deterioration changes by coordinating maneuvers on the network and by directing the intervention squads.

Such a monitoring system increases the *reliability* of the network and that of the utility organization that has the responsibility of the operational functionality of the entire supply and distribution system, especially during crisis events. It also helps to mitigate hazards and *really improve* the quality of the distributed water.

Although the benefits from the use of a remotely controlled, continuous monitoring system of the quality of water appear impressive, not many authorities in charge of the operation of the supply and distribution networks have invested in the new technologies, at least sufficiently, to implement the described monitoring system or perhaps some version of it. There is no single explanation for that. It is, however, apparent that, in societies dominated by market forces, the implementation of automated monitoring has progressed a bit faster. Paris and New York are characteristic representatives of the above philosophy, although the corresponding monitoring networks appear to be far from complete, at least relative to the ideas presented in this work. For further documentation see, for example, [1].

4. Design Considerations

When designing an automated water quality monitoring system for use in water distribution networks, consideration ought to be given to the following aspects:

a) What is the minimum number of parameters (for example, among the 61 included in the EE regulations) that will allow a satisfactory monitoring of the water quality in the distribution system and that will also guarantee the possibility that these parameters become the subject of an *in situ* (chemical or microbial) analysis?

b) What are the characteristics and specifications of the instruments that will be used for continuous and automatic monitoring of the water quality parameters selected in (a)?

c) Is it sufficient to determine, along the distribution network, a limited number of junction sites where the automated and continuous monitoring of the pre-selected water quality indices (along, perhaps, with the data collected from some mobile units) will provide, quality-wise, representativeness of flow conditions and of geographic locations from those encountered in the distribution system?

d) What are the criteria for determining the geographic distribution and the density of the monitoring junction sites?

e) What are the technical characteristics and specifications of a terminal station, built at specific junction sites of the distribution system as a separate part of the junction box, where water is pumped from the distribution system to a small tank for further analysis and proper hardware is installed for sampling, analyzing, storing and transmitting the monitored data to a central control -and- command station?

f) What are the technical specifications of a mobile unit that will perform the crosschecks and calibration tests of the continuously monitoring instruments?

g) What are the physicochemical or other processes that determine the evolution and transformation of the monitored water quality indices along the distribution system; how these processes are modeled and what is the structure of a software that will simulate the water quality across the entire distribution network in an operational mode?

h) How the data, collected at specific sites of the water distribution network, can be used as input, along with a mathematical model of the water quality and in conjunction with an assimilation scheme, to simulate the quality of water in the entire network?

5. Criteria for the Geographic Distribution and Density of Terminal Stations

The selection of the sites along a distribution network, where the small terminal stations are built, may be assisted by the following criteria:

1) Ground morphology;
2) Age of the distribution system;
3) Hydraulic characteristics of the distribution network;
4) The density of distribution network (km of conduits/km^2 network);
5) The spatial and temporal variability of contaminant concentrations in the network;
6) Residence time, or renewal period, of water within particular areas of the distribution system.

As a rule of thumb, it may be argued that, when: ground topography is irregular, the distribution system is aged, the density of network is high (many pipes/km^2 of network), the spatial and/or temporal variability of contaminant concentrations is intense, and the residence time of water is large, then the density of the terminal stations is expected to rise.

6. The Monitoring Sensors

During the last decade, significant contributions were made towards the improvement of the design and manufacturing of a variety of sensors for real time automated monitoring of the environment. During that period of time, various innovative aspects of monitoring the air and water quality were presented by research teams from Universities, small companies, and the respective Industry. These efforts led, initially, to novel ideas for designing instruments and experimental techniques, and later to the production of sensors and processing electronics covering, to some extent, the needs of environmental agencies.

With respect to aquatic environment, there are several sensors, nowadays, appropriate for automated and continuous monitoring of various quality parameters, as specified by the pertinent EE regulations. These parameters are associated with the physicochemical and organoleptic properties of water, the presence of organic pollutants, pesticides and their transformation products, the presence of toxic substances and heavy metals, trace elements, nutrients, the presence of bacteria and, perhaps, of other undesirable substances (as polychlorinated biphenyls in groundwater, and some other substances in waste waters).

For sea water and inland waters, it is also possible to monitor (besides temperature, conductivity, and pH) turbidity, salinity and dissolved oxygen concentrations, attenuation of light with depth at various wave lengths, chlorophyll a and algae concentrations, the radioactivity of surface waters (gamma radiation) and net solar radiation, some nutrient (e.g., orthophosphates, nitrites, nitrates) and hydrocarbon concentrations, suspended solids, sediment and particle concentrations. Extended references for the development of on-stream biosensors for pesticide detection, the ''smart'' sensors used for odor detection, new optical sensors and other sensors that may be used for *in situ* automatic monitoring of various water quality parameters, either from a fixed platform or a vehicle-portable (mobile) platform can be found in [11, 13 and 14].

With regard to drinking water, it is feasible to automatically monitor, in real time, the following parameters: the temperature of water, water conductivity, pH and Redox potential, odor, the dissolved oxygen concentration, the concentrations of residual chlorine and of chlorine disinfection by-products in the network, the concentrations of various heavy metals, of dissolved solids, nutrients, undesirable substances (nitrates and nitrites), and the presence of bacteria. At any rate, the existing capabilities are less than those available for other aquatic environments, as the receiving sensors and the accompanying transducers (and/or other necessary electronics) have to be of miniature type in order to occupy the limited space. This space is usually available inside the terminal stations built at some specific junction sites of the distribution systems. It is worth noting, however, that, due to the interest shown by the industry of designing and producing automated sensors for monitoring (in real time) the water environment, several new sensors appear in the market every year and older versions of existing instruments are upgraded at fast rates. It is hoped that, within the next few years, all of the parameters required to be monitored in water supply and distribution systems will be monitored remotely in a continuous and automated fashion.

In this regard, multiparametric apparatus and sensors for remote monitoring of the quality of drinking water (and of the air) have appeared recently. Some of them are of miniature type and are very appropriate for installation inside the small terminal stations positioned at the pre-selected junction sites of a distribution network. These multiparametric apparatus are equipped with data storage, processing and transmission capabilities, and may be programmed to respond remotely to various commands and pre-selected events. Using the same automatically calibrated sensor, they can monitor simultaneously water temperature, pH, Redox potential, water conductivity, and water depth.

7. Data Transmission

The data collected at each terminal station, in the form of analog or digital time series, are transmitted wirelessly or via leased lines or Internet Service Providers to the control -and-command station. It is clarified that the sampling rates of the monitoring quantities are usually determined by the encountered frequencies in the fluctuating signals, in a way that the Nyquist criterion is satisfied. Usually, a frequency twice the largest frequency encountered in all monitored signals is used to formulate a common sampling rate of all monitored data. The Data Acquisition (DAQ) software, however, is flexible, permitting variable sampling rates, depending on the needs of the monitoring program.

Wireless transmission frequently appears to be more flexible and attractive than the other alternatives. Hybrid solutions are also a possibility. Transmission costs, the quality of data transmission and the possibility of the existence of an independent and autonomous transmission network along the water distribution system are some of the factors that may influence the final choice among the various alternatives.

The presence of a processor at each terminal station facilitates the communication of the data collected by the monitoring instruments, as the latter communicate with the external environment via a GPIB protocol (IEEE Reg., No 488).

There are two communication alternatives among the monitoring instruments and the processor (computer) at each terminal station: either via direct incorporation of the monitoring configuration inside the terminal processor, or via a special BPIB card. That will depend on the specific methodology of monitoring the various chemical indices and the conversion of the monitored quantity to a presentable form. In the first case, a fast, expandable, low cost and highly reliable system of data acquisition is created with direct plug-in of the monitoring instruments to each terminal computer unit. High transmission rates and signals in digital form are also obtained. Similar quality data are obtained when the second alternative is used, although the interconnection among the various units is different.

8. Communication Alternatives

8.1. GENERAL

If a cable communication among the terminal stations and the central control -and- command station is selected, the following possibilities arise: (1) use of leased lines either from a local telecommunication organization or from ISP services (Internet Service Providers); and (2) the use of an autonomous transmission network in the area of interest if such a network exists.

Receiving and transmitting data are accomplished via dial-up access or via simple modems. Such a cable interconnection among the various terminal stations has a low cost of data handling but may present several deficiencies regarding its transmission capabilities when mobile units are used. Reasons that may lead to the selection of either alternative, i.e., cable or wireless transmission, may depend on the specific application characteristics and requirements. Some of the factors that may directly influence the selection process are:

1) The distance between the most remote terminal station and the central control -and- command station;
2) The number of terminal stations;
3) The data volume that must be transmitted to the central control -and- command station;
4) The ground morphology and the environment of H.M. wave transmission;
5) The frequency and the way of possible transportation of monitoring (terminal) stations.

The possible alternatives may be divided into two categories:

a) Creation of a wide area wireless network among all of the terminal station and the central control -and- command station;
b) Serial information transmission from- and -towards the terminal stations and treatment of the respective information by means of appropriate software.

8.2. WIRELESS WIDE AREA NETWORK

Realization of such a network is possible by means of a wireless controller (hub) which is connected to the central control -and- command station. Similar devices are installed at each terminal station, and they communicate with the hub of the central station. With the use of such a controller, all network stations communicate with each other. The flexibility and the ability of communication depend on the supporting software and its particular adjustments at each of the terminal stations.

Controllers of such kind have a data transmission capability of up to 19.2 kbps. Areas that can be covered this way may exceed a range of 50 km, with rates of error propagation in the order of 10^{-8}. These controllers use transmission channels 25 kHz wide, in the areas of 820-960 MHz or 400-512 MHz.

The respective terminal devices are small in size and of low cost, and they can be easily connected to the computers of terminal stations.

8.3. SERIAL TRANSMISSION

Serial data transmission from - and- towards the terminal stations allows their interconnection with the central control -and- command station, maintaining a level in the complexity of the system. Data handling takes place via software execution at both the central and the terminal stations. Data are transmitted within packets as dual information, regardless of their kind, and are decoded with the aid of a special software. This type of interconnection allows for a communication of *half-duplex* type between each terminal station and the central station.

Serial wireless communication and data transmission from various locations of the network may be accomplished by means of (a) radio modems, (b) an existing GSM network, and (c) through a special UHF receiver - transmitter.

8.3.1. *Use of Radio Modems*

Radio modems are RF receivers-transmitters that function in the UHF range. They provide transparent data transmission at rates up to 9600 bps, connected with the computer through an RS-232 or RS-485 gate. They usually use 10-25 kHz bands in frequency ranges of 406-70 MHz. Data transmission through the air may be achieved with antennas of different types as, for example, the dipoles and antennas of Yagi type. Distances covered may vary depending on the type of antenna used and the ground morphology.

Since information transmission is transparent, data gathering and preparation, prior to their transmission, rely on the specific software that is developed for such purposes. This software groups the data that come from the monitoring instruments, along with any other necessary information (e.g., coordinates of the monitoring location, time and date of monitoring, etc.).

8.3.2. *The Use of an Existing GSM Network*

The serial interconnection of terminal stations with the central control -and- command station may be realized with the aid of an existing GSM network of mobile telephone communications. For this, a modem and a mobile phone are needed at each terminal station, as well as in the central control -and- command station. A GSM network allows data transmission through special channels at rates up to 9600 bps. It is, therefore, possible, through the network of mobile telecommunications, to realize a dial-up connection between any two terminal stations and transfer all of the data through such a connection. Such a connection may be energized either from the central station or from a terminal station. Software for data transmission through a GSM network is provided by the mobile phone industry and the makers of special cards for connecting mobile telephones with computers. Data transmission is again transparent, and the data handling and processing are done by a special software.

8.3.3. *The Use of Special UHF Receiver-Transmitter*

As with radio modems, it is possible to use specially fabricated receivers - transmitters. Such a solution helps in the case where the commercially available radio modems impose some serious restrictions on the realization of a system (e.g., limited transmission range, limited zone width, etc.).

9. Water Quality Simulation Model

In the past, water supply and distribution networks were designed and operated without any recourse to water quality aspects because of the difficulties encountered in the analysis of such complex dynamic systems. The huge cost of manually monitoring the water quality from the entire distribution network (i.e., of sampling the water and analyzing it in a conventional way) and the complex reaction of water with the various chemical substances that may exist in the network render mathematical modeling a quite suitable method for predicting the quality of water in distribution networks.

There are several techniques dealing with modeling of the water quality in a water distribution system. Such techniques are known as Eulerian and/or Lagrangian. The distinction depends on the volumetric control approximation used.

Eulerian models divide the network into a number of fixed interconnected control volumes, record changes at the boundaries or within these volumes, and compute changes of concentrations of various substances as the water flows between the various pipes of the network. Lagrangian models track changes in a series of discrete parcels of water as they travel through the network.

These simulation techniques may be time-driven or event-driven. Time-driven simulations update the state of the network at fixed time intervals. Event-driven simulations update the state of the system only at times when a change actually occurs, such as when a new parcel of water reaches the end of the pipe and mixes with water from other connecting pipes.

Some references regarding water quality simulation in water distribution networks may be found in [15, 16, 17, 18 and 19]. In most of these references, a particular software has been used as a tool to conduct several numerical tests and evaluate the performance of the Eulerian and Lagrangian techniques, when the latter are combined with various numerical schemes and used to model the quality of drinking water in distribution systems. This particular simulation software (EPANET) has been developed by the Environmental Protection Agency of USA, and it is often used as a vehicle for further improvements and extensions aiming at the incorporation of more fate and transport processes (physical and biochemical) and of more chemical substances as well. The current capabilities of EPANET still remain limited, and more have to be done before such software becomes operational and meets the needs of real-time water quality monitoring. EPANET has been documented in various references as in Rossman [20].

The software that simulates the quality of drinking water in a distribution system also simulates, in real time, the hydraulic characteristics of the flowing water in the network. More specifically, it tracks the flow of water in each pipe, the pressure at each pipe junction, the water elevation at each storage tank, and the concentrations of various substances through the network. Such substances may be total solids (TS), total dissolved solids (TSS), BOD-5, COD, total coliforms, NH_3-N (ammonium nitrogen), total phosphorus (T-PO_4-P), etc. It is also possible to simulate the concentrations of other toxic substances and heavy metals (e.g., As, Cd, Pb, Cu, Co, Ho, Hg, Zn, Cr), trace elements (e.g., Ca, Mg, K, Na), as well as salts (e.g., Cl^-, SO_4, SO_3, NO_3, NO_2, PO_4).

The number and kind of simulated substances mainly depend on the availability of continuous monitoring of the particular substances. The simulation software takes into account the decay of substances (through first order decay kinetics) and models reactions within the bulk flow, reactions at the pipe walls, and mass transport between the bulk of the flow and the pipe wall. It also considers the variations of concentration in disinfectant residuals, growth of disinfecting by-products, the blending of different sources of waters, and the residence times of water in the system.

The simulation software incorporates an assimilation module that allows assimilation of the monitored data, in a continuous fashion or otherwise, from the specific junction sites where the terminal stations are located. The combination of simulation software and an assimilation module optimizes the simulation precision in the distribution network so that simulations become most reliable, provided that, in all mathematical models, the problem of optimal calibration, validation and verification always exists.

Nowadays, the basic problems associated with the assimilation of data into mathematical models that simulate physical and/or biochemical processes have been understood. Assimilation techniques have also been used with ecosystem models. Bayes Theorem and Monte Carlo-Markov Chain algorithms have been used, for example, to estimate, in an optimal sense, several parameters in a multi-compartment ecosystem model dealing with flows of nitrogen amongst phytoplancton, zooplancton, nitrate, bacteria, ammonium, dissolved organic nitrogen, and detritus [21, 22].

During the last few years, important steps have been taken towards the development of advanced assimilation schemes that handle even strong non-linearities in the relevant equations at reasonable costs, utilizing proper statistical error analysis techniques. Typical methods, such as the "sequence" or the ''optimum interpolation'' technique, or schemes that utilize statistical errors depending on time and showing dynamic consistency, may be used to maximize the precision of water quality simulation predictions. The optimum interpolation technique is a particular choice.

10. Development Stages of a Monitoring System

The design and integration of a remotely (or tele-) controlled, automatic and continuous monitoring system that will serve the needs of a particular distribution network may be realized in the following three phases, although the detailed characteristics of such a monitoring system may differ and depend, in turn, on the characteristics of both the distribution system itself and the geographic area covered.

Phase one: This phase may include two stages. During the first stage, a minimum number of parameters for monitoring the physicochemical characteristics of the quality of drinking water in the specific distribution system is determined. There are about 50 chemical or physicochemical parameters (among the 61) that have to be monitored, but monitoring in a continuous and automatic fashion for all of these parameters is neither feasible nor practical to apply. There are also some microbial parameters that have to be monitored in the system. Thus, the minimum number of parameters selected for monitoring, among the total sum of physicochemical and microbial parameters, will guarantee satisfactory monitoring of the overall quality of drinking water in the distribution system. In the second stage, the telemetry characteristics of the system are designed. More specifically, some of the parameters selected in the previous stage, as for example pH, will be used for continuous monitoring, and the transmission of the respective data will provide guidance for the investigation of any telemetry problems in the particular urban environment.

Phase two: This phase includes six stages. In the first stage, the possibility for automatic and continuous (site) chemical analysis of the quality parameters selected for monitoring is examined. In the second stage, the possibility for automatic and continuous microbial monitoring is examined. Biosensors for environmental monitoring are available today. For example, biosensors for monitoring bacteria and other substances in the water, such as organic pollutants, pesticides, phenols, heavy metals etc., are available [11, 13]. In the third stage, the instruments that will perform the *in situ* automatic and continuous chemical and microbial analyses are selected. In the fourth stage, the physical and biochemical processes that influence the parameters selected for monitoring and to be mathematically modeled are determined. The existence of relevant software, that might be available, and/or the need for modification and further extension of such software is also investigated at this stage. In the fifth stage, the assimilation methodology is selected. In the sixth and last stage, the strategic locations of the terminal monitoring stations are determined among the various junctions sites of the distribution network. Some preliminary simulations are made for pinning down possible locations of unusual changes of the water quality parameters monitored in the entire network.

Phase three: This phase includes two stages. In the first, the specifications of the monitoring terminal stations are completed. In the second stage, the specifications for a vehicle-portable mobile unit are determined. Issues that might be examined are, for example, the kind of energy support this mobile unit may have, the kind and number of working persons in the mobile unit, and the unit equipments.

11. Conclusions and Future Perspectives

It is believed that the most complete and viable alternative for monitoring the quality of drinking water in water distribution systems appears to be that of a system operating in a remotely controlled and continuous fashion. The supporting arguments may be summarized as follows: (a) the strict (EE) regulations and the high drinking water quality standards that have to be met in distribution networks; (b) the uncertainty that frequently exists in the quality of water within the distribution systems, caused by the insufficient knowledge of both the system characteristics and the bio-geo-chemical processes that govern the transport and fate of the various contaminants in the system; and (c) the need: (i) to have reliable and fast information for proper decision making in order to minimize health risks and save lives when extraordinary or other events occur; (ii) to avoid the adverse consequences of possible improper treatment practices that may produce undesirable substances in the system; (iii) to reduce operational costs, (iv) to upgrade the water quality in the system by maintaining its disinfection capabilities at any time; and (v) to increase its reliability.

Today, the instrumentation and computer technologies that may be used for monitoring and control of the water quality in distribution systems are mature and well tested. The specific design of such an automated system for a particular city must take into account the geographic characteristics of the area, the system's characteristics, the regulatory constraints, and the existing socio-economic framework. Intensive research efforts in the design and production of multiparametric apparatus, sensors, biosensors, and mobile analytical instruments are expected to completely cover the existing needs of a fully automated monitoring system to include all of the specified 61 water quality parameters. These new instruments are expected to have the capability of analyzing a large number of compounds in a few seconds. Their smaller size and cost, their energy independence and ease of use, the parallel optimization of special driving software that will be used in conjunction with remote communication capabilities, the further enhancement of the software that will simulate completely the biochemical processes in the distribution system, and the perfection of assimilation schemes will provide unique capabilities for monitoring the quality of drinking water in distribution systems in the near future.

12. Acknowledgments

Particular thanks and appreciation are due to Professors M. Statheropoulos and P. Kottis from the Chemical and Mechanical Engineering Departments of the NTUA, respectively.

13. References

1. Protopapas, A.L. and Nguyen, B. (1995) The command-and-control system for the water supply in Paris, France - A challenge for the New York City Department of Environmental Protection in *Automating to Improve Water Quality*, WEF Specialty Conference Series Proceedings, June 25-28, Minneapolis, Minnesota, pp. 9.

2. Fernandes, C. and Karney, B.W. (1999) Assessing water quality issues in water distribution systems from source to demand in *Water Industry Systems: Modeling and Optimization Applications*, Vol. 2, Research Studies Press Ltd., 231-239.

3. Koutra, M. (2000) Mathematical Modeling of Residual Chlorine in Distribution Networks, Diploma Work (in Greek), National Technical University of Athens, Civil Engineering Dept., Hydraulics Div., pp.100.

4. Theodoulidis, P. (1996) A Supervisory Control and Data Acquisition (SCADA) system for monitoring the water distribution of Cyprus' capital city, (in Greek) in *Management of Water Resources*, International Conference Proceedings, Vol. II, Technical Chamber of Greece (Division of Central and Western Thessaly), Larisa, November 13-16, pp. 6.

5. Haugan, P.M., Evensen, G., Johannessen, J.A., Johannessen, O.M., and Petterson, L. (1991) Modeled and observed mesoscale circulation and wave current refraction during the 1988 Norwegian continental shelf experiment, *J. Geophysical Research* **C6**, 10487-10456.

6. Johannessen, J.A., Roed, L.P., Johannessen, OM Evensen, G., Hackett, B., Petterson, L.H., Haugan, P.M., Sanven, S. and Shuchman, R. (1993) Monitoring and modeling of the marine coastal environment, *J. Photogrammetric Engineering and Remote Sensing* **59-3**, 351-361.

7. Johannessen, J.A., Vachon, P.W., and Johannessen, O.M. (1994) ERS-1 SAR imaging of marine boundary layer process, *ESA, Earth Observation Quarterly* **46**, 1-5.

8. Papadimitrakis, J.A. and Nihoul, J. (1996) Temporal (and spatial) scales and sampling requirements in environmental flows (with emphasis on inland and coastal waters) in N.B. Harmancioglu, M.N. Alpaslan, S.D. Ozkul and V.P. Singh (eds.), *Integrated Approach to Environmental Data Management Systems*, NATO ASI Series, Series 2: Environment-Vo. I31, Kluwer Academic Publishers, Dordrecht, pp.117-132.

9. *National Handbook of Water Quality Monitoring* (1996), U.S. Depart. of Agriculture, National Resources Conservation Service.

10. Soukissian, T.H., Chronis, G.T., Vlachos, D., and Ballas, D. (1999) POSEIDON: A marine environmental monitoring, forecasting and information system for the Greek Seas, in *Abstracts of Proceedings Oceanography of the Eastern Mediterranean and Black Sea: Similarities and Differences of Two Interconnected Basins*, International Conference, Athens, Greece, 23-26 February.

11. M. Ochsenkuhn-Petropulu and K.M. Ochsenkuhn (Eds.) (1999) *Instrumental Methods of Analysis. Modern Trends and Applications*, International Conference Proceedings, Vol. I & II, Chalkidiki, Greece, 19-22, September.

12. Meuzelaar, H.L.C., Dworzannski, J.P., and Amold, N.S. (1999) Advances in field-portable, mobile GC/MS instrumentation in M. Ochsenkuhn-Petropulu and K.M. Ochsenkuhn (eds.), *Instrumental Methods of Analysis. Modern Trends and Applications*, International Conference Proceedings, Vol. II, Chalkidiki, Greece, 19-22, September, 349-356.

13. Biosensors for Environmental Monitoring (1994) Second European Workshop, 17/18 February, Kings College, London, United Kingdom, Technologies for Environmental Protection, Report EUR 15622 EN.

14. Bockreis, A. and Jager, J. (1999) Odour monitoring by the combination of sensors and neural networks, *J. Environmental Modelling and Software* **14-5**, 421-426.

15. Males, R.M., Grayman, W.M., and Clark, R.M. (1988) Modeling water quality in distribution systems, *Ame. Soc. Civil Eng., J. Water Resources Planning and Management* **114-2**.

16. Clark, R.M., Grayman, W.M., Males, R.M., and Hess, A.F. (1993) Modeling contaminant propagation in drinking water distribution systems, *Am. Soc. Civil Eng., J. Environmental Engineering* **119-2**, 349-364.

17. Rossman, L.A., Boulos, P.F. and Altman, T. (1993) Discrete volume-element method for network water-quality models, *Am. Soc. Civil Eng., J. Water Resources Planning and Management* **119**-5, 505-517.

18. Rossman, L.A., Clark, R.M., and Grayman, W.M. (1994) Modeling chlorine residuals in drinking – water distribution systems, *Am. Soc. Civil Eng., J. Environmental Engineering* **120**-4, 803-820.

19. Rossman, L.A., and Boulos, P.F. (1996) Numerical methods for modeling water quality in distribution systems: A comparison, *Am. Soc. Civil Eng., J. Water Resources Planning and Management* **122**-2, 137-146.

20. Rossman, L.A. (1994) *EPANET-Users Manual, EPA-600/R-94/057*, U.S. Environmental Protection Agency, Risk Reduction Engineering Lab., Cincinnati, Ohio.

21. Harmon, R. and Challenor, P. (1977) A Markov Chain Monte Carlo Method for estimation and assimilation into models, *Ecological Modeling* **101**, 41-59.

22. Fasham, M.J.R. and Evans, G.T. (1995) The use of optimization techniques to model marine ecosystem dynamics at the JGOFS station at 470N 200W, *Phil. Trans. R. Soc. Lond. B* **348**, 203-209.

ENTROPY-BASED DESIGN CONSIDERATIONS FOR WATER QUALITY MONITORING NETWORKS

N. B. HARMANCIOGLU and S. D. OZKUL
Dokuz Eylul University, Faculty of Engineering
Tinaztepe Campus, Buca 35160 Izmir, Turkey

Abstract. Assessment of water quality monitoring networks requires potential methods to delineate the efficiency and cost-effectiveness of current monitoring programs. To this end, the concept of entropy has been considered as a promising method in previous studies as it quantitatively measures the information produced by a network. The paper presented discusses an entropy-based approach for the assessment of combined spatial/temporal frequencies of monitoring networks. The results are demonstrated in the case of water quality data observed along the Mississippi River in Louisiana.

1. Introduction

The current status of water quality monitoring networks reflects a wide range of shortcomings despite all the efforts and investment made on monitoring. A major consequence of this situation is the significant gap between information needs on water quality and the information produced by current systems of data collection. In view of this difficulty, researchers have felt the need to focus more critically on design procedures used. Within the last few years, most developed countries have started the redesign of their monitoring programs to revise and/or modify the existing networks [1, 2, 3, 4, 5, 6, 7].

The initial step in the redesign process is to assess the existing network with respect to its basic features, i.e., sampling sites, temporal frequencies, variables sampled, and the duration of sampling. Current literature provides considerable amount of research carried out so far on these four aspects of the design problem [3, 4, 8, 9, 10, 11]. Available studies show that water quality monitoring is recognized basically as a statistical procedure so that both the assessment and the design problems are addressed via statistical methods [12]. Statistical analyses based on regression theory as well as time series analysis, decision theory and optimization methods are used to select the spatial and temporal design features of a network. Spatial-analytical techniques such as kriging and co-kriging are also employed to assess spatial dependencies in data from water quality sampling networks [13, 14]. However, these geostatistical methods are considered to be better suited for the design of groundwater monitoring networks than for the dendritic nature of stream water quality networks, where assumptions of stationarity and spatial continuity may not always be satisfied [10, 15, 16]. On the other hand, some researchers have proposed that the above techniques should be employed in combination with each other for a better assessment of the design problem [17]. This is a plausible approach as none of the methods have yet been widely accepted due to particular deficiencies associated with each technique.

Although it is generally accepted that networks should produce data that permit the application of statistical data analysis techniques, current design methodologies do not fulfill this expectation. Thus, it is often very difficult to assess the information conveyed by existing networks. In general, available methods of network design are deficient because: (a) a precise definition of "information" contained in the data and how it is measured is not given; (b) the value or utility of data is not precisely defined, and consequently, existing networks are "optimal" neither in terms of the information contained in these data nor in terms of the cost of getting the data; and (c) the method of information transfer in space and time is restrictive. Essentially, available design methodologies serve to assess the effectiveness of design decisions and the efficiency of an existing network although each method focuses on the problem from a different perspective, using different criteria. Yet, there is still the question of how one relates such criteria in the assessment process to the value of data [11].

Within this respect, one of the most promising methods for network assessment purposes is based on the entropy concept of information theory, which has been used to evaluate not only water quality but also other hydrometric networks [18, 19, 20, 21, 22, 23]. Entropy is a measure of the degree of uncertainty of random hydrological processes. It is also a quantitative measure of the information content of a series of data since reduction of uncertainty, by making observations, equals the same amount of gain in information [19]. Harmancioglu and Alpaslan [20] claimed that the entropy method allows to quantitatively measure network efficiency in terms of the information produced by the network.

In view of the statistical information expected from a network, i.e., water quality means, extremes and temporal trends, it must be emphasized that entropy measures basically reflect the spatial and/or temporal variability of water quality. Thus, if the primary objective of monitoring is considered as the determination of hydrologic variability, then the entropy principle can be used to evaluate informativeness of data with respect to such changes in time and space dimensions.

The major advantage of the entropy method for network assessment purposes is that it defines information or data utility in quantitative terms and can be used to assess five basic features of a network, i.e., sampling sites, frequencies, combined space/time frequencies, variables to be sampled and the duration of sampling. The method is applicable to cases where a decision must be made to remove existing observation sites, and/or reduce sampling frequencies, and/or terminate collection program. The procedure also indicates needs for network expansion in time and/or space domains [20].

In the following, an entropy-based methodology is presented for the analysis of sampling sites, temporal sampling frequencies, and combined space/time network features. The methodology is demonstrated on water quality data collected along the Mississippi River in Louisiana. As the available data comprise a long observational period with very few missing values, the potential aspects of the method are demonstrated more clearly in comparison with the results of the earlier work by Harmancioglu and Alpaslan [20].

2. Applied Methodology

2.1. DEFINITIONS OF ENTROPY

The definitions of entropy measures are derived in information theory, which analyzes the statistical structure of a series of numbers, signs or symbols that make up a communication signal. The term "information content" refers to the capability of signals to create communication, and the basic problem is the generation of correct communication by sending a sufficient amount of signals, leading neither to any loss nor to repetition of information. The basic principles of the information theory were developed by Shannon, who defined the information content of signals as "entropy" since the mathematical expression for this concept is analogous to that of entropy in statistical mechanics [24].

According to Shannon, information is attained only when there is uncertainty about an event, which implies the presence of alternative results the event may assume. Alternatives with a high probability of occurrence convey little information and vice versa. Here, the probability of occurrence of a certain alternative is the measure of uncertainty (entropy) or the degree of unexpectedness of a signal. According to Shannon, signals sent through a communication channel must be uncertain before they are transmitted; that is, they must have a "surprise value" to create information.

The entropy of a random process can be computed and expressed in specific units (e.g., as bits per symbol) once its statistical structure is identified. The entropy function always assumes positive values, which makes it readily acceptable as an objective criterion for measuring the information content of any random process. Thus, the concept has found a wide area of application in various other fields of science including hydrology [19,23,25,26].

There are four basic information measures based on entropy: marginal, joint, conditional entropies, and transinformation. Shannon and Weaver [24] were the first to define the marginal entropy, H(X), of a discrete random variable X as:

$$H(X) = -K \sum_{i=1}^{N} p(x_i) \log p(x_i) \tag{1}$$

with the constant K=1 if H(X) is expressed in "napiers" for logarithms to the base e. Here, N represents the number of elementary events with probabilities $p(x_i)$ (i=1,..., N).

For a continuous random variable X with probability density function f(x), the total range of X is subdivided into N intervals of width Δx, so that the probability that a value of X is within the *i*th interval, where i=1,2,...,N, is:

$$P_i = P(x_i - \frac{\Delta x}{2} \leq X \leq x_i + \frac{\Delta x}{2}) = \int_{x_i-(\frac{\Delta x}{2})}^{x_i+(\frac{\Delta x}{2})} f(x)\,dx \tag{2}$$

In this case, probabilities $p(x_i)$ of Eq. (1) are approximated as [$f(x_i)\,\Delta x$] for small Δx, with $f(x_i)$ being the relative class frequency and Δx, the length of class intervals or the discretization interval [25]. Then, the definition of entropy is extended to the continuous case in the form of:

$$H(X;\ \Delta x) = -\int_{-\infty}^{+\infty} f(x) \log f(x)\, dx - \log(\Delta x) \tag{3}$$

The total entropy of two independent random variables X and Y is equal to the sum of their marginal entropies:

$$H(X, Y) = H(X) + H(Y) \tag{4}$$

When X and Y are stochastically dependent, their joint entropy is less than the total entropy of Eq. (4). For the continuous case again, the joint entropy, or the total entropy of dependent variables, is:

$$H(X, Y; \Delta x, \Delta y) = -\int_{-\infty}^{+\infty}\int_{-\infty}^{+\infty} f(x, y) \log f(x, y)\, dx\, dy - \log(\Delta x \Delta y) \tag{5}$$

with f(x,y) being the joint probability density function of X and Y. Conditional entropy of X given Y represents the uncertainty remaining in X when Y is known, and vice versa:

$$H(X \mid Y; \Delta x) = -\int_{-\infty}^{+\infty}\int_{-\infty}^{+\infty} f(x, y) \log f(x \mid y)\, dx\, dy - \log(\Delta x) \tag{6}$$

where $f(x \mid y)$ is the conditional probability density function of X with respect to Y.

Transinformation is another entropy measure which measures the redundant or mutual information between X and Y. It is described as the difference between the total entropy and the joint entropy of dependent X and Y:

$$T(X, Y) = H(X) + H(Y) - H(X, Y) \tag{7}$$

or as:

$$T(X, Y) = H(X) - H(X \mid Y) = H(Y) - H(Y \mid X) \tag{8}$$

The above definitions can be extended to the multivariate case with M variables [20]. The total entropy of independent variables X_m (m= 1,..., M) equals:

$$H(X_1, X_2, \ldots, X_M) = \sum_{m=1}^{M} H(X_m) \tag{9}$$

If the variables are dependent, their joint entropy can be expressed as:

$$H(X_1, X_2, \ldots, X_M) = H(X_1) + \sum_{m=2}^{M} H(X_m \mid X_1, \ldots, X_{m-1}) \tag{10}$$

To compute the joint entropy, the multivariate joint probability distribution of M variables is used:

$$H(X_1, X_2, \ldots, X_M; \Delta x_1, \ldots, \Delta x_M) = -\int_{-\infty}^{+\infty} \ldots \int_{-\infty}^{+\infty} f(x_1, \ldots, x_M) \log f(x_1, \ldots, x_M)\, dx_1\, dx_2 \ldots dx_M - \log(\Delta x_1 \Delta x_2 \ldots \Delta x_M) \tag{11}$$

It is sufficient to compute the joint entropy of the variables to estimate the conditional entropies of Eq. (10) since the latter can be obtained as the difference between two joint entropies, e.g.:

$$H(X_M \mid X_1, X_2, ..., X_{M-1}) = H(X_1, X_2, X_3, ..., X_M) - H(X_1, X_2, ..., X_{M-1}) \quad (12)$$

Finally, when the multivariate normal distribution is assumed for $f(x_1, x_2, ..., x_M)$, the joint entropy of **X**, **X** being the vector of M variables, can be expressed as:

$$H(\mathbf{X}) = (M/2)\ln 2\pi + (1/2)\ln|C| + M/2 - M\ln(\Delta x) \quad (13)$$

where $|C|$ is the determinant of the covariance matrix C and Δx, the class interval size assumed to be the same for all M variables.

2.2. ASSESSMENT OF SPATIAL FREQUENCIES

Assessment of sampling sites in a monitoring network has to be carried out separately for each water quality variable observed. The approach here is to assess the reduction in the joint entropy (uncertainty) of two or more variables (i.e., two or more sampling sites where a particular water quality variable is observed) due to the presence of stochastic dependence between them. Such a reduction is equivalent to the redundant information (transinformation) in the series of the same water quality variable observed at different sites. Thus, the objective in spatial orientation is to minimize the transinformation by an appropriate choice of the number and locations of monitoring stations. The combination of stations with the least transinformation reflects the variability of the quality variable along the river without producing redundant information. Such an approach foresees the monitoring of a water quality variable at points where it is the most variable or uncertain. Accordingly, existing sampling sites can be sorted in order of decreasing uncertainty or decreasing informativeness. In this ordered list, the first station is the one where the highest uncertainty occurs about the variable. The following stations serve to reduce this uncertainty further so that the last station in the list brings the least amount of information. This approach conforms to Shannon's definition of entropy, i.e., the more uncertain an event is to the observer prior to sampling, the more is the information obtained when such an event is observed. It is possible here to select a threshold transinformation value as the amount of redundant information to be permitted in the network such that sampling of the particular water quality variable may be quit at stations which exceed the threshold. Here, a priority ordering of stations is accomplished with respect to their degrees of uncertainty to arrive at the best combination of stations where redundancy in the network is kept at a minimum.

The following procedure is applied for each water quality variable separately to select the best combination of stations [27]:

a) It is assumed that there are M monitoring stations in the basin. The data series of the selected water quality variable at each station is represented by X_m, with outcomes $x_{m,i}$, where m (m = 1,...,M) denotes the station identification number and i, the time point along the sample of size N (i = 1,...,N). Here, the sampling duration at all stations is considered to be equal. However, the total number (N) of available data at each station can be different since there are often missing values or gaps within the data series.

b) Next, the type of the multivariate joint probability density function which best fits the distribution of X_m (m=1,...,M) is selected. At present, only normal and lognormal distributions can be used for the computation of entropy measures in the multivariate case since the description of multivariate probability density functions for other skewed distributions is very difficult [23]. If a multivariate normal or lognormal distribution is assumed, the joint entropy of M stations $H(X_1,...,X_M)$ can be calculated by Eq. (13). This joint entropy represents the total uncertainty about the particular water quality variable in the basin, which is to be reduced by sampling at M monitoring stations.

c) In the next step, the marginal entropy $H(X_m)$ (m=1,...,M) of the water quality variable observed at each station is computed again by Eq. (13) where M is replaced by 1. The station with the highest $H(X_m)$ is denoted as the first priority station X_1; this is the location where the highest uncertainty occurs about the variable so that the highest information may be gained by making observations at this site. Note that the station identification number m is now being replaced by the priority index j such that the mth station X_m with the highest entropy is denoted $X_j = X_1$.

d) Later, this station is coupled with every other station in the network to select the pair that leads to the least transinformation. The station that fulfills this condition is marked as the second priority location $X_j = X_2$ such that:

$$\min \{H(X_1) - H(X_1 \mid X_2)\} = \min \{T(X_1, X_2)\} \tag{14}$$

The conditional entropy, which is in the form of:

$$H(X_1 \mid X_2) = H(X_1, X_2) - H(X_2) \tag{15}$$

can be computed by Eqs. (12) and (13). In the next step, the pair (X_1, X_2) is coupled with every other station in the network to select a triple with the least transinformation.

e) The same procedure is continued by considering successively combinations of 3,4,5,...,j stations and selecting the combination that produces the least transinformation by satisfying the condition:

$$\min \{H(X_1, ..., X_{j-1}) - H(X_1, ..., X_{j-1} \mid X_j)\} = \min \{T(X_1, ..., X_{j-1}), X_j\} \tag{16}$$

where X_1 is the 1st priority station and X_j is the station with the jth priority. Conditional entropies and transinformations are calculated as:

$$H(X_1, ..., X_{j-1} \mid X_j) = H(X_1, ..., X_{j-1}, X_j) - H(X_j) \tag{17}$$

$$T((X_1, ..., X_{j-1}), X_j) = H(X_1, ..., X_{j-1}) - H(X_1, ..., X_{j-1} \mid X_j) \tag{18}$$

For the multivariate normal probability density function, transinformation can also be determined by:

$$T((X_1,\ldots,X_{j-1}),X_j) = -(1/2)\ln(1-R^2) \tag{19}$$

where R represents the multiple correlation coefficient. Accordingly, the above procedure assures the selection of a station X_j that has the least correlation with other stations in the network.

f) In carrying out the above procedure, one may evaluate the results at each step by defining the percentage (t_j) of nontransferred information among stations as:

$$t_j = H(X_1,\ldots,X_{j-1} \mid X_j) / H(X_1,\ldots,X_{j-1}) \tag{20}$$

and the percentage of transferred information ($1-t_j$) as:

$$1-t_j = T((X_1,\ldots,X_{j-1}),X_j)) / H(X_1,\ldots,X_{j-1}) \tag{21}$$

Here, the designer may decide how much repeated information he wants to permit in the network. If he specifies this upper limit of redundant information as $(1-t_j)^*$ in percent, he can select the combination of stations that produces this percentage as the one that must be included in the network. Stations that are added to the system after reaching $(1-t_j)^*$ will increase the redundant information further so that one may decide to quit monitoring at such locations.

g) The evaluation explained in step (f) can also be made by defining k_j or the ratio of uncertainty explained by j number of stations to that explained by the total M number of stations in the network:

$$k_j = H(X_1, X_2 \ldots, X_j) / H(X_1, X_2, \ldots, X_M) \tag{22}$$

One may specify here an upper limit k_j^* as the percentage of uncertainty that is to be removed by the network. This upper limit is reached by a certain combination of stations; thus, sampling sites which produce $k_j > k_j^*$ may be discontinued.

In the above procedure, the benefits for each combination of sampling sites are measured in terms of the least transinformation or the highest conditional entropy produced by that combination. To select the best combination of stations, it is sufficient to compare costs (increases in costs by addition of new stations or decreases in costs by deletion of some stations) and benefits represented by t_j or k_j.

The above procedure helps to assess a network with respect to existing monitoring sites. If new stations are to be added to the system, their locations may be selected again on the basis of the entropy method by assuring maximum gain of information. Husain [18] applied this approach to a rainfall gauging network, where he developed a relation between entropy measures and the distance between station pairs.

2.3. ASSESSMENT OF TEMPORAL SAMPLING FREQUENCIES

The analysis of temporal frequencies by the entropy method is based on the assessment of reduction in the marginal entropy of a water quality process due to the presence of serial dependence within its data series. Such a reduction, if any, is equivalent to the redundant information of successive measurements. This analysis has to be carried out separately for each sampling site in case of the particular variable analyzed. Here, the results may indicate either a decrease in time frequencies or a complete termination of data collection at the site if information produced is redundant in the time domain. The steps of the procedure are given as follows [20, 27]:

a) An appropriate probability density function is selected to fit the distribution of the water quality variable X_j at station j, with outcomes $x_{j,i}$, where j represents the station priority index and i denotes the time point along the sample of size N. When the variable X_j is described as subseries $X_{j,k}$ with realizations $x_{j,i-k}$ for time lags k = 0, ..., K, the problem becomes multivariate so that either multivariate normal or lognormal distribution is used to compute the relevant entropy values.
b) Here, one first has to determine the degree of dependence k in the data series. To this end, the marginal entropy of the variable is computed by Eq. (13) where M is replaced by 1. In this case, the covariance matrix C converts directly into the variance σ^2 of the variable X_j and assumes the form of:

$$H(X_j) = (1/2)\ln 2\pi + (1/2)\ln\sigma^2 + 1/2 - \ln(\Delta x) \tag{23}$$

c) In the next step, a maximum lag K is assumed, and considering subseries $X_{j,0}$, $X_{j,1}$, $X_{j,2}$,..., $X_{j,k}$ as separate variables, conditional entropies in the form of $H(X_{j,0} \mid X_{j,1},..., X_{j,k})$ are computed by Eq. (13) and the relation:

$$H(X_{j,0} \mid X_{j,1},...,X_{j,k}) = H(X_{j,0}, X_{j,1},...X_{j,k}) - H(X_{j,1},...X_{j,k}) \tag{24}$$

   The conditional entropies indicate the level down to which the marginal entropy decreases for each lag k. When no further reductions are observed in the marginal entropy at a certain k = K, it is considered that the lags beyond this point do not contribute significantly to the reduction of uncertainty.
d) If, in step (c) above, a significant dependence is detected between successive measurements, this indicates the presence of redundant information between such values. In this case, the sampling frequency may be reduced to decrease the redundancy; otherwise, sampling has to be continued with the already selected frequencies.
e) In general, serial dependence of the first order (k = 1) is significant for monthly observed water quality series [20]. Thus, the highest reduction in marginal entropy occurs at this lag. Here, one may investigate whether the sampling frequency can be extended to bimonthly intervals and determine the change in redundant information. To this end, the transinformation between successive measurements can be computed as:

$$T(X_{j,0}, X_{j,1}) = H(X_{j,0}) - H(X_{j,0} \mid X_{j,1}) \tag{25}$$

   where $H(X_{j,0})$ is the marginal entropy and $H(X_{j,0} \mid X_{j,1})$, the conditional entropy for k = 1. The ratio $T(X_{j,0}, X_{j,1}) / H(X_{j,0})$ shows in percentage the redundant information for bimonthly observations. Accordingly, the loss of redundant information due to the decrease in sampling frequency can be represented by $1-[T(X_{j,0}, X_{j,1})/H(X_{j,0})]$. One may decide upon the appropriate sampling frequency by comparing this level of uncertainty with the reduction in costs accruing from the decreased frequency. Similar evaluations may be made for further lags k = 2, 3, 4, ... which correspond to sampling intervals of every 3, 4, 5, ... months, respectively. For each k, transinformations can be computed as:

$$T(X_{j,0}, X_{j,k}) = H(X_{j,0}) - H(X_{j,0} \mid X_{j,k}) \quad (26)$$

to find the degree of redundancy for each sampling interval.

f) Entropy method can also be used as in Eq. (26) to evaluate time intervals beyond monthly to bimonthly, once in every 3 months, 4 months etc. For this, the procedure explained in (c) is applied for time lags k = 2 (once in every 3 months), k = 3 (once in every 4 months), k = 4 (once in every 5 months), and k = 5 (once in every 6 months). To find the contribution of each successive time lag to the overall uncertainty, transinformations may also be evaluated as:

$$T(X_{j,0},\ldots, X_{j,k}) = H(X_{j,0}) - H(X_{j,0} \mid X_{j,1},\ldots, X_{j,k}) \quad (27)$$

On the other hand, the total entropy (or joint entropy) also increases as k is increased. Thus, one may evaluate the redundant information for decreased sampling frequencies as:

$$T(X_{j,0},\ldots, X_{j,k}) / H(X_{j,0}, \ldots, X_{j,k-1}) \quad (28)$$

The difference 1- $[T(X_{j,0},\ldots, X_{j,k}) / H(X_{j,0}, \ldots, X_{j,k-1})]$ indicates the remaining uncertainty (or the loss of redundant information) due to a decrease in frequency for each time lag k.

In assessing temporal frequencies by the above approach, benefits of a selected alternative frequency can be described by transinformations $T(X_{j,0},\ldots, X_{j,k})$ or $T(X_{j,0}, X_{j,k})$, or the ratio of Eq. (4). When this ratio is high, sampling frequency may be decreased; otherwise, the existing frequency may be preserved as it is. The decision here depends on comparison of benefits and costs accruing from each frequency. Benefits are described in units of information (napier, bit, or as a percentage); costs are expressed as decreases in costs due to decreases in sampling frequency.

It may be followed from the above that the application of the entropy method depends on the already selected temporal frequencies of an existing network. Accordingly, only decreases in the available frequencies may be analyzed [20].

2.4. ASSESSMENT OF COMBINED SPACE/TIME FREQUENCIES

Some network design procedures combine both the spatial and the temporal design criteria to evaluate space-time tradeoffs. The approach in such combined design programs is to compensate for the lack of information with respect to one dimension by increasing the intensity of efforts in the other dimension. An increase in the sampling interval decreases the common information between the stations in a given combination; whereas an increase in the number of stations increases the transinformation for a given time frequency. One would look for the best combination with respect to time and space for reduction of the total uncertainty about a water quality variable [27].

To analyze spatial and temporal frequencies on a joint basis, the best combination of monitoring stations has to be selected first. Next, starting with the first priority station, the number of stations is successively increased by adding to the combination the next station on the priority list. For each number of stations, the temporal frequencies are decreased to identify how much information is provided by those stations at different sampling intervals. Finally, changes in information are plotted on the same graph with respect to both the increases in the number of stations and the decreases in temporal frequencies. The particular information measure used in this analysis is transinformation which represents redundant information in space and time dimensions. The objective is to select a space/time combination that produces the least amount of transinformation. One may specify here a particular level of redundant information which he wants to preserve in the network. Such redundancy is produced by different numbers of stations at different sampling intervals. The most appropriate combination may then be selected by evaluating reductions in costs due to decreases in either the number of stations or the temporal frequencies.

Combined spatial/temporal design features constitute a problem which is multivariate with respect to both time and space. In this case, the joint entropy of M x (K + 1) variables has to be computed, with corresponding transinformations for alternative combinations of numbers and locations of stations versus different Δt sampling intervals. Here, M represents the total number of stations and K the total number of time lags to be considered. The computation of joint and conditional entropies are again realized by Eq. (12) and (13); this time, the C matrix of the latter equation includes both the auto and the crosscovariances and represents stochastic dependence in both the temporal and the spatial dimensions.

To begin with two stations $X_{j,0}$ and $X_{j+1,0}$, transinformations can be obtained in the form of $T(X_{j,0}, X_{j+1,0})$ (for k = 0 or the existing frequency of available observations), $T(X_{j,1}, X_{j+1,1})$ (for k = 1 or the sampling interval increased 1 time unit), or $T(X_{j,k}, X_{j+1,k})$ for any time lag k. Similarly for three stations $X_{j,0}$, $X_{j+1,0}$ and $X_{j+2,0}$, transinformations in the form of $T(X_{j,k}, X_{j+1,k}, X_{j+2,k})$ are needed. For a total number of M stations $X_1, X_2, ...X_j,..., X_M$ with j=1, ... ,M and k = 0, ... , K time lags, the transinformations:

$$T(X_{1,k}, X_{2,k}, \dots, X_{M,k}) \qquad (k = 0, \dots, K) \tag{29}$$

must be computed. These transinformations are to be derived from corresponding joint and conditional entropies. To do this, one may refer back to Eq. (10), which gives the joint entropy of M dependent variables in the general case as:

$$H(X_1,...,X_M) = H(X_1) + H(X_2|X_1) + H(X_3|X_1,X_2) + ... + H(X_M|X_1,X_2,...,X_{M-1}) \tag{30}$$

In the above, each conditional entropy gives transinformations in the form of:

$$T(X_1, X_2, \dots, X_M) = H(X_M) - H(X_M \mid X_1, X_2, \dots, X_{M-1}) \tag{31}$$

where all conditional entropies can be computed by Eqs. (12) and (13). Combining Eqs. (30) and (31) gives:

$$T(X_1, X_2, \ldots, X_M) = H(X_1) + H(X_2) + \ldots + H(X_M) - T(X_1, X_2) - T(X_1, X_2, X_3) - \ldots - T(X_1, X_2, X_3, \ldots X_{M-1}) - H(X_1, X_2, \ldots, X_M) \quad (32)$$

On the other hand, the solution to Eq. (32) requires the computation of several terms on the right-hand side to arrive at a single transinformation value on the left-hand side of the equation. Therefore, a much simpler relationship is followed as:

$$T(X_1, X_2, \ldots, X_M) = H(X_1) - H(X_1 | X_2, X_3, \ldots, X_M) \quad (33)$$

Time lags k may be introduced into Eq. (33) so that one gets:

$$T(X_{1,k}, X_{2,k}, \ldots, X_{M,k}) = H(X_{1,k}) - H(X_{1,k} | X_{2,k}, \ldots, X_{M,k}) \quad (k = 0, \ldots, K) \quad (34)$$

where all terms on the right hand side can be computed via Eqs. (12) and (13) by setting up the appropriate C matrix which includes the selected stations and the selected time lags.

Once all transinformations are obtained by Eq. (34) for M number of stations and k time lags, the final graph is constituted to show the changes in transinformations with respect to both the number of stations and the varying sampling intervals.

The general formulation of Eq. (34) for transinformations of M stations with different sampling intervals considers the same time frequency at all stations. Another more reasonable approach may be to select a base station and apply a rather frequent sampling procedure at this station. Then, the sampling frequencies at other stations may be reduced to reflect how such a reduction affects basinwide information with respect to the base station. In this approach, the best combination of stations is still preserved and stations are added successively to the network in order of priority, as explained in section 2.2. The first priority station is considered as the base station since it represents the highest uncertainty about the variable analyzed. Taking this station as variable $X_{1,0,}$ with k=0 showing the existing sampling frequency, sampling intervals are varied at other stations $X_{2,k}, X_{3,k}, \ldots, X_{M,k}$ (k= 0, ..., K). For each different sampling frequency and each combination of stations, transinformations are computed as:

$$T(X_{1,0}, X_{2,k}, X_{3,k}, \ldots, X_{M,k}) = H(X_{1,0}) - H(X_{1,0} | X_{2,k}, X_{3,k}, \ldots, X_{M,k}) \quad (k=0,\ldots,K) \quad (35)$$

Essentially, this formulation shows how different numbers of stations contribute to reduction of the highest uncertainty at $X_{1,0}$ when different sampling frequencies are applied [27].

3. Application of the Method

3.1. AVAILABLE DATA

The above methodology is applied to the case of the Mississippi River basin in Louisiana, USA for basin segment 07, where the water quality monitoring network comprises 12 stations run by the Louisiana Department of Environmental Quality, Office of Water Resources. Figure 1 shows the locations and the identification numbers of these stations. The total available record at these sampling locations covers a period of

27 years between 1966-1992 with monthly observed values of 26 water quality variables. Almost all data series have regular observations with few missing values, which permit entropy computations in space/time dimensions.

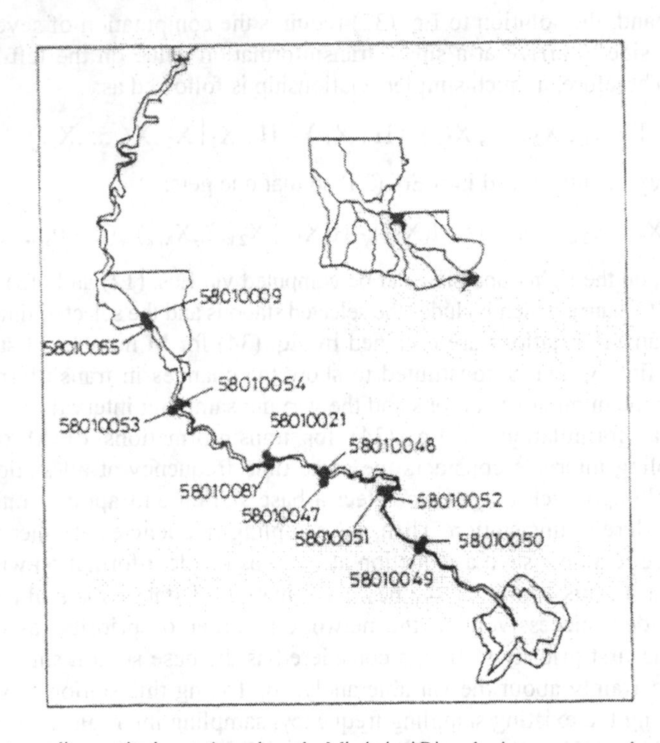

Figure 1. Water quality monitoring stations along the Mississippi River, basin segment number 07, in Louisiana.

3.2. ANALYSIS OF SAMPLING LOCATIONS

The analysis of sampling locations is applied to available DO, EC, Cl^-, TSS, P, COD and NO_3-N data from 12 sampling stations. All variables are assumed to be lognormally distributed except for DO and NO_3-N where the normal distribution gives a better fit. For each variable, the number of stations is increased by fulfilling the condition in Eq.(16), and transinformations are computed for M=2,...,12. The procedure described in Section 2.2 permitted the ordering of existing stations, where the first priority is given to the station with the highest marginal entropy. When new stations are added successively to the list by assuring the least transinformation as in Eq.(16), the best combination of stations is obtained in respect of minimum redundant information. Tables 1 and 2 summarize these computations for the case of Cl^- and DO, respectively, where the level of transinformation corresponding to each combination of stations can be observed.

Table 1 shows that the highest uncertainty about Cl^- occurs at the most downstream station 050; hence, this station appears to be the most informative location. The applied procedure selects the most upstream station of 009 as the second station in the priority list. Accordingly,

050 and 009 constitute the pair with the least amount of redundant information. The third location is station 081 which is right in the middle of the river segment considered. The next stations shift back and forth between 009 and 050; the last station in the list is 049 as expected since it produces highly redundant information with 050 due to its location. Table 1 shows that the addition of 049 to the combination significantly increases the transinformation so that the redundant information increases from 3.7% for 11 stations to 12% for 12 stations. According to Eq.(19), the multiple correlation coefficient, R, for stations 050 and 009 is 46%. It increases to 99.87% for a total of 12 stations, as shown in the last column of Table 1. The same level of correlation is attained with a combination of 10 stations for DO of Table 2.

TABLE 1. Selection of Sampling Stations in Order of Minimum Redundant Information for the Case of Cl^-.

Number of Stations (n)	Station Added	Joint Entropy (napier)	Share in Total Uncertainty, (k_j) (%)	Conditional Entropy (napier)	Trans-information (napier)	Redundant Info.,($1-t_j$) (%)	Multiple Corr. Coef. R (%)
1	050	3.1386	13				
2	009	5.7086	23	3.0197	0.1189	3.79	46.00
3	081	8.1251	33	5.4211	0.2874	5.03	66.12
4	047	10.5304	42	7.7982	0.3269	4.02	69.28
5	021	12.7809	51	10.0796	0.4508	4.28	77.07
6	054	14.8162	60	12.2960	0.4849	3.79	78.79
7	055	16.7687	67	14.2068	0.6095	4.11	83.93
8	053	18.6744	75	16.1168	0.6519	3.89	85.35
9	052	20.8521	84	17.9712	0.7032	3.77	86.89
10	048	22.7457	92	20.0951	0.7571	3.63	88.32
11	051	24.6950	99	21.8966	0.8491	3.73	90.39
12	049	24.8554	100	21.7235	2.9715	12.03	99.87

TABLE 2. Selection of Sampling Stations in Order of Minimum Redundant Information for the Case of DO.

Number of Stations (n)	Station Added	Joint Entropy (napier)	Share in Total Uncertainty, (k_j) (%)	Conditional Entropy (napier)	Trans-information (napier)	Redundant Info., ($1-t_j$) (%)	Multiple Corr. Coef. R (%)
1	052	4.4981	14				
2	009	8.4793	27	4.1252	0.3730	8.29	72.51
3	081	11.8604	37	7.4612	1.0181	12.01	93.24
4	049	15.2607	48	10.8086	1.0518	8.87	93.70
5	047	18.6162	58	14.1930	1.0677	7.00	93.90
6	053	21.6888	68	17.3039	1.3123	7.05	96.31
7	048	24.5327	77	20.1217	1.5671	7.23	97.80
8	021	27.0795	85	22.6950	1.8377	7.49	98.72
9	051	29.6139	93	25.1464	1.9331	7.14	98.95
10	050	31.9070	100	27.4164	2.1975	7.42	99.38

The results can be converted to the transinformation curves for each variable as in Fig. 2. This figure shows that a high transinformation is obtained for DO with a combination of 10 stations, where the percentage of redundant information (ratio of total transinformation to total joint entropy of 10 stations) is almost 7.5%. For other variables, the percentages of redundant information are in the order of 8.5% for EC with 9 stations, 10% for NO_3-N with 11 stations, and only 3.7% for TSS again with 11 stations, 4.6% for P with 12 stations and only 2.6% for COD again with 12 stations.

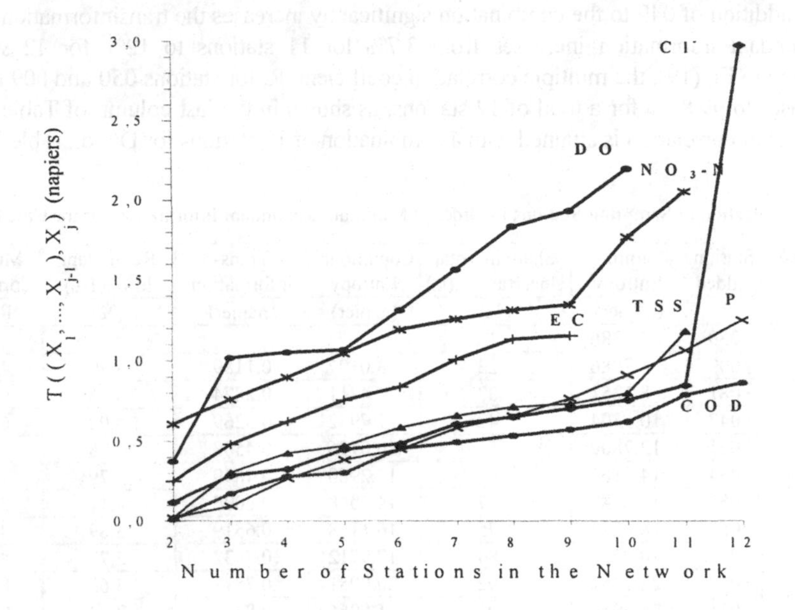

Figure 2. Increases in redundant information with respect to increases in the number of sampling sites for different variables observed.

The priority orders of stations also change from variable to variable as summarized in Table 3. For DO, TSS, and COD, the most uncertain and hence the most informative site is station 052; whereas 050 is the 1st priority station for Cl^- and EC. Station 009 has the 2nd priority for all variables except for NO_3-N. As an overall result, it may be assessed that 9 sites for EC and 10 sampling sites for DO are sufficient to reduce the uncertainty about these variables within the basin segment. To produce the same result, TSS and NO_3-N require 11 sampling stations while Cl^-, P and COD require 12 stations. Essentially, these results agree with the coefficients of variation, C_v, for each variable. Comparison between variables shows that DO and EC have smaller variability and hence require a smaller number of sampling sites. The last two lines in Table 3 show the multiple correlation coefficients, R, and the percentage of redundant information, $(1 - t_j)$, for the given total number of stations for each variable.

3.3. ANALYSIS OF TEMPORAL FREQUENCIES

Analysis of temporal frequencies is carried out for each variable at each sampling site. The method explained in Section 2.3 is used to compute the transinformations between successive observations. Figures 3 through 6 show the percentage of redundant information at different Δt sampling intervals for the case of DO, Cl, EC, and TSS.

TABLE 3. Ordering of stations for water quality variables observed along the Mississippi River.

Priority order of stations	VARIABLES													
	DO		CL		TSS		P		NO_3-N		COD		EC	
	St. no	C_v	St. no	C_v	St. no	C_v	St. no	C_v	St. no	C_v	St. no	C_v	St. no	C_v
1	052	0.27	050	1.61	052	0.97	047	0.55	054	0.40	052	0.86	050	0.57
2	009	0.23	009	0.48	009	0.63	009	0.58	048	0.39	009	0.60	009	0.25
3	081	0.24	081	0.55	054	0.71	051	0.78	009	0.42	081	0.71	052	0.31
4	049	0.26	047	0.32	055	0.64	053	0.59	081	0.39	054	0.62	047	0.25
5	047	0.25	021	0.68	021	0.76	048	0.51	021	0.39	051	0.94	053	0.22
6	053	0.23	054	0.34	047	0.83	021	0.56	050	0.40	021	0.72	021	0.23
7	048	0.25	055	0.36	048	0.78	049	0.48	047	0.37	048	0.57	081	0.23
8	021	0.24	053	0.32	051	0.88	081	0.44	053	0.38	050	0.82	048	0.22
9	051	0.27	052	0.85	081	0.79	055	0.47	051	0.38	055	0.60	054	0.22
10	050	0.27	048	0.41	053	0.70	052	0.57	055	0.39	047	0.60		
11			051	0.76	050	0.88	054	0.44	049	0.39	049	0.63		
12			049	1.59			050	0.48			053	0.63		
R(%)	99.38		99.87		95.20		95.84		99.18		90.68		94.91	
$(1-t_j)$ (%)	7.42		12.03		3.72		4.59		10.08		2.65		8.59	

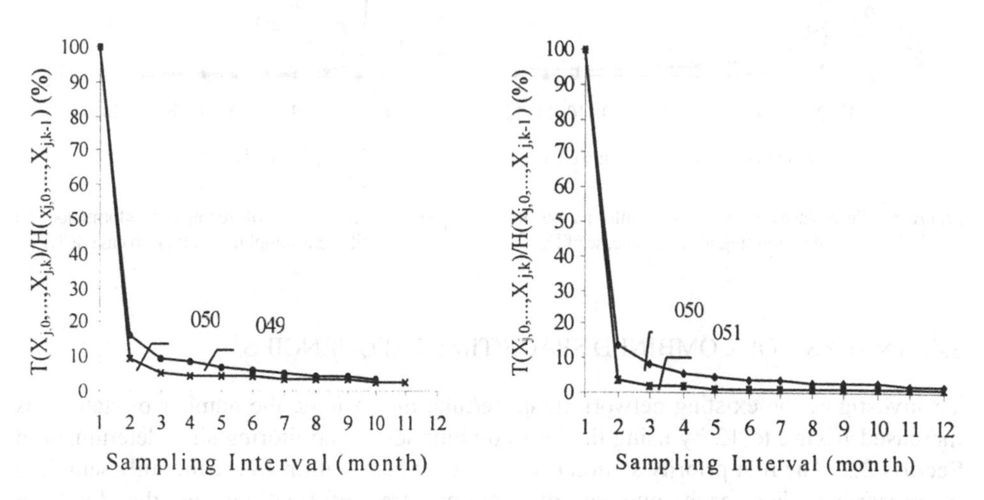

Figure 3. Percentages of redundant information for different sampling intervals in case of DO.

Figure 4. Percentages of redundant information for different sampling intervals in case of Cl^-.

These figures indicate that even the first order serial dependence within the analyzed processes are pretty low. Accordingly, reduction of the sampling frequency from monthly to bimonthly observations causes a loss of redundant information in the order of at least 80% for DO, Cl^-, and EC and more than this level for others. This loss increases further at larger sampling intervals. DO, Cl^-, and EC give similar results, where even the highest percentage of redundant information is less than 20% at a sampling interval of Δt=2 months. This percentage is reduced further as the sampling frequency is reduced.

TSS shows the least amount of dependence among all the variables analyzed. These results indicate that if the monthly sampling frequency is decreased to bimonthly for any of the variables investigated, one must expect at least 80% loss of redundant information. In other words, 80% of the uncertainty represented by the monthly observations still remains in the bimonthly series. Thus, bimonthly series can be considered as informative as the monthly series so that the current practice of monthly observations may be decreased to monitoring every two months. Although Figs. 3 through 6 reflect the situation at only two selected stations for each variable, the above results are found to be valid for all locations and all water quality variables investigated.

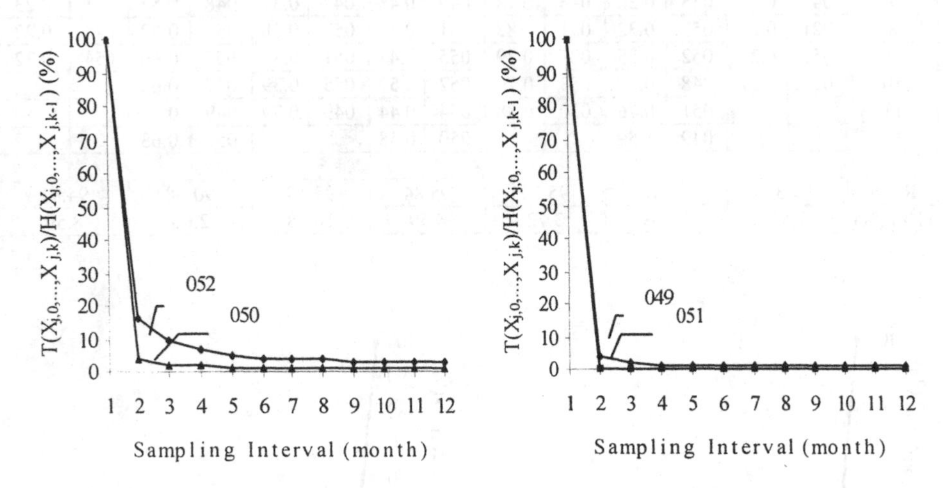

Figure 5. Percentages of redundant information for different sampling intervals in case of EC.

Figure 6. Percentages of redundant information for different sampling intervals in case of TSS.

3.4. ANALYSIS OF COMBINED SPACE/TIME FREQUENCIES

To investigate the existing network in space/time dimensions, the number of stations is increased from 2 to 12 by using the best combination of monitoring sites determined in Section 3.2. The first priority station is taken as the base station where monthly sampling is preserved. For each number of stations, transinformations in the form of $T(X_{1,0}, X_{2,k}, X_{3,k}, \ldots X_{M,k})$ are computed by using Eq. (35). Next, temporal frequencies are changed at all other stations to investigate how these changes affect the reduction of the highest uncertainty in the basin. The results of this application are shown in Figs. 7 through 10 for DO, Cl^-, EC, and TSS, respectively. These figures may be interpreted with respect to three criteria:

a) for a constant sampling interval Δt, redundant information increases as the number of stations is increased;

b) for a particular combination of stations, redundant information decreases as the temporal frequency is decreased;

c) for a constant level of transinformation, a number of space/time alternatives exist such that one may evaluate: 1) whether to increase the number of stations and decrease the temporal frequency, or 2) decrease the number of stations and increase the temporal frequency. The final decision to select among alternatives depends on evaluation of cost reduction with respect to decreases in space or time frequencies.

In Figures 7 through 10, it is observed that Cl^-, TSS and EC produce the smoothest curves. To interpret the results, let's look into Fig. 8 for Cl^-. If one assumes that the level of transinformation has to be kept at 0.10 napiers, monthly sampling intervals must be preserved for a combination of 2 stations. The interval may be increased to bimonthly for a combination of 3 to 7 stations. Beyond 7 stations, Cl^- may be observed once in three months for 8, 9, 10 and 11 stations while still preserving a transinformation value of 0.10 napiers.

Slight variations from smooth curves are observed for DO. Here, the combination with 8 stations produces an unexpected rise in transinformation at $\Delta t = 4$ months. This is essentially a computational result and reflects the effect of missing observations in the data series. On the other hand, the transinformation increases after $\Delta t = 4$ months for all combinations of stations. This is assumed to be due to the seasonal variation in DO which is observed at all stations. However, for purposes of space/time design, it is considered here inappropriate to extend the sampling intervals beyond $\Delta t = 4$ months.

Figures 7 through 10 also show different limiting Δt intervals for each variable. For DO and TSS, the sampling interval cannot be extended beyond 4 months, and for Cl^- and EC beyond 5 months.

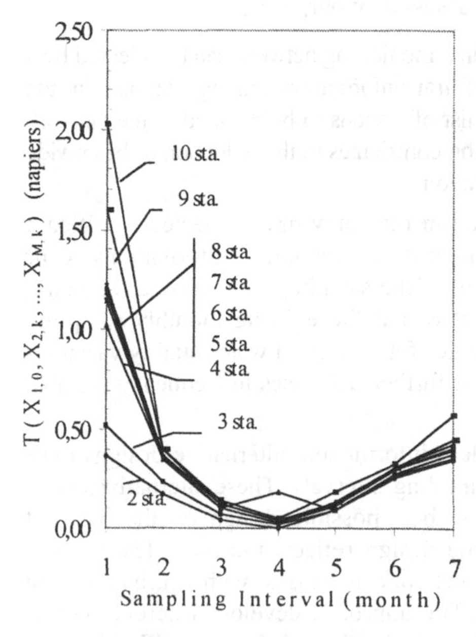

Figure 7. Changes in redundant information for different space/time frequencies in case of DO.

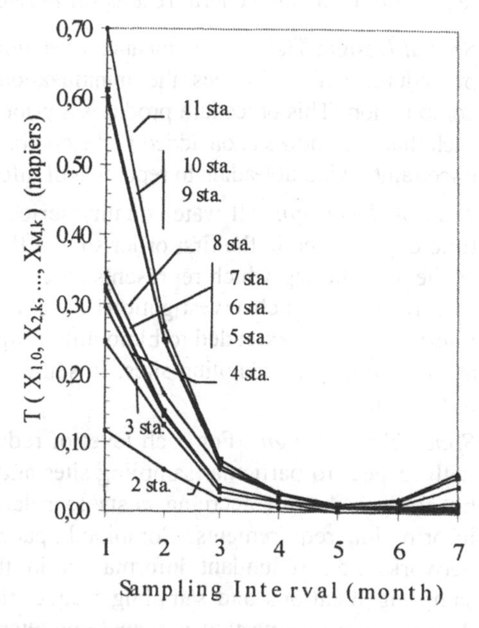

Figure 8. Changes in redundant information for different space/time frequencies in case of Cl^-.

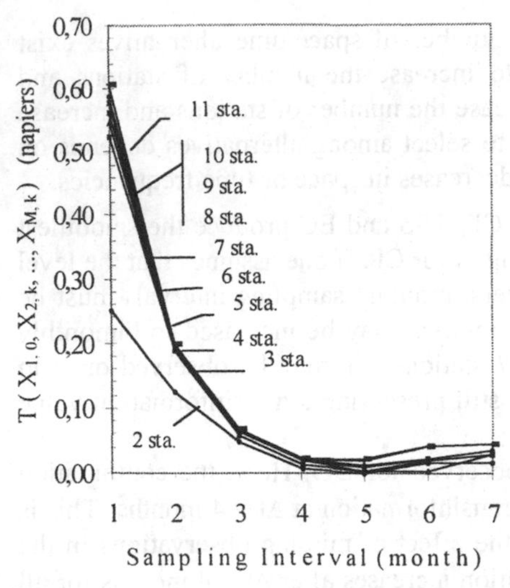

Figure 9. Changes in redundant information for different space/time frequencies in case of EC.

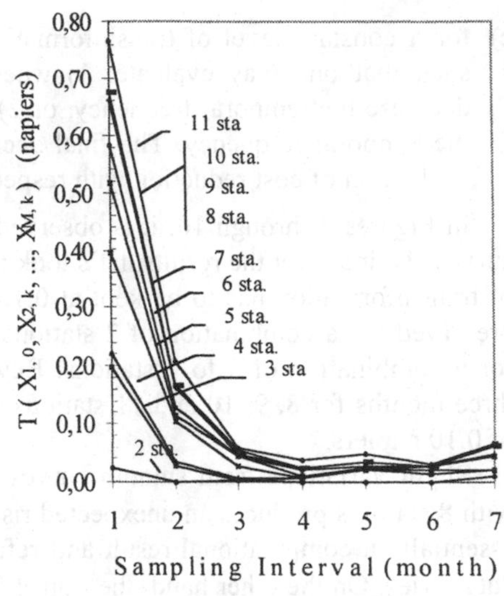

Figure 10. Changes in redundant information for different space/time frequencies in case of TSS.

4. Conclusion

The application of the entropy method to water quality data of the Mississippi River has led to the following general results for network assessment purposes:

Spatial Design: The best combination of stations in a monitoring network can be selected by a procedure, which foresees the minimization of transinformation among stations in the combination. This procedure produces a priority list of stations to be retained in the network such that each new station added to the combination contributes to the reduction of basinwide uncertainty without leading to repetition of information.

Temporal Design: All water quality series used in the study have reflected significant time dependence in the first order so that the major decreases in transinformations occur at the 1st time lag which represents an extension of the sampling interval one time unit. The results of such investigations have indicated that the existing monthly sampling intervals may be extended to bimonthly frequencies for almost all water quality variables at the majority of sampling sites. In some cases, further decreases in frequency are also indicated.

Space/Time Design. For each level of redundant information, alternative designs exist with respect to particular sampling sites and sampling intervals. These alternatives must be compared with accruing costs to select the best possible design on the basis of information requirements. Combined space/time design reflects the basic feature of a network, i.e., redundant information in the network increases with an increase in sampling locations and sampling frequencies. The approach developed herein permits the selection of a particular space/time alternative design feature for a specified level of redundant information to be retained in the network.

Within the broader context of network assessment purposes and information expectations of water quality managers, it must be emphasized that the entropy method is but one of the several available techniques used for network design. Each technique focuses on a specific aspect of the network and uses a different criterion for design. For example, some methods are specifically developed for purposes of trend detection as the major objective of the network (e.g., Schilperoort & Groot [13]); while others serve to design a network that collects data to effectively estimate mean values of water quality variables [8]. In comparison, the entropy method is basically a network assessment procedure that focuses on variability of water quality in time and space. Its basic advantages are that it: (a) provides a quantitative measure of the information content of a sampling site and of an observed time series; (b) assesses, again in quantitative terms, transfer of information in space and time; (c) gives an indication of data utility; and (d) can be used to assess jointly several features of a network, e.g., sampling sites, sampling frequencies, variables to be sampled and sampling duration. Yang & Burn [23] have shown that, in comparison with other measures of association, entropy measures are more advantageous as they reflect a directional association among sampling sites on the basis of information transmission characteristics of each site. On the other hand, the method has the major disadvantage of being sensitive to the selection of the appropriate multivariate probability density function to represent the multivariate nature of a network. Thus, at present, it seems best to use different techniques in combination and to investigate network features from different perspectives before a final decision is made for network assessment and redesign.

5. Acknowledgment

The research leading to this paper has been supported by the NATO Linkage Grant ENVIR.LG.950779. This support is gratefully acknowledged.

6. References

1. *National Water Quality Assessment Program: The Challenge of National Synthesis.* (1994). Committee on U.S. Geological Survey Water Resources Research, Water Science and Technology Board, National Research Council, National Academy Press, Washington, D.C.

2. Ward, R.C. (1996). Water quality monitoring: Where's the beef? AWRA, *Water Resources Bulletin*, **32(4),** 673-680.

3. Ward, R.C., J.C. Loftis and G.B. McBride (1994). *Design of Water Quality Monitoring Systems*, Van Nostrand Reinhold, New York.

4. WMO (1994). *Advances in Water Quality Monitoring - Report of a WMO Regional Workshop (Vienna, 7-11 March 1994).* World Meteorological Organization, Technical Reports in Hydrology and Water Resources, No. 42, WMO/TD-NO 612, Geneva, Switzerland, 332 p.

5. Adriaanse, M.J., van de Kraats, J., Stoks, P.G., and Ward, R.C. (1995). Conclusions monitoring tailor made. In: *Proceedings, Monitoring Tailor-Made*, An International Workshop on Monitoring and Assessment in Water Management, Beekbergen, The Netherlands, Sept. 20-23, pp. 345-347.

6. Niederlander, H.A.G., Dogterom, J., Buijs, P.H.L., Hupkes, R., and Adriaanse, M. (1996). *UN/ECE Task Force on Monitoring & Assessment, Working Programme 1994/1995, Volume:5: State of the Art on Monitoring and Assessment of Rivers,* RIZA report: 95.068.

7. Harmancioglu, N.B. and Alpaslan, M.N. (1997). Redesign of water quality monitoring networks. In: J.C. Refsgaard and E.A. Karalis (eds.), *Operational Water Management*, A.A. Balkema, Rotterdam, pp.57-64.
8. Sanders, T.G., Ward, R.C., Loftis, J.C., Steele, T.D., Adrian, D.D., and Yevjevich, V. (1983).*Design of Networks for Monitoring Water Quality*, Water Resources Publications, Littleton, Colorado, 328p.
9. Tirsch, F.S., & Male, J.W. (1984). River basin water quality monitoring network design. In T.M. Schad (ed.) *Options for Reaching Water Quality Goals, Proceedings of the Twentieth Annual Conference of American Water Resources Association*, AWRA Publications, pp:149-156)..
10. Dixon, W., and Chiswell, B. (1996). Review of aquatic monitoring program design. *Water Research*, **30(9)**, 1935-1948.
11. Harmancioglu, N.B., Ozkul, S.D. and Alpaslan, M.N. (1998). Water quality monitoring and network design. In: N.B.Harmancioglu, V.P.Singh and M.N. Alpaslan (eds.), *Environmental Data Management*, Kluwer Academic Publishers, Water Science & Technology Library, vol.27, ch.4, pp.61-106.
12. Ward, R.C. and Loftis, J.C. (1986). Establishing statistical design criteria for water quality monitoring systems: Review and synthesis. *Water Resources Bulletin, AWRA* **22(5)**, 759-767.
13. Schilperoort, T., and Groot, S. (1983) Design and optimization of water quality monitoring networks. Paper presented at the *International Symposium on Methods and instrumentation for the Investigation of Groundwater Systems (MIIGS*), Noordwijkerhout, the Netherlands, May 1982, publication no.286.
14. Jager, H.I., Sale, M.J. and Schmayer, R.L. (1990). Cokriging to assess regional stream quality in the Southern Blue Ridge Province. *Water Resources Research*, **26(7)**, 1401-1412.
15. Smith, R.A., Schwarz, G.E. and Alexander, R.B. (1997). Regional interpretation of water quality monitoring data. *Water Resources Research*, **33(12)**, 2781-2798.
16. Esterby, S.R. (1986) Spatial heterogeneity of water quality parameters. In: A.H. Shaarawi and R.E. Kwiatkowski (eds.), *Statistical Aspects of Water Quality Monitoring*, Elsevier, pp. 1-16.
17. Moss, M.E. (1989). Water quality data in the information age. In: R.C. Ward, J.C. Loftis, and G.B. McBride (eds.), *Proceedings, International Symposium on the Design of Water Quality Information Systems*, Fort Collins, CSU Information Series No. 61, pp. 8-15.
18. Husain, T. (1989). Hydrologic uncertainty measure and network design. AWRA, *Water Resources Bulletin*, **25**, 527-534.
19. Harmancioglu, N. (1981). Measuring the information content of hydrological processes by the entropy concept. *Ege University, Journal of the Civil Engineering Faculty, Special Issue for the Centennial of Ataturk's birth*, Izmir, pp. 13-40.
20. Harmancioglu, N.B., and Alpaslan, N. (1992). Water quality monitoring network design: A problem of multi-objective decision making, *AWRA, Water Resources Bulletin*, Special Issue on "Multiple Objective Decision Making in Water Resources", **28(1)**, 179-192.
21. Krstanovic, P.F., & Singh, V.P. (1993). Evaluation of rainfall networks using entropy: I.Theoretical development; II. Application. *Water Resources Management*, **6**, 279-314.
22. Moss, M.E. (1997). On the proper selection of surrogate measures in the design of data collection networks. In: N.B. Harmancioglu, M.N. Alpaslan, S.D. Ozkul and V.P. Singh (eds.), *Integrated Approach to Environmental Data Management Systems*, Kluwer Academic Publishers, NATO ASI Series, 2. Environment, vol. 31, pp.79-88.
23. Yang, Y. and Burn, D.H. (1994). An entropy approach to data collection network design, *Journal of Hydrology*, **157**, 307-324.
24. Shannon, C.E., & Weaver, W. (1949). *The Mathematical Theory of Communication*. Urbana, Illinois, The University of Illinois Press.
25. Amorocho, J., and Espildora, B. (1973). Entropy in the assessment of uncertainty of hydrologic systems and models, *Water Resources Research*, **9**, 1511-1522.
26. Chapman, T.G. (1986). Entropy as a measure of hydrologic data uncertainty and model performance. *Journal of Hydrology*, **85**, 111-126.
27. Ozkul, S. (1996). *Space/Time Design of Water Quality Monitoring Networks by the Entropy Method*. Dokuz Eylul University, Graduate School of Natural and Applied Sciences, Izmir, Ph. D. Thesis in Civil Engineering (Advisor: Nilgun B. Harmancioglu).

Part IV

STATISTICAL SAMPLING

UNCERTAINTY IN ENVIRONMENTAL ANALYSIS

V. P. SINGH[1], W. G. STRUPCZEWSKI[2] and S. WEGLARCZYK[3]
[1]*Department of Civil and Environmental Engineering*
Louisiana State University
Baton Rouge, Louisiana 70803-6405, USA

[2]*Water Resources Department, Institute of Geophysics*
Polish Academy of Sciences
Ksiecia Janusza 64, 01-452 Warsaw, Poland

[3]*Institute of Water Engineering and Water Management*
Cracow University of Technology
Warszawska 24, 31-155 Cracow, Poland

Abstract: Questions of safety or reliability in environmental engineering arise principally due to the presence of uncertainty. There are various sources and types of uncertainty due to randomness of physical phenomena and errors in modeling and prediction. The randomness inherent in physical phenomena is normally described using probability distributions and certain moment properties. These properties are central tendency and dispersion or standard deviation. The modeling errors are either systematic or random. The uncertainty in modeling or prediction may be due to sampling error and imperfection of the prediction model, giving rise to both random error and systematic error. This paper describes various types, sources, and measures of uncertainty that apply to environmental data analysis.

1. Introduction

"The only thing certain in life is uncertainty." The history of science is the history of gaining new information for broader and better understanding of the world surrounding us. In a way, it can be considered as the process of limiting uncertainty. In the last century, the views on the possibility of eliminating uncertainty evolved with the progress of science. The deterministic belief originating in the eighteenth century's mechanistic conception, according to which the uncertainty is caused only by our ignorance, has significantly weakened, partly because the quantum mechanics has shed light on chaotic and unpredictable behavior of elementary particles. It is now accepted that chaos is one of the properties of elementary particles. Empirical confirmation of the thesis of quantum mechanics has shown that Einstein was not right in his commonly quoted remark: "God does not cast die."

Over a long period of time, the prevailing opinion was the predictability of the world at macro scale. It was accepted that, by having an adequate mathematical model with appropriate initial and boundary conditions, the prediction problem reduced to a computational problem, which can then be solved using an appropriate numerical method. Although the advent of the theory of deterministic chaos shook this notion, it seems still to be dominant among meteorologists. The heart of the matter can be described as follows: Complex non-linear dynamic systems are, in fact, deterministic, i.e., they can be described by a deterministic system of differential equations; but they are unpredictable, which means the impossibility of the exact prediction of future behavior. It is interesting to learn that researchers from various fields, starting with different premises and using different methodologies, have essentially reached the same conclusion as the unpredictability.

Since the effects of current environmental decisions depend on the future course of hydrometeorological processes and on future socio-economic development, an improvement of the quality of information about the future can improve the quality of decisions. The need for better forecast is obvious, but the expectation to achieve "good forecast" is less than well founded. The thesis on unpredictability should not be considered in the fundamentalist way as the conviction on the total chaos in the universe. It is pragmatic to state that, despite all the progress in science, uncertainty would never be fully eliminated, and any decision based on the best available predictions would contain a certain risk of error. One of the paradigms in science is the necessity of getting reproducible results of experiments, confirming a scientific hypothesis which assumes the possibility of prediction of results. Therefore, it is more appropriate to regard the prediction as limited but not to consider that it is totally impossible. The necessity for improvement of existing models and looking for new prediction models is beyond the scope of discussion here, but one should acknowledge the existence of limits of predictability.

One of the main reasons for disappointment of decision makers and technicians with application of research techniques seems to be due to improper understanding of uncertainty. They can hardly accept that uncertainty is one of the fundamental properties of the world, like the law of energy conservation; and continuing this analogy, the hope for total elimination of uncertainty is illusive. Since even Einstein did not consider uncertainty as one of the principles of physics, the reason for not accepting it is not the lack of education but the deep-rooted conviction with mechanistic and deterministic model of the world that has sprung up particularly in the European civilization. The thesis on limited predictability may be attractive for those dealing with predictions when the forecast happens to be wrong, but it is highly frustrating and disheartening for those trying to make decisions using forecasts. They should learn that uncertainty always exists and that it limits a prediction and cannot be eliminated completely. Limiting uncertainty improves the decision making, but it is difficult to achieve it when prediction is considered entirely as deterministic.

The problems considered in operations research and control theory are usually deterministic, i.e., the availability of full knowledge is assumed to achieve the goal of carrying out an activity (e.g., the maximum gain). In fact, in these problems, decisions are made while the problems are being formulated, i.e., when an objective function and constraints are defined. Optimization, i.e., searching for the extremum of the function, does not involve the necessity of decision making, although the variables with respect to the extremum sought are called decisive variables. It can be shown that the necessity of decision making results totally from uncertainty [1, 2]. That is, if the uncertainty did not exist, there will be no need for decision making. While commenting on uncertain information, decision makers should realize that there is no room for them in the fully predictable world.

Environmental systems are subject to uncertainty, but their planning, design, operation and management are often done without accounting for it. The uncertainty can be inherent or intrinsic, caused by randomness in nature; or it can be epistemic, caused by lack of knowledge of the system or paucity of data. Environmental phenomena exhibit random variability, and this variability is reflected when observations are made and samples analyzed. For example, there is inherent randomness in the climatic system, and it is impossible to precisely predict what the maximum rainfall would be in a given city in a given year, even if there is a long history of data. Similarly, there is no way to precisely predict the amount of sediment load that a given river will carry during a given week at a given location. Likewise, the maximum discharge of a river for a given year cannot be predicted in advance. Because randomness is an inherent part of nature, it is not possible to reduce the inherent uncertainty. There are some local meteorological events of high environmental impact, like spring slight frost, hail and gale (usually discussed by people with respect to possible climate change), which are particularly difficult for prediction.

Epistemic uncertainty is extrinsic and is knowledge-related. The knowledge relates to the environmental system as well as to the data. Thus, this type of uncertainty is caused by a lack of understanding of the causes and effects occurring in the system. If the system is fully known, it may be caused by a lack of sufficient data. For example, using laboratory experimentation or computer simulations, it may be possible to construct a precise mathematical model for an environmental system, but it will be impossible to determine the parameters for the vast range of conditions encountered in nature. Consider the case of flow in a channel. The flow dynamics is reasonably known, but it is not possible to estimate the shear stress for the range of flow and morphologic conditions found in practice. Epistemic uncertainty changes with knowledge and can be reduced with increasing knowledge and longer history of quality data. The knowledge can, in general, be increased by gathering data, research, experience, and expert advice.

Uncertainty is central to decision making and risk assessment. The ultimate goal is to reduce uncertainty and, thereby, risk. The objective of this paper is to provide an overview of uncertainty in environmental analysis, including a discussion of its role in rational decision making and the treatment of different types, sources, and measures of uncertainty.

2. Rational Decision Making

Decisions about environmental systems, such as wastewater treatment plant, soil remediation system, water purification, and so on, are often made under conditions of uncertainty. A question arises: Are these decisions rational? Another question is: What does uncertainty do to decision making? Before proceeding further, one must consider the decision making process itself and especially the meaning of "rational." In other words, what makes a decision rational? There can be more than one criterion of rationality.

One criterion for rationality is that the decision must be aimed at a goal one wants to achieve. For example, we may want to replace an old wastewater treatment plant by a new one. The objective is not just any plant but a plant that will meet the increased wastewater load, that is designed for increasing load due to growing population, and that is economical, durable and more efficient. Failure to meet any of these objectives in a new treatment plant design may not be acceptable and would make the decision irrational. However, there are many ways to the design of a treatment plant that, to a greater or lesser degrees, meet the objective. In other words, there are many acceptable design alternatives from which the decision maker must choose the best design. This leads to a second criterion of rationality: the decision maker must identify and study enough alternatives to ensure that the best alternative is among them. This does not mean that the decision maker must make an exhaustive search of all possible alternatives, which is seldom possible and mostly a waste of time. It does, however, mean that he must have an open mind for alternative solutions. For example, the designer may be a specialist in wastewater treatment, but he should also consider the advantages which a combination scheme, where treatment, use, management and release are combined, offers.

A third criterion for rationality is making the choice between alternatives in accordance with a rational evaluation process. Let us consider an example. Each alternative has consequences some of which are desirable and some not so desirable, and has benefits and costs. Some treatment plant designs have a lower initial cost, some have a longer useful life, some require much maintenance and some require little maintenance, some produce more odor than others, and some can be built faster and with less harm to human health. All these consequences must be taken into account when ranking the alternatives. For comparing alternatives, one may want to express the value of each of the consequences using a common measure of value, say the dollar. Then, one can add up the values of the consequences and compute the relative value of each alternative. These relative values, expressed numerically, can then be compared.

Consider an example to see how it works. The initial cost is expressed in dollars. Maintenance requirements and useful life can also be evaluated in dollars. One can calculate what it will cost to upgrade the plant, renovate the buildings, replace the equipment, and replace the plant. There is, however, a problem with these expenditures, for they are not immediate like the initial cost but occur at specified intervals. The value of dollar at a future data is not the same as it is today. This problem is circumvented by applying interest rates which are established by the money markets, to express the difference in value. The difficulty arises when those aspects, for which there is no market value, need evaluation, as, for example, environmental disruption in plant operation, greater production of odor or injury to human health. The question then arises: what are the people willing to pay for less interruption in plant operation and for less odor production? The answers will, of course, not be precise and are subjective. Nevertheless, it is not too difficult to at least obtain some upper and lower limits of values of each consequence. This will suffice for ranking alternatives in order to find the best alternative.

Thus, a rational decision making involves: (1) defining an objective, (2) identifying alternative means of achieving the objective, and (3) applying a ranking procedure to determine the best alternative. In the real world, often little planning is done, and little rationality is invoked even for important decisions. Errors are commonly made. One of the common errors is the failure to recognize alternatives. All too often, the course of action is determined by precedent, tradition, accident, prejudice or shortsightedness. Another common error is that important decisions are too often postponed till there is hardly anytime for anything but to continue the present practice. A third common error is the presumption in the value judgment as to what is important and what is not. The bottom line should be that the judgment of the client or the community to be served must be paramount.

3. Types of Uncertainty

Following Pat*-Cornell [3], the intrinsic uncertainty can be divided into inherent uncertainty in time and inherent uncertainty in space. The epistemic uncertainty can be divided into statistical uncertainty and model uncertainty. A short discussion of these is now in order.

3.1. INHERENT UNCERTAINTY IN TIME

A stochastic process expressed as a time series of a random variable exhibits uncertainty in time. For example, the time series of annual rainfall at a given station, the time series of the annual instantaneous maximum discharge of a river at a given station, the time series of annual 24-hour maximum rainfall at a gauging station, annual sediment yield of a watershed, the time series of the annual 7-day minimum ozone level in a given city, and so on are examples of the inherent uncertainty in time.

3.2. INHERENT UNCERTAINTY IN SPACE

Environmental systems exhibit uncertainty in space. For example, hydraulic conductivity in an aquifer varies from point to point in location as well as well in direction. Therefore, a space series of the hydraulic conductivity can be expressed along a given transect. The same applies to the hydraulic conductivity of soils. Likewise, the roughness of a river bed varies along the bed and can be expressed in space along the longitudinal direction. Another example is atmospheric pollution in a large urban area, which varies at different points and in different directions and which can be expressed as a function of space. Examples of this kind exhibiting spatial variability abound in environmental analysis.

3.3. STATISTICAL UNCERTAINTY

The statistical uncertainty combines parameter uncertainty and distributional uncertainty [4]. These two types of uncertainties are not always independent and distinguishable. For example, the identification of a correct distribution model depends very much on the accuracy with which its parameters can be estimated. Vrijling and van Gelder [4] proposed to divide the statistical uncertainty into statistical uncertainty in time and statistical uncertainty in space.

3.3.1. *Parameter Uncertainty*

The parameter uncertainty is caused by either lack of data, poor quality of data, or inadequate method of estimation. This type of uncertainty is widely prevalent in environmental analysis.

3.3.2. *Distribution Uncertainty*

It is not always clear which type of a distribution a particular environmental random variable follows. For example, the annual maximum instantaneous discharge of a river can be described by the log-Pearson type 3 distribution, the 3-parameter lognormal distribution, the Pearson type 3 distribution, or the generalized extreme value distribution. In many cases, it is difficult to discern the exact type of the distribution the annual instantaneous maximum discharge follows. Similar is the case with a number of other environmental variables.

3.3.3. *Statistical Uncertainty in Time*

Consider, for example, the time series of the annual maximum instantaneous discharge of a river. In most cases, the time series is short for determining the discharge of a long recurrence interval, say, 500 years. In this case, then, there is the scarcity of information. The same applies to droughts, minimum flows and a host of environmental variables. Although an estimate of a 500-year flood can be made using any of the standard techniques available, this estimate is subject to uncertainty; and this is an example of the statistical uncertainty in time or statistical uncertainty of variations in time. This uncertainty can be reduced by lengthening the database.

3.3.4. *Statistical Uncertainty in Space*

Consider, for example, the spatial mapping of erosion in a large basin. There is very little information available to map spatial variability of erosion. Similarly, the spatial mapping of hydraulic conductivity of an aquifer is very difficult, for there is not enough data available. Such examples abound in environmental analysis. Thus, the spatial mapping of an environmental variable is subject to uncertainty, and this is an example of statistical uncertainty in space or statistical uncertainty of variations in space.

3.4. MODEL UNCERTAINTY

Environmental models are imperfect because of uncertainties. Consider, for example, an air pollution model that describes the pollutant concentration in space and time in the atmosphere. The model is imperfect because there are many gaps in our knowledge about pollutant dispersion in the atmosphere, or the model is simplified for practical reasons. In any case, the model is subject to uncertainty.

4. Sources of Uncertainty

Uncertainties in environmental analysis arise due to: (1) randomness of physical phenomena, or (2) errors in data and modeling. The modeling and data errors are of two types: (1) systematic and (2) random. There is another source of error, called illegitimate error, which results from outright blunders and mistakes. Computational errors are of this type, which can be avoided.

A stochastic or random phenomenon is characterized by the property that its repeated occurrences do not produce the same outcome. For example, when the same rainfall occurs at different times over a watershed, it produces different hydrographs. This means that watershed response is a stochastic variable. In a laboratory experiment, when a measurement is repeatedly made, the measured values do not identically match; the deviations between the values are called random fluctuations. Essentially, these are random errors and are also called statistical errors in common parlance. Roughly, random errors tend to be higher and lower than the true value about the equal number of times. Furthermore, estimators subject to only random errors tend to be consistent. Thus, random errors tend to exhibit a statistical regularity, not a deterministic one. Examples of such errors are numerous.

Systematic errors are characterized by their deterministic nature and are frequently constant. For example, a raingauge located near a building tends to produce biased rainfall measurements. A gauge operator tends to operate the gauge in a biased manner. The efficiency of observation is often a source of systematic error. Changes in experimental conditions tend to produce systematic errors. For example, if the location of a raingauge is changed, it will produce biased measurements.

In order to better define constraints for environmental modeling, let us assume that certain mathematical variables are functionally related [5]. In general, we have a set of variables $X_1, ..., X_k$ related in p functional forms:

$$f_j(X_1,...,X_k;\alpha_1,...,\alpha_l) = 0, \quad j = 1, 2, ..., p \tag{1}$$

depending on the l parameters α_r, $r = 1, 2, ..., l$. If the variables are random variables, the functional relation reduces to a structural relation. In environmental modeling, one often deals with time-space distributed variables.

The problem commonly posed in natural sciences is to estimate α_r from a set of observations. If we were able to observe values of $\boldsymbol{X}$ without error, there would be no statistical problem here at all, and the problem would merely be a mathematical one of solving the set of equations. However, our mathematical variables are unobservable, and what we observe (or measure) are the indices of mathematical variables, i.e., $(\xi_1,...,\xi_k)$, being often lumped in space or in time.

The problem of the α_r estimation from a sample of observable variables $(\xi_1,...,\xi_k)$ needs several assumptions with respect to the properties of errors, and it is merely an academic one. However, one can ask about the practical need for Eq. (1) and for the values of its α_r-parameters if observable variables $(\xi_1,...,\xi_k)$ differ much from their respective unobservable counterparts (X_1, ..., X_k.). The interest in environmental modeling is in the relation between observable variables, which is statistical in character. It clearly shows the gap between experimental physics and environmental approaches. If observed variables differ much from physical unobservable variables, then the knowledge of the actual functional forms f_j may be of little use, even if it is gained. An application of statistical models is constrained to the same kind of data as used for their calibration. Therefore, in case of a substantial change in an observation network or in measurement techniques of input (independent) variables, the model should be recalibrated. Moreover, the model should be tuned with data range interesting for its operational use, e.g., flood protection.

5. Analysis of Errors

Errors are determinate if a logical procedure, whether experimental or theoretical, can be employed to evaluate them. Random errors can be determined using statistical tools.

Systematic errors are often determinate because they can be evaluated using subsidiary experiments or other means. Sometimes, some determinate errors can be removed using correction factors. For example, the U.S. National Weather Service pan is used to determine evaporation in an area by applying a correction factor to the pan measurement. There are other examples of this kind in environmental analysis.

If experimental data are subject to small errors, then the terms "precision" and "accuracy" are often employed. Consider an experiment. It is said to have high precision if it has small random error. It is said to have high accuracy if it has small systematic error. This means that precision relates to the repeatability of measurements and accuracy to the deviation between the true value and the estimator. Thus, there are four possibilities for characterizing experiments: (1) precise and accurate, (2) precise and inaccurate, (3) imprecise and accurate, and (4) imprecise and inaccurate. The objective is to reduce both systematic and random errors as much as possible. However, for economy of effort, one must try to strike a balance between the sources of error, giving greater weight to the larger of the two.

There is another important but difficult aspect of data errors, relating to mistakes in the data and rejection of data. When data of natural phenomena are collected, anomalies are seemingly found. These anomalies and their causes should be carefully analyzed. The causes of these anomalies may be random or systematic. Under no circumstances should they be discarded, unless there is very strong compelling reason to do so. For example, in frequency analysis of the annual instantaneous maximum discharge of a river, the so-called outliers or inliers are usually encountered. They must be dealt with, not discarded away. Most often, these anomalies are expected. If the normal probability law (or chi-square criterion) indicates that these anomalous values are expected, they must be retained. Consider another case. If the deviation of an anomalous value is too large and has a small chance of occurring, the Chauvenet criterion may be used as a guide for accepting or rejecting the value, or more appropriately, flagging the suspicious situation. According to this criterion, if the probability of the value deviating from the mean by the observed amount is 1/2N or less, where N = number of observations, then there is reason for suspicion or rejection.

6. Measures of Errors

The experimental data errors can be investigated using the criterion of repeatability of measurements. In observing natural phenomena, however, there is no way to repeat the measurement. In laboratory experimentation, if a measurement is repeated, say N times, and the measured values do not exactly match the "true" value, the differences are analyzed. It should, however, be noted that the true value is never known, and only an estimator can be obtained. If N becomes large, the arithmetic average of the measured values approaches a constant value; and if this is the case, the estimator approaches a constant value and is qualified as a "consistent" estimator. Thus, consistency is one measure of experimental data error and is tied to the sample size. Ideally, the estimator should be consistent and without bias. However, the estimator, although consistent, may be biased if N is too small, i.e., the estimator may be either too large or too small. Thus, bias is another measure of experimental error, which is not tied to the sample size.

Not all statistics are unbiased estimators. Some are consistent estimators because they converge to the parent population as sample size increases; but, for finite sample size, they need correction to become unbiased best estimate. For example, for N identical independent measurements, only the sample mean and sample variance are consistent statistics, but only the sample mean is an unbiased estimator. To obtain an unbiased estimate of the variance, the sample variance is multiplied by the factor N/(N-1).

7. Extraction of Information

Data constitute a source of information. They are the only media of communication with nature. The purpose of data analysis is therefore to extract the maximum information. Statistical concepts used to analyze data are threefold: (1) aggregate characteristics, (2) variation of individual values from aggregate properties, and (3) frequency distribution of individual values. In the first case are the mean (arithmetic, median, mode, harmonic, and geometric), deviation (mean and standard), variance, coefficient of variation, and higher order (such as skewness, kurtosis, etc.) or other types of moments (such as probability weighted, linear, and geometric). The moments and frequency distribution are interconnected. While computing these statistics, issues relating to rounding off and truncation have to be dealt with.

8. Analysis of Uncertainty

An evaluation of safety and reliability requires information on uncertainty, which is determined by the standard deviation or coefficient of variation. Questions of safety or reliability arise principally due to the presence of uncertainty. Thus, an evaluation of the uncertainty is an essential part of the evaluation of engineering reliability. The uncertainty due to random variability in physical phenomena is described by a probability distribution function. For practical purposes, description may be limited to: (a) central tendency, and (b) dispersion (e.g., standard deviation) or coefficient of variation.

To deal with uncertainty due to prediction error (estimation error or statistical sampling error and imperfection of the prediction model), consider the mean annual rainfall for Baton Rouge to be 60.00 inches. Conceivably, this estimate of the true mean value would contain error. If the rainfall measurement experiment is repeated and other sets of data obtained, the sample mean estimated from other sets of data would most likely be different. The collection of all the sample means will also have a mean value, which may well be different from the individual sample mean values, and a corresponding standard deviation. Conceptually, the mean value of the collection of sample means may be assumed to be close to the true mean value (assuming that the estimator is unbiased). Then, the difference (or ratio) of the estimated sample mean (i.e., mean value of 60 in.) to the true mean is the systematic error, whereas the coefficient of variation or standard deviation of the collection of sample means represents a measure of the random error.

In effect, random error is involved whenever there is a range of possible error. One source of random error is the error due to sampling, which is a function of the sample size. Taking a sample standard deviation of, say, 15 in., the corresponding coefficient of variation (CV) is, therefore,

$$CV = \frac{15\,in.}{60\,in.} = 0.25$$

Assuming the number of observations in the sample to be, say 25, the random sampling error (expressed in terms of *CV*) would be:

$$\Delta_1 = \frac{0.25}{\sqrt{25}} = 0.05$$

The systematic error or bias may arise due to factors not accounted for in the prediction model that tends to consistently bias the estimate in one direction or the other. For example, the mean rainfall estimated by the arithmetic mean method may be about, say, 5% to 10% higher than the true mean, say, yielded by the isohyetal method. With this information, a realistic prediction of the mean rainfall requires a correction from 90% to 95% of the corresponding mean (arithmetic) rainfall. If a uniform probability density function (pdf) between this range of correction factors is assumed, then the systematic error in the estimated arithmetic mean rainfall of 60 in. will need to be corrected by a mean bias factor of:

$$e = \frac{1}{2}(0.9 + 0.95) = 0.925$$

whereas the corresponding random error in the estimated mean value, expressed in terms of *CV*, is

$$\Delta_2 = \frac{1}{\sqrt{3}}\left(\frac{0.95 - 0.90}{0.90 + 0.95}\right) = \frac{1}{\sqrt{3}}\left(\frac{0.05}{1.85}\right) = \frac{0.027}{1.73} = 0.016$$

The total random error in the estimated mean value is, therefore:

$$\Delta = \sqrt{\Delta_1{}^2 + \Delta_2{}^2} = \sqrt{(0.05)^2 + (0.016)^2} = \sqrt{0.0028} = 0.053$$

The systematic error is a bias in the prediction or estimation and can be corrected through a constant bias factor. The random error, called standard error, requires statistical treatment. It represents the degree of dispersiveness of the range of possible errors. It may be represented by the standard deviation or coefficient of variation of the estimated mean value. An objective determination of the bias as well as the random error will require repeated data on the sample mean (or medians), which are hard to come by.

For a random phenomenon, prediction or estimation is usually confined to the determination of a central value (e.g., mean or median) and its associated standard deviation or coefficient of variation. The uncertainty associated with the error in the estimation of the degree of dispersion is of secondary importance, whereas the uncertainty due to error in the prediction of the central value is of first-order importance. To summarize, through methods of prediction we obtain:

$\bar{x}$ = estimate of the mean value
σ_x = estimate of the standard deviation

An assessment of the accuracy or inaccuracy of the above prediction for the mean value is made to obtain:

e = bias correction for error in the predicted mean value $\bar{x}$
Δ = measure of random error in $\bar{x}$

For quantification of uncertainty measures, we confine ourselves to the error of prediction or the error in the estimation of the respective mean values, i.e., the systematic and random errors will refer to the bias and standard error, respectively, in the estimated mean value of a variable (or function of variables). It is important that the uncertainty measures are credible, for the validity of a calculated probability depends on credible assessments of the individual uncertain measures. Methods for evaluating uncertainty measures depend on the form of the available data and information.

For a set of observations, the mean value is:

$$\bar{x} = \frac{1}{n} \sum_{i=1}^{n} x_i \tag{2}$$

and the variance is:

$$\sigma_x^2 = \frac{1}{n-1} \sum_{i=1}^{n} (x_i - \bar{x})^2 \tag{3a}$$

The uncertainty associated with the inherent randomness is given by:

$$CV = \frac{\sigma_x}{\bar{x}} \tag{3b}$$

The above estimated mean value may not be totally accurate relative to the true mean (especially for small sample size n). The estimated mean value given above is unbiased as far as sampling is concerned; however, the random error of the above $\bar{x}$ is the standard error of $\bar{x}$, which is:

$$\sigma_{\bar{x}} = \sigma_x / \sqrt{n} \tag{4}$$

Hence, the uncertainty associated with random sampling error is:

$$\Delta_x = \sigma_{\bar{x}} / \bar{x} \tag{5}$$

This random error in $\bar{x}$ is limited to the sampling error only. There may, however, be other biases and random errors in $\bar{x}$, such as the effects of factors not included in the observational program.

Often, the information is expressed in terms of the lower and upper limits of a variable. Given the range of possible values of a random variable, the mean value of the variable and the underlying uncertainty may be evaluated by prescribing a suitable distribution within the range. For example, for a variable x, if the lower and upper limits of its values are x_l and x_u, the mean and CV of x may be determined as follows:

$$\bar{x} = \frac{1}{2}(x_l + x_u) \tag{6}$$

$$CV = \frac{1}{\sqrt{3}}\left(\frac{x_u - x_l}{x_u + x_l}\right) \tag{7}$$

where the probability density function (pdf) of the variable is uniform between x_l and x_u.

Alternatively, if the pdf is given by a symmetric triangular distribution prescribed within the limits x_l and x_u, the corresponding CV would be:

$$CV = \frac{1}{\sqrt{6}}\left(\frac{x_u - x_l}{x_u + x_l}\right) \tag{8}$$

With either the uniform or the symmetric triangular distribution, it is implicitly assumed that there is no bias within the prescribed range of values for x. On the other hand, if there is bias, the skewed distributions as shown above may be more appropriate. If the bias is judged to be toward the higher values within the specified range, then the upper triangular distribution would be appropriate. In such a case, the mean value is:

$$\bar{x} = \frac{1}{3}(x_l + 2x_u) \tag{9}$$

and CV is:

$$CV = \frac{1}{\sqrt{2}}\left(\frac{x_u - x_l}{2x_u + x_l}\right) \tag{10}$$

Conversely, if the bias is toward the lower range of values, the appropriate distribution may be a lower triangular distribution, with the mean value as:

$$\bar{x} = \frac{1}{3}(2x_l + x_u) \tag{11}$$

and CV as:

$$CV = \frac{1}{\sqrt{2}}\left(\frac{x_u - x_l}{x_u + 2x_l}\right) \tag{12}$$

Another distribution may be a normal distribution where the given limits may be assumed to cover $\pm 2\sigma$ from the mean value. In such cases, the mean value is:

$$\bar{x} = \frac{1}{2}(x_u + x_l) \tag{13}$$

and CV is:

$$CV = \frac{1}{2}\left(\frac{x_u - x_l}{x_u + x_l}\right) \tag{14}$$

The seemingly different types and sources of uncertainty can be analyzed in a unified manner. Consider a variable x (true value) and its prediction is given as $\hat{x}$. Let there be a correction factor N to account for error in $\hat{x}$. Therefore, the true x may be expressed as:

$$x = N\hat{x} \tag{15}$$

For a random variable x, the model $\hat{x}$ should be a random variable. The estimated mean value $\hat{x}$ and variance σ_x^2 (e.g., from a set of observations) are those of $\hat{x}$. Then $CV = \sigma_x / \bar{x}$ represents the inherent variability. The necessary correction factor N may also be considered a random variable, whose mean value "e" represents the mean correction for systematic error in the predicted mean value $\bar{x}$, whereas CV of N, Δ, represents the random error in the predicted mean value $\bar{x}$. Assuming N and $\hat{x}$ to be statistically independent, the mean value of x, is:

$$\mu_x = e\bar{x} \tag{16}$$

The total uncertainty in the prediction of x then becomes:

$$\Omega_x \cong \sqrt{CV_x^2 + \Delta_x^2} \quad , \qquad CV_x = \sigma_x / \bar{x} \tag{17}$$

The above analysis pertains to a single variable.

If Y is a function of several random variables $x_1, x_2, ..., x_n$, i.e.,

$$Y = g(x_1, x_2, ..., x_n) \tag{18}$$

the mean value and associated uncertainty of Y are of concern. A model (or function) $\hat{g}$ and a correction N_g may be used so that

$$Y = N_g \hat{g}(x_1, x_2, ..., x_n) \tag{19}$$

Thus, N_g has a mean value of "e_g" and CV of Δ_g. Using the first-order approximation, the mean value of Y is:

$$\mu_y \cong e_g \hat{g}(\mu_{x1}, \mu_{x2}, ..., \mu_{xn}) \tag{20}$$

where e_g is the bias in $\hat{g}(.,.,\_\_\_\_,.)$ and $\mu_{xi} = e_i \bar{x}_i$.

Also, the total CV of Y is:

$$\Omega_y^2 = \Delta_g^2 + \frac{1}{\mu_g^2} \sum_i \sum_j \rho_{ij} c_i c_j \sigma_{xi} \sigma_{xj} \tag{21}$$

in which $c_i = \frac{\partial g}{\partial x_i}$ evaluated at $(\mu_{x1}, \mu_{x2}, ..., \mu_{xn})$, ρ_{ij} = correlation coefficient between x_i and x_j.

9. Propagation of Errors

Consider a function z = f(x, y, ...) of two or more quantities x, y, ... If there are errors in x, y, ..., then there will be error in z, whose amount is denoted as dz. Let the errors in x, y, ... be denoted as dx, dy, ... The errors dz, dx, dy, ... must, of course, be of the same kind, i.e., all average errors, all standard deviations, etc. For propagation of errors, two cases can be distinguished: (1) independent errors, and (2) nonindependent errors. In the case of independent errors, there is the possibility of compensation or counterbalancing. For example, if the error in x causes z to be large, the error in y may cause it to be small. Thus, the net result would be that the total error in z would be less than the sum of individual errors. This result does not hold in the case of nonindependent errors where the errors actually add up algebraically.

Let the error be specified by variance whose square root yields the standard deviation. It is assumed that the function varies sufficiently slowly so that it can be represented by the first few terms in the Taylor series expansion. Let $\bar{z}$, $\bar{x}$, $\bar{y}$, ... be the mean values of z, x, y, ..., respectively. Their variances are defined respectively as $\sigma_z^2 = (z - \bar{z})^2$, $\sigma_x^2 = (x - \bar{x})^2$, $\sigma_y^2 = (y - \bar{y})^2$, ... Let a point P be defined by ($\bar{x}$, $\bar{y}$, ...). Expanding z in Taylor series at P,

$$z = f(\bar{x}, \bar{y}, ...) + \left.\frac{\partial f}{\partial x}\right|_P (x - \bar{x}) + \left.\frac{\partial f}{\partial y}\right|_P (y - \bar{y}) + ... \tag{22}$$

where the derivatives of f are evaluated at pint P. Equation (22) yields $\bar{z} = f(\bar{x}, \bar{y}, ...)$, which is expected.

Consider the variance of z, which can be expressed with the use of Eq. (22) as:

$$\sigma_z^2 = (\left.\frac{\partial f}{\partial x}\right|_P)^2 (x - \bar{x})^2 + (\left.\frac{\partial f}{\partial y}\right|_P)^2 (y - \bar{y})^2 + ... + 2(\left.\frac{\partial f}{\partial x}\right|_P)(\left.\frac{\partial f}{\partial y}\right|_P)(x - \bar{x})(x - \bar{y})^2 + ... \tag{23}$$

or

$$\sigma_z^2 = (\frac{\partial f}{\partial x}\bigg|_P)^2 \sigma_x^2 + (\frac{\partial f}{\partial y}\bigg|_P)^2 \sigma_y^2 + \ldots\ldots\ldots\ldots + 2(\frac{\partial f}{\partial x}\bigg|_P)(\frac{\partial f}{\partial y}\bigg|_P)\sigma_{xy} + \ldots\ldots\ldots\ldots \tag{24}$$

where σ_{xy} is the covariance of x and y.

Equation (24) can be generalized for independent and nonindependent or correlated errors. Let $z = f(x_1, x_2, \ldots\ldots\ldots\ldots, x_n)$. Then, the variance of z follows:

$$\sigma_z^2 = \sum_{i,j=1}^{n} (\frac{\partial f}{\partial x_i}\bigg|_P)(\frac{\partial f}{\partial x_j}\bigg|_P)\sigma_{x_i x_j} \tag{25}$$

It should be noted that covariances are always symmetric, i.e., $\sigma_{x_i x_j} = \sigma_{x_j x_i}$ and vanish for independent errors.

It may be worthwhile to illustrate the propagation of errors using a few examples. First, consider the case of two independent variables x and y, and $z = f(x, y)$. The errors are independent. Then,

$$\sigma_z = \sqrt{(\frac{\partial f}{\partial x})^2 \sigma_x^2 + (\frac{\partial f}{\partial y})^2 \sigma_y^2} \tag{26}$$

If $z = xy$, then

$$\sigma_z = \sqrt{(\bar{y})^2 \sigma_x^2 + (\bar{x})^2 \sigma_y^2} \tag{27}$$

Equation (27) is more meaningfully expressed in terms of the coefficient of variation (standard deviation divided by the mean), ε, as:

$$\varepsilon_z = \sqrt{\varepsilon_x^2 + \varepsilon_y^2} \tag{28}$$

If $z = x/y$, then Eq. (28) also holds.

If $z = x + y$ or $z = x - y$, then

$$\sigma_z = \sqrt{\sigma_x^2 + \sigma_y^2} \tag{29}$$

Equation (29) shows why the method of computation in environmental analysis, based only on the water balance equation are not popular. An example is the significant error obtained when computation of evaporation from a lake or a watershed is based on water balance.

Another case is $z = x^m y^n$. In this case,

$$\varepsilon_z = \sqrt{m^2 \varepsilon_x^2 + n^2 \varepsilon_y^2} \tag{30}$$

Equation (30) shows that the error multiples in a nonlinear case. If $z = x^m$, then $\varepsilon_z = m\varepsilon_x$. If $z = cx$, where c is constant, then $\varepsilon_z = \varepsilon_x$.

It is now possible to determine the best value of a quantity x from two or more independent measurements whose errors may be different. Intuitively, the measurement with less error should carry more weight. However, how exactly the weighting should be done is not quite clear. To that end, the principle of least squared error may be invoked. Consider two independent measurements of x as x_1 and x_2, with their respective (plus or minus) errors as σ_1 and σ_2. It may be reasonable to assume an estimate of x as:

$$\bar{x}_{12} = a x_1 + (1-a) x_2 \tag{31}$$

Equation (31) is similar to the Muskingum hypothesis used in flow routing. Then, it can be shown that:

$$x_{12} = \frac{\dfrac{x_1}{\sigma_1^2} + \dfrac{x_2}{\sigma_2^2}}{\dfrac{1}{\sigma_1^2} + \dfrac{1}{\sigma_2^2}} \tag{32}$$

$$\sigma_{12}^2 = [\frac{1}{\sigma_1^2 + \sigma_2^2}]^{-1} \tag{33}$$

Equations (32) and (33) can be generalized as:

$$\bar{x} = \frac{\sum_{i=1}^{n} (\dfrac{x_1}{\sigma_i^2})}{\sum_{i=1}^{n} (\dfrac{1}{\sigma_i^2})} \tag{34}$$

and

$$\sigma_z = \{\sum_{i=1}^{n} (\frac{1}{\sigma_i^2})\}^{-1/2} \tag{35}$$

Now, consider the case when errors are correlated. Let $E = A/(A + B)$, where A and B are independent measurements, with their mean and variances, respectively, denoted as $\bar{A}, \sigma_A^2$ and $\bar{B}, \sigma_B^2$. It can be shown that:

$$\sigma_E^2 = \frac{(\bar{B})^2}{(\bar{A}+\bar{B})^2}\sigma_A^2 + \frac{(\bar{A})^2}{(\bar{A}+\bar{B})^2}\sigma_B^2 \tag{36}$$

Equation (36) can also be derived by expressing $E = A/U$, where $U = A + B$, and then applying the Taylor series.

If z = x + y or z = x – y, then it can be shown that:

$$\sigma_z = \sqrt{\sigma_x^2 + \sigma_y^2 \pm 2\sigma_{xy}} \tag{37}$$

Equation (37) contains the covariance term. Similarly, if $z = xy$, then:

$$\varepsilon_z = \sqrt{\varepsilon_x^2 + \varepsilon_y^2 + 2\frac{\sigma_{xy}}{(\bar{x}\,\bar{y})}} \tag{38}$$

If $z = x/y$, then:

$$\varepsilon_z = \sqrt{\varepsilon_x^2 + \varepsilon_y^2 - 2\frac{\sigma_{xy}}{\bar{x}\,\bar{y}}} \tag{39}$$

Equations (38) and (39) are similar, except for the sign of the covariance term.

If $z = x^m\, y^n$, then it can be shown that:

$$\varepsilon_z = \sqrt{m^2 \varepsilon_x^2 + n^2\, \varepsilon_y^2\, 2\frac{mn\sigma_{xy}}{(\bar{x}\,\bar{y})}} \tag{40}$$

Equations (37) - (40) contain covariance terms and should be calculated.

10. Dealing with Uncertainty

Now, some thought must be given to the vexing problem of uncertainty in the decision making process. It is not possible to deal with every kind of uncertainty. If people cannot make up their minds, if there are delays in the transmission of pertinent information, or if there is mismanagement, then this may create considerable uncertainty in the decision making process. But, it is not the kind of uncertainty one will deal with here for the simple reason that it is virtually impossible to quantify the degree of uncertainty. Here, only the kind of uncertainty that can be measured quantitatively, at least in principle, has been dealt with.

For example, if a coin is tossed, it is not known in advance whether head will turn up. But, it is known that the event "head will turn up" has a probability of ½ or 50% each time the coin is tossed. Similarly, it is possible to assign a probability to the event that the peak flow in the Red River in any given year will exceed 3000 m³/sec. One can also determine the probability that the compressive strength of concrete, manufactured in accordance with given specifications, will exceed 25,000 kN/m² or that the maximum number of cars that have to wait at a railway crossing will exceed 20 on any given working day. The kind of uncertainty involved here is the uncertainty associated with the randomness of the event.

To investigate the uncertainty in the conclusions reached about uncertain events is beyond the scope here. For example, one may calculate the probability p that, in any given year, the peak flow in the Red River exceeds 3000 m^3/sec and conclude that p is 3.3%. But, there is an element of uncertainty in estimation of p, which depends very much on the length and quality of data used for its determination. The question thus arises: what is the probability that p lies within a given range p + or – Δp? In each case, one can talk about events that can be analyzed to a degree that reasonable people, using reasonable procedures, come up with reasonably close probability assessments. If that sounds rather vague, it should be remembered that the goal is only to deal rationally with uncertainty, not to eliminate it completely.

Now, the question is: how to deal with uncertain events in the decision making process? In principle it is quite simple. Events that are certain to occur, or conclusions that are certainly true, must, of course, be fully taken into account. These can be given a weight of 1, corresponding to the probability assigned to all certain events, which is 1. Impossible events, on the other hand, are disregarded in decisions, and these are given the weight of 0, corresponding to the probability of 0 assigned to all impossible events. Any event in between is given a weight equal to the probability of its occurrence. Thus, the more likely an event, the more the weight it gets, and the greater its relative effect on the outcome or the decision.

11. Conclusions

The following conclusions are drawn from this study: (1) environmental data are subject to both intrinsic and epistemic random errors which can be random or systematic; (2) the sources of errors are many and are the root cause of uncertainty; (3) two main measures of errors are bias and consistency; (4) errors propagate in analyses, and the rate of propagation depends primarily on the nature of the function describing the system and the dependence among errors; and (5) errors can be minimized but cannot be eliminated completely.

12. References

1. Klir, G. J. (1991) Measures and principles of uncertainty and information, *Information Dynamics, Proceedings of NATO Advanced Study Institute on Information Dynamics,* Plenum Press, New York.

2. Shackle, G. L. S. (1961) *Decision, Order and Time in Human Affairs*, Cambridge University Press, Cambridge.

3. Pat*-Cornell, M. E. (1996) Uncertainties in risk analysis: six levels of treatment, *Reliability Engineering and System Safety*, No. 54, Elsevier Science Limited, Northern Ireland.

4. Vrijling, J. K. and van Gelder, P. H. A. J. M. (2000) Policy implications of uncertainty integration in design in Wang and Hu (ed.), *Stochastic Hydraulics 2000*, Balkema, Rotterdam, pp. 633-646.

5. Kendall, M. G. and Stuart, A. (1973) *The Advanced Theory of Statistics, V.2. Inference and Relationship*, Ch.29. Functional and structural relationship, Charles Griffin & Company Ltd, London.

PHYSICS OF ENVIRONMENTAL FREQUENCY ANALYSIS

W. G. STRUPCZEWSKI[1], V. P. SINGH[2] and S. WEGLARCZYK[3]
[1] *Water Resources Department, Institute of Geophysics*
Polish Academy of Sciences
Ksiecia Janusza 64, 01-452 Warsaw, Poland

[2] *Department of Civil and Environmental Engineering*
Louisiana State University
Baton Rouge, Louisiana 70803-6405, USA

[3] *Institute of Water Engineering and Water Management*
Cracow University of Technology
Warszawska 24, 31-155 Cracow, Poland

Abstract. A multitude of environmental processes embody both the elements of chance and the descriptive laws of physics. Excessive process description at one scale is lost through the processes of integration in time and space and through averaging. This justifies simplification in representation of the processes. It is hypothesized that if an environmental process is described by a linear or linearized governing equation, then the solution of this equation for a unit impulse (or Dirac delta) function can be interpreted as a probability density function for describing the probabilistic properties of the process. This hypothesis is tantamount to mapping from the unit impulse response function (UIR), $h(t)$, to the probability density function (PDF), $f(x)$, where h is UIR as a function of time or space variable denoted by t and f is the PDF as a function of the random variable of the process. For example, the impulse response of a diffusion equation for pollutant transport described by space-time variation of concentration can be used as a probability distribution for pollutant concentration in a medium, such as a river, lake, tube, storm water, soil, or saturated geologic formation. Likewise, the impulse response of a linearized diffusion model of channel flow can be interpreted as a probability distribution for frequency analysis of extreme values (such as floods, droughts, hurricanes, earthquakes, and so on). Similarly, the impulse response of a linear reservoir can be used as an exponential probability distribution model. The impulse response of a cascade of linear reservoirs is the gamma distribution which has a number of applications in environmental data analysis. In this vein, a number of impulse responses of physically-based equations which apply to environmental processes and data are discussed and illustrated using field or laboratory data.

1. Introduction

Environment can be defined as a continuum of three components: air, water and land. The processes dealing with these components and their dynamic interactions constitute environmental processes. Examples of such processes are solute transport in a river, lake, storm water, soil or aquifer; flood; drought; rainfall; erosion; sediment transport by storm water or in a river; air pollution; depletion in ozone layer; glacial movement and melting; climate change; occurrence of an epidemic; sea water rise; salt water intrusion; and so on. A quantitative description of these processes involves the determination of either the space, time or the space-time history of flux, concentration, the average or the volume of the process variable. For example, for describing the transport of a pollutant in a river, the pollutant flux or concentration as a function of space and time may be selected. One may also select the pollutant load passing through a given point on the river over a selected period of time, say a month or a year. Similarly, for describing the quality of air in an urban area, one may select the ozone level and determine its variability in time.

For quantitatively describing environmental variables, it is frequently observed that environmental processes embody both the elements of chance and the descriptive laws of physics. In other words, these variables cannot be completely described either by deterministic means or by stochastic means alone. Rather, the approach has to be a combination of both the stochastic and the deterministic means. Viewing the variables deterministically, considerable simplification is usually needed. This can be justified in light of the observation that excessive process description at one scale is lost through the operations of integration in time and averaging. Furthermore, the governing equations themselves have inherent limitations with regard to accuracy. Even more stark is the state of data acquisition and processing.

When the stochastic aspect of the environmental variables is considered, the element of chance is attributed to a complex mix of factors, such as inherent stochasticity in environmental processes due to interactions of the environmental continuum components, interaction between man and the environment, and our inability to observe and quantify spatial and temporal variability of environmental variables. A stochastic description usually includes a time series analysis, analysis of variance or frequency analysis. Each type of analysis is needed, depending on the demand of a problem. For example, a frequency analysis is needed for planning and design. A time series analysis is needed for operation and management. A regression analysis is needed for prediction, extrapolation, or interpolation. Analogous to a deterministic description, a stochastic analysis also involves simplifications which can be justified on the lack of data, lack of adequate knowledge of processes to be modeled as well as of the methodologies for modeling of nonlinear and non-Gaussian processes, requirement of simplicity, parsimony of parameters, and so on.

An environmental process in nature exhibits itself in a way that it should, and a variable selected to describe some aspect of this process must obey the commands of the process. The variable is unaware of deterministic and stochastic issues and analyses thereof. This means that there must be an inherent connectivity between these analyses. In environmental science and engineering, this connection is frequently observed. For example, a regression analysis without error analysis in statistics is no different from curve fitting techniques in mathematics. Indeed, regression analysis techniques are often employed to find a best-fit curve for a given set of data, and the connection between these types of techniques is well known. Another example is the autoregressive (AR) technique in the time series analysis. When an AR technique is applied to, say, daily, monthly or annual river flows, the coefficients associated with the autoregressed variable are nothing but the ordinates of linear kernel of the flow variable. Since the AR technique is linear, it is equivalent to the impulse response function of a linear flow process. In hydrology, this is known as the unit hydrograph method. One also finds a connection between the unit hydrograph method and the spectral analysis. This means that certain linear time series analysis techniques are equivalent to linear response functions of environmental processes. However, the connection between frequency analysis methods and deterministic methods is not clear yet. This may be because frequency by definition is stochastic in nature, and finding a deterministic equivalent seems somewhat contradictory in terms. However, our objective here is to find a connection through techniques of analysis, not through concept. This constitutes the subject of this paper.

2. Hypothesis

It is hypothesized that, if an environmental variable is described by a linear or linearized governing equation, then the solution of this equation for a unit impulse (or Dirac delta) function (UIF) can be interpreted as a probability density function (PDF) for describing the probabilistic properties of the random variable, say X. The solution for UIF can be characterized as the unit impulse response (UIR) or $h(t)$. If UIR is a function of time, t, then PDF is a mapping from the (h, t) plane to the (f, x) plane, where x is the value (or quantile) of X.

There are many environmental variables which can be linearly described reasonably well. If some of the variables cannot be described linearly in the real domain, then they can be described linearly in the logarithmic domain or in an appropriate transformed domain. Examples of linear approximation are surface runoff due rainfall excess, river flow, monthly sediment discharge, solute concentration in a tube or soil, and so on. Thus, their UIRs can be considered as their PDFs. It is not surprising that several probability distributions have found their niche in linear environmental analyses. This hypothesis will be explored in what follows.

3. Impulse Responses of Linear Systems

3.1. LINEAR RESERVOIR

The simplest linear system is probably a linear reservoir described by the spatially lumped form of the continuity equation:

$$\frac{dS(t)}{dt} = I(t) - Q(t) \tag{1}$$

and a storage-discharge type relation:

$$S = k\,Q \tag{2}$$

where $I(t)$ is the rate of inflow to the reservoir at time t, $Q(t)$ is the rate of outflow from the reservoir at time t, $S(t)$ is the storage in the reservoir at time t, and k is the storage coefficient or average travel (or residence or lag) time. The linear reservoir has been used by itself or as an element of a network model for rainfall-runoff modeling. If $I(t)$ is denoted by a unit Delta function, $\delta(t)$, then the UIR of the linear reservoir, $h(t)$, is:

$$h(t) = \frac{\exp(-t/k)}{k} \tag{3}$$

In hydrology, $h(t)$ is known as the instantaneous unit hydrograph (IUH). Then, the PDF of a variable described by a linear reservoir becomes:

$$f(x) = \frac{\exp(-x/k)}{k} \tag{4}$$

where x is the quantile of the variable X described by the linear reservoir, and k is a parameter. Thus, it is seen that $h(t)$ is mapped onto $f(x)$. Equation (4) is an exponential density function and is widely used in environmental and water resources analyses. For example, if an environmental process is described by the Poisson process, then the interarrival times follow an exponential distribution. Interarrival times of floods can be modeled using Eq. (4). Rainfall depth, intensity, and duration have been modeled using Eq. (4). It should be noted that k in $f(x)$ represents the average of X, and hence, its interpretation from Eq. (3) remains unchanged under mapping of $h(t)$ onto $f(x)$.

Another modification of the linear reservoir is when the unit Delta function, $\delta(t)$, is modified as $\delta(t-t_0)$, where t_0 is the time at which the function occurs. In that case, $h(t)$ of Eq. (3) becomes:

$$h(t) = \frac{\exp[-(t-t_0)/k)}{k} \tag{5}$$

Equation (5) is the UIR of a lag and route linear reservoir system in which t_0 is the amount of lag time before water is released from the reservoir. Mapping Eq. (5) onto the probability plane, the PDF becomes:

$$f(x) = \frac{\exp[-(x-x_0)/k)}{k} \tag{6}$$

where x_0 is the threshold of X, $x \geq x_0$. The threshold is the minimum value of X. This is useful in frequency analysis of environmental data.

3.2. MUSKINGUM MODEL

The Muskingum model is described by Eq. (1) and the Muskingum hypothesis:

$$S(t) = K[\alpha\ I(t) - (1-\alpha)Q(t)] \tag{7}$$

where K is the average reach travel time, and α is a parameter or a weighting coefficient. The unit impulse response of the Muskingum model is given by:

$$h(t) = -\frac{\alpha}{1-\alpha}\delta(t) + \frac{1}{K(1-\alpha)^2}\exp\left[-\frac{t}{K(1-\alpha)}\right] \tag{8}$$

It has been shown that modeling flood routing along a short reach of a lowland river may result in the negative value of the α and

$$0 \leq \left(-\frac{\alpha}{1-\alpha} \right) \leq 1 \tag{9}$$

Denoting $\frac{1}{1-a} = \beta$ and $\frac{\beta}{K} = \gamma$ and renaming t as x, one gets a two-parameter probability distribution function:

$$f(x) = (1-\beta)\delta(x) + \beta\,\gamma\,\exp(-\gamma x) \tag{10}$$

The PDF given by Eq. (10) is a weighted sum of two functions: a delta function and an exponential function. It is interesting to note that, in this function, parameter β is a weighting factor and parameter K–β/γ becomes the average of X. Thus, the original expressions of the weighting factor and the average travel time are modified under mapping, but the conceptual meaning of the modified expressions remains more or less intact.

3.3. CASCADE OF LINEAR RESERVOIRS

If an environmental system is represented by a cascade of n-equal linear reservoirs, then its UIR becomes:

$$h(t) = \frac{1}{k\Gamma(n)} \left(\frac{t}{k}\right)^{n-1} \exp(-t/k) \tag{11}$$

where k is the storage parameter of each reservoir and $\Gamma(n)$ is the gamma of n or the number of reservoirs. Since there are n reservoirs, nk represents the total lag time (or the average residence time) of the system. Mapping onto the probability plane, the PDF becomes:

$$f(x) = \frac{1}{k\Gamma(n)} \left(\frac{x}{k}\right)^{n-1} \exp(-x/k) \tag{12}$$

where k and n are parameters. Equation (12) is the gamma probability density function. The gamma distribution results from the sum of exponentials, where n will be the number of exponentials. In deterministic parlance, each exponential represents a linear reservoir. Thus, the deterministic interpretation of parameters is carried over under mapping. The gamma distribution is one of the most commonly used probability distributions for environmental frequency analysis.

If an environmental system satisfies the requirement that $h(t) > 0$ if $t \geq t_0$, then the UIR becomes:

$$h(t) = \frac{1}{k\Gamma(n)} \left(\frac{t-t_0}{k}\right)^{n-1} \exp[-(t-t_0)/k] \tag{13}$$

The interpretation of t_0 is that it takes the cascade of equal linear reservoirs t_0 to retain water before it starts to release it. Mapping on the probability plane, Eq. (13) becomes:

$$f(x)=\frac{1}{k\Gamma(n)}(\frac{x-x_0}{k})^{n-1}\exp[-(x-x_0)/k) \tag{14}$$

which represents the 3-parameter Pearson type 3 probability density function. This is one of the most widely used frequency distributions in hydrology and environmental resources. Here, parameter x_0 is the lowest value or the threshold of the variable X. Although these parameters, k, n, and x_0, can be interpreted using the deterministic analogy, their optimal values are better found by curve fitting. This means that under mapping onto the probability plane, the meaning of the parameters is somewhat distorted.

3.4. LINEAR CHANNEL DOWNSTREAM MODEL

One of the most important problems in one-dimensional flood routing analysis is the downstream problem, i.e., the prediction of flood characteristics at a downstream section on the basis of the knowledge of flow characteristics at an upstream section. Using the linearization of the Saint-Venant equation, the solution of the upstream boundary problem was derived by Deymie [1], Masse [2], Dooge and Harley [3], Dooge *et al.* [4, 5], among others; a discussion of this problem is presented in Singh [6]. The solution is a linear, physically based model with four parameters dependent on the hydraulic characteristics of the channel reach at the reference level of linearization. However, the complete linear solution is complex in form and is relatively difficult to compute [6]. Two simpler forms of the linear channel downstream response are recognized in the hydrologic literature and are designated as linear diffusion analogy model (LD) and linear rapid flow model (LRF). These correspond to the limiting flow conditions of the linear channel response, i.e., where the Froude number is equal to zero [7, 8] and where it is equal to one [9].

The complete linearized Saint Venant equation is of hyperbolic type, and it may be written as:

$$a\frac{\partial^2 Q}{\partial x^2}+b\frac{\partial^2 Q}{\partial x\partial t}+c\frac{\partial^2 Q}{\partial t^2}=d\frac{\partial Q}{\partial x}+e\frac{\partial Q}{\partial t} \tag{15}$$

where Q is the perturbation of flow about an initial condition of steady state uniform flow, x is the distance from the upstream boundary, t is the elapsed time, and a, b, c, d and e are the parameters being a function of channel and flow characteristics at the reference steady state condition. A number of models of simplified forms of the complete St. Venant equation have been proposed in the hydrologic literature.

If all three of the second-order terms on the left-hand side of that equation are neglected, we obtain the linear kinematic wave model. Expressing the second and the third second-order term in terms of the first on the basis of the linear kinematic wave approximation leads to a parabolic equation [8] in contrast to the original Eq. (15), which is a hyperbolic equation. Its solution for a semi-infinite channel, known under the name of the linear diffusion analogy (LDA) or the convective-diffusion solution, has the form:

$$h(x,t)=\frac{x}{\sqrt{4\pi D t^3}}\exp\left[-\frac{(x-u\cdot t)^2}{4D\cdot t}\right] \tag{16}$$

where x is the length of the channel reach, t is the time, u is the convective velocity and D is the hydraulic diffusivity. Both u and D are the functions of channel and flow characteristics at the reference steady state condition. Besides flood routing, it was applied by Moore and Clarke [10] and Moore [11] as a transfer function of a sediment routing model.

The function given by Eq. (16) is rarely quoted in statistical literature. It was derived by Cox and Miller [12] as the probability density function of the first passage time T for a Wiener process, starting at 0 to reach an absorbing barrier at the point x, where u is the positive draft and D is the variance of the Wiener process. Tweedie [13] termed the density function of Eq. (16) as an inverse Gaussian PDF; Johnston and Kotz [14] summarized its properties, and Folks and Chhikara [15] provided a review of its development. The function in Eq. (16) has been applied by Strupczewski *et al* [16] as a flood frequency model.

If the diffusion term is expressed in terms of two other terms, using the kinematic wave solution, one gets the rapid flow (RF) equation, which is of parabolic-like form [17]. Therefore, it filters out the downstream boundary condition. It provides the exact solution for a Froude number equal to one and, consequently, can be used for large values of the Froude number. If the alternative approach is taken by expressing all the second-order terms as cross-derivatives, one gets the equation representing the diffusion of kinematic waves [9, 18]. The kinematic diffusion (KD) model, being of the parabolic-like form, satisfactorily fits the solution of the complete linearized Saint Venant equation only for small values of the Froude number and slow rising waves.

Although RF and KD models correspond to quite different flow conditions, the structure of their impulse response is similar [19, 20]. In both cases, the impulse response is:

$$h(x,t) = P_0(\lambda)\,\delta(t-\Delta) + \sum_{i=1}^{\infty} P_i(\lambda)\,h_i\left(\frac{t-\Delta}{\alpha}\right).1(t-\Delta) \tag{17}$$

where

$$P_i(\lambda)=\frac{\lambda^i}{i!}\exp(-\lambda) \tag{18}$$

is the Poisson distribution and

$$h_i\left(\frac{t}{\alpha}\right) = \frac{1}{\alpha\,(i-1)!}\left(\frac{t}{\alpha}\right)^{i-1}\exp\left(-\frac{t}{\alpha}\right) \tag{19}$$

is the gamma distribution and $1(x)$ is the unit step function. Parameters α, λ *and* Δ are functions of both channel geometry and flow conditions, which are different for both models. Furthermore, there is no time lag (Δ) in the impulse response function of the KD model.

Both models can be considered as hydrodynamic and conceptual. Note that the solution of both models can be represented in terms of basic conceptual elements used in hydrology, namely, a cascade of linear reservoirs and a linear channel in case of the RF model. The upstream boundary condition is delayed by a linear channel with time lag, Δ, divided according to the Poisson distribution with mean λ, and then transformed by parallel cascades of equal linear reservoirs (with time constant K) of varying lengths. Note that λ is the average number of reservoirs in a cascade. Strupczewski *et al.* [19] and Strupczewski and Napiorkowski [9] have derived the distributed Muskingum model from the multiple Muskingum model and have shown its identity to the KD model. Similarly, the RF model happens to be identical to the distributed delayed Muskingum model [20].

Einstein [21] introduced the function given by Eq. (17) to hydrology as the mixed deterministic-stochastic model for the transportation of bed load. For the second time, it has appeared as the probability distribution function of the total rainfall depth obtained from the assumption of Poisson process for storm arrivals and the exponential distribution for storm depths [22]. The function in Eq. (17) is considered to be a flood frequency model.

3.5. DIFFUSION EQUATION

Many natural processes are diffusive in nature and can therefore be modeled using diffusion-based equations. Examples of such processes are dye transport in a container of water, contaminant mixing in rivers and estuaries, transport of sediment in rivers, migration of microbes, to name but a few. Thus, probability distributions based on random diffusion processes may be better suited to represent the data of diffusion driven processes.

To illustrate this concept, a dye diffusion equation is considered to describe the concentration distribution produced from an injection of a mass of dye, that is introduced as a plane source located at coordinate x_0 at time zero into a liquid-filled, semi-infinite tube of cross-section A. The tube is closed on the left end and extends on the right to infinity. The one-dimensional diffusion equation for a mass M of dye introduced at time $t = 0$ into a liquid-filled tube of cross-section A that extends from $x = 0$ to infinity is:

$$\frac{\partial C}{\partial t} = D\frac{\partial^2 C}{\partial x^2} + \left(\frac{M}{A}\right)\delta(x - x_0)\delta(t) \tag{20}$$

where D is the diffusion coefficient; $\delta(x - x_0)$ is a Dirac delta function of $(x - x_0)$; $\delta(t)$ is a Dirac delta function of t; and x_0 is the location where the mass is inserted at time $t = 0$. A Dirac delta function has the property that it is equal to zero if the argument is nonzero; when the argument is zero, the Dirac delta function becomes equal to infinity. A Dirac delta function has units which are (units of the function=s argument)$^{-1}$ [23].

The first boundary condition states that there is no diffusion of dye through the closed left end of the tube at $x = 0$:

$$\frac{\partial C}{\partial x} = 0, \quad at\ x = 0 \tag{21}$$

The second boundary condition states that the concentration and the concentration flux are zero at infinity; more generally, all of the terms in the Taylor series expansion of the concentration are zero at infinity. The second boundary condition is stated for the terms of the Taylor series of concentration as:

$$C = 0, \frac{\partial C}{\partial x} = 0, \ldots, \frac{\partial^n C}{\partial x^n} = 0, \ldots \; as \; x \to \infty \tag{22}$$

$$C(x,0) = 0 \tag{23}$$

The initial condition states that there is no dye in the tube at time zero.

Using the integral transform method, the solution of Eq. (20) subject to (21)-(23) is found to be [24, 25]:

$$C(x,t) = \frac{M}{A} \frac{1}{\sqrt{4\pi Dt}} \left(\exp\left[-\frac{(x-x_0)^2}{4Dt} \right] + \exp\left[-\frac{(x-x_0)^2}{4Dt} \right] \right) \tag{24}$$

We now reduce the number of terms in Eq. (24), and normalize the equation so that it represents a unit mass injected over a unit area, and map onto the probability plane by introducing a frequency term instead of concentration. These changes make Eq. (24) resemble a probability distribution. The term "$4Dt$" appears together in Eq. (24). We define a new term:

$$\sigma^2 = 2Dt \tag{25}$$

In addition, the mass and cross-sectional area are combined with concentration, σ is held constant so it is treated as a parameter, and a new term $f(x;x_0,N)$ is introduced, so that $f(x;x_0,N) = AC(x,t)/M$, which has units length^{-1}. The result is the equation:

$$f(x;\sigma,x_0) = \frac{1}{\sqrt{2\pi}\sigma} \left\{ \exp\left[-\left(\frac{x-x_0}{\sqrt{2}\sigma} \right)^2 \right] + exp\left[-\left(\frac{x+x_0}{\sqrt{2}\sigma} \right)^2 \right] \right\} \tag{26}$$

which is now interpreted as a probability distribution that is bounded by $x = 0$ on the left side and extends to infinity on the right. The term σ represents the spread of the probability distribution; and x_0 usually represents the location of the peak frequency, although it is possible that, if σ is large, the peak frequency may not be located at x_0 but may be located at $x = 0$. The distribution is called a two parameter semi-infinite Fourier distribution as it was developed from the diffusion (Fourier) equation using semi-infinite Fourier transforms.

Equation (26) is limited to application to data that are distributed along the positive x axis. However, if the restriction on x only being able to represent distance is relaxed, so that x can represent any dimension that is appropriate for a frequency distribution, then the number of applications of Eq. (26) can increase. For example, if one is interested in the frequency distribution of a chemical, such as manganese concentration in a river, then x could have units of milligrams per liter. The units of σ are the same as the units of x.

4. Application

In the frequency analyses of environmental (say, hydrologic) data in arid and semiarid regions, one often encounters data series that contain several zero values while zero is the lower limit of the variability range. From the viewpoint of the probability theory, the occurrence of zero events can be expressed by placing a nonzero probability mass on a zero value, i.e., $P(X=0)\neq 0$, where X is the random variable, and P is the probability mass. Therefore, the distribution functions from which such hydrological series were drawn would be discontinuous, with discontinuity at the zero value having a form:

$$f(x) = (1-\beta)\delta(x) + f_c(x;\mathbf{h}).1(x) \tag{27}$$

where (1-β) denotes the probability of the zero event, i.e., $1-\beta = P(X=0)$; $f_c(x;\mathbf{h})$, is the continuous function such that $\int_0^{\infty} f_c(x;\mathbf{h})\,dx = \beta$; $\mathbf{h}$ is the vector of parameters; $\delta(x)$ is the Dirac delta function, and $1(x)$ is a unit step function.

The estimation procedures for hydrologic samples with zero events have been the subject of several publications. The theorem of total probability has been employed to model such series [26, 27, 28]. Then, Eq. (27) takes the form:

$$f(x) = (1-\beta)\delta(x) + \beta f_1(x;\mathbf{g}).1(x) \qquad \beta \notin \mathbf{g} \tag{28}$$

where $f_1(x;\mathbf{g})$ is the conditional probability density function (CPDF), i.e., $f_1(x;\mathbf{g}) \equiv f_1(x;\mathbf{g}|X>0)$, which is to be continuous in the range $(0,+\infty)$ with a lower bound of zero value. Wang and Singh [28] estimated β and the parameters of CPDF separately, considering the positive values as a full sample for the purpose. Having estimated **g** and β, the conditional distribution can be transformed to the marginal distribution, i.e., to $f(x)$, by Eq. (28). Among several PDFs with zero lower bound recognized in FFA (e.g., Rao and Hamed [29], the Gamma distribution given by Eq. (12) was chosen by Wang and Singh [28] as an example of CPDF. Four estimation methods were applied, i.e., the maximum likelihood method (MLM), method of moments (MOM), probability weighted moments (PWM), and ordinary least squares method. Using monthly precipitation and annual low-flow data from China, and annual maximum peak discharge data from the United States, the suitability of the distribution and the estimation methods was assessed. The histogram and the estimated PDF of all three series indicated a reverse J-shape without mode, while the value of the coefficient of variation of $f_1(x;\mathbf{g})$ was close to one, pointing out on a good fit of data to Muskingum origin PDF given by Eq. (10).

Among positively skewed distributions, it is the lognormal (LN) distribution which, together with the Gamma, is most frequently used in environmental frequency analysis. LN has been found to describe the distribution of hydraulic conductivity in a porous media, annual peak flows, raindrop sizes in a storm, and other hydrologic variables [30]. Chow [31] reasoned that this distribution is applicable to hydrologic variables formed as the product of other variables since, if $X = X_1 \cdot X_2 \cdot ... X_n$ then $Y = \log X$ tends to the normal distribution for large n provided that the X_i are independent and identically distributed.

Kuczera [32] considered six alternative PDFs and found the two-parameter LN to be the most resistant to an incorrect distributional assumption in at-site analysis and also while combining site and regional flood information. Strupczewski *et al.* [33] fitted seven two-parameter distribution functions, namely, normal, gamma, Gumbel (extreme value type I), Weibull, log-Gumbel and log-logistic to thirty nine 70-year long annual peak flow series of Polish rivers. The criterion of the maximum log-likelihood value was used for the best model choice. From the above competing models, lognormal was selected in 32 cases out of 39, gamma in 6 cases, Gumbel in one case, and the remaining four were not identified as the best model even in one case.

The LDA model shows a similarity to the LN model [16]. It is only for large values of the coefficient of variation that the LDA lines deviate apparently from straight lines on a lognormal probability paper. A comparison of the maximum likelihood values of both distributions of the thirty-nine 70-year long annual peak flow series of Polish rivers has shown in 27 out of 39 series a better fit of LDA to the data than did the LN model [16]. For Polish rivers, the average value of the ratio of the skewness coefficient (c_s) to the coefficient of variation (c_v) equals 2.52. This value is only just closer to the ratio of LDA model, where $c_c / c_v = 3$, than to the respective of the LN model, where $c_s = 3c_v + c_v^3$. Moreover, it is interesting to learn that, the LDA model represents flood frequency characteristics quite well when the LDA model is likely to be better than the other linear models, i.e., for lowland rivers (Fig. 1).

Comparing the potential for applicability of the two distributions, one should also take into account real conditions, where the hypothetical PDF differs from the true one. Applying the ML method, one gets unbiased moment estimates if, instead of the LN model, the LD model is used for the LN-distributed data, while in the opposite case, the ML-estimate of the variance is biased. Therefore, if both models show an equally good fit to the data, it seems reasonable to select the LD model if the ML method is to be applied. One should be also aware that the LN distribution is not uniquely determined by its moments [34].

An assessment of the suitability of the kinematic diffusion (KD) model given by Eq. (17) for environmental samples with zero values and the development of estimation methods are in progress [35]. Examples of its application for annual peak flow series of the U. S. arid regions are being worked out. In general, a shortage of long series with zero values is observed. An examination of function given by Eq. (17) reveals that, for $\lambda<2$ (which corresponds to the probability of the zero event being equal to $P_0(\lambda=2)=\exp(-2)=0.135$) the PDF has a reverse J-shape without mode. Modeling longer time series, it may be reasonable to introduce a third parameter to the model of Eq. (17), making the shape of the continuous part of the distribution independent on the probability of zero event:

$$f(x) = \beta \cdot \delta(x) + \frac{1-\beta}{1-e^{-\lambda}} \sum_{i=1}^{\infty} P_i(\lambda) \cdot h_i\left(\frac{x}{\alpha}\right) \cdot 1(x) \qquad (29)$$

where $P_i(\lambda)$ and $h_i\left(\frac{x}{\alpha}\right)$ are defined by Eqs. (18) and (19).

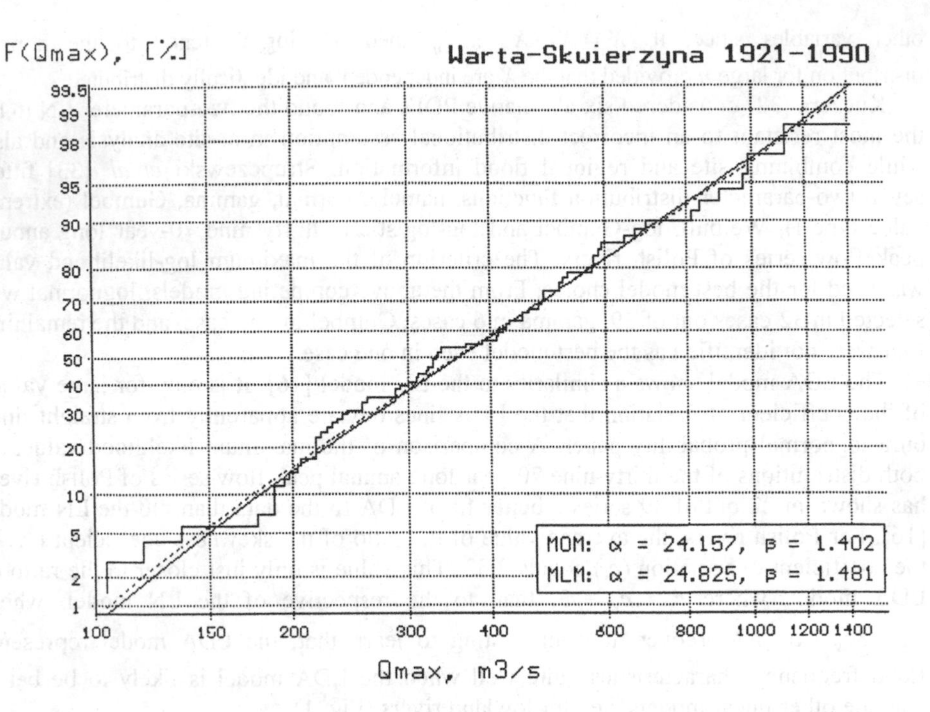

Figure 1. Empirical and two theoretical LD cumulative distribution functions (CDF's) for the Warta River, Skwierzyna cross-section data. MOM and MLM estimated parameters are shown. Solid line: MOM estimated CDF, dotted line: MLM estimated CDF.

5. Conclusions

The following conclusions are drawn from this study: (1) the unit impulse response (UIR) functions of linear or linearized physically based models form suitable models for environmental frequency analysis; (2) many of the unit response functions are found to be the same as have been used in statistics for a long time; (3) the use of the unit impulse response functions can provide a physical basis to many of the statistical distributions; and (4) the UIR approach provides a hope for linking deterministic and stochastic frequency models.

6. Acknowledgment

This study has been partly financed by the Polish Committee for Scientific Research, Grant no. 6 P 4D 056 17.

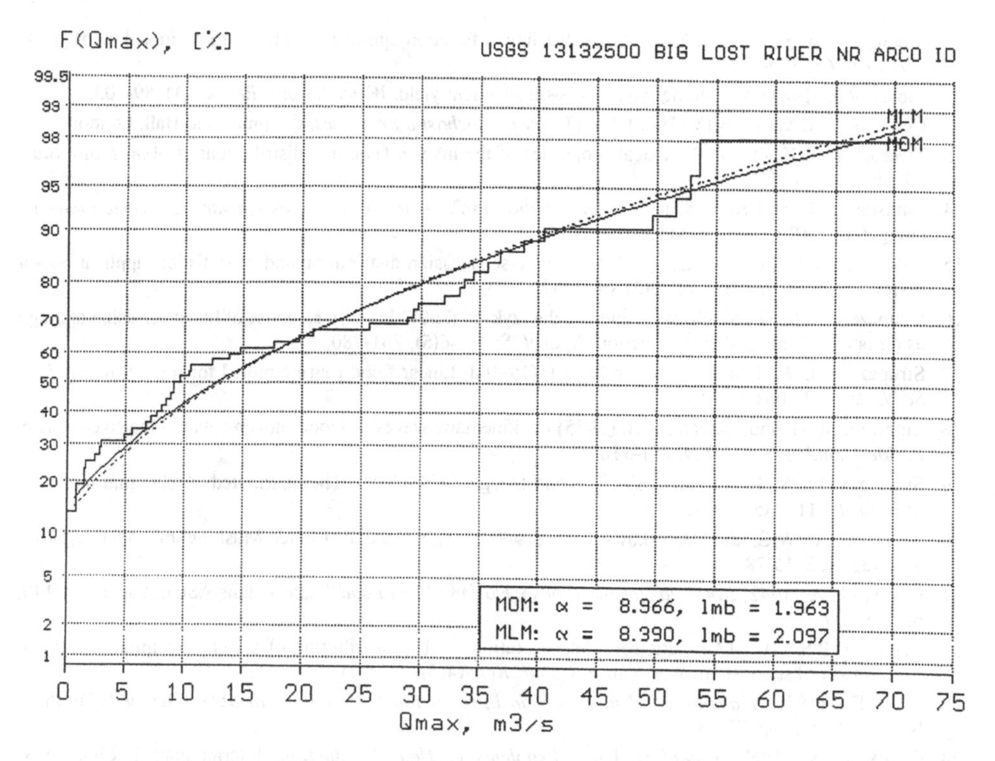

Figure 2. Empirical and two theoretical KD cumulative distribution functions (CDF's) for the Big Lost River, Arco, Id. MOM and MLM estimated parameters are shown. Solid line: MOM estimated CDF, dotted line: MLM estimated CDF.

7. References

1. Deymie, P. (1939) Propagation d'une intumescence allongee (Problem aval) [Propagation of an elongated intumescence], *Proc. 5th International Cong. Appl. Mech.*, New York, pp.537-544.
2. Masse, P. (1939) Recherches sur la theorie des eaux courantes [Researches on the theory of water currents], *Proc. 5th Internatl. Cong. Appl. Mech.*, New York, pp. 545-549.
3. Dooge, J.C.I. and Harley, B.M. (1967) Linear routing in uniform open channels, *Proc. Int. Hydrol. Symp. Fort Collins, Co.*, Sept.1967, Paper No.8, 1, 57-63.
4. Dooge, J.C.I., Napiorkowski J.J., and Strupczewski, W.G. (1987 a) The linear downstream response of a generalized uniform channel, *Acta Geoph.Pol.* **35**, 277-291.
5. Dooge, J.C.I., Napiorkowski, J.J.,and Strupczewski, W.G. (1987 b) Properties of the general downstream channel response, *Acta Geoph.Pol.* **35**, 405-418.
6. Singh, V.P. (1996) *Kinematic Wave Modeling in Water Resources: Surface Water Hydrology*, John Wiley & Sons, New York.
7. Hayami, S. (1951) On the propagation of flood waves, *Kyoto Univ. Japan, Disaster Prevention Res. Inst. Bull.* **1**, 1-16.
8. Dooge, J.C.I. (1973) *Linear Theory of Hydrologic Systems, Tech. Bull.*, 1468, Agricultural Research Service, Washington.
9. Strupczewski, W.G. and Napiorkowski, J.J. (1989) Properties of the distributed Muskingum model, *Acta Geoph. Pol.* **V.XXXVII, 3-4**, 299-314.

10. Moore, R.J. and Clarke, R.T. (1983) A distributed function approach to modeling basin sediment yield, *J. Hydrol.* **65**, 239-257.
11. Moore, R.J. (1984) A dynamic model of basin sediment yield, *Water Resour. Res.* **20 (1)**, 89-103.
12. Cox, D.R. and Miller, H.D. (1965) *The Theory of Stochastic Processes*, Chapman and Hall, London.
13. Tweedie, M.C.K. (1957) Statistical properties of the inverse Gaussian distributions, *I. Ann. Math. Stat.* **28**, 362-377.
14. Johnston, N.L. and Kotz, S. (1970) *Distribution in Statistics: Continuous Univariate Distributions 1.*, Houghton-Mifflin, Boston, Mass.
15. Folks, J.L. and Chhikara, R.S. (1978) The inverse Gaussian distribution and its statistical application – a review, *J.R.Stat. Soc. Ser. B.* **40(3)**, 263-289.
16. Strupczewski, W.G., Singh, V.P., and Weglarczyk, S. (2001) Impulse response of linear diffusion analogy as a flood probability density function, *Hydrol. Sc. J.* **46(5)**, 761-780.
17. Strupczewski, W.G. and Napiorkowski, J.J. (1990 a) Linear flood routing model for rapid flow, *Hydrol. Sc. J.*, **35, 1, 2**, 49-64.
18. Lighthill, M. H. and Witham, G.B. (1955) On kinematic waves. I. Flood movements in long rivers, *Proc. R. Soc., London, Ser. A* **229**, 281-316.
19. Strupczewski, W.G., Napiorkowski, J.J., and Dooge, J.C.I. (1989) The distributed Muskingum model, *J. of Hydrol.* **111**, 235-257.
20. Strupczewski, W.G. and Napiorkowski, J.J. (1990 b) What is the distributed Muskingum model?, *Hydrol. Sc. J.* **35, 1, 2**, 65-78.
21. Einstein H.A. (1942) *Formulas for the Transportation of Bed Load*, Transactions ASCE, Paper No.2140, pp.561-597.
22. Eagleson, P.S. (1978) Climate, soil and vegetation. 2. The distribution of annual precipitation derived from observed storm sequences, *Water Resour. Res.* **14(5)**, 713-721.
23. Scott, E.J. (1955) *Transform Calculus with an Introduction to Complex Variables*, Harper & Brothers, New York, NY, pp. 71-73.
24. Özisik, M.N. (1968) *Boundary Value Problems of Heat Conduction*, International Textbook Co., Scranton, PA, pp. 48-79.
25. Cleary, R.W. and Adrian, D.D. (1973) New Analytical Solutions for Dye Diffusion Equations, *Journal of the Environmental Engineering Division, ASCE* **99 (EE3)**, 213-227.
26. Jennings, M.E. and Benson M.A. (1969) Frequency curve for annual flood series with some zero events or incomplete data, *Water Resour. Res.* **5(1)**, 276-280.
27. Woo, M.K. and Wu, K. (1989) Fitting annual floods with zero flows, *Can. Water Resour. J.* **14(2)**, 10-16.
28. Wang, S.X. and Singh, V.P. (1995) Frequency estimation for hydrological samples with zero value, *J. of Water Resour. Planning and Management, ASCE* **121(1)**, 98-108.
29. Rao, A.R. and Hamed, K.H. (2000) *Flood Frequency Analysis*, CRC Press, Boca Raton, Florida.
30. Freeze, R.A. (1975) A stochastic-conceptual analysis of one-dimensional groundwater flow in nonuniform homogenous media, *Water Resour. Res.* **11(5)**, 725-741.
31. Chow, V.T. (1954) The log-probability law and its engineering applications, *Proc. Am. Soc. Civ. Eng.* **80**, 1-25.
32. Kuczera, G. (1982) Robust flood frequency models, *Water Resour. Res.* **18(2)**, 315-324.
33. Strupczewski, W.G., Singh, V.P., and Weglarczyk, S. (2002 a) Asymptotic bias of estimation methods caused by the assumption of false probability distribution, *J. of Hydrol.* **258**, 122-148.
34. Kendall, M.G. and Stuart, A. (1969) *The Advanced Theory of Statistics, V.1. Distribution Theory*, Charles Griffin&Comp. Limited, London.
35. Strupczewski, W.G., Singh, V.P., and Weglarczyk, S. (2002 b) Physically based model of discontinuous distribution for hydrological samples with zero values, in Singh, Al-Rashed & Sherif (eds.), *Proc. of the Int. Conf. on Water Resources Management in Arid Regions*, A.A. Balkema Publishers, Swets & Zeitlinger, Lisse, 523-537.

ASSESSMENT OF OUTLIERS IN STATISTICAL DATA ANALYSIS

B. ONOZ and B. OGUZ
Istanbul Technical University
Faculty of Civil Engineering
80626 Istanbul, Turkey

Abstract. The first step in any statistical data analysis is to check whether the data are appropriate for the analysis. In such analyses, the presence of outliers appears as an unavoidable important problem. Thus, in order to manage the data properly, outliers must be defined and treated. Tests for outliers are well established in the statistical literature. Methods for the processing of outliers take on an entirely relative form, that is, relative to the basic model so that an examination of the outlier allows a more appropriate model to be formulated. One of the two extreme choices in the analysis of outliers is either to reject them with the risk of loss of genuine information or to include them with the risk of contamination. Outliers can be treated in four possible ways: they can be rejected as erroneous, identified as important, tolerated within the analysis, or incorporated into the analysis. This treatment can be performed by two different approaches. One of these approaches is called the *accommodation* procedures, which make use of 'robust' methods of inference, employing all the data but minimizing the influence of any outliers. The second type of statistical method for handling outliers is *discordancy* procedures, namely those of 'testing' an outlier with the prospect of rejecting it from the data set or of 'identifying' it as a feature of special interest. It is possible to perform outlier analysis not only for univariate cases but also for multivariate cases. In this study, different techniques for the statistical analysis of outliers are considered, and some examples are given.

1. Introduction

In Barnett and Lewis [1], an outlier in a set of data is defined as "an observation (or subset of observations) which appears to be inconsistent with the remainder of that set of data". The main point in this description is the "appearance" of an observation to be "inconsistent". This is an informal and subjective judgment. It is important to discriminate whether such an observation is a genuine member of the main population, or whether it is a contaminant arising from a different population. Contaminants may lead us to make incorrect inferences about the original population. Sometimes, the existence of contaminants is not evident; thus, in such cases, the inference process is not significantly distorted. When there are observations strange to the main population, the attempt to represent the population becomes difficult. Such observations can greatly distort estimates (or tests) of parameters in the basic model for the population.

Before starting to process the principal mass of data, the data set must be examined for the presence of outliers. The question is what the reaction towards outliers should be. Should they be rejected, adjusted, or remain unaltered? Clearly, the answer depends on the form of the population; therefore, methods used in the processing of outliers become totally relative to the basic model. The rejection, or the accommodation by special treatment of the apparently unreasonable observations, for which no tangible explanation can be found, is invalid.

The question whether an observation is an outlier or not is related to the selected distribution. For example, an observation which stands out as an outlier when the normal distribution is accepted may not be an outlier if log-normal distribution were selected. Accordingly, an examination of the outlier allows a more appropriate model to be formulated, or it enables one to assess any danger that may arise from basing inferences on the normality assumption.

Reed [2] interprets Barnett and Lewis' suggestion for four possible treatments of outliers as follows. Outliers can be:

- Rejected as erroneous. This is appropriate if the outlier results from a mistaken reading, but inappropriate if it is simply poorly measured;
- Identified as important. This treatment may be appropriate where a series includes several outliers. If these are held to be the most relevant to the problem, the analyst concentrates on the outliers and pays less attention to the main data;
- Tolerated within the analysis. This is called "accommodating" the outlier. This tactic assumes that the procedure is robust against outliers, i.e., that the analysis will not be seriously distorted by their presence;
- Incorporated into the analysis. In this case, the analysis defers to the outlier. The procedure is varied in a manner that is consistent with the outlier being not as extreme as implied by the standard analysis. For example, a different distribution or fitting method may be specially adopted.

The treatment of outliers can be performed by two different approaches. One of these approaches is named as *accommodation* procedures, which make use of 'robust' methods of inference, employing all the data but minimizing the influence of any outliers. The second type of statistical method for handling outliers is *discordancy* procedures, namely that of 'testing' an outlier with the prospect of rejecting it from the data set, or of 'identifying' it as a feature of special interest. The subjective basis of the outlier concept and the long history of its study have tended in the past to encourage an informal attitude to 'outlier rejection'.

Outstanding observations can be categorized as extreme observations, outliers or contaminants; outliers may or may not be contaminants; contaminants may or may not be outliers. There is no way of knowing whether or not an observation is a contaminant. Knowing that outliers are possible manifestations of contamination, attention will be focused on the analysis of outliers.

In this paper, different techniques for dealing with outliers will be considered, and the difficulties encountered in their application will be mentioned. Two case studies, a univariate flood analysis and a multivariate low flow example, will be discussed.

2. Methods of Accommodation

In this section, some standard methods which implicitly provide certain protection against outliers will be mentioned. Though the primary objective of the following methods is to decrease the effect of the tail behavior of the distribution on estimation of the mean, these methods also serve as a protection against the presence of outliers. Here, two simple methods of accommodation will be given, namely trimming and Winsorizing. The aim is to control the effect of r lowest $x_{(1)},\ldots,x_{(r)}$ and s highest $x_{(n-s+1)},\ldots,x_{(n)}$ values within a sample of magnitude n. If these $r+s$ observations are not included in the assessment, the trimmed mean can be calculated as follows:

$$x_{r,s}^{T} = \left(x_{(r+1)} + \ldots + x_{(n-s)}\right)/(n-r-s) \tag{1}$$

On the other hand, if the r lowest and s highest observations are not omitted but replaced by the neighboring observation, this technique is called Winsorization. The (r,s)-fold Winsorized mean can be calculated by the following equation:

$$x_{r,s}^{w} = \left(r.x_{(r+1)} + x_{(r+1)} + \ldots + x_{(n-s)} + s.x_{(n-s)}\right)/n \tag{2}$$

In case r is taken equal to s, then the estimation of location is referred to as r-fold symmetrically trimmed and Winsorized means, $x_{r,r}^{T}$ and $x_{r,r}^{w}$. It is natural to apply a symmetrical trimming or Winsorization if the underlying distribution is symmetrical or vice versa.

Instead of considering the number of observations r and s, the ratio of the number of omitted observations to the total number of observations can be used as α. α being selected, the number of α n observations supposed to be trimmed at each end may not be an integer. Suppose its integer part is r, so that $\alpha\, n = r + f\ (0 \le f \le 1)$. Then, the r observations at each end are omitted and the nearest retained observations, $x_{(r+1)}$ and $x_{(n-r)}$ are included with reduced weight $1\text{-}f$. The α-trimmed mean can be expressed as:

$$x^{T}(\alpha,\alpha) = \frac{\left((1-f)x_{(r+1)} + x_{(r+2)} + \ldots + x_{(n-r-1)} + (1-f)x_{(n-r)}\right)}{n(1-2\alpha)} \tag{3}$$

Since there is no need for any fractional weighting for α-Winsorized means $x_{(\alpha,\alpha)}$, its expression can be given as follows:

$$x_{(\alpha,\alpha)}^{w} = \left(r.x_{(r+1)} + x_{(r+1)} + \ldots + x_{(n-r)} + r.x_{(n-r)}\right)/n \tag{4}$$

Clearly, the 0-trimmed and 0-Winsorized means are both the same as the sample mean $\bar{x}$, and the ½ -Winsorized mean is the same as the sample median $\tilde{x}$. The ½ -trimmed mean, by a suitable limiting argument, can also be taken to be $\tilde{x}$. The ¼ -trimmed mean, $x^{T}(¼, ¼)$, is called the mid-mean. In literature, it is possible to find several altered trimming and Winsorization techniques, such as modified trimming, modified Winsorization, and semi-Winsorization.

It is well-known that, in statistical analyses, the dispersion parameter is the second important parameter following the location parameter. A robust estimation of dispersion can be derived by using a suitably trimmed or Winsorized sample. An example of such an estimation of dispersion, i.e., a Winsorized variance, can be given as follows:

$$w_{s^2_{r,r}} = \frac{1}{n-1}\left\{r\left(x_{(r+1)} - m\right)^2 + \sum_{j=r+1}^{n-r}\left(x_{(j)} - m\right)^2 + r\left(x_{(n-r)} - m\right)^2\right\} \quad (5)$$

In the above equation, m is the (r,r) Winsorized mean, $x_{r,r}$.

Another method for the robust estimation of dispersion may be the L-estimation procedure:

$$s_L = \sum_{i=1}^{n} b_i . x_{(i)} \quad (6)$$

In the above equation, weights b_i must be suitably chosen for robust estimation against outliers. For this purpose, one would seek weights b_i to be lower in the extremes than in the body of the data set.The semi-interquartile range Q is a simple and useful L-estimator of dispersion, where q_1, q_2 are the nearest integers to (n/4, 3n/4), respectively [1]:

$$Q = 1/2\left\{x_{(q_2)} - x_{(q_1)}\right\} \quad (7)$$

3. Discordancy Measures

In literature, discordancy tests are often referred to as tests for the rejection of outliers; but, as it is mentioned before, rejection is not the only way to be adopted if an observation is detected as foreign to the main data set. For the measurement of discordancy, a simple and widely used statistic of the form N/D is used. For example, for testing of the largest value in a univariate ordered sample of size n, N is the measure of separation of $x_{(n)}$ from the remainder of the sample, and D is a measure of the spread of the sample. In discordancy analysis, there is a wide choice of test statistics. For most outlier examinations, it is not possible to obtain globally uniform powerful tests. Here, the problem is to decide on which tests should be preferred in any particular situation and how they can be constructed. How the significance level will be obtained and how the performance tests will be assessed must be determined. The following four aspects must be considered for discordancy analysis: the basic model and the assessment of significance; the alternative (outlier-generating) hypothesis; the assessment of the test performance and the power concept; and the desirable properties a discordancy test should have.

There are numerous common types of test statistics in discordancy analyses. Some of these are more appropriate than others under different situations, e.g., in examining a single upper outlier, or two lower outliers, and so on. Seven common types of statistics are: excess/spread statistics, range/spread statistics, deviation/spread statistics, sums of squares statistics, extreme/location statistics, higher-order moment statistics and W-statistics. In literature, there are several discordancy tests using the above mentioned test statistics for particular samples such as Gamma, exponential, normal, truncated exponential, Poisson, Binomial etc. and particular cases such as a single upper outlier, single lower outlier, upper outliers, lower outliers, a lower and upper outlier pair, and so on [1].

The handling of outliers for multivariate data is more complicated than it is for univariate data. Incorrect data values, outliers, trends and specific shifts in the mean of the sample can all be reflected in the L-moments of the sample. Hosking and Wallis [3] found that useful information can be obtained by comparing the sample L-moment ratios for different variables in the multivariate analysis. Given a multivariate data set, the aim is to identify as discordant the variables which stand aside with the group as a whole.

In hydrology, the following method has been adopted by Hosking and Wallis [3] for regional analysis. They have proposed a measure of the discordancy between the L-moment ratios of a site and the average L-moment ratios of a group of similar sites. Considering the sample L-moment ratios (L-CV (t), L-skewness (t_3), L-kurtosis (t_4)) of a site as a point in three-dimensional space, a group of sites will yield a cloud of such points. Any point with a large distance measured from the center of the cloud will be flagged as discordant. The large distance measure is interpreted in such a way as to allow for correlation between the sample L-moment ratios.

The necessary formal definition of the above mentioned method is as follows. If, for example, there are N sites in the group, $u_i = \left[t^{(i)}\ t_3^{(i)}\ t_4^{(i)}\right]^T$ will be a vector containing the t, t_3 and t_4 values for site i with:

$$\bar{u} = N^{-1} \sum_{i=1}^{n} u_i \tag{8}$$

where $\bar{u}$ is the unweighted group average. The sample covariance matrix is defined as:

$$S = (N-1)^{-1} \sum_{i=1}^{n} (u_i - \bar{u})(u_i - \bar{u})^T \tag{9}$$

The discordancy measure is defined for site i in the following way:

$$D_i = (1/3)\,(u_i - \bar{u})\, S^{-1} (u_i - \bar{u}) \tag{10}$$

The site i for which D_i is large is declared as discordant. The definition of "large" depends on the number of sites in the group. It is suggested that a site be flagged as discordant if its D_i value exceeds the critical value at a significance level of 10% given in Table 1. The critical value of max D_i has been given by a test statistic derived by Wilks. An approximate critical value of max D_i for a significance level α is $(N-1)\ Z\ /\ (N-4+3Z)$, where Z is the upper $100\alpha/N$ percentage point of an Fischer distribution with 3 and N-4 degrees of freedom. If N is equal or less than 3, D_i cannot be calculated.

The methodology described above is applied to a hydrologic problem in Gediz River Basin in Western Anatolia, along the coast of the Aegean Sea [4]. The total drainage area of the basin is about 18,000 km^2. There are 12 flow gaging stations, one of which is affected by the regulation of the reservoirs upstream and is therefore not considered in this study.The discordancy analysis is performed to flag as discordant the sites where the data stand out from those of the other sites. The data used are 7-day annual average minimum flows. Three of these have more than 60% zeros and are not included in the regional analysis. Data of the remaining 8 stations are used in the regional analysis. Record lengths vary from N=17 to 33 years. Discordancy is measured in terms of the L-moment (L CV, L skewness and L kurtosis) of the data at these sites. Discordancy measure D_i is computed at each site (Table 2). All D_i values are smaller than 2.14, which is the critical value for 8 sites (Table 1), and cannot be rejected as discordant outliers by the criterion for multivariate discordancy analysis.

TABLE 1. Critical values for the discordancy test statistic D_i [3]

Number of sites in region	Critical value	Number of sites region	Critical value
5	1.333	10	2.491
6	1.648	11	2.632
7	1.917	12	2.757
8	2.140	13	2.869
9	2.329	14	2.971
		≥15	3

TABLE 2. Characteristics of flow gaging stations

N(years)	Station no	A(km^2)	D_i
17	501	5675	1.18
33	509	902	1.25
24	510	3185	0.22
30	514	690	0.95
30	515	740	0.96
25	523	3272	1.37
22	524	176	0.98
19	525	64	0.92

4. Water Resources Council Method

US federal agencies have recommended some procedures for flood frequency analysis in Bulletin 17 B issued by the Water Resources Council [5].

Bulletin 17 B defines outliers to be "data points which depart significantly from the trend of the remaining data" [6]. The original Bulletin 17 recommends the log Pearson type 3 distribution for analysis of flood frequency. In this method, the thresholds of high and low outliers in log space are defined as:

$$\bar{Y} \pm K_n S_y \tag{11}$$

where y and s_y are the mean and the standard deviation of logarithms of the flood peaks. K_n is a critical value for sample size n. K_n can be computed by using the following equation:

$$K_n = -0.9043 + 3.345\sqrt{log(n)} - 0.4046\, log(n) \qquad (12)$$

Thus Eq. (11) is a one-sided outlier test at 10% significance level [7].

As an example, this method has been applied to the 27-year long flood data of flow gaging station number 601 on K. Menderes River in Turkey. The mean, standard deviation and the coefficient of skewness of the data in log space are respectively:

$$\bar{Y} = 2.146; \quad S_y = 0.250; \quad C_s = -0.082 \qquad (13)$$

By Eq. (11), y_{high} = 2.776, which gives the threshold value for high outliers as 597 m^3/sec. The largest record value with a magnitude of 693 m^3/sec remaining significantly above this threshold is detected as a high outlier. After the elimination of the high outlier, the parameters are found as $\bar{Y} = 2.119$, $S_y = 0.212$ and $C_s = -1.252$. Similarly, the y_{low} value and the corresponding threshold value for low outliers are obtained as 1.516 and 32.8 m^3/sec, respectively. Since the lowest value of the record (36 m^3/sec) remains above the threshold value, no low outliers are detected in this example.

A flood frequency analysis has been performed for these data for the two cases, retaining the detected outlier and rejecting the detected outlier. This analysis was made by using the log Pearson type 3 distribution as is recommended by Bulletin 17. Flood estimates at various return periods T = 5, 10, 25, 50 and 100 years were made for the two cases, and the results are given in Table 3. In Table 3, Q_T and Q'_T stand for the estimated flood discharges for the complete data case and the high outlier rejected case, respectively.

TABLE 3. Results of frequency analysis using the Water Resources Council method

Return Period	K_T	K'_T	Q_T (m^3/sec.)	Q'_T (m^3/sec.)
5	0.8453	0.84105	228	198
10	1.27238	1.07547	291	222
25	1.72208	1.26127	377	243
50	2.00945	1.35171	445	255
100	2.26586	1.41605	516	263

Inspection of Table 3 shows that, as the return period is increased, the estimated discharge decreases. After the elimination of the high outlier, the standard deviation reduces, and the distribution becomes even more negatively skewed. Together with negative skewness, K_T decreases and, related to this, the flood discharge also decreases. However, in case the standard deviation becomes very small due to the elimination of the low outlier, it is possible to detect a decrease in the discharges.

Since the Water Resources Council Method adopts the logarithmic transformation of data, the advantages and disadvantages of this kind of transformation may be discussed.

On one hand, logarithmic transformation is a reasonable tool for normalizing values and for occasionally preventing the domination of large values on the calculation of product-moment estimators. On the other hand, logarithmic transformations have the danger of giving greatly increased weights to low outliers [7].

5. Conclusion

Before starting the statistical analysis of any data, the data must be checked to see whether they are appropriate for the hydrologic analysis. In fact, the search for outliers may be considered as a search for the homogeneity of data. No matter whether the data are univariate or multivariate, the quest for and the treatment of the outliers lead us towards more homogeneous data. Among the two approaches employed for the treatment of outliers, accommodation procedures make use of 'robust' methods of inference, employing all the data but minimizing the influence of any outliers. In other words, robust models tolerate the existing outliers more broadly. Thus, accommodation techniques retain the outstanding data instead of eliminating them. Yet, the alternative technique known as the discordancy procedures are used for the identification and rejection of outliers. These methods reject the outlier after identification instead of keeping it within the data series. The recent trend is to prefer accommodation techniques rather than discordancy techniques to retain the complete data set.

6. References

1. Barnett, V. and Lewis, T. (1994) *Outliers in Statistical Data,* John Wiley & Sons, Chichester.
2. Reed, D. (1999) *Flood Estimation Hand Book,* Volume 1, Wallingford.
3. Hosking, J.R.M. and Wallis, J.R. (1997) *Regional Frequency Analysis*, Cambridge University Press.
4. Onoz, B., Bayazıt, M., and Oguz, B. (2000) Methodology for regional low flow analysis and an application, Hydrology of the Mediterranean Regions, Montpellier, France.
5. United States Water Resources Council (1976) *Guidelines for Determining Flood Flow Frequency*, Washington, D.C.
6. Chow, V.T., Maidment, D.V., and Mays, L.W. (1988) *Applied Hydrology*, McGraw Hill, Singapore.
7. Stedinger, J.R., Vogel, R.M., and Foufoula-Georgiou, E. (1992) *Handbook of Hydrology,* Frequency Analysis of Extreme Events, Chapter 18, Editor in Chief, Maidment, D.V., McGraw-Hill, USA.

Part V

PHYSICAL SAMPLING AND PRESENTATION OF DATA

MODERN DATA TYPES FOR ENVIRONMENTAL MONITORING AND WATER RESOURCES MANAGEMENT

G. A. SCHULTZ
Ruhr University Bochum
Institute of Hydrology, Water Resources Management and Environmental Techniques
D-44780 Bochum, Germany

Abstract. In the field of water resources management, we observe presently a significant change of philosophy. The introduction of the principle of sustainable development by the United Nations and the principle of integrated river basin management postulated by the European Union play a major role in this context. Thus, new planning tools have to be developed, which are more sophisticated than those hitherto and, therefore, require a much better data basis than presently available. This paper deals with new data types partly available already now and partly expected to be developed in the medium range future. Furthermore, it is shown how the integration of remote sensing and other data leads to new types of information, allowing integrated planning of water resources systems in a more suitable way. The potential of real time data is highlighted, particularly in the context of real time operation of water resources systems, especially for flood control in rural and urban areas. Also, the potential of large-scale data schemes in the context of regional and even continental water management schemes is discussed.

1. Introduction

Hydrologic processes, e.g., precipitation, infiltration, soil water storage, evaporation, transpiration, etc., are processes which occur on and below the earth's surface area. In addition, water management activities like integrated river basin management, as required by the new "European Union Water Framework Directive", are supposed to be carried out in consideration of the whole drainage basin area. The description of such processes and activities, as well as mathematical modeling thereof, would be necessary to be carried out with the aid of areal information. The present status of measurement techniques allows, in principle, only point measurements. Transformation of these point measurements into areal information can be presently done only with rather high inaccuracies, even if relatively dense networks exist. This fact causes, not only uncertain knowledge about hydrologic processes in river basins, but it also leads to deficits in the efficiency of water management, in planning as well as in real-time operation of water resources systems. These problems lead to the necessity of collecting areal information instead of point measurements. Following the title of this paper, only the acquisition and the use of areal hydrometeorological data are considered here as "modern data types". Thus, this paper discusses mainly the use of digital thematic maps, digital terrain models, and remote sensing data within the framework of geographic information systems (GIS). In modern environmental monitoring, as well as in information production for water resources management purposes, the data coming from different sources are not used separately, but all these different data types have to be combined so that they can be used in an integrated way, as required by modern mathematical hydrologic or water management models.

In this paper, various types of areal data obtained from rather different sensors or other information sources are discussed, and then various examples are given, demonstrating their integrated use with the aid of integrated technologies within a GIS framework.

"Integrated", in the context of river basin management, means also the integration of information from different disciplines, e.g., technical disciplines (like hydrology and water management), ecology, economy and social sciences. These types of integrated techniques are also discussed in the following sections.

2. Monitoring and Modeling in Hydrology

While most conventional hydrologic catchment models are single input - single output models of the black-box type, the new data types allow development and application of distributed system models with high resolution in space. This high resolution is relevant for both the hydrologic model parameters as well as the model input [1]. While model parameters are mostly based on catchment characteristics, the model input is usually an external forcing, e.g., rainfall, snowmelt, temperature, radiation, etc. Besides catchment characteristics (forming model parameters), the model input information often comes from remote sensing sources, too. A further feature of these new models, based on the new types of data, is the fact that it now becomes feasible to use time varying model parameters, while the model parameters in conventional models do not change over time. An example for this situation is given in Section 3.2.

2.1. INTEGRATION OF REMOTE SENSING AND OTHER DATA FOR ESTIMATION OF SOIL STORAGE CAPACITY

The coupling of remote sensing data from various sources for hydrologic modeling with other relevant information is of increasing importance. Model parameters are often estimated with the aid of satellite data (e.g., land use classification from Landsat or Spot and vegetation indices from NOAA-AVHRR), while model input can be derived, e.g., from ground based weather radar (precipitation) or from satellite data combined with airborne information (snowmelt). Often, remote sensing data are combined with other information like Digital Elevation Models (DEM), digital maps, or field observations, thus forming an integrated data set.

One problem in the application of gridded data (RS, DEM, digital maps) with rather high resolution in space lies in the necessity to aggregate pixels (picture elements) of similar hydrologic behavior into the so-called Hydrological Response Units (HRU), Representative Elementary Areas (REA), or Hydrological Similar Units (HSU) in order to avoid excessive computation time in the rather complex, distributed models. This is true particularly if such models are applied to larger catchment areas. At the author's institute, one such aggregation technique was developed, which takes care of spatial heterogeneity of a drainage basin in a specific way. The integration of remote sensing data with a digital map is used to define the area distribution of soil storage capacity within a drainage basin [2]. Figure 1 shows this principle in detail. From Landsat data, land use classification is derived (top left map), which allows the estimation of root depth. A digital soil map (top right map) is used to derive information on soil porosity, which is merged with the root depth in order to produce soil storage capacities (bottom left map). This diagram shows that the soil storage capacity varies strongly across the catchment area (basis: 50 x 50 m pixels). This information is highly important for hydrologic modeling. It is not necessary to work with a very high resolution in space of Landsat data (30x 30 m) since, in modeling, one is mainly interested in homogeneous areas of similar behavior. Therefore, pixels of similar soil storage capacity must be aggregated to larger area elements in order to keep the computation efforts for modeling within reasonable limits. In the mentioned case, these larger area elements for computational purposes are generated on the basis of the entropy concept. In Fig. 1, the bottom right diagram shows the application of this concept for the aggregation of pixels in a way suitable for a distributed model where the soil storage capacity plays an important role. The computations are then carried out for each such area unit within the hydrologic model. This procedure is particularly useful for modeling flood hydrographs within a river basin.

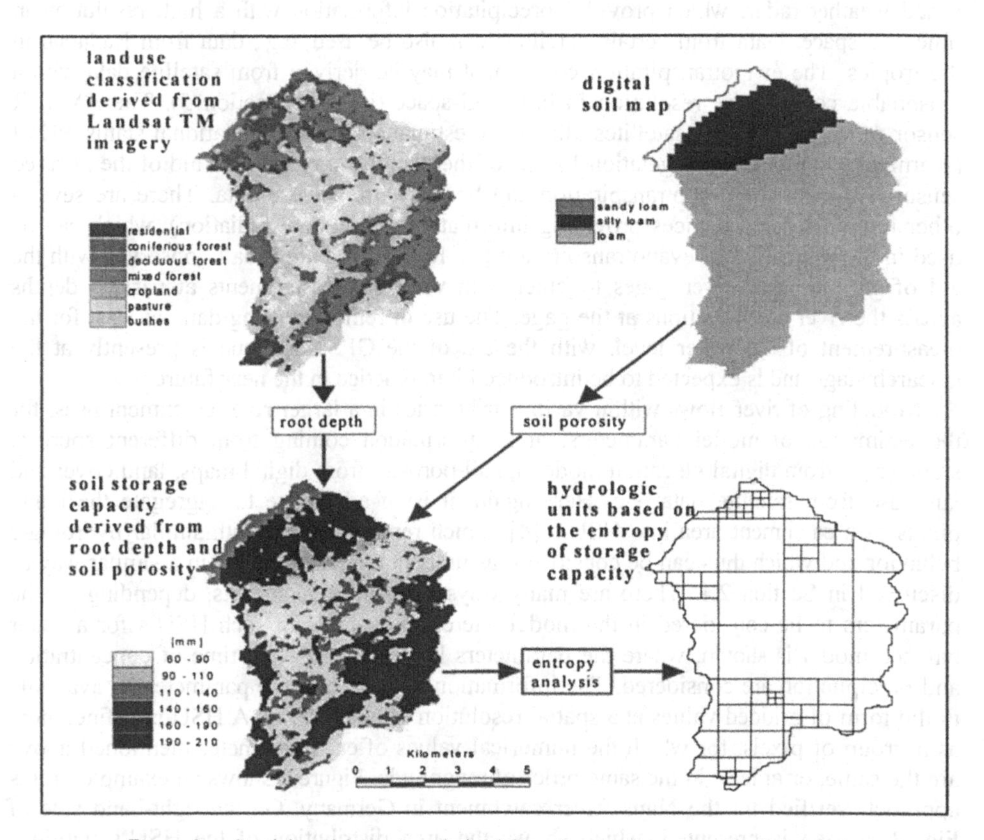

Figure 1. Spatial model structure derived from Landsat imagery and a digital soil map (Pruem River Basin, Germany, 53 km^2)

2.2. WATER BALANCE MODELING

In water balance modeling, the whole terrestrial part of the water cycle is represented in a mathematical form. According to the conventional water balance equation, the water cycle comprises the hydrologic components of precipitation, runoff, evapotranspiration and soil water storage changes. The water balance equation in its conventional form is:

$$P(t) = E_t(t) + Q(t) + dS(t)/dt \quad (1)$$

with: $P(t)$ = precipitation, $E_t(t)$ = evapotranspiration, $Q(t)$ = runoff, $dS(t)/dt$ = soil water storage change in time, t = time.

For the mathematical simulation of each component of the hydrologic cycle, various areal data are required, usually coming from various different sources. This means that they have to be integrated. The overall computation of the water balance then requires the integration of all these components into one equation. Data from numerous different sources have to be integrated for the required information production.

The precipitation component can be obtained from conventional precipitation observations with the aid of some geo-statistical techniques or with the aid of ground based weather radar, which provides precipitation information with a high resolution in time and space. Data from certain satellites can also be used, e.g., data from Radarsat in the tropics. The evapotranspiration component may be derived from satellite data with a reasonable compromise resolution in time and space (see also Section 5). The AVHRR sensor of various NOAA satellites allows the estimation of the vegetational status NDVI (Normalized Difference Vegetation Index) of the plant canopy with the aid of the infrared sensor. Estimates of evapotranspiration can be made from these data. There are several other remote sensing devices providing information (e.g., about radiation), which can be used in the estimation of evapotranspiration [3]. Runoff data are usually observed with the aid of ground-based river gages together with velocity measurements at various depths across the river cross-sections at the gage. The use of remote sensing data, at least for the measurement of the water level, with the aid of the GPS-technique is presently at the research stage and is expected to be introduced into practice in the near future.

Modeling of river flows within various tributaries in a larger river catchment uses, for the estimation of model parameters, areal information coming from different sources: slope (e.g., from digital elevation models), soil porosity from digital maps, land cover and land use from satellite data, etc. Here again, it is of advantage to aggregate the many pixels of a catchment area into HSU's [4], which represent areas with similar hydrologic behavior and which thus can be considered as units in modeling efforts in a similar way as discussed in Section 2.1. There are many ways to identify such units, depending on the parameters to be considered in the model. Here, an example of such HSU's for a water balance model is shown, where the parameters land use, soil type, time of concentration and precipitation are considered. The information on each of these parameters is available in the form of gridded values at a spatial resolution of 50 by 50 m. A HSU is defined here as a group of pixels, for which the numerical values of each parameter mentioned above are the same, or at least in the same order of magnitude. Figure 2 shows an example of this approach verified for the Nims River catchment in Germany. On the right-hand side of Fig. 2, a map is presented, which shows the area distribution of the HSU's within a drainage basin. Altogether 293 combinations of parameter value groups (forming HSU's) were identified, and they are shown on the map in different shades. The distributed hydrologic model then uses the HSU's as area elements for which the hydrologic model computations are carried out separately. The advantage of such a technique is not so much an improvement of model accuracy; but it rather provides the potential to compute runoff at points along the river network where no gages exist, and the possibility to quantify the impact of potential future land use changes of planned hydraulic structures on the hydrologic regime (see Section 3.2).

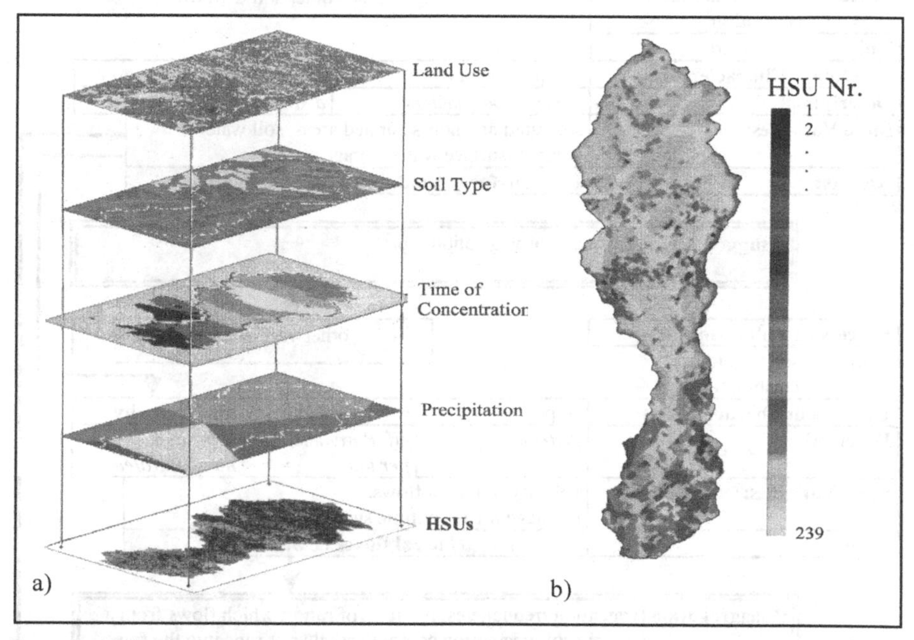

Figure 2. Derivation of Hydrologically Similar Units (HSU's) by overlay operation: (a) the principle; (b) generated (HSU's) in the Nims River catchment (267 km^2, Mosel tributary), Germany [5].

2.3. RAINFALL-RUNOFF MODELS

Rainfall-runoff models are distinguished from water balance models not only by their purpose, but also by a difference in time scale. Rainfall-runoff models, especially in urban areas, have to work with very small time increments of a few minutes; and the whole flood often does not last much longer than a few hours or a few days. On the other hand, water balance models deal with long time increments of e.g., a day, considering long-term processes occurring over years. This difference in time scale requires different model structures. Here, a conceptual rainfall-runoff model is presented, which is valid for the spatial meso-scale considering statistical descriptions of distributed catchment characteristics to account for spatial heterogeneity [2]. The model contains 3 semi-distributed modules. These 3 modules are combined to form a hydrologic model including feedback components between surface flow and infiltration and between subsurface return flow and surface flow in saturated areas. Figure 3 shows the structure of this model and its 3 components. For each of these components, a spatial differentiation with regard to one physical characteristic, which is seen as the most relevant for a specific hydrologic process, is combined with a lumped description of the heterogeneity of other characteristics which are also important for this process.

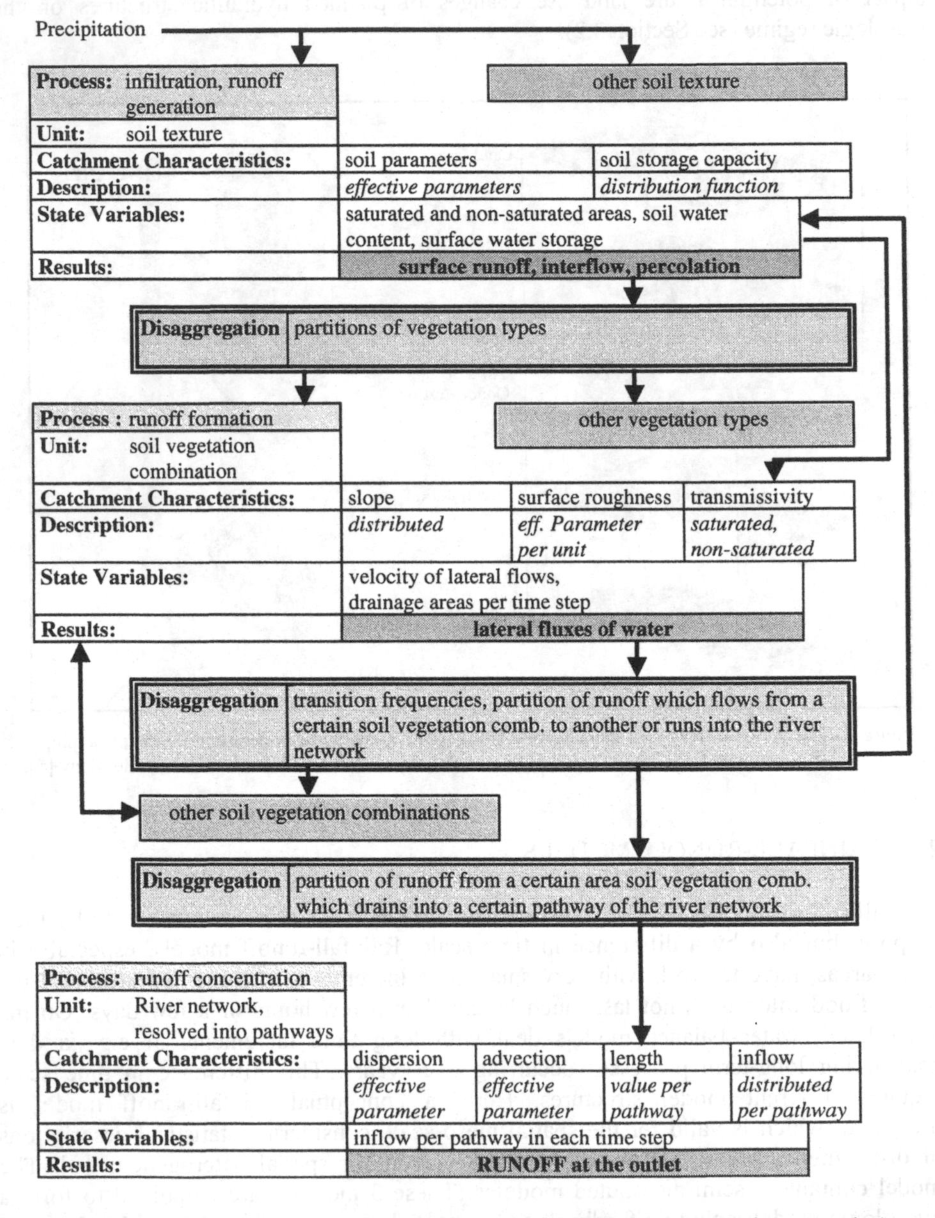

*Figure 3.*Coupling three process modules (runoff generation, runoff formation and concentration) to form a rainfall-runoff model.

The model is based on a GIS with a data resolution of 50 m x 50 m, in which the following catchment characteristics are stored: soil type, land cover, slope, elevation, and drainage network estimated from the digital elevation model. The main advantage of this approach consists in the reduction of the number of conceptual parameters. The model structure is based on the classical approach of 3 different hydrologic processes represented as modules: runoff generation (vertical hillslope processes) (upper part of Fig. 3), runoff formation of hillslopes (lateral hillslope processes), presented in the central part of Fig. 3, and runoff concentration within the river network presented in the bottom part of Fig. 3. Within the flow chart of Fig. 3, many parameters relevant for the model are given. Many of these parameters are obtained from remote sensing data, digital maps, and digital elevation models. Thus, it can be seen from Fig. 3 that modern model types require a lot of information distributed in space, and they are integrated in the models in order to allow an efficient computation of runoff hydrographs.

2.4. NEW MODELS WITH SOIL MOISTURE AS THE CENTRAL COMPONENT

Hydrologic models of both the water balance type and the rainfall-runoff type are structured such that precipitation (or snowmelt) serves as model input and one or several other parameters, preferably runoff, represent the model output. Evapotranspiration is modeled and supported with point measurements on the ground. However, one of the most important parameters, namely, the soil water storage component (Eq. (1)), is usually not modeled at all and is considered as a residual. This is rather unfortunate since, in this model component, not only the actual soil water storage is represented, but also all errors of the other components are included due to the fact that it is usually not measured and, thus, the errors cannot be identified. If, in the future, it becomes possible to measure soil moisture directly, not only at a few points but rather as distributed over the basin area, this would allow the development of completely new model structures, where soil moisture and soil water storage would form a central component and would be supported by measurements. Such models would be more efficient and less erroneous than the present ones.

They require, however, areal soil moisture measurements. This can be presently done by remote sensing with airborne sensors. Figure 4 shows a sequence of soil moisture measurements at 8 consecutive days in the catchment area “Little Washita” in Oklahoma, USA. This time sequence represents a drying process in the catchment after strong rainfall at the beginning. The soil moisture values were derived from data of the ESTAR sensor (Electronically Steered Thinned Area Radiometer), which was flown on a NASA airplane. The spectral band relevant here is passive microwave (PMW) in the L-band range. It can clearly be seen how the soil becomes dryer from day to day. These promising images can presently be taken, however, only from airplanes, and one should not forget that the accurate determination of soil moisture via remote sensing is still a field in the state of research. It can be hoped, however, that, in the not too distant future, such PMW sensors will be flown on satellites, allowing the use of their data in river basins all over the world.

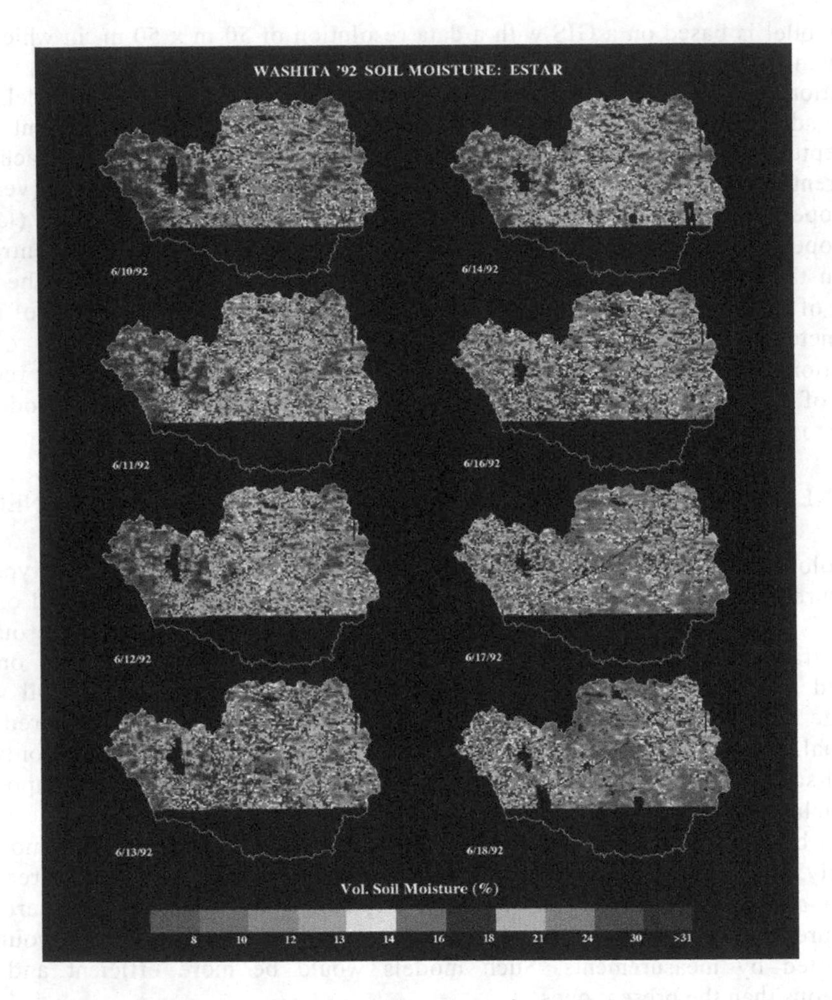

Figure 4. Sequence of daily surface soil moisture for the Little Washita Watershed, Oklahoma, USA. Measurements from the airborne ESTAR instrument (passive Microwave, L-band). (Courtesy of P.O'Neil, NASA, GSFC).

3. Medium Range Forecasting and Long-term Prediction in Hydrology and Water Management

For many purposes in hydrology and water management, it would be of great advantage to have medium range forecasts in some cases and long-term predictions in others. Medium range forecasts would be very valuable for medium range operation of water resources systems, while long-term prediction would be useful in the planning of new water management systems. For both cases here, only one example will be given.

3.1. HYDROPOWER FORECASTING AND SCHEDULING (MEDIUM RANGE FORECASTING)

Hydropower plants transform potential energy (represented by the difference of two water levels) into kinetic energy via turbines and electrical generators. The power production over time depends on water availability in storage located upstream of the power plant. In colder regions of the world, storage does not only mean water stored in reservoirs, but also the water quantity stored in the snow cover accumulated during the winter months. Water stored in the snow pack is made available for power production during the melting period of the snow in spring and early summer. For the power production within a medium time range (e.g., several months), it is important to know the expected water availability during this period to ascertain economic efficiency. Generally, it is obvious that the more water is stored in the snow cover, the more kilowatt-hours can be produced. Remote sensing data coupled with hydrologic models allow quantifying this qualitative information. Snow cover can be measured with the aid of satellites (e.g., Landsat, SPOT, NOAA etc.). This information is, however, not sufficient since the water content within the snow cover varies considerably due to the status of the snow. Therefore, it is necessary to know the snow water equivalent (SWE). This can be detected by sensors flown on airplanes or, more efficiently, by sensors on satellites. Although SWE was operationally observed from airplanes, e.g., by a Norwegian hydropower company, this is cumbersome and expensive. Much more favorable would be the acquisition of snow water equivalent by satellites. In recent times, it became possible to get such information from the military satellite DMSP (USA) with the aid of the passive microwave sensor SSM/I. Figure 5 shows such a passive microwave derived snow water equivalent for March 1, 1997, which happened to be a flood year on the Canadian prairies. Although this information looks good in Fig. 5, one has to consider, however, that the figure covers a huge area. If smaller catchment areas were to be observed, the rather coarse resolution in space of this sensor would be recognized as disadvantageous. Nevertheless, one can hope that future passive microwave sensors on satellites will be able to provide a better resolution in space.

Based on the known snow cover (acquired, e.g., by NOAA or Landsat) and the known SWE (obtained either from airplane sensors or satellites according to their distribution in space), it is possible to compute the total SWE for a whole river basin upstream of a hydropower plant. What is not known yet, but is necessary for hydropower scheduling, is the time distribution of the snowmelt runoff relevant for hydropower production. Here, another hydrologic model can be applied. The so-called "Snowmelt Runoff Model (SRM)" [6] is a model using remote sensing and other information for snowmelt runoff computations. The model uses the "degree day approach" in estimation of the melting snow cover in a basin. This version of SRM has been tested on over 80 basins in 25 countries worldwide. For each day, the water produced from snowmelt and from rainfall is computed, superimposed on the calculated recession flow and transformed into daily discharge from the basin with the aid of a certain equation [3, 6]. The model uses, as further information snow cover, depletion curves, which may be derived from Landsat data for different elevation zones. The degree-day version of SRM can be obtained by visiting the SRM worldwide Web homepage (http://hydrolab.arsusda.gov/cgi-bin/srmhome). The technique described above for real-time scheduling of hydropower production represents a highly integrated technology since information obtained from satellites (Landsat, NOAA, DMSP), airplane monitoring, digital maps, snow depletion curves etc., have to be integrated in a GIS and be processed in a hydrologic model.

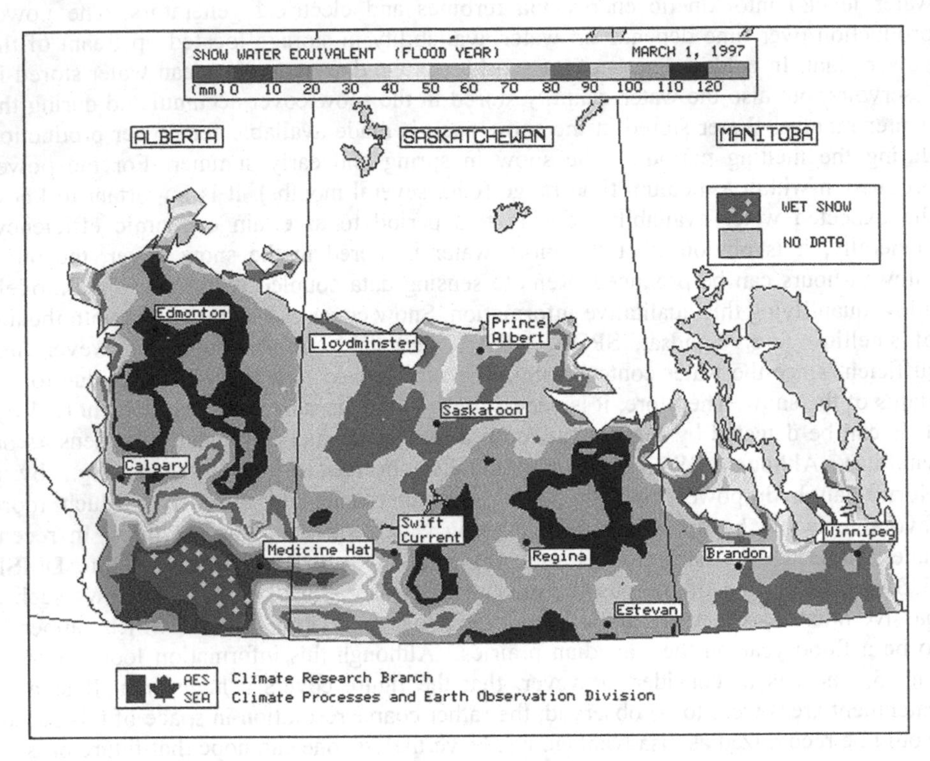

Figure 5. Passive microwave-derived snow water equivalent for 1 March 1997 (flood year) on the Canadian prairies. Derived from passive microwave sensor SSM/I on Satellite DMSP [3].

3.2. INFLUENCE OF LAND USE CHANGES ON FLOOD INTENSITIES (LONG-TERM PREDICTION)

In almost all river basins in the more densely populated areas of the earth, one can observe long-term land use changes, e.g., the increase of large urban settlements (mega-cities), forest diseases, changes in utilization of agricultural areas (use of agricultural areas in a more intense or more extensive way), clearance of woodland for cultivation etc. Such changes have a significant influence on hydrologic processes in the river basin and, thus, on water management schemes. Therefore, it is important to create timely information on such expected long-term changes and their impact. As an example, Fig. 6 shows at the top 3 maps of the catchment of the Nims River, a tributary of the international Mosel River for 3 different conditions. On the left map, land use conditions derived from a Landsat satellite image of the year 1989 are presented. It shows great variability of the various land uses, e.g., forest, cropland, pasture, build-up areas

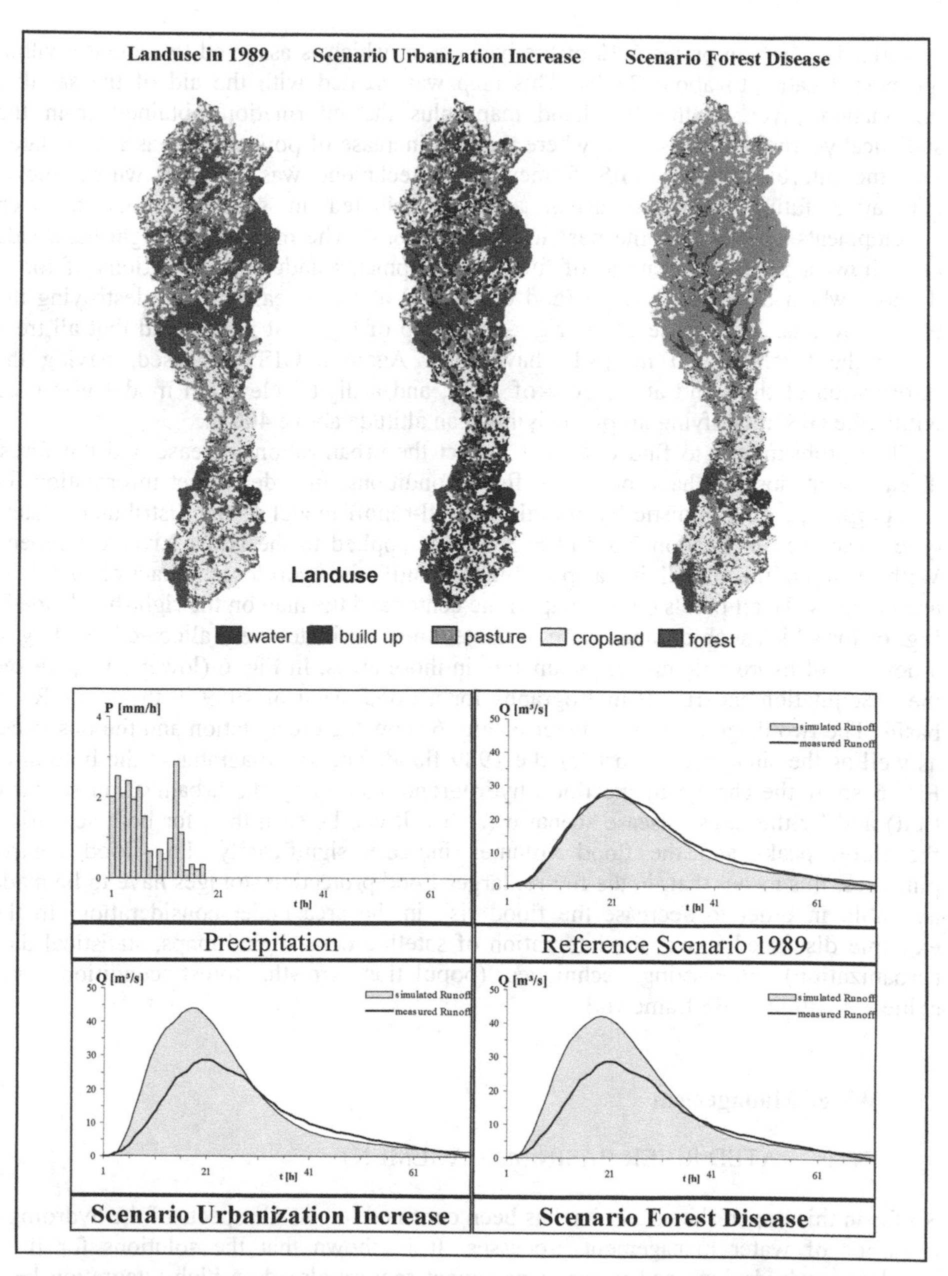

Figure 6. Land use classification (1989) of the Nims River catchment as a reference case (6% of area urbanization), scenario of increased urbanization (24%), and scenario of forest disease (all trees above 400 m a.S.L. assumed dead), and the corresponding flood hydrographs (observed 1989) and predicted.

and water. The map in the center shows the same catchment area for hypothetical future conditions, where it is assumed that the urban areas would grow significantly. Presently, the urbanization comprises 6 % of the basin area, which is assumed to increase within the next decades to about 24 %. This map was created with the aid of the satellite information given in the left hand map, plus the information obtained from the statistical yearbook of this area, where the past increase of population was extrapolated into the future. Within a GIS framework, a technique was applied, which allows generating future increased urban areas (distributed in space) similar to such developments observed in the past in the real world. The map on the right-hand side also shows a potential scenario of future developments under the conditions of forest disease, which occurs in Europe (and elsewhere) at an increasing rate, destroying the trees of forests. In the case of the right-hand map of Fig. 6, it is assumed that all trees above the level of 400 m a.s.L. have died. Again a GIS was used, having the information of the Landsat imagery of 1989; and a digital elevation model was used within the GIS, identifying all pixels lying at an altitude above 400 m.

The problem was to find out what impact the urbanization increase and the forest disease scenarios will have on future flood conditions. In order to get information for this purpose, a deterministic hydrologic rainfall-runoff model of the distributed system type (discussed in Section 2.3 and Fig. 3) was applied to the Nims River catchment. With the aid of this model, it was possible to identify the hydrologic impact of such land use changes. To all pixels on the map in the center and the map on the right-hand side in Fig. 6, for which such changes were predicted, new attributes were allocated, leading to a new set of hydrologic model parameters in those areas. In Fig. 6 (lower part), we see the precipitation and runoff hydrographs for a flood event of 1989 in the Nims River basin. The two diagrams in the center of Fig. 6 show the precipitation and the observed, as well as the simulated, runoff for the 1989 flood. The two diagrams at the bottom of Fig. 6 show the change of the flood hydrograph, caused by the urbanization scenario (left) and by the forest disease scenario (right). It can be seen that, for both scenarios, the flood peaks and the flood volumes increase significantly. For flood control purposes, this means that, in the future, larger flood protection storages have to be made available in order to decrease the flood risk in the area under consideration. In the example discussed above, the integration of satellite data, digital maps, statistical data (urbanization), forecasting techniques (population growth, forest condition) was achieved within a GIS framework.

4. Water Management

4.1. INTEGRATED RIVER BASIN MANAGEMENT

So far in this paper, the discussion has been confined to partial aspects of the hydrologic cycle or of water management processes. It is shown that the solutions for these problems in hydrology and water management require already a high integration level of data from different sources, as well as of data combined with hydrologic or water management models.

In the future, we will be faced with demands for integrated technologies and information production (particularly for decision processes) at a much higher and a more demanding integration level. In Europe, the Water Framework Directive to be followed in all EU countries requires Integrated River Basin Management. This means the integration of different spatial scales, different segments of the water cycle (surface water, ground water), integrated consideration of different properties (water quantity and quality) and of different functions of water (ecological, economical, social), as well as the integration at different time horizons (development in the past, requirements of the present, and objectives for the future).

Furthermore, integrated river basin management requires that, for each river and river basin (national or international), a catalogue of intended measures for improvement of the water quality has to be produced along with a water resources and catchment management plan for the whole river basin. This requires, among other things, the development and application of multi-disciplinary evaluation systems combined with multi-dimensional Decision Support Systems (DSS). Most of the theoretical and technical tools to meet those requirements do not exist yet. Thus, the implementation of the principle of integrated river basin management in Europe (and elsewhere) will request the development of specific, high level integrated technologies and information production systems. Therefore, an increased attention will be paid to the topic of this workshop in the future in Europe and elsewhere.

4.2. FLOOD MANAGEMENT IN REAL-TIME

Hydrologic processes of high temporal and spatial dynamics, e.g., flood events in urban areas, are difficult to handle by means of the operation of hydraulic structures, since the information needed for flood management is usually not available early enough in order to reduce the flood danger. Since remote sensing data can be transferred relatively easily, and without delay, electronically to a flood control center, it is attractive to use such data as a basis for flood management decisions, e.g., the operation of a group of flood retention reservoirs. Often, such reservoir groups are operated according to fixed rules which are not changed from flood event to flood event. It is, however, known that the efficiency, i.e., reduction of flood damages and loss of lives, can be improved considerably if a flood forecast is known in due time before the flood event occurs. This means that early flood forecasting in real-time is needed in the form of a forecast flood hydrograph for the whole duration of the flood event, or at least for several hours in an urban system.

In order to achieve a reasonably long lead time of the flood forecast for efficient flood management, it is necessary to produce such flood forecasts very early, i.e., during the storm event. A suitable technique for this purpose would use a rainfall-runoff model, which transforms the precipitation input into runoff in real-time. By precipitation input, we mean not only the rainfall, which is already observed, but also a Quantitative Precipitation Forecast (QPF). Since the spatial distribution of the storm event across the catchment area is of importance for the shape of the forecast flood hydrograph, it is recommendable to use areal rainfall measurements as model inputs. For this purpose, the data obtained by ground-based weather radar is of great use. The measured radar echo (active microwave) is transformed first into areal rainfall by some calibration technique, and then, with the aid of a distributed system rainfall-runoff model, into a forecast flood hydrograph to be expected in the immediate future. On the basis of such a forecast, existing dams or flood protection reservoirs can be operated in an optimum way with the aid of a mathematical optimization technique. Figure 7 shows a sequence of 4 consecutive radar images during a storm. In the 4 pictures, an urban catchment area of the city of Bochum, Germany, can be recognized. These 4 radar images were calibrated to represent rainfall intensities. The sequence of images shows the time and space distribution of a storm event moving across the urban catchment area. The high variability of rainfall intensities in time and space can be seen, which has a strong influence on the resulting flood hydrograph. With the aid of a distributed system rainfall-runoff model, the storm is transferred into a forecast flood hydrograph. The forecast can be improved if, besides the observed information, which is already available when the forecast is made, a QPF for the next hours is also computed. This can be done with the aid of consecutive radar images or on the basis of a Numerical Weather Prediction model, as provided by several national meteorological services. The output of the hydrologic model consists in real-time forecasts of the inflow hydrographs into one or several flood retention reservoirs within a river system or an urban drainage system. In order to secure the best possible flood control, an operation mode for the reservoirs has to be computed in real-time with the aid of some optimization technique on the basis of the computed flood forecast. Flood events in an urban system do not cause only water quantity problems, but also water quality problems because, during the beginning of a flood, the pollution load increases considerably. Thus, flood management here means integrated flood management, considering water quantity

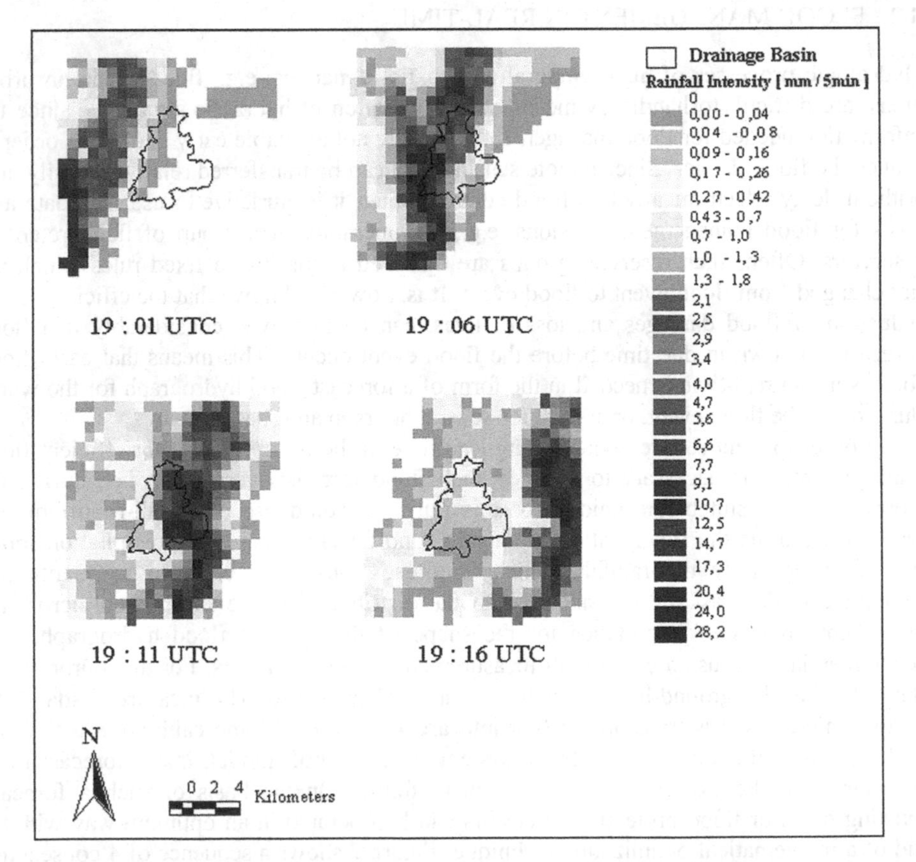

Figure 7. Sequence of consecutive radar images (C-band, Δt = 5 min, Δx = 1 km) during a storm (July 23, 1996) over an urban catchment in NW Germany (Courtesy German Weather Service (DWD))

and water quality parameters (e.g., nitrogen, phosphorus, chemical oxygen demand). Since the flood process in an urban environment shows a highly dynamic behavior, it is necessary to work with time increments as short as 5 min. (Fig. 7). Figure 8 shows schematically how the prediction and simulation of the hydraulic load, as well as the pollution load, in an urban sewer network has to be done. It also shows the role of the sewage treatment plant.

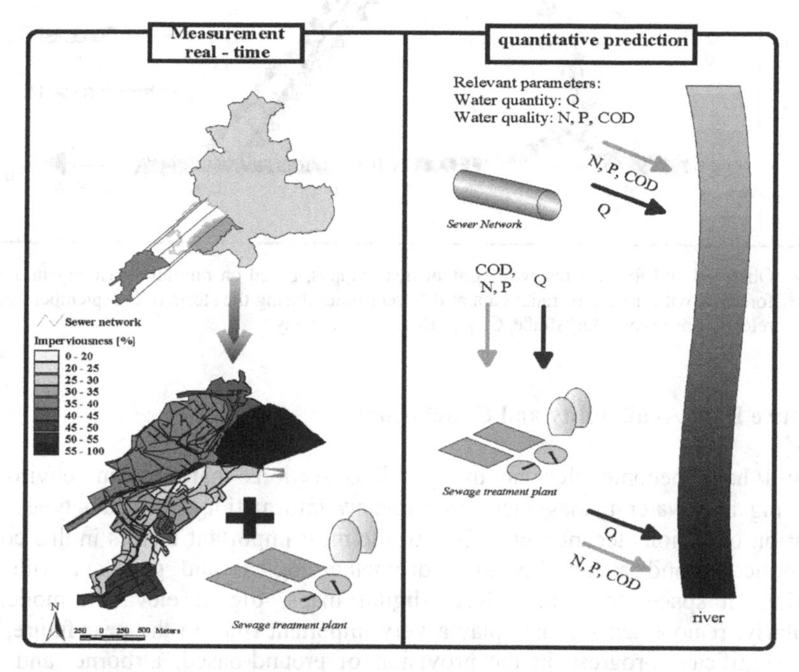

Figure 8. Prediction and simulation of the hydraulic load and the pollution load in the sewer network and the sewage treatment plant

In a research project at the author's institute, together with two other institutes, the whole flood process, comprising radar rainfall measurement, rainfall forecasting, flood hydrograph computation, routing of flood hydrograph and pollution load hydrograph through the urban drainage system and simulation of the operation of the sewage treatment plant, was considered [7]. The relevant pollution load forecasts are computed as a function of a forecast runoff hydrograph. Figure 9 shows the observed and forecast reservoir content hydrographs in an urban flood retention storage, based on rainfall measurements by radar and rainfall forecasts (with the aid of radar data) at different times during a storm in 1998. It can be seen that the forecast made at 6:05 h, i.e., at the start of the storm about an hour before the observed flood peak, is already very good. On the basis of such forecasts, it is possible to achieve an efficient flood management concerning both water quantity and quality. The main problem associated with such flood management is the relevant information production (areal rainfall measurement, rainfall forecast, flood forecast, operation strategy of urban flood protection reservoirs, operation of the sewage treatment plant) in real-time as early as possible and with very short time increments.

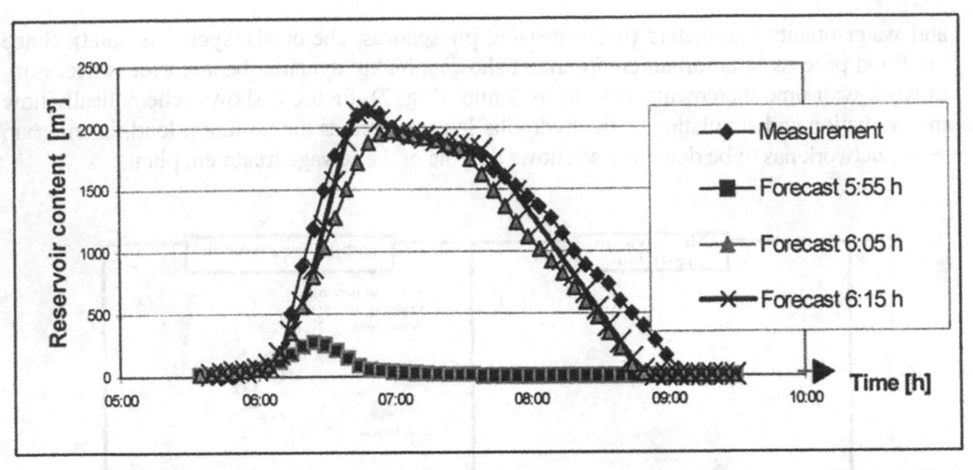

Figure 9. Observed and forecast reservoir content hydrographs, based on rainfall measurement and rainfall forecasts with the aid of radar data at different times during the storm of 2 September 1998. Flood retention reservoir Markstraße, City of Bochum, Germany.

5. Future Data Availability and Conclusions

It should have become clear in the previous sections that efficient environmental monitoring and water management require new information, new data types and the integration of various technologies. One of the most important factors in this context is the production and availability of hydrometeorological and geo-data with a high resolution in space and time. Here, digital maps, digital elevation models, and, particularly, remote sensing data play a very important role. In the near future, we can expect significant progress in the provision of ground-based, airborne, and satellite remote sensing systems. In the field of ground-based sensors, large scale integrated radar networks are under development in Western Europe including the UK (C-band), in North America (NEXRAD (S-band)) and in Japan (two different weather radar systems). In many other parts of the world, similar efforts are in progress. In the field of airborne sensors, we can observe an evolution in several directions. The development of light and small instruments, e.g., multi-spectral cameras or radiometers, are of particular interest for real-time water management. Probably the most important platforms and sensors will be in the field of satellite remote sensing. At present, about 80 satellites are in space or will be launched in the near future relevant to hydrologic applications [3]. Table 1 shows a selection of existing and other satellites to be launched in the near future and the corresponding sensors along with their spectral bands, their spatial resolution and revisiting capability (repetition rate). Some of these sensors have a large number of spectral channels (e.g., MERIS on ENVISAT: 15 channels, MODIS on EOS AM-1 and PM-1: 36 channels). Some sensors have a rather good resolution in space (e.g., IKONOS (1 m and 4 m), OrbView-4 (1 m, 4 m, 8 m)) but a less sufficient temporal resolution (1 or several days). Other satellites have a very good temporal resolution (e.g., METEOSAT and other geo-stationary satellites (0,5 h)) but a less satisfying spatial resolution.

TABLE 1. Specifications for sensors on existing and future RS satellites relevant to hydrologic applications

Sensor	Mission(s)	Spectral Bands	Spatial Resolution	Revisit Capab.
AMSR Advanced Microwave Scanning Radiometer	ADEOS II	P. Microwave: 6,9, 10.65, 18.7, 23.8, 36.5, 50.3, 52.8 and 89 GHz	5-50 km (dependent on frequency)	4 days (57 revisit)
ASAR Advanced Synthetic Aperture Radar	ENVISAT-1	C-band	Image, wave and altern. polariz. modes: 30m x 30m, Wide swath mode: 100m x 100m, global monit. modes: 1km x 1km	35 days
ASTER Advanced Space-borne Thermal Emission & Reflection Radiometer	EOS PM-1	VIS-NIR: 3 channels (0.5-0.9 μm), SWIR: 6 channels (1.6-2.5 μm), TIR: 5 channels (8-12 μm)	VNIR: 15 m, stereo: 15 m, horiz. and 25 m vertic. SWIR: 20 m, TIR: 90 m	16 days
CMIS Conical MW Imager Sounder	NPOESS Nation. Polar Orb. Env.	more channels at the high frequencies	will replace AVHRR	
HYDROSTAR	HYDROSTAR ESSP, NASA	L-band, H-Polarization	30 km	3 days
IKONOS	IKONOS 1	Panchrom.: 0.45 -0.90 μm, Multispectr.:0.45-0.52, 0.52-0.60, 0.63-0.69, 0.76-0.90	Panchromatic: 1 m Multispectral: 4 m	1.5 days at 1 m resolution
MERIS Medium Resolution Imaging Spectrometer	ENVISAT 1	VIS-NIR: 15 bands select. range: 0.4 -1.05 μm (bandwidth between 0.0025 and 0.03 μm)	300 m or 1200 m at SSP	3 days
MODIS Moderate Resolution Imaging Spectro Radiometer	EOS AM-1 EOS PM-1	VIS-TIR: 36 bands in range 0.4-14.4 μm	cloud cover: 250 m (day) and 1000 m (night)	Daylight reflect. and day/night emiss. spectr. imag. at least every 2 days
OrbView-4 (OrbView 2 and 3)	OrbView-4	Panchrom.: .450-.900 μm, Multispectr.: .450-.520 μm,.520 -.600 μm, .625-.695 μm, .760-.900 μm Hyperspectr.: .450 -2500 μm (200 channels)	Panchromatic: 1 m Multispectral: 4 m Hyperspectral: 8 m	< 3 days
PR Precipitation Radar	TRMM	MW: 13.796 and 13.802 GHz	Range Resol.: 250 m Horiz. Resol.: 4,3 km at Nadir	~ 5 days
QuickBird 1	QuickBird	Panchrom.: .450-.900 μm, Multi-spectr.: .450-.520 μm, .520-.600 μm, .630-.690 μm, .760-.890 μm	Panchromatic: 1 m Multispectral: 4 m	1-5 days depending on latitude
SMOS	New Explorer ESA	L-band, HV-polarization	50 – 60 km	3 – 4 days
VCL Vegetation Canopy Lidar	ESSP-1A	Visible-NIR	5.25 m footprints	No period. repeat. Decay orbit f. ground sampling

We can expect that environmental monitoring and information production in the form of integrated technologies for hydrology and water resources management have a fascinating future. This will be based on the new remote sensing devices mentioned above and an expected new hydrologic model generation suitable for these new information fields. We also hope for 3D-GIS and multi-temporal GIS (4D-GIS) for data processing and management. The development of integrated river basin management techniques as discussed above, including multi-disciplinary evaluation systems and multi-dimensional decision support systems, should lead to highly efficient integrated water management in the future.

6. References

1. Kite, G.W. (1997) Simulating Columbia River flows with data from regional-scale climate models, *Water Resources Research* **33**, 6.

2. Schumann, A.H., Funke, R., and Schultz, G.A. (2000) Application of a geographic information system for conceptual rainfall-runoff modeling, *Journal of Hydrology* **240**, 45-61.

3. Schultz, G.A. and Engman, E.T. (eds.) (2000) *Remote Sensing in Hydrology and Water Management*, Springer Verlag Berlin, New York, London, Singapore, Tokyo.

4. Schultz, G.A. (1996) Remote sensing applications to hydrology: Runoff, *Hydrological Sciences Journal* **41(4)**, August 1996.

5. Su, Z. (1996) Remote sensing applied to hydrology: The Sauer river basin study, in G.A. Schultz (ed.), *Schriftenreihe Hydrologie/Wasserwirtschaft*, No. 15, Ruhr University Bochum, Germany.

6. Martinec, J., Rango, A., and Roberts, R. (1998) *Snowmelt Runoff Model (SRM) User's Manual*, Geographica Bernensia, Department of Geography, University of Berne.

7. Quirmbach, M., Schultz, G.A., and Frehmann, T. (1999) Use of weather radar vor combined control of an urban drainage system and sewerage treatment plant, in *Impacts of Urban Growth on Surface Water and Groundwater Quality*, IAHS publ. No. 259.

ASSESSING THE APPLICABILITY OF HYDROLOGIC INFORMATION FROM RADAR IMAGERY

F. P. De TROCH, N. E. C. VERHOEST, V. R. N. PAUWELS
and R. HOEBEN
Laboratory of Hydrology and Water Management
Ghent University
Coupure links 653, B9000 Ghent, Belgium

Abstract. During the last two decades, the potential of radar remote sensing in the retrieval of the water content of the near-surface unsaturated soil zone has been explored. This water content is usually referred to as soil moisture. The inversion of radar observations into soil moisture values has been hampered by an insufficient characterization of the soil roughness. However, some studies have focused on relating temporal changes of the radar signal to hydrologic relevant information. One result is presented here, where we show that variable source areas, which are mainly responsible for runoff in a catchment, can be visualized through a principal component analysis. A second part of this paper shows how soil moisture information obtained from radar imagery can be incorporated into hydrologic models. A first example uses an extended Kalman filtering technique, which adjusts the state variables from a hydrologic model. Through this technique, one-dimensional soil moisture profiles are retrieved with high accuracy. In a second example, we show a data-assimilation method which uses both the statistics and the spatial distribution of radar-retrieved soil moisture values, to adjust the modeled soil moisture profile. This methodology enables a better modeling of the rainfall-runoff behavior of the catchment.

1. Introduction

Near surface soil moisture is a major control on hydrologic processes at both the storm event scale and at longer temporal scales. It influences the partitioning of precipitation into infiltration and runoff. Near surface soil moisture also affects the evapotranspiration as it controls the water availability to plants [1]. Improved understanding of the spatial distribution of soil moisture is useful for a range of applications in hydrology [2]. As topography, soil texture, vegetation and meteorological conditions all have some influence on soil moisture, it will exhibit spatial variability in a wide range of scales [3].

Unlike *in situ* measurements, remote sensing offers the potential for spatial observations of soil moisture at basin and regional scales. A variety of infrared and microwave (both active and passive) sensors have been tested for soil moisture retrieval, and the limitations of these techniques are well documented [4].

In this paper, data from synthetic aperture radar (SAR) are employed for the retrieval of hydrologic relevant information. Several backscattering models, which relate the backscattered intensity of the incident microwave to soil moisture, soil roughness and vegetation properties have been presented. For bare soil, the integral equation model (IEM) [5, 6] and the model of [7] have been tested intensively and show good agreement with the field measured values. The main problem encountered with the retrieval of soil moisture from radar backscattering intensities is the characterization of soil roughness. Recently, some studies have tried to bypass this characterization, based on several assumptions such as by assuming an a priori knowledge of the soil roughness or by applying multiple backscattering models [8], and by using temporal radar data, assuming a constant surface roughness in the retrieval of soil moisture [9].

Three examples which retrieve hydrologic relevant information from radar remotely sensed data are presented. A first example studies the changes occurring in the backscattered signal during a winter period and relates this to soil moisture changes. Therefore, a principal component analysis is used. The obtained results show the saturated areas in catchment, which mainly contribute to the runoff production during rainfall events. A second example uses retrieved soil moisture information of the top soil layer to estimate the soil moisture profile in the unsaturated zone through assimilation of the radar retrieved soil moisture into a hydrologic model through extended Kalman filtering. This technique allows an accurate estimation of the soil moisture profile from top soil moisture measurements. A last example assimilates spatial soil moisture estimates in a distributed hydrologic model in order to improve the rainfall-runoff processes observed in a catchment.

2. Maps of Variable Source Areas

The main difficulty with SAR imagery is that soil moisture, surface roughness, and vegetation cover all have an important and nearly equal effect on radar backscatter. These interactions make retrieval of soil moisture possible only under particular conditions such as bare soil or surfaces with low vegetation density [10]. It should be possible to separate the vegetation, topography and soil moisture effects on the radar response, using multi-frequency and/or multi-polarization measurements; but, currently, operational satellites are not equipped with sensors that provide such data. A related concept that can be applied to existing satellite imagery is multi-temporal analysis, which is based on statistical techniques such as filtering, differencing, and rationing [11]. Although, on its own, it will not produce direct or absolute measurements, multi-temporal analysis could be useful for deriving wetness indices and for monitoring temporal changes or elucidating spatial patterns in soil moisture. Here, a principal component analysis (PCA) is performed on a wintertime series of European Remote Sensing (ERS) satellites 1 and 2 images taken over the Zwalm catchment, Belgium. As one can assume that, during winter, soil roughness and vegetation changes are minimal, the changes in the backscattered signal can be related to soil moisture changes during the spanned period.

The principal components transformation is a standard tool in image enhancement, image compression, and classification. It linearly transforms multi-dimensional data into a new coordinate system in which the first component contains the largest percentage of the total variance (i.e., the maximum or dominant information), the second component contains the second largest percentage, and so on. Images transformed by PCA may make evident features that are not discernible in the original data, changes and trends in multi-temporal data, which typically show up in the intermediate principal components.

Applying the PCA to eight ERS images (for a detailed description of the analysis, the reader is referred to [12]) leads to a separation of the information contained in the original images into several principal components (PCs) that can be attributed to different factors influencing the backscatter. The first principal component can be assigned to the effects on the backscattered signal of a changing local incidence angle due to topography. This topographic effect could be expected, as hillslopes facing towards the spacecraft will consistently reflect more microwave energy than the slopes turned away from the sensor. The second principal component (shown in Fig. 1(a)) displays a strong spatial organization, with the highest values grouped along the drainage network of the catchment. To test the hypotheses that the information contained in this image is related to the drainage conditions of the catchment, the image can be compared to the drainage map of the Zwalm catchment, as given in Fig. 1(c). [12] showed that the radar response, brought out in this second principal component, was caused by the soil moisture patterns that result from the drainage characteristics of the basin. Areas with a low drainage capability are mostly found near the rivers and saturate very quickly after a rainfall event. These areas, therefore, can be defined as variable source areas, as they will lead to overland flow once they saturate. The patterns retrieved in the second principal component cannot be attributed to any single event in the time series but reflect the overall response of the soil to the rainfall and inter storm periods spanned by the images. The third principal component can be assigned to land use and land cover. The fourth and sequent PCs do not seem to reveal significant physical features. They are mainly characterized by noise (including speckle). A detailed analysis of the obtained second principal component with respect to variable source areas is given in [12].

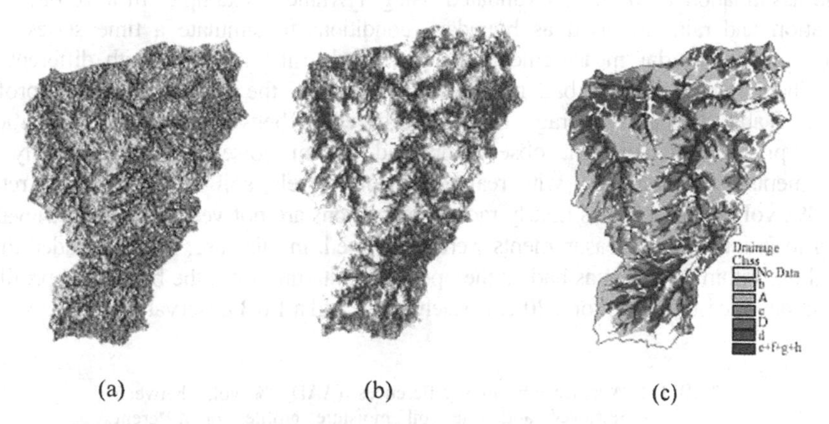

Figure 1. (a) Not filtered second principal component image, after [12]; (b) Wavelet filtering applied to (a); (c) Drainage map of the Zwalm catchment (b, D: well-drained soils; c, D, d: intermediate drained soils; c, f, g, h: badly drained soils). The river network is draped on top of the three images.

The obtained second principal component image is still corrupted with noise, which can partly be related to the speckle observed in the SAR images. In order to get noise reduced images, which are more appropriate in hydrologic modeling schemes, we apply a recently developed wavelet-based image denoising technique [13] that preserves the spatial patterns and observed edges, while increasing the signal-to-noise ratio significantly. This technique applies the classical Bayes estimation theory to estimate the wavelet coefficients using prior knowledge of their spatial distribution. The result of this filtering is shown in Fig. 1 (b). Immediately apparent is the large noise reduction obtained by this filtering technique, which results in a better visual correspondence with Fig. 1 (c). [14] found that the wavelet-based filter of [13] is able to reduce this noise considerably, while conserving spatial detail in the valley regions. [14] showed that the filtered second principal component image corresponds better to the observed drainage characteristics of the catchment than the original PC.

3. Soil Moisture Profiles

Radar measurements offer the possibility to retrieve surface soil moisture. In order to obtain soil moisture profile information, data assimilation of these radar measurements into a hydrologic model is necessary. In this case, the extended Kalman filter is used. A Kalman filter merges information from a system model and an observation model. The system model comprises a 1D Richard's equation based hydrologic model, which can successfully model soil moisture profile evolution based on meteorological data. The observation model simulates the radar measurement for which we use the Integral Equation Model (IEM) [6]. The Kalman filter compares the simulated observation and the radar measurement and adjusts the state vector of the system model in a statistical optimal way, taking into consideration error levels of both the system and the observation model. The reader is referred to [15] for a full description of the applied methodology.

The assimilation algorithm is validated using a synthetic example. In a 16-day period, evaporation and rain are used as boundary conditions to simulate a time series of soil moisture profiles. Radar measurements are generated, using the IEM with different noise levels. Then, starting from a bad initial moisture profile, the true soil moisture profile is retrieved. Table 1 shows average absolute differences between retrieved and true soil moisture profiles for different observation and system noise levels and hourly radar measurements. It is seen that, with real-world noise levels, soil moisture profile retrieval within 4% vol. is possible. As hourly radar observations are not very realistic, retrieval runs with up to 2-daily radar measurements were performed. In this case, it is concluded that the retrieval algorithm performs as bad as the open loop (starting from the bad initial profile and not using any measurement) for a 20 cm system noise and a 1 dB observation noise.

TABLE 1. Average absolute differences (AAD, % vol.) between retrieved and true soil moisture profile for different observation and system noise levels and hourly updates

Observation noise [dB]	System noise [cm]			
	1	5	10	20
0.0	0.025	0.025	0.025	0.025
0.1	0.325	0.433	0.534	0.638
0.2	0.559	0.826	1.064	1.586
0.5	0.629	1.226	1.823	3.234
0.75	0.850	1.397	2.090	3.785
1.0	1.100	2.099	3.126	3.853

The assimilation technique is also applied to a data set obtained at the European Microwave Signature Laboratory (EMSL). The reader is referred to [16] for a full description of this experiment. The data set contains soil moisture profile measurements in a soil sample together with radar observations of this soil sample. Four observations in a 9-day period were used. Figure 2 shows measured and retrieved soil moisture at 5 depths during the entire 9-day period. For a real-world configuration and noise levels and starting from a good initial guess, one is able to retrieve the entire soil moisture profile at the end of the time period within 4% vol. The use of a wildly guessed initial soil moisture profile results in the same accuracy for the retrieved profile.

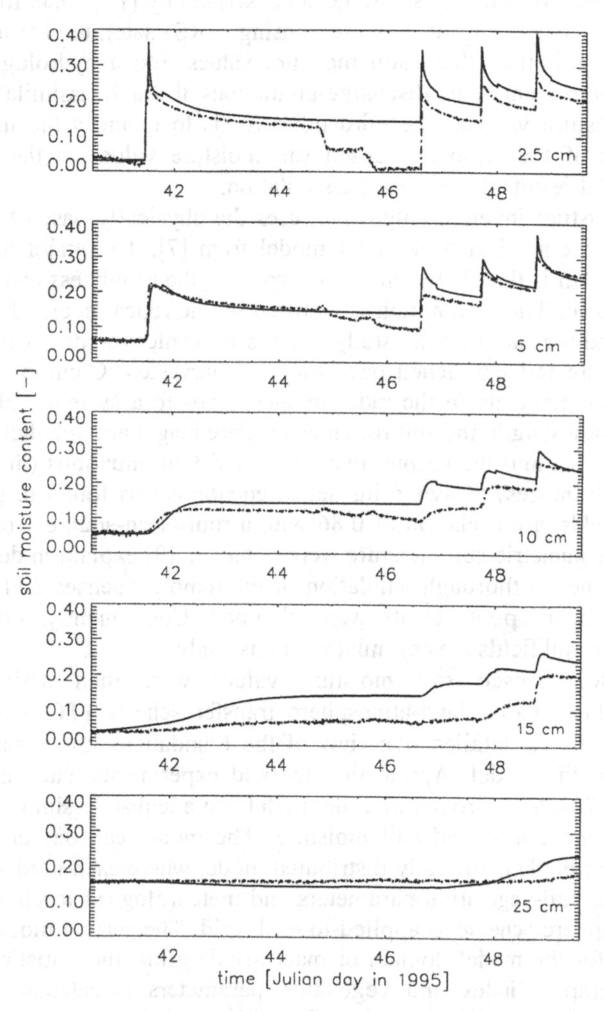

Figure 2. Retrieved soil moisture at 5 depths using radar measurements at 23° incidence angle [System noise: 20 cm, observation noise: 1.13 dB; dash-dotted line: TDR-measured, solid line: assimilated values]

4. Rainfall-Runoff Modeling

The purpose of this study is to assess the improvement of model-predicted fluxes, i.e., discharge, through assimilation of remotely sensed soil moisture values. The study was done using data obtained over the Zwalm catchment in Belgium. Troch *et al.* [17] give a detailed overview of the Zwalm catchment. Three specific objectives are addressed in this study. The first objective is to develop a soil moisture estimation method, by the use of radar data, with, as a major characteristic, the possibility to estimate soil moisture based on radar observations without having to measure *in situ* soil roughness parameters. The concept of the soil moisture estimation method used is similar to the estimation of effective roughness parameters described by [18]. Data from the European Space Agency (ESA) European Remote Sensing (ERS) satellites are used. The second objective is to assimilate these soil moisture values into a hydrologic model and to quantify the improvement in discharge-predictions through assimilation of remotely sensed soil moisture values. The third objective is to quantify the importance of the spatial patterns of the remotely sensed soil moisture values in the improvement of hydrologic model results through data assimilation.

The soil moisture inversion algorithm uses the physically based backscatter model from [5] and the empirical backscatter model from [7]. The major assumption in the inversion algorithm is that, between two overpasses, the roughness characteristics of the soil do not change. The assumption is justified by the repeat cycle of the ERS-satellite (five weeks) and the fact that the study focuses on winter months, when most fields in the study area are left untouched once they are harvested. Combination of these two models, for each pixel inside the radar images, leads to a system with four unknowns (the soil roughness length, the soil root mean square height and the dielectric constant of the soil for the first and the second overpass), and four equations (the two backscatter models for both images). Solving this set of equations was found to give good results for bare soil fields: a correlation of 0.86 with a root-mean-square error of 5.6% and a bias of 0.47% volumetric soil moisture were obtained. [9] explain in detail the inversion algorithm and show a thorough validation of the remotely sensed soil moisture values. For vegetated fields, poor results were obtained. Consequently, only soil moisture values over bare soil fields are assimilated in this study.

The remotely sensed soil moisture values were then assimilated into the TOPMODEL [19], based land-atmosphere transfer scheme (TOP-LATS) [20]. [20], [21], and [22] give a detailed overview of the foundations, the assumptions, and the development of the model. Application to field experiments, such as FIFE [21] and BOREAS [22, 23], have proven that the model can adequately simulate surface energy fluxes, soil temperatures, and soil moisture. The model can be run in two different modes. The first mode is the fully distributed mode, where each model unit (a grid) has its own specific soil-vegetation parameters and meteorological forcing data, and where the land-atmosphere scheme is applied to each grid. The second mode is the statistical mode, where, for the model domain or macro-scale grids, the statistical distribution of the soil topographic index and vegetation parameters is calculated; and the land-atmosphere scheme is applied to each statistical interval.

First, for all catchments inside the Zwalm catchment, a baseline run (without assimilation) was established. Then, the statistics (mean and standard deviation) of the remotely sensed soil moisture data were assimilated into the lumped version of TOPLATS through a modified version of the statistical correction assimilation method [24], where not only the moisture content of the upper soil layer, but also the moisture content of the lower soil layer is adapted. The improvement in the discharge peaks was quantified. [25] further describe this assimilation study.

As a second step, the statistics of the remotely sensed soil moisture data were assimilated into the distributed version of TOPLATS through the same assimilation method. Finally, both the statistics and the spatial distribution of the remotely sensed soil moisture data were assimilated into the distributed version of TOPLATS through the nudging technique [26]. These last two assimilation studies are described in more detail in [9, 27].

Figure 3 shows the results of the model applications for the Wijlegemse Beek catchment inside the Zwalm catchment. For the other catchments, similar results are obtained. Table II gives an overview of the results of the assimilation procedures. One can observe an initially higher RMSE between the observed and the simulated discharge for the lumped version than for the distributed version. However, the statistical correction assimilation method reduces the RMSE to a similar value for both model versions. Assimilating the spatial patterns of the remotely sensed soil moisture values, on top of the statistics (through the nudging technique), leads to only slightly better results than those obtained by only assimilating the statistics. As a consequence, it is suggested that it is sufficient to assimilate the statistics of the remotely sensed soil moisture data into the lumped model to improve discharge predictions, and that assimilating the spatial patterns of these remote sensing data as well, into the distributed model, has only a secondary effect on the modeled discharge.

5. Conclusions

In this paper, it is shown that radar remote sensing is able to improve hydrologic modeling, as it is able to provide spatial information of the moisture content of the top soil layer in a catchment. In the first part, it is shown that, through studying the temporal behavior of the backscattering coefficient by a principal component analysis, one is able to visualize the variable source areas in a catchment. As these areas are prone to generate runoff, their spatial organization in the catchment determines the rainfall-runoff behavior of the river basin. A second part discusses the possibilities of using radar retrieved soil moisture for estimating the moisture content in the unsaturated zone, using a Kalman filter technique. The moisture deficit in the unsaturated zone determines the amount of water which can be stored before saturation excess runoff occurs. Therefore, a correct estimation of the spatial distribution of the moisture deficit in the catchment will improve the hydrologic modeling. The experiments conducted in this part only consider a 1D profile for testing the potential of updating a hydrologic model using radar satellite data. The third example given in this paper deals with applying spatial soil moisture values obtained in bare soil or low vegetated areas for updating a spatially distributed model. Therefore, a simple data-assimilation method, different from the Kalman filter technique, is applied. It is shown that it is sufficient to assimilate the statistics of the remotely sensed soil moisture into the lumped version of the TOPLATS model for improving discharge predictions.

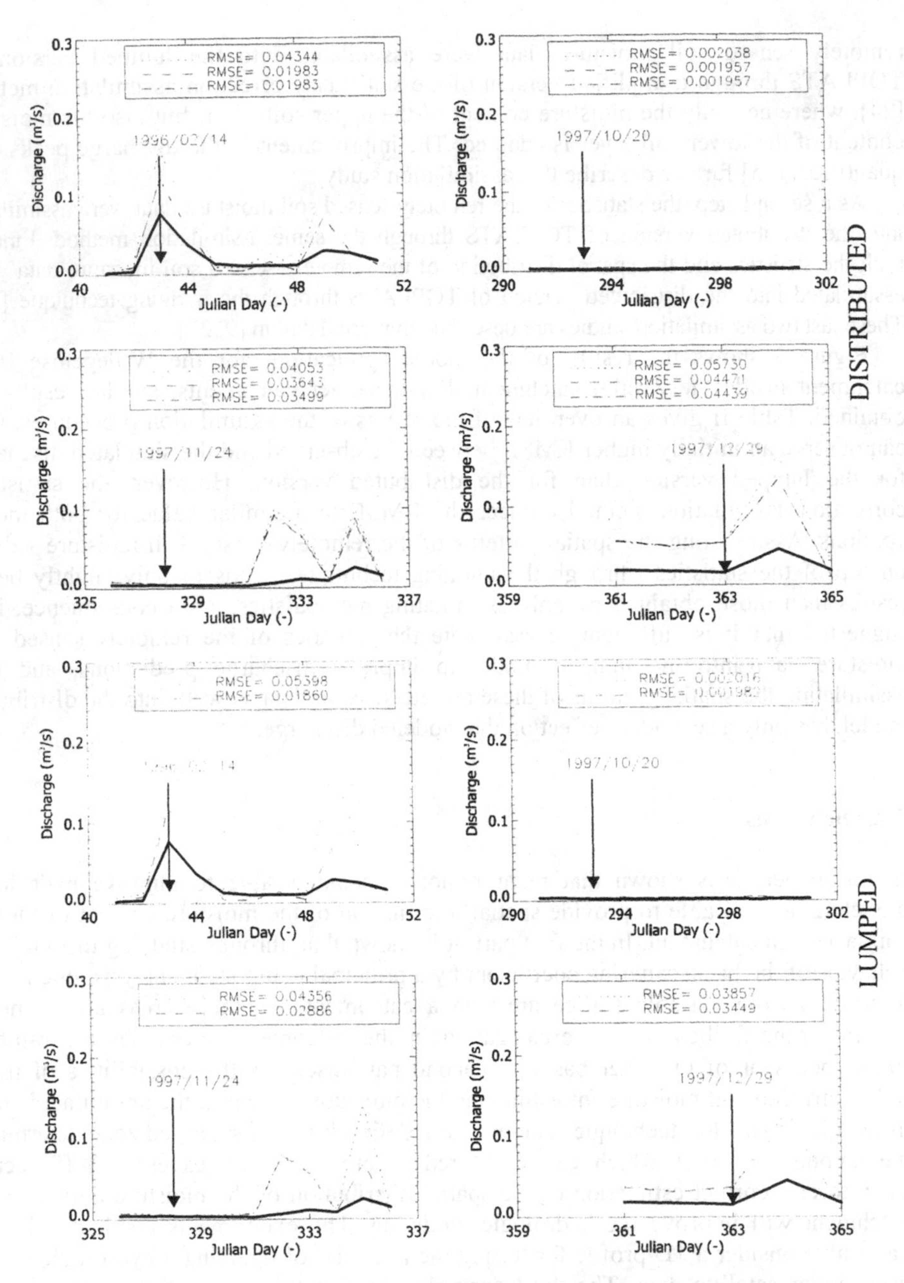

Figure 3. Results (daily values) of the model applications run for the Wijlegemse Beek catchment. Observations are in thick solid lines, results from the baseline run are in dotted lines and results from the soil moisture assimilation runs are in dashed lines. The results from the distributed runs with the statistical correction assimilation method are in dot-dashed lines. The top 4 plots are the results from the distributed runs; the bottom 4 plots are the results from the lumped runs. The arrows indicate the time of the ERS overpass. Validation data for 1994, 1995,and 1998 are not available.

TABLE 2. The average RMSE between observations and simulations of the discharge from the baseline runs and the soil moisture assimilation runs for the four modeled catchments and their average. Units are in m^3s^{-1}

Catchment	RMSE Distributed Run			RMSE Lumped Run	
	Baseline	Nudging	Statistical Correction	Baseline	Assimilation
Peerdestok	0.0654	0.0656	0.0712	0.101	0.0797
Sassegem	0.0109	0.0113	0.0118	0.0118	0.0124
Passemare	0.0296	0.0194	0.0183	0.0330	0.0134
Wijlegemse	0.0358	0.0257	0.0253	0.0345	0.0210
Average	0.0354	0.0305	0.0316	0.0452	0.0316

With the advent of new multi-polarized satellites, such as ENVISAT-1, more accurate soil moisture inversions are expected. Through acquiring data in HH and VV mode of the same area, field measurements of soil roughness will not be needed any more, as one will be able to retrieve it from the remotely sensed data itself. Therefore, the soil moisture estimates are expected to be more reliable. This will have profound implications on the hydrologic modeling using spatial remotely sensed soil moisture information through data assimilation.

6. References

1. Grayson, R.B. and Western, A.W. (1998) Towards areal estimation of soil water content from point measurements: time and space stability of mean response, *J. Hydrol.* **207(1-2)**, 68-82.
2. Grayson, R.B., Western, A.W., Chiew, F.H.S., and Blöschl, G. (1997) Preferred states in spatial soil moisture patterns: local and nonlocal controls, *Water Resour. Res.* **33(12)**, 2897-2908.
3. Western, A.W., Grayson, R.B., Blöschl, G., and Willgoose, G.R. (1999) Observed spatial organization of soil moisture and its relation to terrain indices, *Water Resour. Res.* **35(3)**, 797-810.
4. Engman, E.T. (1995) Recent advances in remote sensing in hydrology, in R.A. Pielke Sr. and R.M. Vogel (eds.), *U.S. National Report to International Union of Geodesy and Geophysics 1991-1994: Contributions in Hydrology*, American Geophysical Union, Washington, DC, pp. 967-975.
5. Fung, A.K., Li, Z., and Chen, K.S. (1992) Backscattering from a randomly rough dielectric surface, *IEEE Trans. Geosc. Rem. Sens.* **30(2)**, 356-369.
6. Fung, A.K. (1994) Microwave Scattering and Emission Models and their Applications, Artech House.
7. Oh, Y., Sarabandi, K., and Ulaby, F.T. (1992) An empirical model and an inversion technique for radar scattering from bare soil surfaces, *IEEE Trans. Geosc. Rem. Sens.* **30(2)**, 370-381.
8. Davidson, M.W.J., Le Toan, T., Mattia, F., Satalino, G., Manninen, T., and Borgeaud, M. (2000) On the characterization of agricultural soil roughness for radar remote sensing studies, *IEEE Trans. Geosc. Rem. Sens.* **38(2)**, 630-640.
9. Pauwels, V.R.N., Hoeben, R., Verhoest, N.E.C., De Troch, F.P., and Troch, P.A. (2002) Improvement of TOPLATS-based discharge predictions through assimilation of ERS-based remotely sensed soil moisture values, *Hydrol. Proc.* **16(5)**, 995-1013.
10. Altese, E., Bolognani, O., Mancini, M., and Troch, P.A. (1996) Retrieving soil moisture over bare soil from ERS-1 synthetic aperture radar data. Sensitivity analysis based on a theoretical surface scattering model and field data, *Water Resour. Res.* **32(3)**, 653-661.

11. Singh, A. (1989) Digital change detection techniques using remotely-sensed data, *Int. J. Rem. Sens.* **10(6)**, 989-1003.
12. Verhoest, N.E.C., Troch, P.A., Paniconi, C., and De Troch, F.P. (1998) Mapping basin scale variable source areas from multitemporal remotely sensed observations of soil moisture behavior, *Water Resour. Res.* **34(12)**, 3235-3244.
13. Pizurica, A., Philips, W., Lemahieu, I., and Acheroy, M. (1999) Image denoising in the wavelet domain using prior spatial constraints, in: *Proc. IEE Conf. on Image Processing and its applications, IPA'99*, Manchester, pp. 216-219.
14. Verhoest, N., Pizurica, A., Philips, W., and De Troch, F. (2000) The application of wavelet-based filtering techniques for retrieving bio-physical parameters from multi-temporal ERS-images, in: *Proc. of the ERS-Envisat Symposium: Looking down to Earth in the New Millenium*, Gothenburg, Sweden, on CDROM, 182verho.pdf, 8 pp.
15. Hoeben, R. and Troch, P.A. (2000) Assimilation of active microwave observation data for soil moisture profile estimation, *Water Resour. Res.* **36(10)**, 2805-2819.
16. Mancini, M., Hoeben, R., and Troch, P.A. (1999) Multifrequency radar observations of bare surface soil moisture content: a laboratory experiment, *Water Resour. Res.* **35(6)**, 1827-1838.
17. Troch, P.A., Smith, J.A., Wood, E.F., and De Troch, F. P. (1994) Hydrologic controls of large floods in a small basin: Central Appalachian case study, *J. Hydrol.* **156**, 285-309.
18. Su, Z., Troch, P.A., and De Troch, F.P. (1997) Remote sensing of soil moisture using EMAC/ESAR data, *Int. J. Rem. Sens.* **18(10)**, 2105-2124.
19. Beven, K.J. and Kirkby, M.J. (1979) A physically-based, variable contributing area model of basin hydrology, *Hydrol. Sci. Bull.* **24(1)**, 43-69.
20. Famiglietti, J.S. and Wood, E.F. (1994) Multiscale modeling of spatially variable water and energy balance processes, *Water Resour. Res.* **30(11)**, 3061-3078.
21. Peters-Lidard, C., Zion, M., and Wood, E.F. (1997) A soil-vegetation-atmosphere transfer scheme for modeling spatially variable water and energy balance processes, *J. Geophys. Res.* **102(D4)**, 4303-4324.
22. Pauwels, V. and Wood, E.F. (1999) A soil-vegetation-atmosphere transfer scheme for the modeling of water and energy balance processes in high latitudes, 2, Application and validation *J. Geoph. Res.* **104(D22)**, 27,823-27,839.
23. Pauwels, V. and Wood, E. F. (2000) The importance of classification differences and spatial resolution of land cover data in the uncertainty in model results over boreal ecosystems, *J. Hydromet.* **1(3)**, 255-266.
24. Houser, P.R., Shuttleworth, W.J., Famiglietti, J.S., Gupta, H.V., Syed, K.H., and Goodrich, D.C. (1998) Integration of soil moisture remote sensing and hydrologic modeling using data assimilation, *Water Resour. Res.* **34(12)**, 3405-3420.
25. Pauwels, V.R.N., Verhoest, N.E.C., and De Troch, F.P. (2001 a) The effect of changing stream levels on the water table configuration through an analytical solution of the Boussinesq-equation, Advances in Water Resources, submitted.
26. Stauffer, D.R. and Seaman, N.L. (1990) Use of four-dimensional data assimilation in a limited-area mesoscale model, Part I: experiments with synoptic-scale data, *Monthly Weather Rev.* **118(6)**, 1250-1277.
27. Pauwels, V.R.N., Hoeben, R., Verhoest, N.E.C., and De Troch, F.P. (2001 b), The importance of spatial patterns of remotely sensed soil moisture in the improvement of discharge predictions for small-scale basins through data assimilation, *J. Hydrol.* **251(1-2)**, 88-102.

INTEGRATED SATELLITE – AIRBORNE TECHNOLOGY FOR MONITORING ICE COVER PARAMETERS AND ICE-ASSOCIATED FORMS OF SEALS IN THE ARCTIC

V. V. MELENTYEV[1], V. I. CHERNOOK[2] and L. H. PETTERSSON[3]
[1]*Nansen International Environmental and Remote Sensing Center (NIERSC)*
B. Monetnaya Str. 26/28, St. Petersburg, 197101, Russia

[2]*Polar Research Institute of Marine Fisheries and Oceanography (PINRO)*
Knippovich Str. 18, Murmansk, 183763, Russia

[3]*Nansen Environmental and Remote Sensing Center (NERSC)*
Edv. Griegsv. 3A, Bergen, N-5059, Norway

Abstract. Airborne observations of the Arctic Ocean have become an indispensable part of fish industry and marine ecology. An application of satellite data for monitoring sea mammals has been attractive for a long time. However, the practical use of satellite information was restrained by the low spatial resolution of satellite data. Additional limits are related to daytime illumination and the effects of clouds at the polar regions.

After the launching of ERS Synthetic Aperture Radar (SAR) in 1991, which provides the global all-weather sounding of the Earth with a spatial resolution of 20-25 m, the situation has changed to a certain extent. Satellite SAR provides the possibility to identify the types of ice, to recognize the outline and the shape of ice floes, and to document many other parameters. The use of ERS SAR signatures of the ice, as an abiotic factor of the ecology of harp seals *Pagophillus groenladicus* and other sea mammals in the Arctic and sub-Arctic, is a new approach in marine biology. The first experience with the use of SAR data was undertaken in the White Sea and the contiguous Barents and Pechora Seas in 1996. These investigations were supported by ESA in 1998-2000 as part of the A03.440 Project entitled "Application of ERS SAR Data for Studying Migration of White Sea Population of Greenland Seals".

SAR application studies demonstrate that satellite radar can be used now as the main and, sometimes, the only information source about different environmental processes and ice phenomena. Unfortunately, for the time being, the available resolution cannot help to investigate the individual marine mammals that have on average a length of 1-2 m or more.

1. Introduction

Airborne observations of the Arctic Ocean have become an indispensable part of fish industry, ice navigation, and marine ecological investigations. An application of satellite data for monitoring sea mammals has been attractive for marine biologists for a long time. However, the practical use of satellite information to study seals, polar bears, whales and other ice-associated marine mammals was restrained by the low spatial resolution of satellite data. Additional limits are related to the daytime illumination and the effects of clouds at the polar regions.

After the launching of ERS Synthetic Aperture Radar (SAR) in 1991, which provides the global all-weather sounding of the Earth with a spatial resolution of 20-25 m, the situation has changed to a certain extent. High transparency of dry snow and relatively deep penetration of radio signals into the sea ice constitute the basis for SAR sub-surface sounding. The penetrability and radar return signals are strongly dependent on salinity and other ice parameters, i.e., its roughness, deformation and compactness. Satellite SAR provides the possibility to identify the types of ice, to recognize the outline and the shape of ice floes, and to document many other parameters.

Using ERS SAR signatures of the ice, as an abiotic factor of the ecology of harp seals *Pagophillus groenladicus* and other sea mammals in the Arctic and sub-Arctic, is a new approach in marine biology [1, 2, 3, 4].

The first experience with the use of SAR data was undertaken in the White Sea (northwest of Russia) and contiguous Barents and Pechora Seas, the inhabited areas of harp seals, in 1996 [1]. The synoptic acquisition of ERS SAR images and sub-satellite field experiments onboard the nuclear icebreaker "Taymir" provided a validation program for sea ice description; its conditions and drift patterns were evaluated in connection with the migration features of seals. Since then, satellite investigations of the migration of sea mammals, in relation with peculiarities of ice regime of the White Sea and the surrounding waters, have been provided systematically each winter season. Research aircraft PINRO Antonov-26 "Arktika" was used to carry out multi-spectral measurements. Helicopters performed ice reconnaissance and registered ice-associated sea mammals.

These investigations were supported by ESA in 1998-2000 as part of the A03.440 Project entitled "Application of ERS SAR Data for Studying Migration of White Sea Population of Greenland Seals" [2, 3, 4].

2. Specific Objectives

The specific objectives of the A03.440 Project on "Application of ERS SAR Data for Studying Migration of White Sea Population of Greenland Seals" were:

- Multi-level investigations of the White Sea as the inhabited area of harp seals: an integrated approach to monitoring (including climatic aspects);
- Classification of radar signatures of different environmental objects (open water, sea and saltish ice, brackish and fresh water ice, snow cover, coastal zones and shores);
- Study of water masses and ice dynamics; regional exchange and water catchment input (ice drift as a tracer of weather conditions and winter hydrology);
- Interdisciplinary study of ice as an abiotic factor of seal ecology; SAR charting of whelping rookeries and unfit zones;
- Development of integrated satellite-airborne technology for identifying ice cover parameters; population assessment and carthography of whelping migration.

3. Methodological Aspects, Instruments and Database

3.1. NORTHERN SEA ROUTE AND ROUND-THE-YEAR NAVIGATION SUPPORTS

The Northern Sea Route (NSR), as a part of the Arctic Ocean, is very important for marine transportation to Siberia, as well as for future transportation between Europe and the Pacific countries. Recent opening of the gigantic oil and gas deposits on the Siberian shelf will require the development of special integrated technologies for ice navigation and environmental monitoring, including wild nature control.

The present study is based on the availability of ERS SAR images archive for the western part of the NSR, collected by NERSC/NIERSC within the framework of different international and national research programs since 1992. Since 1993, ice information derived from ERS-1/2 radar images has been transmitted regularly to the Russian nuclear icebreakers (I/B) operating along the Siberian coast and to the Marine Operational Headquarters in Dikson [5, 6, 7].

The above studies have the overall objective of validation, decoding and thematic interpretation of ERS-1/2 SAR data to provide real time satellite monitoring of the ice cover parameters in the Kara, Barents and Pechora Seas. They aim to plan and assess risks of round-the-year ice navigation in the Russian Arctic. Studies were accomplished through cooperation with Icebreaker Fleet Service of the Murmansk Shipping Company (MSC). Since the winter navigation of 1995/96, this study has been continued as the cooperative Project between the European Space Agency and the Russian Aerospace Agency.

The SAR application study demonstrates that satellite radar can be used now as the main and, sometimes, the only information source about different environmental processes and ice phenomena. Unfortunately, for the time being, the available resolution cannot help to investigate the individual marine mammals that have on average a length 1-2 m or more.

3.2. HARP SEALS POPULATION ECOLOGY

Seals are representative of high-level fodder chains in the ocean, and the assessment of their population and migration tendency is a very significant problem for marine biology, ecology, fishery, and for many other applications (in particular for the assessment of human hunting capacity). Such assessments could be realized more precisely in the winter season at separate regions of the Arctic at the time of mass accumulation of ice-associated forms of seals, that is the whelping and molting period. In order to estimate the population of seals, marine biologists perform annual inspections of the Russian sector of the Arctic Ocean, extending from the Barents and White Sea to the Bering and Okhotsk Sea. Number of adults and pups are checked separately. However, an aerial observation is a very difficult task because different representatives of seals are associated with different types of pack or fast ice.

Ice reconnaissance and visual inspection of sea mammals depend on weather conditions. Holes in the ice, ice drift, shadows from hummocks, false targets, etc., create many unfavorable circumstances. Sometimes harp seals cannot be found at the open sea. In counting the sea mammals, an additional problem may be the color of their fur: females have gray-white coat with dark dirt-pits, but the snow-white color of pups is masked on the ice. Therefore, the pups are distinguished vaguely in aerial photography images.

Whelping and molting processes of ice-associated forms of seals are studied only in a general way [8]. Let's describe briefly the host data about harp seals which have the largest population among seals. This knowledge helps to formulate monitoring of sea mammals via instrumental satellite-airborne technology.

There are three populations of harp seals (or Greenland seals) in the Arctic: White Sea, Newfoundland, and Jan-Majen populations of Greenland seals. These marine mammals are important and unique parts of the wild nature; each population strongly depends on environmental conditions and is stressed by anthropogenic factors. The whelping of the White Sea population of Greenland seals begins every year during the end of February and early March. The whelping period includes proper whelping and the lactation period. A major component of whelping is the changing of the fur of pups. Only after the change of the uterine fur, pups are ready to carry out self-independent living. These processes continue for about three weeks. The contact of pups with water at this time is fatal for them. Laws of the population ecology require that harp seals select the safest ice conditions to protect and save the newborn generations. Seeking by SAR the optimal ice features for whelping is the fundamental objective of an integrated monitoring technology. The initial stage of the second event, or the mass molting of adults, starts at the beginning of April.

3.3. DATA SET

Within the framework of this study, NERSC/NIERSC archive of ERS SAR and RADARSAT data were used, including more than 30 PRI images of the inhabited areas of harp seals in the White Sea. Part of them was received almost in real time, synchronously with airborne observations. The interval between SAR surveys and airborne and *in situ* measurements amounts to 5-6 hours. Composite ice maps based on SAR and other information were prepared. Separate scenes of visual and infrared satellite data for cloudless surveys were used. The archive of SMMR/SSMI daily scenes were available for assessment of the ice climate. Ice boundaries were determined by visual and NOAA/DMSP satellite microwave data.

ERS SAR signatures for mass congestion areas of seals were verified by *in situ* observations and compared with radar contrasts of pack and fast ice zones that are avoided by seals for whelping. ERS SAR images were received before the airborne observations at the airport of Arkhangelsk, and, after verification, they were used as the principal document for ice patrol and reconnaissance of whelping patches. Aircraft PINRO provided IR, visual and UV airborne measurements, which were used for computation of the seal population (Fig. 1). Multi-spectral information avoids the influence of daytime (shadows) and weather conditions (false heat targets) and increases the accuracy in counting the harp seals (including taxation of adults and pups, separately). It has been proved that SAR short period scenes can be used for assessment of ice drift parameters (direction and velocity). Long-term SAR series are used to study the relation between the arrangement of whelping rookeries and the annual modification of hydrological features, weather conditions and ice dynamics in the southern and central parts of the White Sea (so-called the "Basin" and the "Neck") [9, 10, 11, 12].

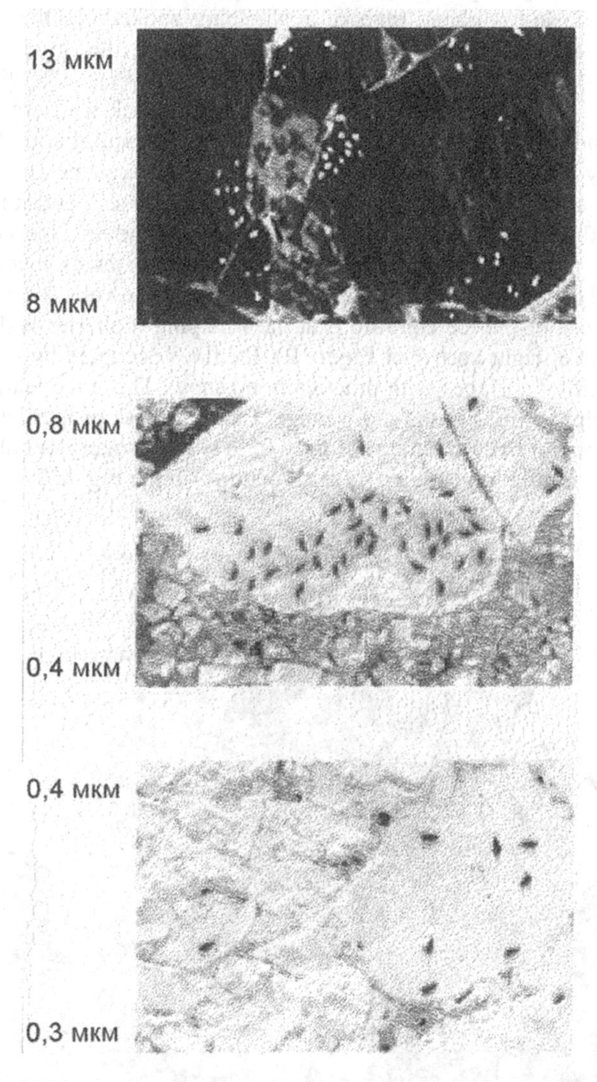

Figure 1. Multi-spectral technology of an aerial survey: a) IR – high thermal contrast seal-ice; b) visual – identification; c) UV – high contrasts of pups with high resolution and filtration.

4. Whelping Migration of Greenland Seals (White Sea Population): Results of the Validation Program

Figures 2 a and b show, respectively, an ERS-2 Synthetic Aperture Radar (SAR) image of 12 March 1997 and a photograph of a whelping rookery in the White Sea. The photograph was taken simultaneously with the sub-satellite image at location **A** indicated in the image. The satellite image covers the "Neck", "Funnel" and the southwestern part of the Mezen Bay. Harp seals select pack ice here traditionally for mass whelping and molting. The Island Morzgovets is the natural boundary between the whelping rookery zone and the mass molting area. As it can be seen from the SAR image, there is a radar contrast between these two zones, and there is a signal heterogeneity within the bounds of the whelping area. Studying ice cover parameters and the behavior of seals inside the mass accumulation zones of marine mammals by using satellite radar portraits of the sea is the main task of the validation program.

A typical feature of the ice conditions at the whelping rookeries is the thin first-year level pack ice (lower right corner of Figure 2). Ice floes occupied by seals for whelping have ideally unruffled surfaces with practically no snow. These ice features can be used as tracers of whelping areas in the radio range. Exactly, these properties are responsible for a specific character of signatures, not only for a single rookery but also for the whole region of mass accumulation of seals (areas A_1 and A_2 in the Figure 2).

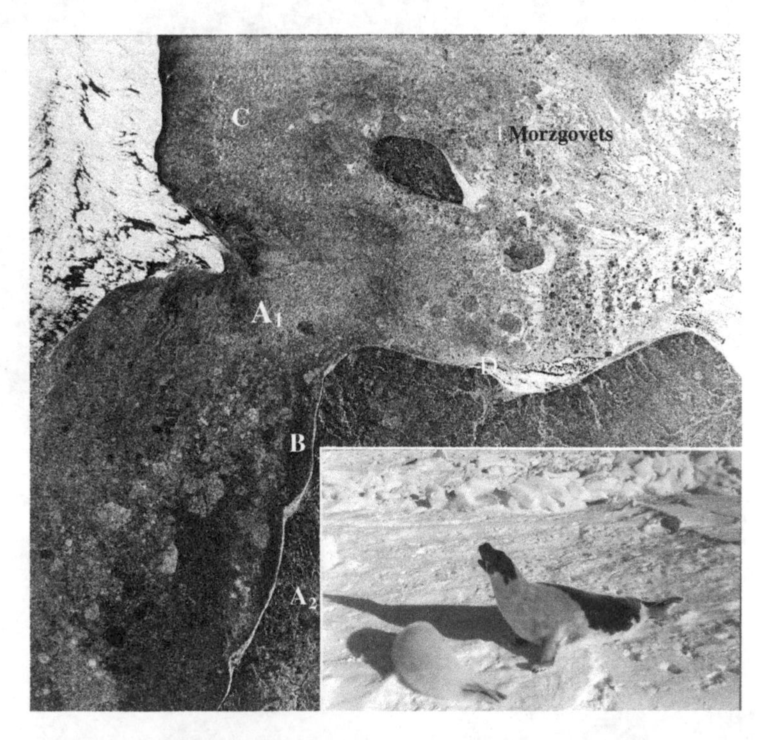

Figure 2. (a) ERS SAR image from the White Sea on 12 March 1997; 08.39 GMT; and (b) photograph of the whelping rookery obtained on the same day.

The integrated monitoring technology provides the thematic decoding of airborne instrumental measurements. IR, UV and visual airborne data are processed automatically in an operational mode. Ice maps and images of seals (first version) are composed during the flight. Final versions are prepared after the flight. The presence of experts on ice and seals (sometimes from different countries) onboard the aircraft "Arktika" is a permanent condition of validation programs. Helicopter Myl'-2 is employed for reconnaissance, collection of ice cores and water samples, and for registering harp seals and other ice-associated sea mammals.

Comprehensive analyses showed the wind speed at the time of SAR survey of 12 March 1997 to be 8-10 m/s (NW). As it has been proved, SAR fixes the position and the displacement of the ice massif that is selected by harp seals for whelping distinctly. Earlier, the wind was in the SE direction; it divided the rookery zone into two parts $\mathbf{A_1}$ and $\mathbf{A_2}$.

SAR signatures of the mass congestion areas of sea mammals were verified by *in situ* observations. These interdisciplinary investigations allow relating the radar contrasts of whelping zones with the type of ice and with the solidity and the salinity of this ice. An important part of the technology is to develop the relation between the whelping ice floes and the origin of water masses forming these saline or brackish water ice types. The variation of the salinity of sea waters at different parts and gulfs of the White Sea, formation of level lubricious ice zones, regional features of snow parameters, arrangement of fractures, cracks and other holes in the ice are studied with respect to the variation of the solidity ice floes. ERS SAR signatures for the mass congestion areas of seals are compared with pack ice zones that are avoided by seals as uncomfortable (Figure 3 a and b).

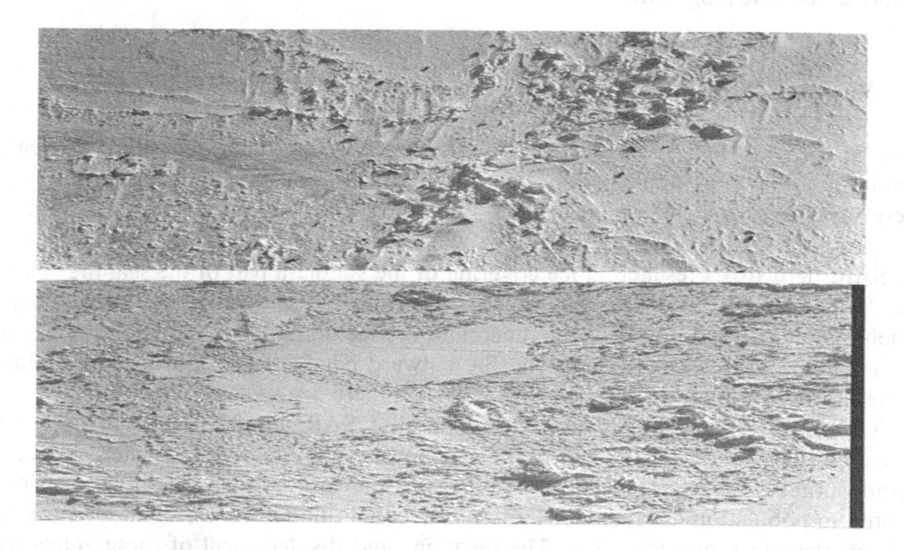

Figure 3. (a) Ice convenient and (b) inconvenient for whelping, that has the contrast of signatures: dark and bright white signals, correspondingly. Airborne photo survey, White Sea, 19 March 2000.

Ice floes more suitable for whelping are situated within the outer limits of the "Basin" and inside the "Neck". Low saline, strong and solid pack ice zones are formed exactly in these areas.

White Sea is the inland water area, and possibilities of ice cover contouring and the identification of fast ice and pack ice zones are the advantages of SAR surveys. They help to improve the organization of airborne inspections because the Greenland seals avoid fast ice zones for whelping. The elicited fact is the methodological basis of satellite-airborne technology for seal reconnaissance.

Chemiluminiscence analysis of samples allows relating the rookery location to certain types of water masses. Ice commonly used for whelping is thin, second stage, first-year FY ice (30 - 70 cm). Originating from brackish-waters, it is solid, strong, and resistant against any mechanical pressing, breaking, and abrasion. Freshly salted and snowless, it has special electrical and radiophysical parameters.

It is identified that the ice most suitable for whelping has deep dark SAR signatures (zones $\mathbf{A_1}$ and $\mathbf{A_2}$ in Fig. 2 a). These areas have a contrast with the surrounding waters and the ice areas with brighter radar return signals. Multi-level investigations demonstrate that seals avoid ice with gray, gray-white and bright white signatures as unfit for whelping (zone **B** in Fig. 2). Bright white signals correspond to wavy open water surface (zone **C** in Fig. 2) and broken ice floes where harp seals are wrecked. The land surface has dark signatures differing from the ice cover (zone **D** in Fig. 2).

Classification of SAR signatures for different types of ice and identification of the mass accumulation areas of seals allow suggesting SAR imagery for optimal shipping and ice routing. Murmansk Shipping Company (MSC) and other prominent shipping companies from Russia and foreign countries work in the White Sea round the year, and they need to pay the penalties for involuntary damage to the representatives of wild nature at the whelping period.

5. Whelping Migration: Population Assessment for Different Winter Severity

The wild life in the Arctic Ocean strongly depends on winter severity and weather conditions [8]. In the framework of this study, different ice conditions in the White Sea were investigated, including the 1998/1999 winter season that was severe at the north-western part of Russia and the 1999/2000 winter season that was warm. As it was fixed by SAR, hard frosts resulted in ice covering of the southern part of the sea, the "Neck", the "Funnel", and the Mezen Bay. It created problems for seals at the initial stage of whelping and caused it to penetrate deep into the ice massif. This situation implies that the mammals selected ice floes along the outward edges of the "Basin" for whelping, using the recurring cracks and openings in the ice there.

The other parameters retrieved from satellite SAR are the total and partial ice concentrations of ice. Multi-level data analysis for the warm winter conditions demonstrates that the arrangement of floe size at warm winter conditions is closely related to population destruction. Harp seals avoided thin ice zones, small ice floes (20-100 m diameter), and ice cakes. The locations and displacement of these zones were detected using SAR series; small floes and ice cakes have the bright radar contract.

Another important aspect of the integrated monitoring technology is the detection of ice drift vectors and the migration tendency of seals. Opportunity to fix, by SAR, the blocked-up pack ice zones inside the "Basin" and "Neck" can be used for predicting the decrease in seal population.

Field experiments to check the reduction of seals in summer-fall seasons were organized periodically. For example, in 1999, they were carried out at the "Neck", Dvina and Mezen Bays (March-April), at Dvina Bay and Solovky Archipelago (June), at the "Neck", Dvina, Chekh, Pechora and Khaypudirskaya Bays (September-October).

The field campaign of 2000, which had similar aims, was organized at the "Neck", Dvina, and Mezen Bay. The incoming stream of Barents Sea waters along the Kola Peninsula (Murmansky Coast) was also studied.

The central part of the White Sea is distinguished as a region of very dynamic hydrologic processes. Here, the monitoring of peculiar properties of floating whelping floes by SAR and the determination of drift vectors and seal migration are very important problems. The analysis of ERS SAR images for the "Neck" and "Basin" on 3 consecutive days was used for retrieving the spatial and temporal displacements of the aforementioned dark zones A.

SAR allows fixing the adverse N/NE winds which block up pack ice inside the "Basin" (or "Neck") for a long time, sometimes till the end of April. These ice conditions create a crucial situation when harp seals cannot reach finny water areas at the northern part of the White Sea and at the border with the Barents Sea. According to [8], enclosure of the direction of the traditional migration leads to destruction in seal population. Mass loss of pups was fixed in March-April in 1999. Later, in the summer season, the research vessel "Ekolog" met adults in unusual regions of the White Sea.

The winter of 1999/2000 was abnormally warm in Russia. Similar situations happened only eight times within the last 120 years. The average temperatures in February and March in Arkhangelsk exceeded the normal degrees of +4° and +3°C, respectively. Relatively thick ice, sufficient for organization of whelping rookeries, began forming only at the third quarter of February when air temperatures in the White Sea region fell down to – 30-35°C. Ice-free conditions dominated at the main part of the White Sea, practically the whole of March-April period. In this situation, it was difficult even to find the rookeries without satellite data.

Identification, by SAR documentary, the crucial ice and weather conditions that are related to the mass destruction of seal populations makes it possible for environmental institutions to suggest different proactive steps for effective mitigation of ecological catastrophes. Some of these steps are conservation actions, active influence on the ice, lowering the annual shoot off quotas of sea mammals, etc. Prediction of ecological catastrophes and mass destruction of seals in the White Sea by SAR was done for the first time in March 1999. Then, a similar warning was provided in May 2001 (the scale of the second event was weaker).

Monitoring of ice and harp seals via airborne and satellite observations help to define schemes for monitoring other seals. For example, the study of the multiyear ice cover parameters in connection with whelping migration of hooded seals allows the modification of the integrated technology for monitoring these sea mammals. According to a preliminary assessment, hooded seals can be studied in association with this type of drift ice. Ringed seals can be investigated in relation to fast ice zones at seas and lakes;

SAR satellite charting of the hummock arrangement is the basis for the identification of ringed seals through rookeries.

6. Results of the Application of Satellite-Airborne Technology for Monitoring of Ice and the Number of Ice-Associated Seals

The results of the thematic decoding of ERS SAR observations and airborne enumeration of the number of harp seals in the White Sea at the whelping period (pups and adults were checked separately) are described in the following.

The observed area covers Dvina Bay, northeastern part of the "Basin", "Neck", Onegsky, Zimny, and Tersky Coasts (Figure 4). The majority of the brackish-water ice originated at the Dvina Bay is broken by E/SE winds and displaced westward; thin first-year (FY) ice is dominant. Eastern winds affected Dvina Bay and a considerable part of the "Neck" close to the Zimny Coast, acting jointly with Zimnegorsky stable current fractured ice along the outer border of the "Basin". Compacted ice edge (bright signature) restricted this area for mammals.

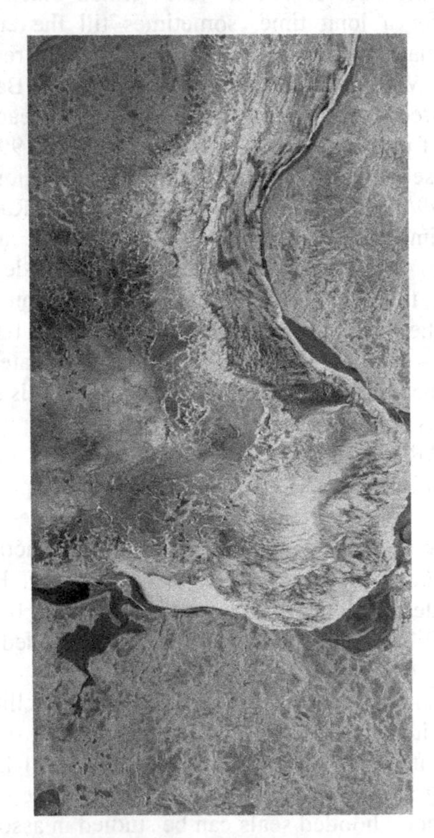

Figure 4. The initial stage of whelping: ERS SAR image 4 March 1999 08.41GMT.

The ice situation was appropriate for whelping of 4 March'99, when strong southern winds pressurized ice massif and resulted in additional ice breaking. Winds opened the fractures between the eastern borders of the "Neck"; ice originated in Dvina Bay was displaced northwestward. Polynyas at the "Neck" and Dvina Bay are covered with slush, grease ice. Southern winds were propitious for ice drift toward to the Barents Sea, and this was a favorable situation for seal migration to the North. Harp seals were expected to accumulate along the border of "Basin" and inside the central part of "Neck".

Polynya along the Kanin Cape was covered by level gray/gray-white ice (Figure 5). Anti-cyclone circulation prevailed, and thin fragmented FY ice was dominant. N/NE winds ruffled the openings at the Mezen and Chekh Bay. Freezing of fractures allows the evaluation of air temperatures to be minus 10-12 °C. The Zimniberezny stable current was traced as consecutive dark and gray ice blocks elongated from SE to NW. SAR showed unfavorable ice features for whelping migration; sea mammals were still trapped inside the limits of "Basin" and "Neck".

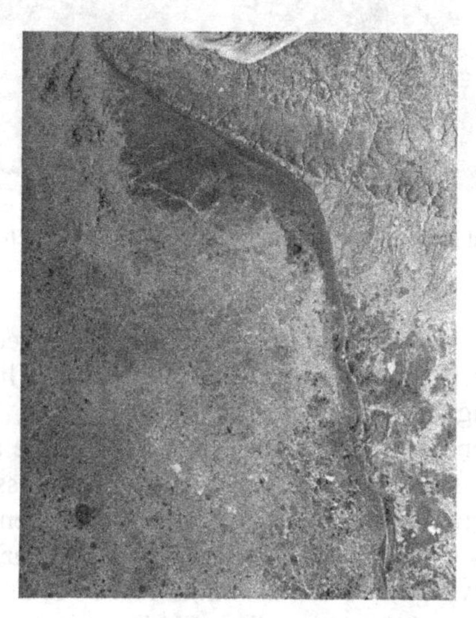

Figure 5. The first stage of whelping: ERS SAR image 17 March 1999 08.32GMT.

Figure 6 shows the observed areas of the Mezen Bay, "Neck" and "Funnel"; Murmansky, Tersky and Zimny Coasts. N/NE winds continued to hinder ice drift towards the Barents Sea. Thin FY ice was dominant. Separate ice floes were marked, respectively at 17 and 20 March SAR scenes; ice drift parameters were assessed (speed and direction). Floe with the initial coordinates of 67°10'N/42°E changed the position to 67°N/41°50'E. This meant that the ice drift was directed to the SSW. 3-days drift distance was 25-30km. Ice drift to the south disturbed the migration features and continued to confine the mammals to the inner part of the White Sea. Ice breaking and fracturing resulted in rookery destruction. SAR detected new holes in ice and rafting processes, and air temperatures were evaluated as minus 10-12 °C.

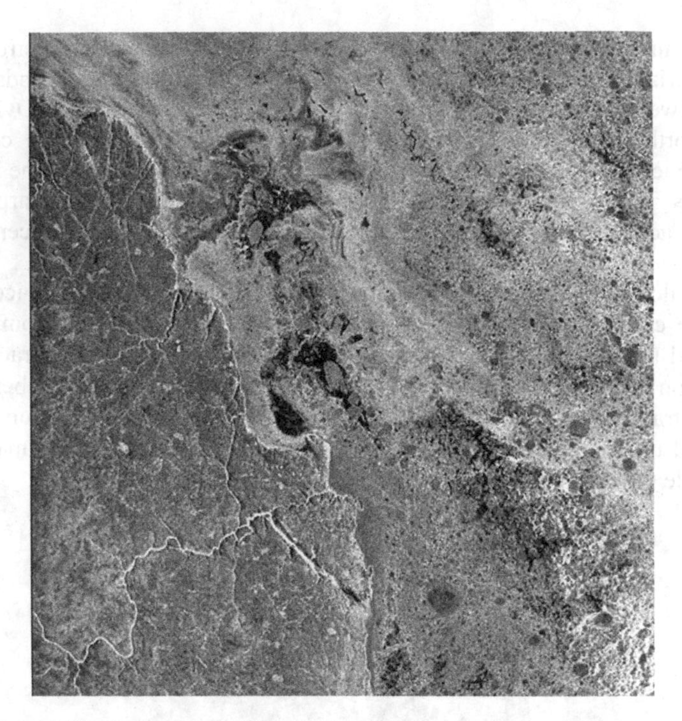

Figure 6. The final stage of whelping: ERS SAR image 20 March 1999 08.38GMT.

At the first quarter of April, most parts of the White Sea became commonly ice-free; ice floes with the newborn generation reached the southern limits of the Barents Sea. Mammals started molting at this time. According to SAR, the spring of 1999 was extremely unfavorable for Greenland seals. Almost the whole area was covered by ice; ice concentration was close to 10 tenths, and freezing processes continued. NE winds were weak (2-3 m/s), but thickening of the ice was evident. Ice covered even the Polynya along the Tersky Coast and saline Barents waters near Cape Svyatoy Nos (dark nilas). Air temperatures were evaluated as minus 15-20 ^{0}C. Different types of first-year ice were determined; thin and thin to medium FY ice was dominant. Ice cover got very fragmented; the melting processes did not start at the region.

The modifications of the whelping conditions in the 1999/2000 winter season are demonstrated in Fig. 7. SAR series show that ice conditions at the beginning of the winter were unfavorable for whelping, but strong frost during the third quarter of February saved the situation. Ice conditions were quite suitable for whelping migration at the beginning of March when passive ice drifting stayed stable in the direction of the Barents Sea.

Figure 8 presents the results of airborne investigations of whelping migrations in 2000. The accumulation area of harp seals was determined by aircraft PINRO; mass rookery location during the second quarter of March'2000 was situated close to the Tersky Coast. These rookeries were separated from another suitable ice zone by winds and by the incoming flux of Barents Sea's stable current.

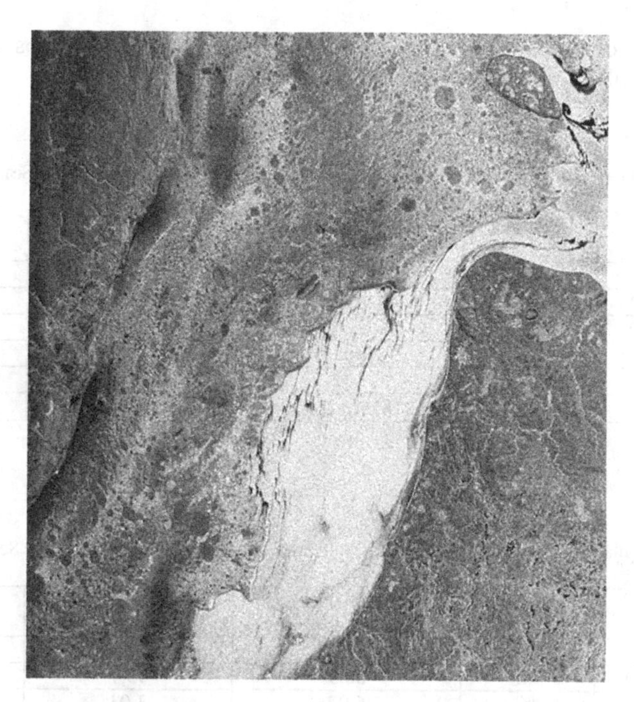

Figure 7. Inter-annual variability of seal migration features: ERS SAR image 20 March 2000 08.35GMT.

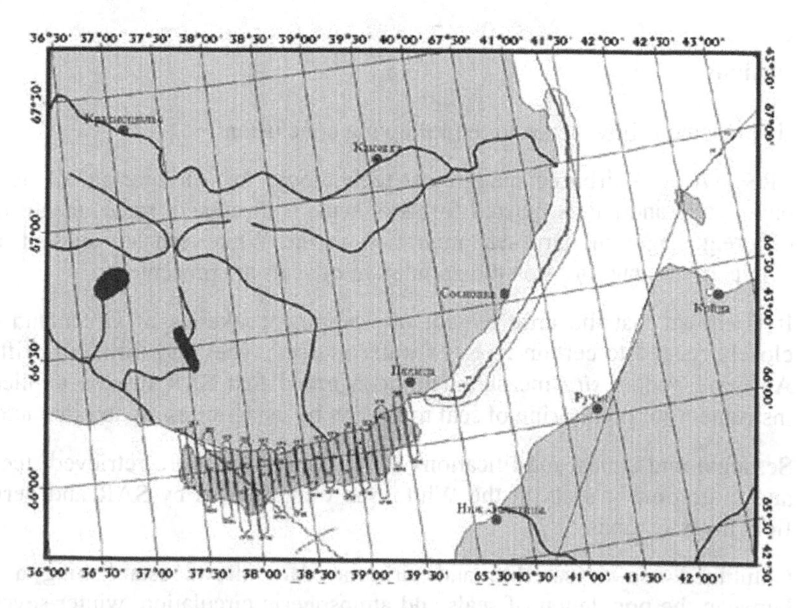

Figure 8. Results of IR charting of seal population: aircraft "Arktika", White Sea, 19 March 2000.

The results of computations on the population of harp seals, pups and adults, are given in the Tables 1 and 2, respectively.

TABLE 1. The population of Greenland seals (pups); aircraft "Arktika" measurements, White Sea, 18 March 2000.

Parameters	**Units**	**Rookeries (compact zones)**	**Rookeries (scattered zones)**	**Rookeries (total)**
Area	km^2	1345	3089	4434
Area (IR survey)	km^2	80,3	93,7	174
Area (IR survey)	%	5,97	3,03	3,92
Number of pups	pieces	13024	3546	16570
Density	pieces/km^2	162	38	76
Numbers	pieces	218200±20600	116900±24900	335100±32300

TABLE 2. The population of Greenland seals (adults); aircraft "Arktika" measurements, White Sea, 18 March 2000.

Parameters	**Units**	**Rookeries (compact zones)**	**Rookeries (scattered zones)**	**Rookeries (total)**
Area	km^2	1345	3089	4434
Area (IR survey)	km^2	80,3	93,7	174
Area (IR survey)	%	5,97	3,03	3,92
Number of pups	pieces	12780	6151	18923
Density	pieces/km^2	159	66	94
Numbers	pieces	214100±13600	202900±33400	417000±36100

7. Conclusions

The study presented above leads to the following conclusions:

- ERS SAR – airborne integrated technology for monitoring of ice cover parameters and ice-associated forms of seals is discussed; radar images of ice at different stages of harp seal migration in the White Sea are studied, and the results of computations on the population of seals are presented;
- It is shown that the arrangement of whelping rookeries of Greenland seals is closely related to certain types of water masses, types and forms of drifting ice. Airborne and *in situ* measurements confirmed that SAR can be applied as an instrument for monitoring of seal migration by using ice as an abiotic factor;
- Seasonal and annual modifications of ice parameters were retrieved. Ice regime and migration of seals in the White Sea were studied by SAR and verified by field investigations;
- Combined use of satellite and airborne data allows establishing a relation between the population of seals and atmospheric circulation, winter severity and

ice type arrangement. It was concluded that the number of seals in the White Sea and the neighboring regions was significantly reduced in 1999;

- Preliminary studies show that the SAR airborne technology can be used for monitoring the migration of different representatives of harp seals and other ice-associated marine mammals, i.e., ringed and hooded seals, polar bears, etc.;
- The management of sea mammals in the Arctic is realized via ENVISAT that provides a high spatial resolution of ice in the order of 500 km. Spaceborne SAR in combination with airborne IR, UV and visual surveys are the most effective instruments for investigating wild nature.

8. Acknowledgment

The authors would like to thank ERS Help and Order Desk for supplying the SAR data. This work is supported by ESA; 20 ERS SAR images of the White Sea and the neighboring regions were obtained within the framework of the A03.440 Project.

9. References

1. Chernook, V.I., Johannessen, O.M., and Melentyev, V.V. (1997) Connection between distribution of Harp Seals and ice cover parameters determined using ERS-2 SAR imagery, *Proc. of the ICES/NAFO Workshop on Survey Methodology for Harp and Hooded Seals*, Copenhagen, 28.08-02.09.97, 22-27.
2. Melentyev, V.V., Chernook, V.I., and Johannessen, O.M. (1998) Analysis of ice dynamics of Arctic Seas for research of the Harp Seals migration in the White Sea using satellite data, *Earth Obs. Rem. Sens.* **5**, 76-93, (in Russian).
3. Chernook, V.I., Kuznetsov, N.V., Melentyev, V.V., Yegorov, S.A., Meisenheimer, P., Innes, S., and Timochenko, Y.K. (1999) Experience of synergetic use of visual, IR and high resolution SAR data for investigation the number population and migration of harp seals in the White Sea region, in *Proc. of the Fourth Int. Airborne Remote Sensing Conference and Exhibition / 21st Canadian Symposium on Remote Sensing*, Ottawa, Ontario, Canada, 21-24 June 1999, Vol. 1, pp. 574–580.
4. Melentyev, V.V., Pettersson, L.H., and Chernook, V.I. (2001) ERS SAR data application use for studying sea ice parameters and retrieving of Greenland seals migration, in *Proc. of the ERS-ENVISAT Symposium*, Goethenburg, Sweden,. 16-20 October 2000, pp. 567-578.
5. Johannessen, O.M., Sandven, S., Pettersson, L.H., Miles, M., Melentyev, V.V., and Bobylev, L.P. (1997) Northern Sea route ice monitoring by satellite radar data, in H. Yamagushi (ed.), *Proc. of the 16th Int. Conference on Offshore Mechanics and Arctic Engineering*, Book No. HO 1083, pp. 9-17.
6. Johannessen, O. M., Sandven, S., and Melentyev, V.V. (1996) ICEWATCH-Ice-SAR monitoring of the Northern Sea Rout, in *Proc. of the Second ERS Applications Workshop*, London, UK, 6-8 December 1995, ESA SP-383, European Space Agency, Paris, pp. 291-296.
7. Johannessen, O.M., Sandven, S., and Melentyev, V.V. (1997) ERS-1/2 SAR monitoring of dangerous icve phenomena along the western part of Northern Sea Route, *Int. J. Remote Sensing*, **18(12)**, 2477-2481.
8. Chernook, V.I. and Kuznetsov, N.V. (1995) *Methods of Airborne Surveys of Greenland Seals*, PINRO Publ. House, Murmansk. (in Russian).

9. *Hydrometeorology and Hydrochemistry of the Seas of the USSR,* Vol. 2 *"The White Sea"* (1991), Gidrometeoizdat, St.Petersburg. (in Russian).

10. Melentyev, V.V., Johannessen, O.M., Chernook, V.I., and Arkhipov, A.V. (1998) Investigations of ice regime and hydrology of the White Sea in conditions of 1994-95 abnormally warm winter, *Proc. of the Russian Geography Society,* **130(4)**, 47-64 (in Russian).

11. Melentyev, V.V., Kloster, K., Chernook, V.I., and Arkhipov, A.V. (1998) Study of ice regime and winter hydrology of the White Sea using ERS SAR surveys, *Earth Obs. Rem. Sens.* **4**, 73-92, (in Russian).

12. Kondratyev, K.Ya., Melentyev, V.V., and Nazarkin V.A. (1992) *Space Remote Sensing of Waters and Catchment Areas (Microwave Methods)*, Gidrometeoizdat, St. Peterburg. (in Russian).

Part VI

ENVIRONMENTAL DATABASES

INTEGRATED APPLICATION OF UNITED KINGDOM NATIONAL RIVER FLOW AND WATER QUALITY DATABASES FOR ESTIMATING RIVER MASS LOADS

I. G. LITTLEWOOD
Centre for Ecology and Hydrology
Wallingford, OX10 8BB, United Kingdom

Abstract. Temporally consistent time series of river mass loads are required to assist with the rational management of the environment (e.g., annual loads of nutrients to lakes, estuaries and coastal sea areas). Issues concerning the integrated application of the United Kingdom national databases for river flows and water quality for calculating river mass loads are discussed. The paper focuses on the range and nature of error sources in river mass load estimates and warns against assigning unrealistically high precision to river loads estimated from the national datasets. Annual mass loads from the freshwater monitored area of Great Britain, 1974 to 1994, are presented for suspended solids, "total" nitrogen (NO_3-N + NO_2-N + NH_4-N) and zinc (dissolved and suspended).

1. Introduction

Environmental managers in the United Kingdom, Europe, and more widely, are becoming increasingly active in gathering and analyzing data to assist with monitoring contaminant fluxes from the land to the sea. One of the key objectives of the UK Land-Ocean Interaction Study (LOIS), for example, was "To estimate the contemporary fluxes of momentum and materials (sediments, nutrients, contaminants) into and out of the coastal zone, including transfers via rivers ..." [1]. International research programmes with similar needs for good estimates of fluvial contaminant loads are: (i) the European Land-Ocean Interaction Studies (ELOISE) of the European Commission [2], and (ii) the Land-Ocean Interactions in the Coastal Zone (LOICZ) scheme of the International Geosphere-Biosphere Programme (IGBP) [3]. Monitoring contaminant inputs to the North Sea from the land and the atmosphere is an international concern in northwest Europe [4].

Any rational assessment of the material transported by rivers (dissolved and suspended constituents) must deal with: (a) institutional aspects of data management (for both water quantity and water quality data), and (b) uncertainties in river mass load estimates (i.e., statistical aspects). Rational management of the environment must be based on an understanding of the processes that act as "sources" and "sinks" of material in rivers, lakes, estuaries, coastal waters and seas, and the incorporation of that understanding into operational models to assist decision making. Indeed, the quality of river mass load estimates (statistical uncertainties) and improving the understanding of processes are highly complementary aspects of modeling the land-ocean transfer of material by rivers.

Within their lifetimes, research programmes like LOIS, ELOISE and LOICZ undoubtedly generate new data of great value, often in vast volumes, e.g., [5]. Furthermore, such studies stimulate the use of other river flow and concentration datasets in order to estimate river mass loads over longer timeframes and for wider geographical areas than the studies themselves might allow. This paper is primarily concerned with the systematic estimation of annual river mass loads discharged from the freshwater-monitored area of Great Britain, from 1975 to 1994. The paper summarizes some of the work undertaken for the UK Department of Environment, Transport and Regions (DETR) [6] and for LOIS [7].

A prerequisite to the analysis was the bringing together of the national river flow and river quality databases in the same computing framework to allow systematic manipulation of those data for load estimation. The Harmonised Monitoring Scheme (HMS) water quality data were installed on the same relational database system used for the National River Flow Archive (NRFA). Special software (CORAL) was then prepared to access the HMS and NRFA datasets simultaneously and to apply a selection of different algorithms for estimating annual loads [8]. This paper discusses some of the uncertainty issues involved and presents selected results in a format not published previously.

2. River Mass Load Estimation in Practice

In principle, a good estimate of the load of a substance (determinand) that passes a non-tidal river cross-section over a period of time is given by the sum of products of frequent measurements of flow and determinand concentration at the river cross-section in question. The uncertainty associated with such an estimate, i.e., a combination of its accuracy and precision (see Fig. 1 for working definitions of these terms), depends on many factors, e.g.:

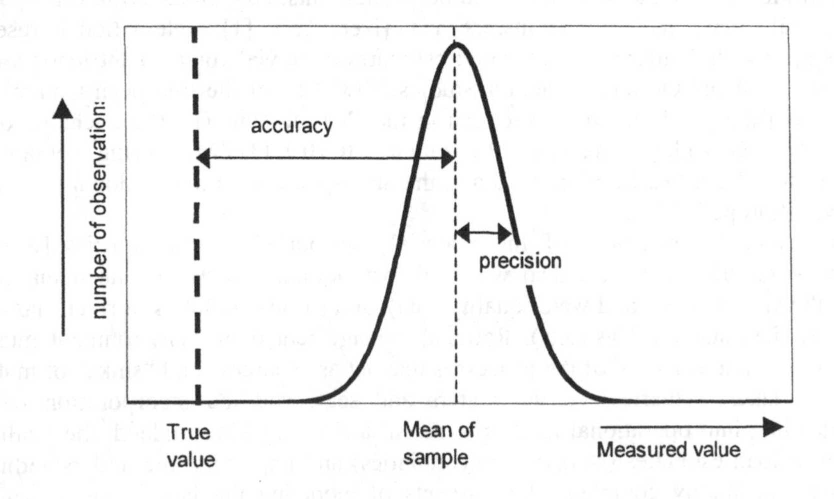

Figure 1. Definition sketch for accuracy and precision.

- river flow and determinand concentration measurement errors (including the representativeness of any bank-side sampled concentration with respect to the mean concentration for the cross-section);
- the closeness with which the actual temporal variations in flow and concentration can be reproduced from available flow and concentration data (the frequency of concentration measurements is often much less than that for flow); and
- the estimation algorithm employed for calculating the mass loads (e.g., some algorithms are inaccurate, and some yield better precision than others).

These and other factors[1] mean that the uncertainty in some estimates of river mass loads can be both large and variable in time, confounding comparisons between river mass loads for different catchments, the detection and interpretation of trends, etc.

When the focus of interest is on deriving a temporally consistent time series of river mass load for a group of monitored catchments (e.g., annual nitrate loads carried by all of the major UK rivers draining towards the North Sea), other factors can become important. For example, it is necessary to pay attention to any year-on-year changes in the network of catchments used for the calculations and to any years when the sampling frequency is very low. Adjustment for such factors may be necessary to provide a consistent time series with which to assess the combined effects of natural and anthropogenic influences over time (further analyses leading hopefully to resolution of these effects).

When the estimate required is the amount of material leaving the lower reaches of a river system, or entering the sea, and where routine measurements of flow and concentration at that point are not available, further factors still must be borne in mind. In the lower reaches of many major rivers in the UK, there are point discharges to the river or estuary from urban and industrial areas. Furthermore, within-channel processes in rivers and estuaries can act as a "source" or "sink" for a given determinand. The mass load will not then be simply the (measured) mass load at the upstream end of the reach plus an amount to account for "direct" or "point discharge" inputs in between, even when the variable concentration and quantity of those point discharges are known. Where routine within-channel measurements of flows and determinand concentrations are not available at such downstream locations, the estimation of the seaward flux of material requires modeling of the in-channel source and sink processes [9].

The most downstream HMS sites in the UK are above but usually close to the tidal limit, where the flow comprises freshwater and is uni-directional. Sampling-points in some other European countries are located further downstream, i.e., in tidal river reaches or in saline estuary waters [10], making meaningful comparisons of river mass load estimates for different countries (e.g., for the North Sea) problematical in some cases.

Below the most downstream sampling point, whether in a non-tidal river reach, tidal river reach or estuary, there will also be waterborne transfers of material from the land

[1] For example, values recorded as "less than the limit of detection, LOD" (large proportions of some HMS site/concentration datasets comprise < LOD values, and the LOD can change with time). Or, the relatively large uncertainty associated with measurements of high river flows (when the transport of material, e.g., suspended sediment, can be a large proportion of the total load).

by "diffuse" processes. These diffuse inputs are not amenable to calculation as essentially the product of flows and concentrations measured in open channels. For such cases, and including non-monitored (usually small) rivers draining coastal catchments directly to the sea, recourse has to be made, therefore, to estimating by other means, e.g., modeling the contributions of diffuse-source material to the total land-sea transfer of material. This may involve seeking relationships between specific river mass loads (e.g., tons/km^2) and rainfall, geology, soil type, land-use, etc. for catchments where measurements of flow and concentration in open channels are available, and applying such relationships to non-monitored areas.

It is evident, therefore, that, in the UK, a comprehensive approach to monitoring the flux of waterborne material from the land to the sea over large regions requires a complex blend of:

- frequent measurements of flows and concentrations in non-tidal river channels,
- an inventory of point discharge inputs below the most downstream sampling points (with records of their variable quantity and quality), and
- modeling of outputs from non-monitored catchments and of source/sink processes in rivers and estuaries.

During projects like LOIS, it may be possible, over the duration of the project (a few years at most), to co-ordinate the necessary measurement, database and modeling activities undertaken by the various institutions involved. Importantly, such studies can indicate where, and how, future efforts related to measurements, data management and analysis might be best directed. However, sustaining a comprehensive approach for monitoring the waterborne discharge of materials from the whole of Great Britain is not a viable option at the present time. But, by using the relatively long datasets from the NRFA and the HMS database, an attempt has been made to provide temporally consistent time series of annual river mass loads for the aggregated freshwater monitored area of Great Britain, extending over more than 20 years [7].

3. The Databases

Most of the data held in the HMS database and the NRFA have been derived from measurements by the Environment Agency (EA) for England and Wales and the Scottish Environment Protection Agency (SEPA). The NRFA, which is managed and maintained by the Centre for Ecology and Hydrology (CEH), formerly known as the Institute of Hydrology (IH), includes essentially unbroken time series of daily mean flows for over 1100 sites throughout the UK (more than 15 million values). Details of the NRFA and its data are available on the Internet[2]. The NRFA plays a major role as a source of hydrometric data and related information to assist with many research aspects of water resources management in the UK.

The HMS commenced in 1974 to provide a coherent source of water quality data to assist with estimating river pollutant loads discharged by UK rivers to the sea [11].

[2] http://www.nwl.ac.uk/ih/nrfa/index.htm

Since 1990, estimates of annual river mass loads have been supplied annually by the Department of the Environment, Transport and the Regions (DETR) to the Oslo[3] and Paris[4] Commission (OSPAR), who collate such information from countries bordering the north-east Atlantic. The NRFA and the HMS databases are the best-organized UK national datasets for river flow and river water quality data, respectively. However, additional river flow data and other water quality datasets exist (e.g., sub-daily river flows and non-HMS river quality data held by the EA and SEPA). In 1998, responsibility for the management and maintenance of the HMS database was transferred from the DETR to the EA.

There are about 200 HMS sites throughout Great Britain located at, or near, an NRFA river gauging station (but in some cases the HMS and NRFA sites are several kilometers apart). The areas draining to the lowest HMS sites throughout Great Britain, grouped by coastal regions, are shown in Fig. 2. For each sample of water taken for analysis, more than 100 determinands can be accommodated in the HMS database. Usually, however, about 25 determinands are recorded at a given site, many of which are "standard" (e.g., temperature, suspended solids, nitrate) while others are measured and recorded on the basis of local importance. The assemblage of determinands recorded at a site can vary with time.

Although the sampling frequency at an HMS site might be nominally weekly, monthly, etc., in practice, HMS concentrations are typically irregular in time (i.e., aperiodic), varying from about four to 60 days apart. The HMS database holds mainly river water quality data. However, to enable estimation of river mass loads without recourse to any other source of data or information, it also holds river flow data for the time (instantaneous flow) or day (daily mean flow) of each sample. A simple "mean of the sum of products" algorithm can be applied to estimate river mass loads. Unfortunately, the HMS instantaneous flow datasets are not always complete with respect to the corresponding concentration datasets. Furthermore, it was found [6] that both the instantaneous and the daily mean flow data in the HMS database often refer to the flow at the nearest NRFA site rather than at the HMS site which, as noted already, might be some distance away.

Previous work [12] has employed the HMS database for estimating river mass loads but, given the computing technology available at the time (1983), the estimation methodologies that could be applied systematically then were of limited efficacy compared to methods which can be applied systematically today. It is now possible to handle and manipulate very large databases in ways that were not a practical option before the 1990s. Databases like the NRFA and HMS can now be installed in the same system and accessed simultaneously, enabling the implementation of methodologies for river mass load estimation, which yield better results than can be derived solely from the HMS database. It has been shown [6, 7, 8] that, when the HMS database and the NRFA are employed together, rather than the HMS database alone, there is a synergy of information in terms of improved quality in the estimates of river mass loads.

---

[3] The (Oslo) Convention for the Prevention of Marine Pollution from Ships and Aircraft.

[4] The (Paris) Convention for the Prevention of Marine Pollution from Land-Based sources, 1974 (implemented 1978). The Oslo and Paris Commissions combined in 1994 (OSPAR).

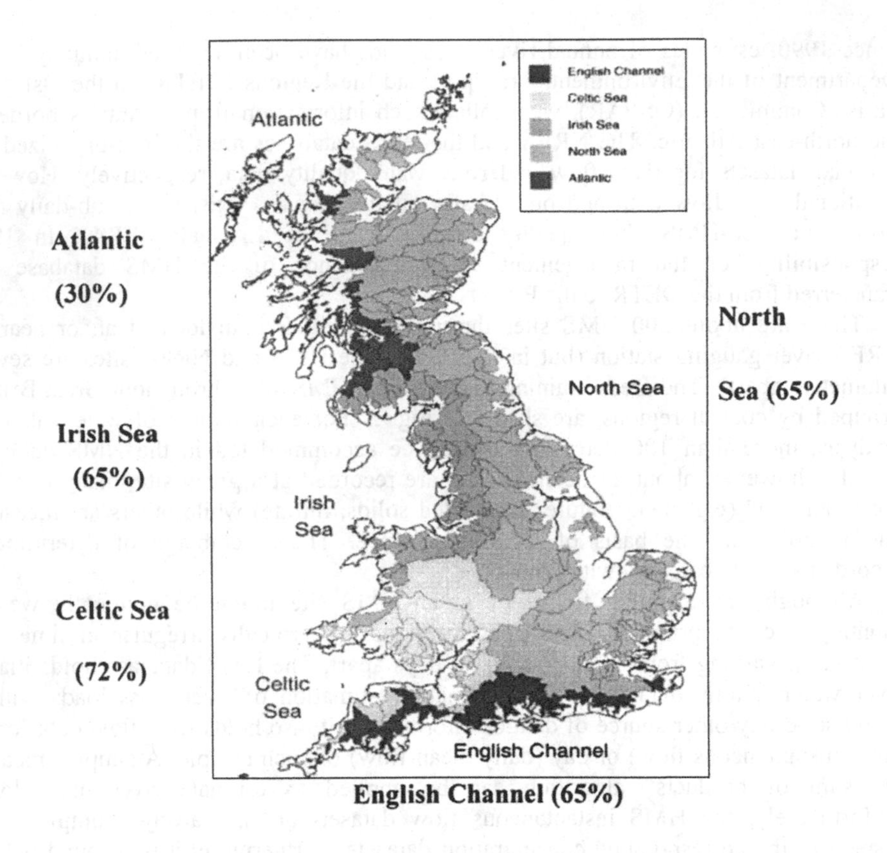

Figure 2. Areas of Great Britain draining to HMS sites (the percentages indicate the areal coverage in each of the five coastal zones).

4. Estimates of Annual River Mass Loads, 1975 to 1994

In the study referred to earlier [6, 7], annual mass load time series, from 1975 to 1994, for the spatial aggregation of all HMS catchments[5] located close to the tidal limit of major rivers in Great Britain (HMS_{GB}), were estimated for the 12 determinands listed in Table 1.

TABLE 1. Selected determinands

total nitrogen (NH_3-N + NO_2-N + NH_4-N)	cadmium
orthophosphate	nickel
suspended solids	mercury
zinc	lindane
copper	arsenic
lead	total phosphorus (dissolved and suspended)

[5] In this paper, aggregations or groupings of catchments exclude HMS basins nested within other HMS catchments.

Figure 3 shows: (i) the HMS_{GB} mean annual mass loads over the 20-year period between 1975 and 1994 for suspended solids, total nitrogen (NO_3+NO_2+NH_4), orthophosphate and zinc (dissolved + suspended); and (ii) a breakdown by groups of HMS catchments draining to five sea areas around Great Britain. Mass load estimates for the other 8 determinands in Table 1 were judged [7] to be relatively unreliable because of problems arising from one or more of several factors (low sampling frequencies, missing concentration data for some catchments in some years, <LOD values, and extreme outlier concentrations). Figure 4 shows the HMS_{GB} annual specific loads (tons/km^2) for the four determinands considered to be estimated the most reliably, again for sea area groups.

5. Concluding Remarks

The effective merger of the NRFA and the HMS databases discussed in this paper has yielded new information in the form of annual mass load time series (HMS_{GB}), from 1975 to 1994. Previously, only values from 1990 onwards were available.

Improved precision in river mass load estimates for some determinands can be achieved by applying the NRFA and the HMS databases simultaneously, rather than the latter database in isolation. However, it should be noted that some estimation algorithms (including the one employed to derive the estimates presented in this paper – see [7] for details) assume implicitly that the concentration data are periodic (i.e., spaced regularly in time). Such algorithms are not strictly suitable for application with HMS data, which are typically aperiodic. Some work has been undertaken, that takes account of aperiodicity in concentration data [7]. Where there is a useful relationship between flow and concentration, the concentration at each time-step between sample times can be estimated from continuous flow records, leading to calculation of load as the sum of products of flow and estimated concentration (with occasional measured concentrations) at each time-step. In order to make the best possible use of available data for river mass load estimation, a site/determinand-specific approach is required. Different estimation algorithms are 'best' for different groups of combination of site and determinand.

Where concentration frequency is very low, or a large proportion of the concentration data are less than the limit of detection (<LOD), the uncertainty in the estimated mass load (e.g., setting <LOD values to LOD/2) may be so large that it can be misleading even to give an estimate. River mass loads, when reported, should always be given with indicative uncertainties, remembering that the level of precision implied by the number of significant figures used to express the estimate is a fundamental issue. In a paper that presents river mass loads for the LOIS area [13], estimates for some determinands are tabulated to 11 significant figures, implying a precision of about 1 part in 10^{10}. This level of precision, rare even in controlled experiments in physics or chemistry laboratories, cannot be attained using real hydrometric and water quality data. Realistically, given the number and the nature of error sources involved (as discussed in Section 2), river mass loads with a precision of about +/-1% (3 significant figures) can be considered to be very good indeed. River mass loads given to more than 3 significant figures lack credibility, unless accompanied by clear justification of the higher implied precision.

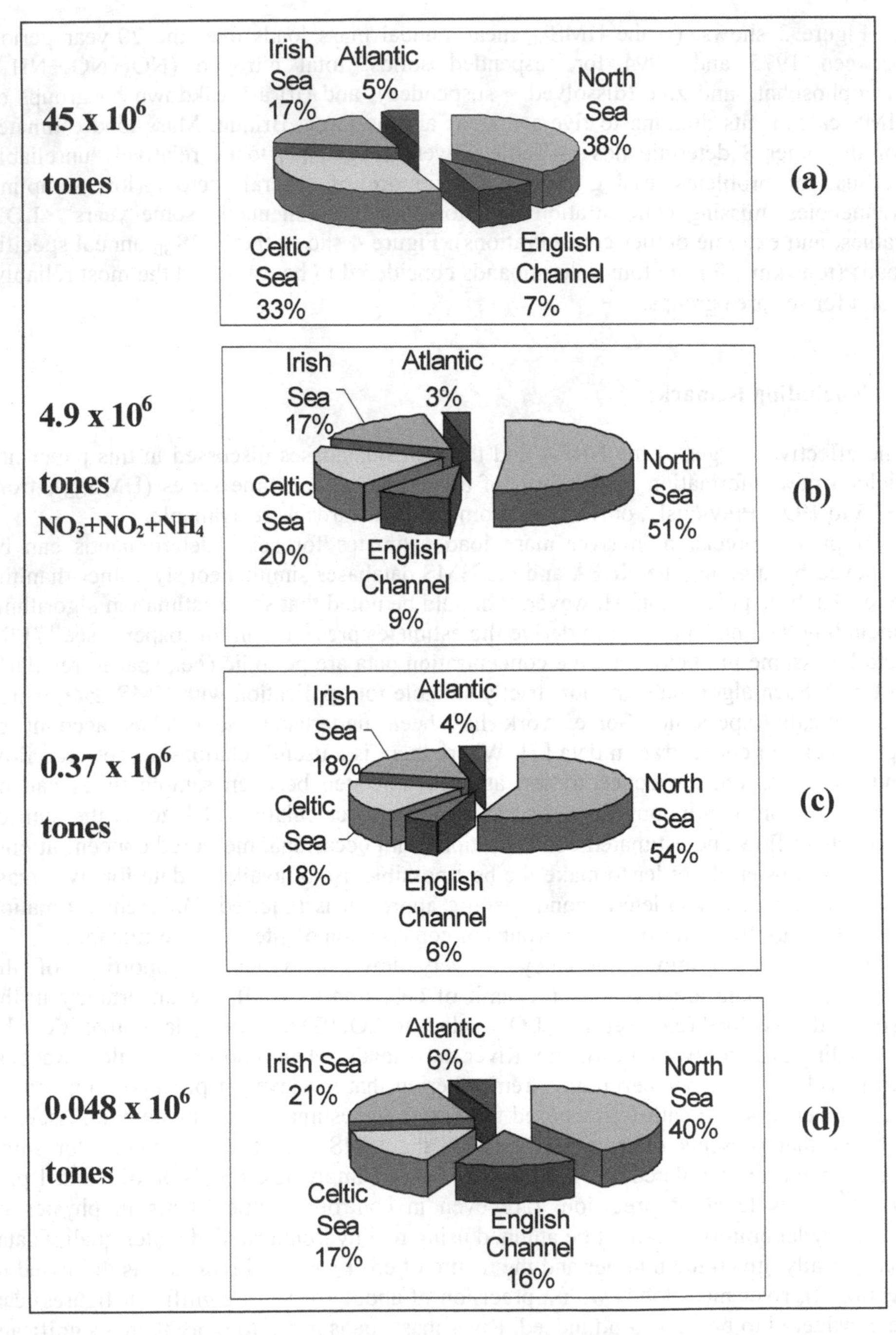

Figure 3. HMS_{GB} mean annual river mass loads by coastal zones, 1975-1994
(a) suspended solids, (b) total nitrogen
(c) orthophosphate, (d) zinc (dissolved + suspended)

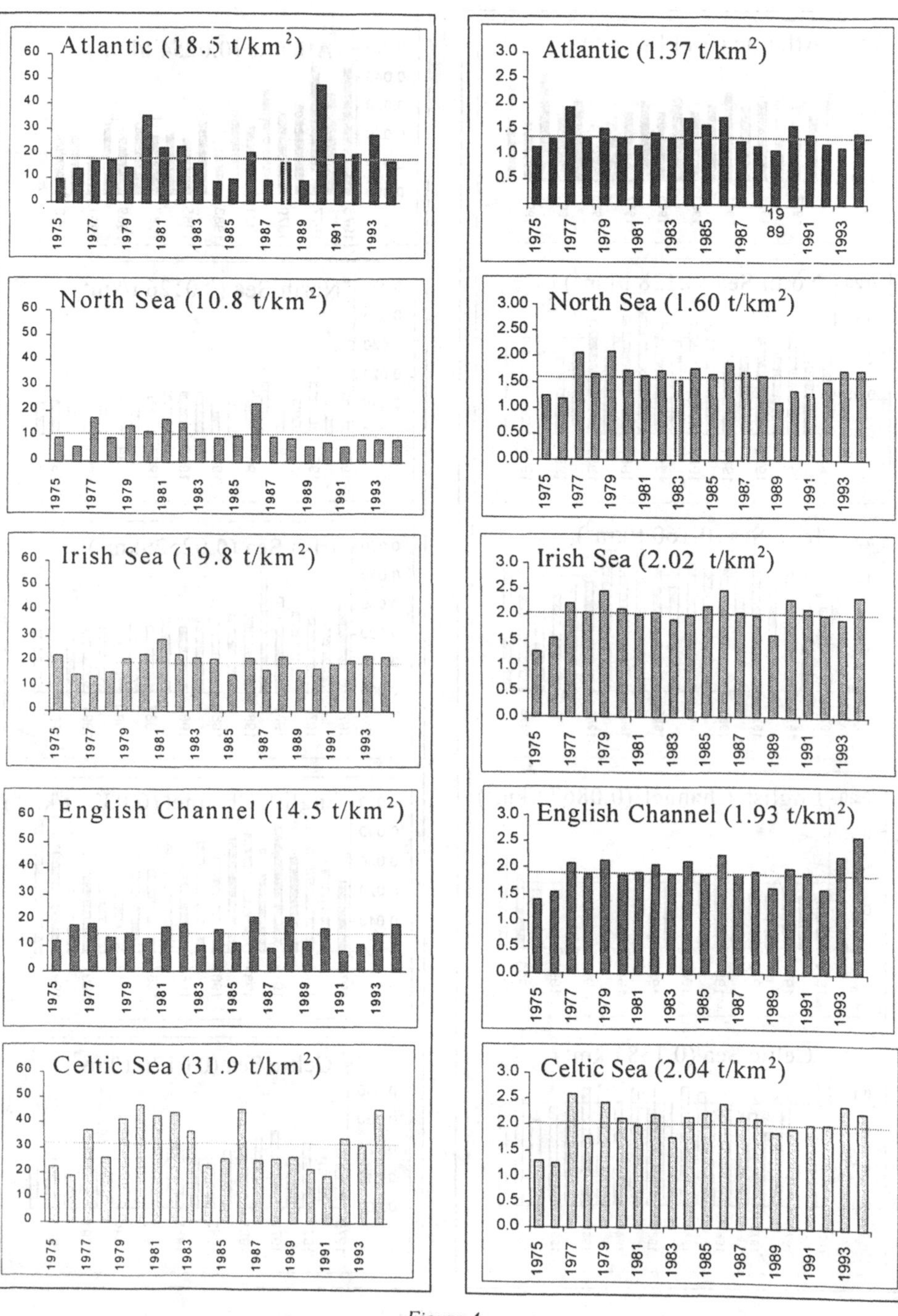

Figure 4.

(a) Annual specific suspended solids specific loads (1975-1994).

(b) Annual specific "total N" specific loads (NO_3+NO_2+NH_4), (1975-1994).

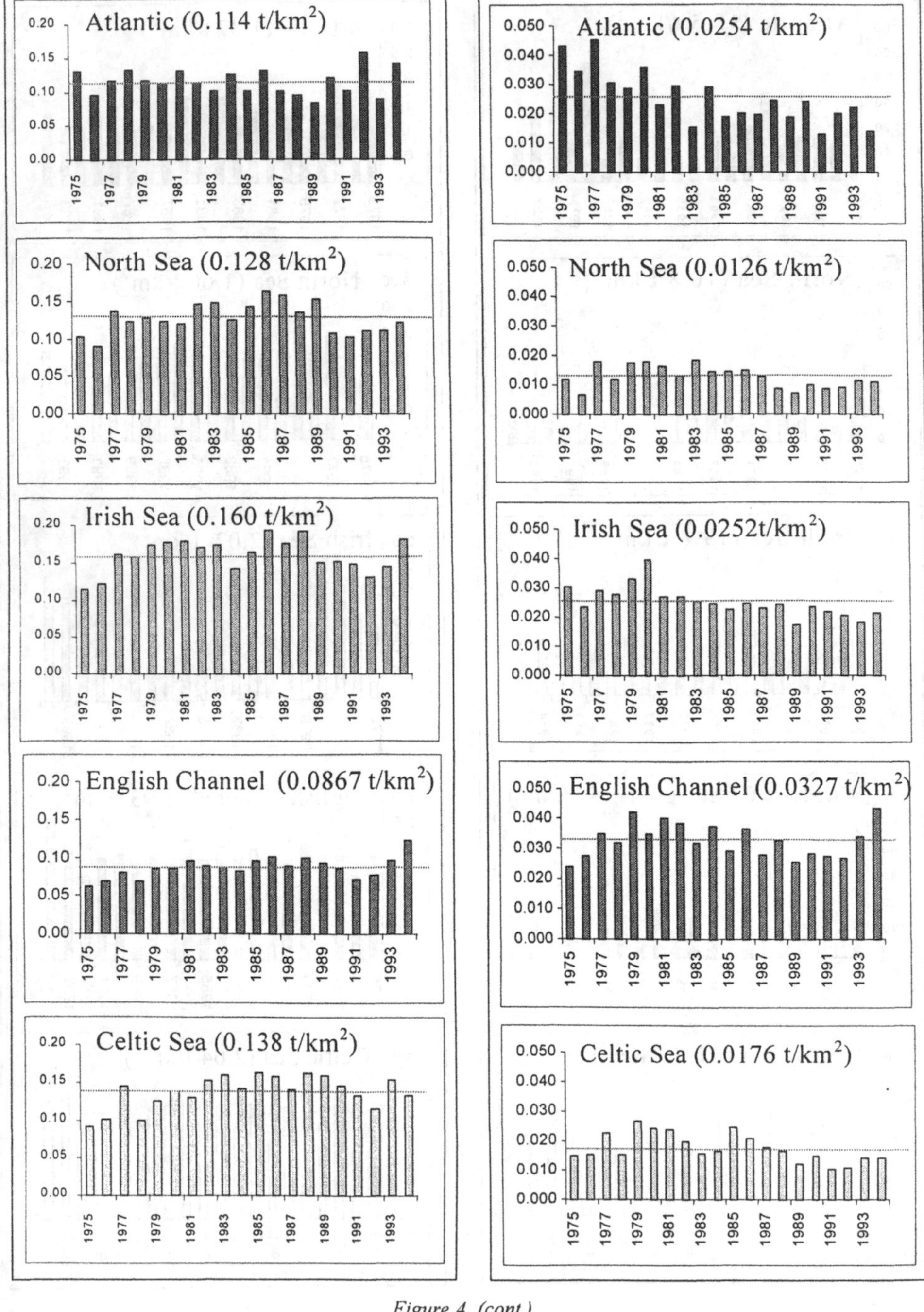

Figure 4. (cont.)

(c) Annual specific orthophosphate specific loads, 1975-1994.

(d) Annual specific zinc (dissolved + suspended) specific loads, 1975-1994.

Further harmonization of the UK river flow and quality monitoring networks, and their databases, is required. Across Europe, the provision of good-quality river mass load estimates should be placed high on the list of outputs required from combined river flow and water quality monitoring networks. Attention should be directed, in the first instance, to those determinands most amenable to measurement in rivers (e.g., excluding those determinands for which <LOD values at a given site comprise a large proportion of the total number of measurements). There is a need to move towards consistent monitoring of European rivers and the systematic estimation of river mass loads. For some determinands, it appears that the detection of mass load trends and the comparison of trends between regions will remain problematical due to large and temporally variable uncertainties in field measurements (especially for concentrations). Regarding uncertainties in river mass load estimates, it is better to be realistic about what can be achieved than to publish results with indicative precisions that cannot be justified (whether the uncertainties are stated explicitly or whether they are implied by the number of significant figures given).

6. References

1. Wilkinson, W.B., Leeks, G.J.L., Morris, A., and Walling, D.E. (1997) Rivers and coastal research in the Land Ocean Interaction Study, *The Science of the Total Environment* **194/195**, 5-14.

2. Cadée, N., Dronkers, J., Heip, C., Martin, J-M., and Nolan, C. (eds.) (1994) *European Land-Ocean Interaction Studies (ELOISE): Science Plan*, European Commission, Directorate-General XIII, Luxembourg, 52pp.

3. Pernetta, J.C. and Milliman, J.D. (eds.) (1995) *Land-Ocean Interactions in the Coastal Zone (LOICZ)*, International Geosphere-Biosphere Programme, Global Change Report No. 33, Stockholm, 215pp.

4. Andersen, J. and Niilonen, T. (eds.) (1995) *Progress Report: 4th International Conference on the Protection of the North Sea*, Esbjerg, Denmark, 8-9 June 1995, Ministry of the Environment and Energy, Danish Environmental Protection Agency, Strandgade 29, DK-1401, Copenhagen, Denmark, 247 pp.

5. Tindall, C.I. and Moore, R.V. (1997) The Rivers Database and the overall data management for the Land Ocean Interaction Study programme, *The Science of the Total Environment* **194/195**, 129-135.

6. Littlewood, I.G., Watts, C.D., Green, S., Marsh, T.J., and Leeks, G.J.L. (1997) Aggregated river mass loads for Harmonised Monitoring Scheme catchments grouped by PARCOM coastal zones around Great Britain, Institute of Hydrology contract report to the Department of the Environment, 91pp + Appendices.

7. Littlewood, I.G., Watts, C.D., and Custance, J.M. (1998) Systematic application of United Kingdom river flow and quantity databases for estimating annual river mass loads (1975-1994), *The Science of the Total Environment* **210/211**, 21-40.

8. Watts, C.D. and Littlewood, I.G. (1998) CORAL: a system for calculating river mass loads from UK databases, in: H.S. Wheater and C. Kirby (eds.), *Hydrology in a Changing Environment, British Hydrological Society International Symposium*, Exeter, 6-10 July 1998, John Wiley & Sons, 559-571.

9. Huntley, D., Leeks, G.J.L., and Walling, D.E. (eds) (in press). Land Ocean Interaction: Measuring and Modelling Fluxes from River Basins to Coastal Seas, IWA Publishing, London.

10. Jarvie, H.P., Neal, C., and Tappin, A.D. (1997) European land-based pollutant inputs to the North Sea: an analysis of the Paris Commission data and review of monitoring strategies, *Science of the Total Environment* **194/195**, 39 - 58.

11. Simpson, E.A. (1980) The harmonization of the monitoring of the quality of rivers in the United Kingdom, *Hydrological Sciences Bulletin* **25**, 13-23.

12. Rodda, J.C. and Jones, G.N. (1983) Preliminary estimates of loads carried by rivers to estuaries and coastal waters around Great Britain derived from the Harmonized Monitoring Scheme, *Journal of the Institution of Water Engineers and Scientists* **37**, 529-539.

13. Webb, B.W., Phillips, J.M., and Walling, D.E. (2000) A new approach to deriving 'best estimate' chemical fluxes for rivers draining the LOIS study area, *The Science of the Total Environment* **251/252**, 45-54.

INTEGRATED MULTIDISCIPLINARY MARINE ENVIRONMENTAL DATABASES

OceanBase System – An Effective Tool to Manage Integrated Databases

V. L. VLADIMIROV, V.G. LYUBARTSEV and
V.V. MIROSHNICHENKO
Marine Hydrophysical Institute
National Ukrainian Academy of Sciences
2 Kapitanskaya St., Sevastopol, Crimea 99000, Ukraine

Abstract. Special environmental database management systems were developed to fulfill the multipurpose multidisciplinary management of marine and coastal environment data using the Windows platform. The main objective of these systems is to provide for easy, quick and effective work with marine environmental data in oceanography, ecology and environmental management. The system provides many possibilities for data loading, processing, analysis, selection, access, preview, and export. It has a customized user-friendly multi-windows interface. Data, information, and graphics in different windows are cross-linked and synchronized. This system was successfully used in recent years for development of databases and data management systems in several international projects and institutions.

1. Introduction

It is necessary to use large sets of interdisciplinary data of the marine and coastal environment for integrated coastal zone management or for the establishment of regional historical environmental databases. These data may contain a lot of variables from different sources and may come with different structures (oceanographic, meteorological, biological, etc.). Moreover, the most interesting scientific results can be obtained at the boundaries between different disciplines.

Yet, it is a very complicated task to combine these multidisciplinary data sets, to verify them, and to analyze them jointly. The industrial standard database management systems (DBMS) are not suitable enough to process such combined data sets and to perform this task efficiently.

Research has been conducted at the Database Laboratory of the Marine Hydrophysical Institute (MHI) since 1989 to facilitate the above task. As a result of this work, some special marine database management systems were developed at MHI to perform the multi-purpose multidisciplinary management of such data, using the Windows platform [1, 2, 3]. The main objective of these systems is to allow easy, quick and effective work with marine environmental data in oceanography, ecology, and environmental management for scientists and experts from different disciplines, e.g., oceanographers, biologists, chemists, and ecologists. These systems are also used for the calibration and interpretation of remote sensing data and images.

2. OceanBase System

A special unique database management and processing system (*OceanBase*) has been developed to compile large sets of multi-variable interdisciplinary oceanographic and environmental data. The system operates under Windows 95/98/NT/2000 as a Local Area Network (LAN) application, has effective Administrator and Data Manager modules, and can be used simultaneously by several users. The structure of database management in the *OceanBase* system is presented in Fig.1, and LAN possibility for the Database exploitation is shown in Fig. 2.

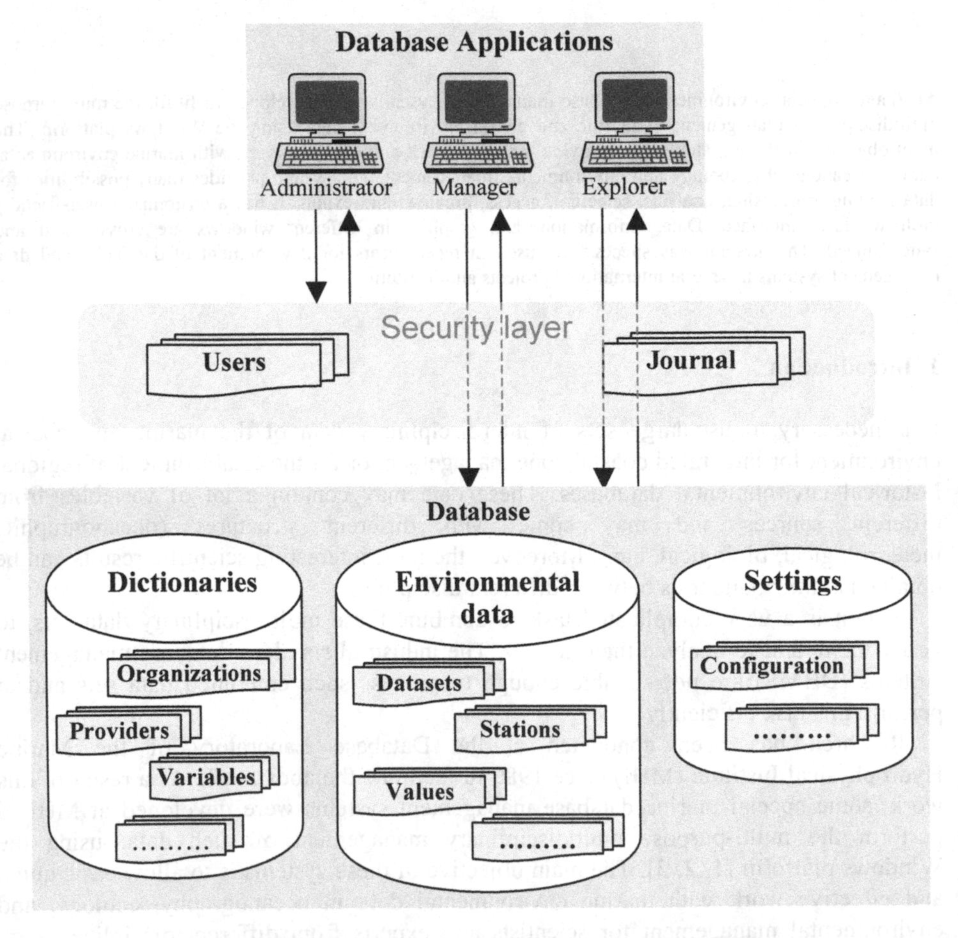

Figure 1. Database Management Structure in the OceanBase.

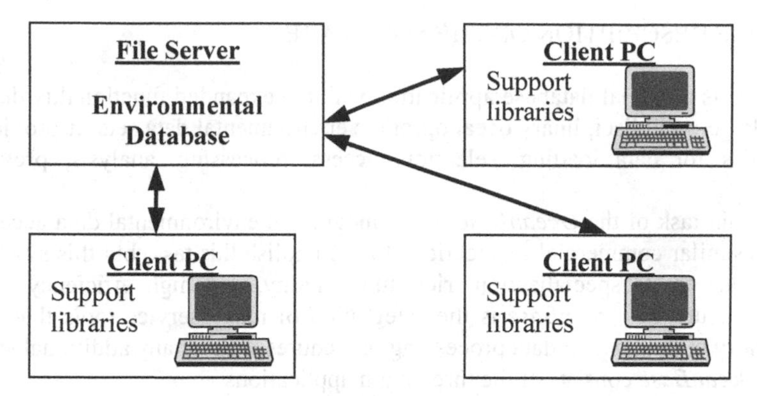

Figure 2. OceanBase in LAN.

2.1. BRIEF DESCRIPTION OF THE DATABASE STRUCTURE

OceanBase works with its own database, which is a relational database in Borland® Paradox™ format. To avoid unauthorized access, database files are protected with passwords, and they are not accessible through the Paradox or any other software.

All collected data are grouped into *data sets.* A data set is usually a set of *stations*. It is described by its *start* and *end* date, has a *name,* and belongs to an *organization.* Usually, a data set contains data of a research vessel cruise. The division of data into *data sets* and *stations* is arbitrary and is determined by *data providers.* Generally speaking, a *data set* may not belong to only one *organization*; for example, it may come from a joint *cruise.* The *data provider* can differ from the *organization;* it can be a department, a scientist, or a data center.

A *station* is a point located in space and time, where some oceanographic environmental measurements are carried out. It is characterized by its *name*, *date* and *time*, geographic *co-ordinates*, and *water depth.* It contains at least one *data profile* with at least one value. The station *name* is a text string containing not more than 20 characters. Time and water depth may be unknown. A station can belong to only one data set.

The initial list of *variables* (*names*, *full names*, *formats*, and *units*) was created by the *OceanBase* developers and can be extended by the Administrator and Data Managers.

Each data value is supplied with a quality flag. For this purpose, the scheme employed for data reported in real-time by GTSPP is selected [4]. The main advantages of this scheme are its universality and applicability to any type of data loaded into the database. It uses one character field with the following interpretation:

0 = data are not checked
1 = data are checked and appear to be correct
2 = data are checked and appear to be inconsistent but correct
3 = data are checked and appear to be doubtful
4 = data are checked and appear to be wrong
5 = data are checked and the value has been altered
6 to 9 - reserved for future use.

2.2. BRIEF DESCRIPTION OF THE SOFTWARE

OceanBase is a special database application with the expanded functionality designed to explore large multidisciplinary oceanographic environmental data sets. It provides many possibilities for data loading, selection, access, processing, analysis, preview, and export.

The main task of the *OceanBase* is oceanographic environmental data access. There are some similar commercial applications to accomplish this task, but this system makes it better due to its specialization, rich functionality, and high efficiency. The most valuable feature of *OceanBase* is the integration of many service tools that allow the user to carry out almost all data processing procedures without any additional software.

The *OceanBase* consists of the three main applications:

- ***OceanBase Administrator***
- ***OceanBase Data Manager***
- ***OceanBase Data Explorer***

The ***OceanBase Administrator*** is the main application of the *OceanBase*, and it allows running all main modules and tasks of the system, except the *OceanBase* Explorer. The ***OceanBase Data Manager*** permits to organize the work of the Administrator and the Data Managers for data entry, editing, and quality control. The ***OceanBase Data Explorer*** (former *OceanBase* itself) is the main module for end-users of the system. It provides the possibilities for data selection, access, preview, export, and processing, but it does not allow to change data loaded into the database.

Borland Delphi was used to develop the database application. This is one of the most comprehensive development platforms for creation of the PC–based database applications with high performance.

2.3. DATABASE USERS

To provide more efficient and secure access to the database, the *OceanBase* comprises five types of Database users:

- ***Master*** (system developer), with full access to all system possibilities. After system installation, nobody has this status permanently;
- ***Administrator***, with full access to the system except the eligibility to change the internal system features. Usually, only one person has the Administrator rights for a particular database;
- ***Data Manager(s)***, with the capacity to add new data and information and to change the data and information previously entered by this data manager. In principle, the number of data managers is unlimited. However, it is recommended to limit it, taking into account that they are responsible for the quality of data loaded into the database;

- ***End-user(s)***, who can work with all data and information loaded into the database. They cannot make changes in data or system information. The number of the end users is unlimited.
- ***Disconnected***, This status can be set up for all types of users (except the master) to prevent their access to the *OceanBase*, while preserving their references on the entered data and information.

2.4. DATABASE SECURITY LEVELS

The *OceanBase* has been designed to work in the Local Area Network. That is why it has the security features to prevent unauthorized access to the system and to data, and to track any activity related to the system.

All files (tables) of the database are encrypted. There are two passwords, which are the same for each table: master and secondary passwords. The master passwords allow all types of operations on the tables (read, write, restructures, etc.). The secondary password provides the "read" operation on all tables and "write" operations on tables related to the user registration. These two passwords are also encrypted and stored in the *OceanBase* configuration file. Thus, it is impossible to read any information from the database without *OceanBase* software and passwords. The two passwords are used to provide additional protection of the database and to give flexibility to the Administrator in database security management. The common user opens the database with an auxiliary password, and he/she can change only the registration tables, but not the tables containing real data.

All users have their own private passwords. These passwords are necessary to register and run the *OceanBase* software. Moreover, they are used to track the user activity so that all serious actions are tracked and authorized.

The Data Managers can create, edit, and delete records in the database tables. Almost all records comprise information on their development and modification. Only the Data Manager who has entered the data is permitted to modify or delete them, and only the Administrator can change this reference. By default, the Data Manager who has created the record is responsible for its content.

The Administrator maintains the list of all users. Data Managers and Users cannot change any registration information.

3. OceanBase User Interface

The *OceanBase* has a customized user-friendly multi-windows interface. A user can create his own desktops and can open and arrange as many windows as he/she needs (Fig. 3). All settings can be stored in the configuration files to create task-oriented desktops for routine tasks. Data, information, and graphics in different windows are cross-linked and synchronized. Main features of the system user interface are as follows:

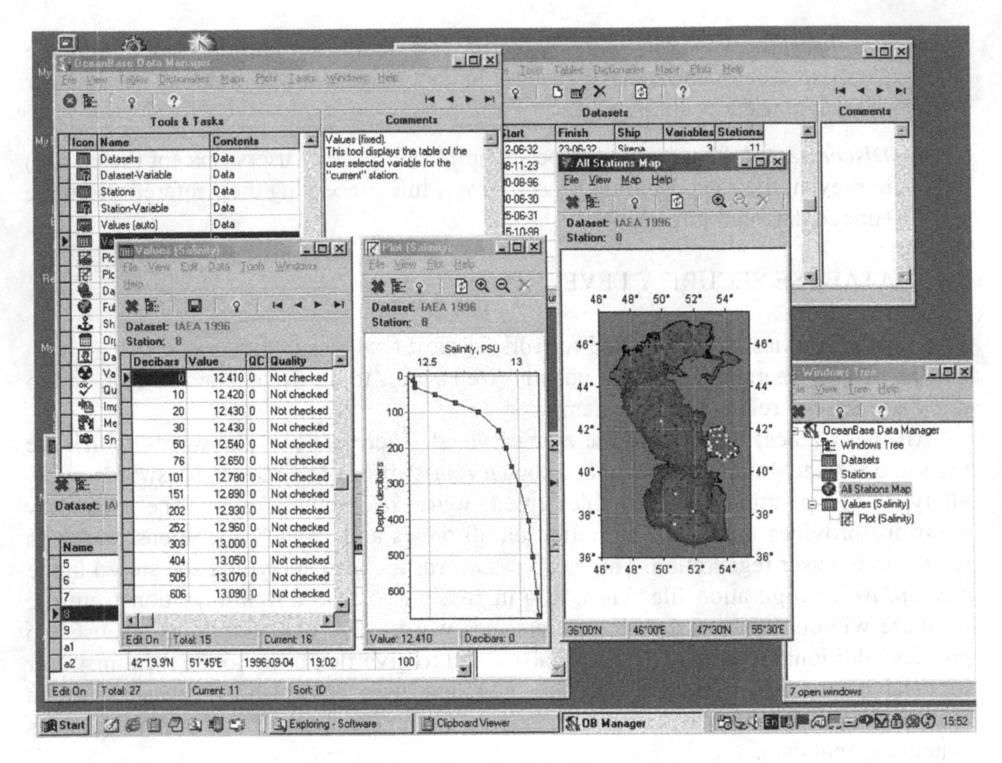

Figure 3. Example of the Data Manager desktop.

- Multi-windows approach, which allows the user to display only the necessary tools and to arrange them on the screen in a convenient manner;
- Ability to save/load the most important settings to the configuration file; auto save/load option is available;
- Easy navigation through the data sets, stations, and data values;
- Cross-links between windows, which synchronize the tool operations and provide convenient access to referenced information;
- A rich set of tools to display original data and the results of data processing (various kinds of maps, plots and histograms);
- Auto scaling, auto stepping, auto formatting which reduce user's time and efforts;
- Many controls allowing customization of windows, tools, algorithms, etc.;
- Long operations are accompanied by a progress bar, indicating the percent of completion; there is also the ability to cancel these operations and preview their preliminary results;
- Context-sensitive help which provides users with all the necessary information.

Selected screenshots presented in Fig. 4 give examples of working desktops created by users.

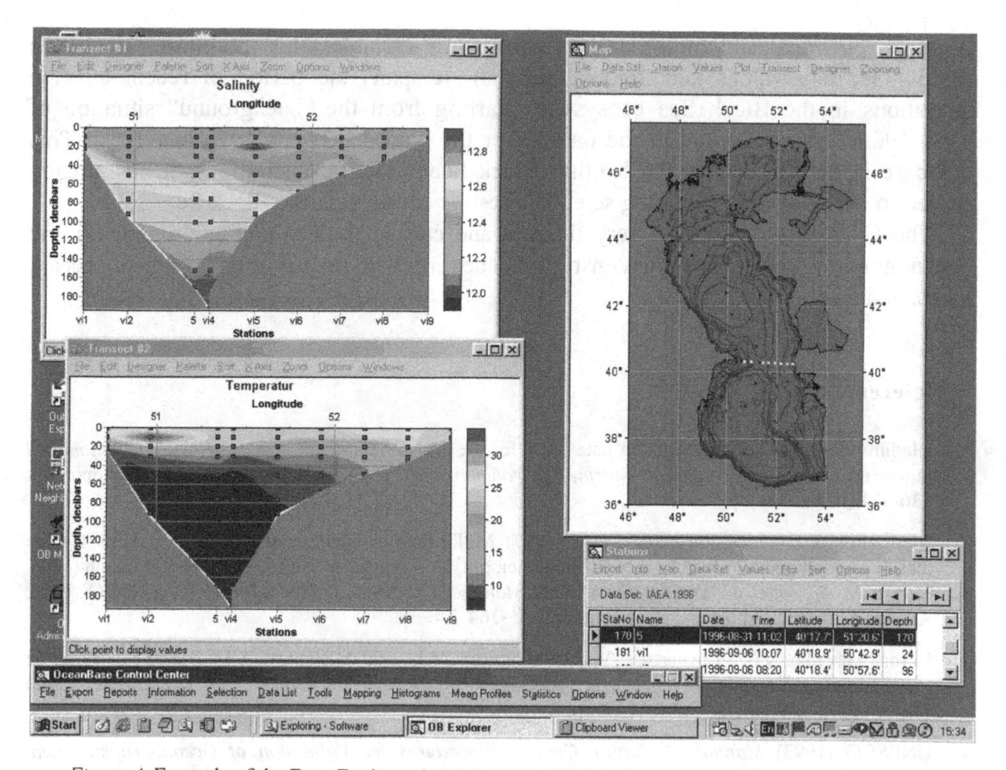

Figure 4. Example of the Data Explorer desktop (analysis of data along the oceanographic transects).

4. Conclusions

The *OceanBase* system has been successfully used in recent years for development of databases and data management systems in several international projects and by different institutions [5,6] such as:

- NATO TU Black Sea Project (http://sfp1.ims.metu.edu.tr/TU-BlackSea)
- NATO SfP ODBMS Project (http://sfp1.ims.metu.edu.tr)
- Caspian Environment Programme
- INTAS Project UA-95-80
- Darwin Initiative Project No. 162/8/251 (http://www.pml.ac.uk/diocean)
- IBSS NASU, Ukraine
- IMS METU, Turkey
- NatMIRC, Namibia

An example for the successful use of this system is the Black Sea interdisciplinary historical database [5]. It was created, using *OceanBase*, in 1997 within the framework of the NATO TU-BLACK SEA project, for which prominent regional oceanographic institutions provided more than 13,000 data files. This database includes all the basic physical, chemical and biological variables (116 variables) for the entire Black Sea (8,364,731 data values and 26,035 stations). It spans the period of recent adverse alterations in the Black Sea ecosystem, starting from the "background" situation of 1950-1960s to 1996. Most of the data cover the period between 1976 and 1996. This database is now in use by the leading Black Sea scientific institutions. It has already served to obtain some interesting scientific results concerning the Black Sea.

The *OceanBase* system is very flexible and can be used in any environmental and oceanographic project (or by any institution) dealing with the water environment (ocean, lakes, and rivers).

5. References

1. Vladimirov, V.L. (1992) Integrated data bank for the cruise data of the research vessel, in *Automated Systems for the Monitoring of the Marine Environment*, Marine Hydrophysical Institute, Sevastopol, pp. 126-131 (in Russian).

2. Vladimirov V.L., Miroshnichenko V.V. (1997) Multipurpose database management systems for the marine environmental research, in N.B. Harmancioglu, M.N. Alpaslan, S.D. Ozkul and V.P. Singh (eds.), *Integrated Approach to Environmental Data Management Systems*, Proceedings of the NATO ARW, Kluwer Academic Publishers, Dordrecht, pp. 355-364.

3. Miroshnichenko, V.V., Luybartsev, V.G., Vladimirov, V.L., and Mishonov, A.V. (1998) Multipurpose Multidiscipline Database Management Systems for the Ocean Data, *Ocean Data Symposium, Dublin Castle 1997, Full papers, CD-ROM*, 8 pp.

4. UNESCO (1993) *Manual of Quality Control Procedures for Validation of Oceanographic Data*, Prepared by: CEC: DG: XII, MAST and IOC/IODE, Manual and Guides No.24, p.310.

5. Vladimirov, V., Besiktepe, S., and Aubrey, D. (1998) Database and Database Management System of the TU Black Sea Project, in L. Ivanov and T. Oguz (eds.), *NATO TU-Black Sea Project Ecosystem Modeling as a Management Tool for the Black Sea, Symposium on Scientific Results*, Kluwer Academic Publishers, Volume 1, pp.1-10.

6. Mishonov, A.V, Vladimirov, V. L., Miroshnichenko, V.V., and Luybartsev, V.G. (1998) Software Systems for Access to Oceanographic Data and Information, *Proc. "Oceanology International 98, The Global Ocean"* Exhibition and Conference, 10-13 March 1998, Brighton, UK, Conference Proc. Vol. 1 (Published by Spearhead Exhibitions Ltd. ISNB: 0 900254 20 3), pp. 171-179.

REGIONAL ENVIRONMENTAL CHANGES: DATABASES AND INFORMATION

K. A. KARIMOV and R. D. GAINUTDINOVA
Institute of Physics, National Academy of Sciences
265-A Chui Prosp., Bishkek 720071, Kyrgyz Republic

Abstract. The investigation of regional environmental changes is necessary to forecast global environmental changes. On the basis of long-term measurements of thermodynamic parameters of the lower atmosphere the background fluctuations are determined, and the estimations of low-frequency trends are carried out. The study has allowed the identification of climatic trends for temperature. It is shown that the method of maximum entropy is one of the effective techniques to identify periodic components in long-term environmental data.

1. Introduction

The retrieval and interpretation of experimental data of large spatial and temporal scales require standardization of both the measuring systems and the methods of analysis. The lack of standardization results in heterogeneous data that are incompatible and difficult to compare. Presentation of retrieved data is also significant in the selection of algorithms to process large numbers of data and further in the development of models based on experimental data.

Recent research has shown that investigations are needed on the development of the theory and methods for forecasting regional environmental changes. In particular, there is the need to forecast the changes in environmental parameters at various time scales (e.g., monthly, annual, long-term, etc.) and to evaluate their sensitivity to natural and anthropogenic changes. These changes are insufficiently reproduced by modern dynamic models, which often have low spatial resolution. Therefore, the estimation of regional changes in environmental parameters is highly significant as is the correct choice of methods for processing experimental data.

In the presented work, the regional changes of the thermal regime in Kyrgyzstan are investigated under conditions of constantly amplifying anthropogenic loadings, using long-term temperature data. A quantitative estimation of temperature changes in the lower atmosphere is derived, and the general trend in temperature increase is evaluated with respect to the elevation of the region analyzed. The spectral structure of long-term variations of temperature in Kyrgyzstan is investigated.

2. Data and Methods of the Analysis

The spectral analysis of temporal data is usually applied to derive information about the periodic behavior of environmental parameters. However, it is often difficult to obtain statistically significant results. Thus, the traditional spectral analysis of Blekman and Tiuki may be inapplicable under these conditions due to data limitations.

The method of maximum entropy is the most appropriate for the analysis of short duration data; it is superior to the traditional spectral analysis of time series with respect to resolution. For a short series of data, the resolution of spectral peaks is lost. With a correct choice of the length of the filter, the method of maximum entropy allows identifying the latent periodicity and shows peaks at close frequencies, which, as a rule, are impossible to detect by other spectral methods.

The method of maximum entropy allows the detection of fluctuations in the data, reflecting the periodic structure of the processes. Such an approach may be used to detect the probable influence of one periodic natural process on another.

The above method was used with long-term temperature data collected by a network of meteorological stations in Kyrgyzstan. The heliogeophysical parameters describing solar activity were also used.

3. Main Results

3.1. SOME PECULIARITIES OF LONG-TERM TEMPERATURE FLUCTUATIONS IN THE LOWER LAYER OF THE ATMOSPHERE

Kyrgyzstan, surrounded by Tien Shan Mountains and situated at the central part of Central Asia, reflects specificities with respect to fluctuations of environmental parameters in the region [1].

The long-term data of temperature in the lower layer of the atmosphere were used to investigate changes in parameters and the regularities of a thermal regime. The results were obtained on the basis of long-term monitoring at stations situated in the middle-altitude region (Chui Valley) and the high-altitude region (Issyk Kul region). The data of three meteorological stations were used for the analysis: Frunze (760 m above sea-level), Cholpon Ata (1640 m above sea-level), and Tien Shan (3700 m above sea-level), for the periods between 1920-1990, 1929-1990 and 1930-1990, respectively.

On the basis of monthly average values of temperature, the "background" fluctuations in temperature and trend components were analyzed. The long-term changes of temperature were considered for winter and summer seasons.

For detection of low-frequency trends in interannual variations of temperature, i.e., the so-called "background" variations, measurements for each consecutive 11 years were averaged. Such an approach allows the exclusion of the effects of an 11-year cycle of solar activity.

The value of background temperature was calculated as follows:

$$T_{\Phi} = \frac{1}{n} \sum_{i=1}^{n} T_{i+k} \tag{1}$$

where T_i are the monthly average values of temperature in the given season with, n=4 and k = 0,1,2..., m- 4,where m is the sample size. For station Frunze, m=70.

Figure 1 presents an example of the smoothed interannual variations of temperature at Frunze, Cholpon Ata and Tien Shan meteorological stations for the winter period. The thin line shows seasonal variations of temperature for the winter season with average values for each year. The thick line presents "background" changes of temperature, derived as explained above. The deviations of seasonal temperature values from the "background" smoothed variations are sharply decreased in accordance with increasing elevation of the monitoring station analyzed.

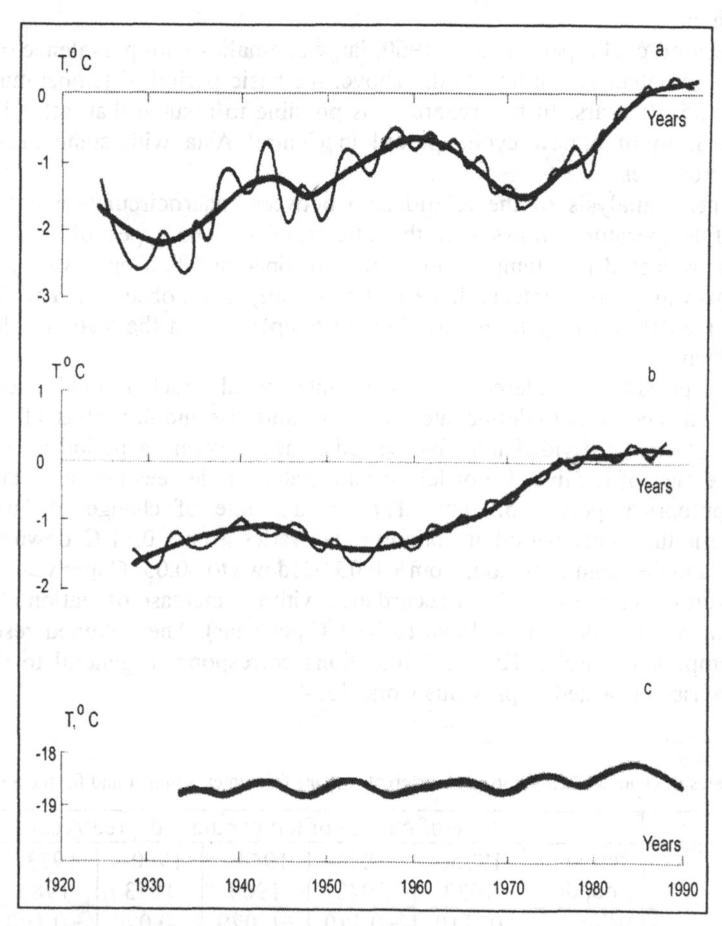

Figure 1. Interannual variations of temperature in the lower atmosphere: a) Frunze; b) Cholpon Ata; c) Tien Shan.

The analysis of results has shown that, in various intervals of time, it is possible to detect the separate periods of the relative rise and fall of temperature, especially for station Frunze. It is necessary to note that, during the same cyclic period, the changes of temperature in different areas of Central Asia do not occur synchronously. According to the data of station Frunze, which is on open valley, the lowest winter temperatures have repeated in 40 years: in 1929-1935 and in 1972-1979. In the closed valleys or orographic regions, these effects did not show up. The last cold period of 1972-1979 was replaced by the warmest period between 1981-1990, which was practically observed in the whole Central Asia region.

In [2, 3, 4], peculiarities of atmospheric circulation in Central Asia for the same period are analyzed. The authors [2] refer to two cyclic periods, before and after 1959-1960. In the first period, which began in 1930, latitudinal forms of circulation prevail. In this period, the relative stability of temperature was observed at stations Cholpon Ata and Tien Shan.

In the second cyclic period since 1960, large anomalies with prevalence of meridian processes dominated. According to the above, the basic period of temperature changes accounts to 35-40 years. In this regard, it is possible to assume that, after 1990, there was a formation of a new cyclic period in Central Asia with some prevalence of latitudinal processes.

The general analysis of the relationship between macrocirculation processes and changes of temperature shows that the effects of various types of circulation will basically be reflected in a temperature regime of open and half-open valleys. For high mountainous valleys and plateau, this effect practically is not observed. This is shown to be true by the data on long-term variations of temperature at the stations Cholpon Ata and Tien Shan.

In the separately considered periods of interannual "background" fluctuations of temperature, it is possible to define rates of temperature rise and fall. Thus, all temperature measurements can be divided into five periods, each covering a period of 10 years. In Table 1, the rates of relative rise or fall of temperature in degrees per one year are given for the appropriate period of time. The average rate of change of "background" temperature in the winter period at station Frunze varies from + 0.11°C down to –0.11 °C per year, and in the summer period, from + 0.05 °C down to –0.05 °C per year.

These rates decrease sharply in accordance with the increase of station elevation (at station Tien Shan, it decreases down to 0.01°C per year). The obtained results on the rates of temperature rise at Kyrgyzstan stations correspond in general to the data on temperature rise presented in previous works [3, 4].

TABLE 1. Rates of change of temperature at different stations for winter, summer and for the all year

Monitoring station	Rate of change of temperature, degrees/year						
	Period (month)	1925–1933	1933–1943	1944–1960	1960–1973	1973–1985	1925–1990
Frunze	Winter	-0. 110	+0.110	+0.070	-0.070	+0.100	
	Summer	+0.010	+0.010	-0.050	+0.050	+0.050	
	Year	*-0.060*	*+0.070*	*0.000*	*-0.070*	*+0.058*	*+0.011*
Cholpon-Ata	Winter	-	+0.060	-0.050	+0.050	+0.050	
	Summer	-	-0.010	+0.020	+0.020	+0.050	
	Year	-	*0.000*	*-0.015*	*+0.046*	*+0.054*	*+0.021*
Tien Shan	Winter	-	0.000	0.000	0.000	+0.030	
	Summer	-	0.000	-0.010	+0.010	+0.010	
	Year	-	*0.000*	*-0.015*	*+0.007*	*0.000*	*+0.007*

It is necessary to note the general tendency of temperature rise at stations Frunze and Cholpon Ata since 1920 and 1929, respectively. At station Frunze, for the winter season, the average background temperature for the whole period has increased till 1990 from -2.5 °C up to 0 °C ; on station Cholpon Ata from –1.5 °C up to 0 °C ; and on station Tien Shan from –19.1 °C up to –18.7 °C.

The rather large changes of temperature in winter and summer at station Frunze for different years of the examined period are related to anthropogenic factors and natural fluctuations of the climate.

As a whole, it is possible to note that the winter and summer variations of temperature for the whole observed period are characterized by a positive tendency, mainly in winter seasons. In the spring and autumn, it is possible to note only a weak tendency towards temperature rise.

To estimate the trends, i.e., the rates of growth of "background" temperature, it is necessary to take into account the period in which a particular trend was observed. The period since 1960-1962 till 1990-1992 is the most interesting for Kyrgyzstan. The beginning of the so-called "industrial period" (period of intensive industrial development) in Kyrgyzstan can be approximately related to 1960-1962. Since this period, the increase of temperature at stations Frunze and Cholpon Ata has been observed for the last 30 years. The increase of temperature since the "pre-industrial period" has been on average between 0.8-1.2 °C since 1990.

The most probable rates of temperature rise, i.e., the annual rise of temperature for different seasons during the "post industrial" period, have been: -0.027 °C per year (for winter and summer) at Frunze, -0.040 °C per year (for winter) at Cholpon Ata, and 0.027 °C per year (for summer), and –0.013 °C per year (for winter) and 0.020 °C per year (for summer) at Tien Shan.

The average rate of temperature changes in winter and summer seasons on all stations of Kyrgyzstan since the "pre-industrial" period till the present time accounts to 0.025 °C per year and, obviously, is caused by increased anthropogenic loadings. These data do not contradict estimates on the global rise of temperature [5, 6].

3.2. SPECTRUM OF PERIODIC COMPONENTS IN SEASONAL VARIATIONS OF TEMPERATURE

For statistical processing of the available experimental data to investigate the latent periodic temperature, the method of maximum entropy (MME) [7, 8] was used. This method surpasses the traditional spectral analysis of time series in resolution.

In contrast to the traditional spectral analysis [9], where the function of a spectral window plays an essential role, the type of the function is not fixed in MME. In estimating spectral density at a certain frequency, the type of the function is self-corrected so that the influence of spectral density on other frequencies is minimal [10]. The MME surpasses the traditional spectral analysis in resolution and is especially advantageous for the analysis of short data series.

It is known that the entropy of a continuous stationary Gaussian process is related to its spectral density. In [11], the expression for the entropy of a discrete stationary Gaussian process is given as:

$$H = \frac{1}{2}\log\{\det[C\ (M)]\} \tag{2}$$

where C (M) is the determinant of a semi-positive matrix of autocovariances:

$$C(M) = \begin{pmatrix} \rho_0 & \rho_1 & \cdots & \rho_M \\ \rho_1 & \rho_2 & \cdots & \rho_{M-1} \\ \rho_M & \rho_{M-1} & \cdots & \rho_0 \end{pmatrix} \tag{3}$$

To estimate a spectrum by MME, the unknown autocovariances ρ_{M+1}, ρ_{M+2} are selected so that at each step *k*, the entropy of the analyzed process is maximal. That is, the known autocovariance function ρ_M is extrapolated for limits of the maximal shift so that the entropy of the appropriate probability density function is maximal at each step of extrapolation. Thus, the density spectrum is defined by the principle of maximum entropy. The expression for spectral density with maximum entropy is as follows [11]:

$$P(f) = \frac{P_{M+1}}{\left[1 + \sum_{k=1}^{M} a_{Mk} \exp(-i2\pi f \kappa)\right]^2} \tag{4}$$

where, $-f_N \le f \le f_N = 1/2\Delta t$, (t: discrete time steps; f_N: Nyquist frequency; a_{Mk} : a coefficient to reflect the effect of prediction errors on the filter). The algorithm of the above calculation is presented in [8].

The autocovariances ρ_M, the power P_{M+1} and coefficients of the filter a_{Mk} are connected by the matrix equation [12]:

$$\begin{pmatrix} \rho_0 & \rho_1 & \cdots & \rho_M \\ \rho_1 & \rho_2 & \cdots & \rho_{M-1} \\ & & \cdots & \\ \rho_M & \rho_{M-1} & \cdots & \rho_0 \end{pmatrix} \begin{pmatrix} 1 \\ a_{M_1} \\ \vdots \\ a_{MM} \end{pmatrix} = \begin{pmatrix} P_{M+1} \\ 0 \\ \vdots \\ 0 \end{pmatrix} \tag{5}$$

which is usually solved by an iterative method. The algorithm for calculation of the coefficients a_{Mk} and P_{M+1} is presented in [8].

MME allows the identification of latent periodicity and of peaks at close frequencies in the spectrum when data are of limited length, which, as a rule, is impossible to be realized by other spectral methods. For the practical use of MME, a number of difficulties related to separation and displacement of spectral peaks is overcome by the selection of the length of data, N; the model accounts for the appropriate choice of the length of the filter or for the order of autoregression M.

The effect of the length of filter on positions of spectral peaks is significant when using the method of maximum entropy. This issue was investigated by many authors, and it was shown that the best spectrum occurs at the order of autoregression M within limits of 1/5 up to 1/2 length of the sample. Numerous calculations have shown that M should not be more than N/2, N being the length of data record.

The calculations of the spectrum for temperature at various M, ranging from 1/3N up to 3/4N, have shown that, at rather wide filters where M>1/2N, there is an instability in the spectrum with splitting of spectral peaks. At narrow filters, the low-frequency part of the spectrum is poorly specified. One of the factors resulting in a change of the position of spectral peaks is related to the rather weak stationarity of the analyzed process. In [13], it was shown that the displacement of spectral peaks is related to this factor and is observed at the low-frequency part of the spectrum.

Time series temperature obtained at three stations with duration from 720 to about 840 months were exposed to spectral analysis. The estimation of confidence intervals of the allocated peaks was realized by the use of standard methods for calculation of spectral density. For an estimation of the reliability of the received spectra, the criterion χ^2 was used. In Tables 2a and 2b, the results of the spectral analysis for confidence intervals of $0.69 < P < 0.95$ and $P \geq 0.95$ are given.

Figure 2 presents the spectrum of temperature variations for three stations for the analyzed period. On x-coordinate, the time in months is presented. For convenience of presentation, the scale of the x-coordinate is divided into three intervals: 1st from 0 to 100 months (8.5 years); 2nd from 100 to 200 months (16.5 years); and the 3rd from 200 to 700 months (58 years). On the axis of ordinates, the natural logarithm of spectral density *lnP(f)* is presented.

The peaks at large confidence levels ($P \geq 0.95$) are detected in the period ranging from 0.5 to 2.0 years. The decrease of the noted peaks is observed for the periods ranging from 2.5 to 4.5-5.5 years at stations of Frunze and Cholpon Ata. Next, an increase of peaks is found in an interval between 5.5 and 7.5 years at station Frunze. At the other two stations, the power of peaks in this range remains constant.

TABLE 2a. Results of spectral analysis of temperature data obtained by maximum entropy method (for a period ranging from 0.5 to 6.5 years).

Monitoring station	Period (years)						
	0.5-1.0	1.5-2.0	2.5-3.5	3.5-4.0	4.0-5.0	5.0-6.0	6.0-6.5
Frunze (1920-1990)	0.69< P< 0.95						
	-	1.5	3.1	3.8	4.2-4.8	-	-
	P ≥ 0.95						
	0.5-1.0	2.0	2.5	-	-	5.7	6.5
Cholpon-Ata (1929-1990)	0.69< P< 0.95						
	-	2.0	-	3.5-3.7	4.2	5.5	-
	P ≥ 0.95						
	0.5-1.0	1.5	3.1	-	4.8	-	6.3
Tien Shan (1930-1990)	0.69< P< 0.95						
	-	1.5-2.0	2.5	3.5-3.8	-	-	-
	P ≥ 0.95						
	0.5-1.0	-	3.1	-	4.6	5.4	6.3

TABLE 2b. Results of spectral analysis of temperature data obtained by maximum entropy method (for a period ranging from 7 to 50 years).

Monitoring station	Period (years)						
	7-8	9-11	12-14	14-15	16-18	20-25	28-50
Frunze (1920-1990)	$0.69 < P < 0.95$						
	-	9.7	-	15	-	-	-
	$P \geq 0.95$						
	7.9	-	12.7-13.2	-	18	-	37.0-39.6
Cholpon-Ata (1929-1990)	$0.69 < P < 0.95$						
	8	-	12.5	-	16	-	49.0
	$P \geq 0.95$						
	7.4	9.2-10.8	-	-	-	-	28.0
Tien Shan (1930-1990)	$0.69 < P < 0.95$						
	-	-	-	-	-	22.5	-
	$P \geq 0.95$						
	7.8	10.7	12.5-13.7	-	-	-	-

The increase of selected peaks at all three stations corresponds to the periods between 10 to 15 years, which is apparently connected with the period of change of the basic 11-year cycle of solar activity. This period is not strictly constant and, for the analyzed 70-year record, it may change between 9.5 years and 13.5 years with an average value of about 11 years [14].

The peaks at 16, 18 and 22.5 years are less intensive maximums. The periodicities at 16 and 18 years, obviously, are a superposition of 11- and 15-year cycles. It is known that a 22-year cycle is present at a magnetic cycle of the sun and is connected with an integrated magnetic field and its polarity. The cycle of 22.5 years is detected with a small probability of confidence and is present only at temperature variations of the high-mountainous station Tien Shan.

The next group of peaks occurs at periods of 28 years and 37-39 years. This group is probably a superposition of the low-frequency cycles of 12-16 and 18 years. It is necessary to note that, in variations of temperature at Cholpon Ata station, one more low-frequency component with a cycle of 49 years is observed.

It is possible to divide the detected cycles into five intervals:

1) an interval of the spectrum from 0.5 to 1.5 years, with a maximum at 1 year, at all three stations;
2) part of the spectrum from 2.5 to 3.5 years, with a maximum close to 3 years at all stations;
3) part of the spectrum from 5.5 to 7 years, with a maximum close to 6.5 years, only at stations Frunze and Cholpon Ata;
4) part of the spectrum from 8 to 15 years, with a maximum close to 12.5 years at all stations;
5) part of the spectrum from 28 to 49 years, with a maximum close to 37 years only at stations of Frunze and Cholpon Ata.

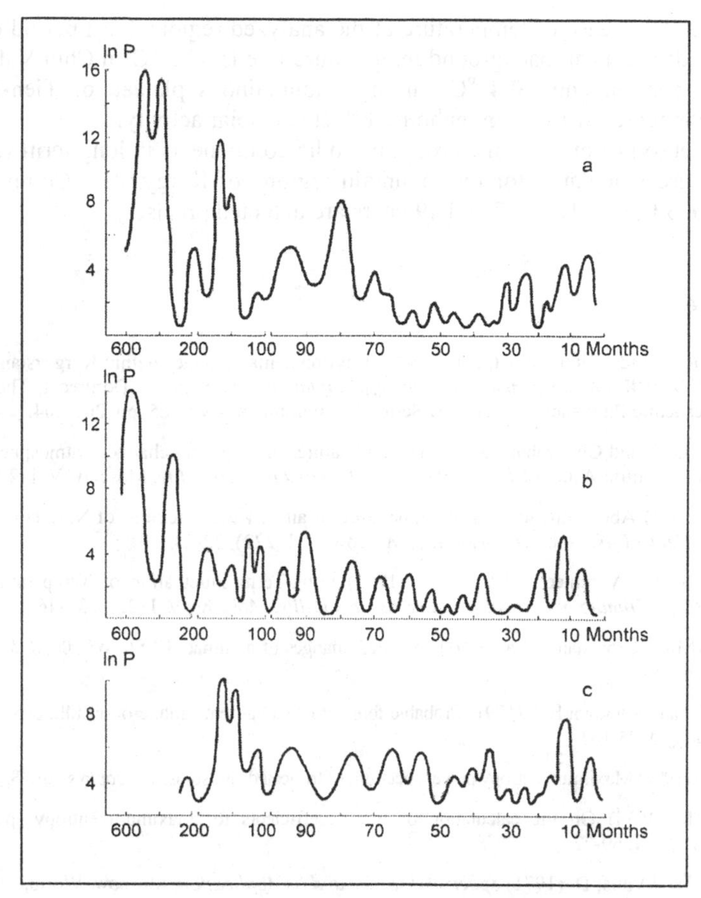

Figure 2. Temperature spectrum for three stations:
a - station Frunze, b - station Cholpon Ata, c - Tien Shan.

The low-frequency periodicity from 28 years up to 39 years explains the rise of temperature in Chui Valley and the Issyk Kul region. These periods are related to the variability of climate in the Central Asia region [4,15].

The periodicity at 49 years detected at the Cholpon Ata station is caused by fluctuations of the general circulation of the atmosphere. All data can be used for the long-term forecast of a temperature regime, not only in Kyrgyzstan, but also in the northern part of the Central Asian region.

4. Conclusions

In the presented study, the basic features of the regional trends in temperature in Kyrgyzstan are investigated. The quantitative characteristics of a temperature rise in the lower layer of the atmosphere in areas of middle- and high-altitude are noted.

The general increase of temperature at the analyzed regions for a period of 70 years is investigated: the mean background temperature rise is: 2.5 °C in Chui Valley, 1.5 °C in Isyk Kul region, and 0.4 °C in high-mountainous plateau of Tien-Shan. This increase is connected with the greenhouse effect and solar activity.

By the method of maximum entropy, periodic components in long-term variations of temperature are calculated for the mountain regions of Kyrgyzstan. Components with cycles of 1.0; 3.0; 6.5; 12.5; 37 and 49 years are detected precisely.

5. References

1. Karimov, K.A. and Gainutdinova, R.D. (1999) Environmental changes within Kyrgyzstan, *Proceedings of the NATO ARW on Environmental Change, Adaptation And Security*, Dordrecht, The Netherlands, Kluwer Academic Publishers, NATO ASI Series 2: Environment - Vol. 65, pp. 201-204.
2. Subbotina, O.N. and Chevychalova T.M. (1991) Features of long-term changes - atmospheric circulation in territory of Central Asia, *A Climate of Central Asia and its Variability*, Moscow, V. I (222), 12-21.
3. Kim, I.S. (1991) About fluctuations of temperature of air in various regions of Northern Hemisphere, *A Climate of Central Asia and its Variability*, Moscow, V. I (222), 22-28.
4. Chanysheva,S.G., Veremeeva T.L., et.al. .(1991) Long-Term Fluctuation of Temperature Regime of Central Asia, *A Climate of Central Asia and its Variability*, Moscow, V. I (222). 32-46.
5. Budyko, MM and Groisman, P.Ya. (1991) Expected changes of a climate USSR to 2000, *J. Meteorology and Hydrology* **4**, 74-83.
6. Gruza, G.V. and Rankova E.P. (1991) Probable forecast of global temperature of air till 2005, *J. Meteorology and Hydrology* **4**, 95-103.
7. Ulrych T.J. (1972) Maximum entropy power spectrum of truncated sinusoids, *J. Geophys. Res.* **8**, 1396-1400.
8. Andersen R. (1974) On the calculation of filter coefficients for maximum entropy spectral analysis, J. *Geophys. Res.* **1**, 69-72.
9. Jenkins, R. and Vatts, D. (1971) *Spectral Analysis and its Application*, Moscow, World, 316 p.
10. Tsvetkov, A.V. (1981) Results of the analysis of the data of a gradient of potential by a method of maximum entropy, *Proceedings of GGO*, Leningrad, Gidrometeoizdat, **458**, 43-48.
11. Smyle, D.E., Clarke, G.K., and Ulrych, T.J. (1973) Analysis of irregularities of the Earths rotation, *J. Methods in Computational Phys.*, 13-22.
12. Ulrych, T.J., Bishop, T.N.(1975) Maximum entropy spectral analysis and autoregressive decomposition, *J. Rev. Geophys. Space Phys.*, 183-200.
13. Rotanova, N.I., Papitashvili, N.E., and Pushkov, A.N. (1979) About the spectral analysis of temporary numbers of a geomagnetic field by method of maximum entropy, *J. Geomagnetism and Aeronomy*, 1091-1096.
14. Vitinsky, Yu.I., Ol, A.I., and Sazonov, B.I. (1979) *Sun and Earth Atmosphere*, Leningrad, Gidrometeoizdat.
15. Karimov, K.A. and Gainutdinova, R.D. (2000) *Changes of Regional Climate Caused by Natural and Antropogeneous Factors, Ecology of Kyrgyzstan: Problems, Forecasts, Recommendations,* Bishkek, Ilim, pp. 66-81.

ENVIRONMENTAL HEALTH INDICATORS IN EUROPE: A PILOT PROJECT

D. DALBOKOVA and M. KRZYZANOWSKI
WHO – European Centre for Environment and Health, Bonn Office
Hermann Ehlersstrasse 10, D - 53113 Bonn, Germany

Abstract. The presented paper describes an ongoing project aimed at the development of a pan-European environmental health indicators system. It should serve each individual member state, as well as supporting multinational analyses using comparable data and information. The system is being tested in 14 countries of the European region of WHO. The evaluation comprises data availability, quality, and usefulness of the indicators in monitoring policy progress to reduce EH (Environmental Health) impacts. The paper explores the main problems and challenges in setting indicator-based reporting and the need for methodological refinements to better address policy questions.

1. Introduction

Decision making requires information support to identify priorities and to track policy progress. This information is often presented in the form of indicators, integrating data and assessments on relevant aspects in a quantitative form. Well-designed indicators focus public attention on significant issues, while harmonizing and streamlining data collection. European environmental indicator-based reporting, allowing for policy-oriented monitoring and benchmarking, has advanced considerably [1]. Recently, the first sectorial indicator report on transport has provided a model for the integration of environmental and sectorial policies [2]. A better integration of health concerns in environmental policies is still an open issue. Efforts to monitor public health and its determinants, including environmental hazards, are increasing across Europe. This requires consistent reporting and appropriate indicators to support evaluation and adjustment of public/ environmental health action plans for reducing and preventing these risks.

The present project aims at the establishment of a common environmental health (EH) indicators system for regular reporting and integrated assessment, using comparable data and information across the European region of the WHO. The focus is on the development of a methodology to support decision making in the rational management of EH risks, serving each member state and, at the same time, enabling multinational analyses. In addition, the indicators should aid in bridging the gaps between local, state, international agencies, and the programs addressing public health and environmental quality.

2. Development Phase of EH Indicators: Definitions, Attributes and Proposed Use

The issues considered for indicator development should provide a comprehensive policy-oriented monitoring of the most important environmental health issues in terms of both "scale" (of widespread significance at the national as well as at the multinational level) and "size" (a clear public health impact). Ten EH issues were selected according to the guidelines of the multidisciplinary Steering Committee: air quality, housing and settlements, traffic, noise, radiation, waste, water and sanitation, food safety, chemical emergencies, and workplace.

The scope of the indicators system was determined, based on the current commonly accepted cause-effect framework: **D**riving Force – **P**ressures – **S**tate – **E**xposure – **E**ffect – **A**ction (Fig. 1). The focus was on the development of indicators that provide information about specific health outcomes and environmental exposures, taking into account the existing evidence for a valid exposure – health effect relationship according to the WHO guideline document [4]. A number of action indicators, showing the implementation of programs and official policies to manage environmental risks and health protection have also been included. Indicators on the state of the environment, pressures, and driving forces are usually provided by the environmental and statistical agencies. They were included in the present EH indicator set when they could be considered as "proxy" of exposure and also when they allow for some communalities with the sectorial reporting mechanisms. The resulting set of EH indicators is shown in Table 1. It should be noted that there is no clear-cut distinction between the EH issues: they are interrelated, depending on the type of sources or economic sectors generating the EH risks as well as on the settings in which these factors exert effects. Therefore, the indicators form and should be presented as "clusters" or "chains".

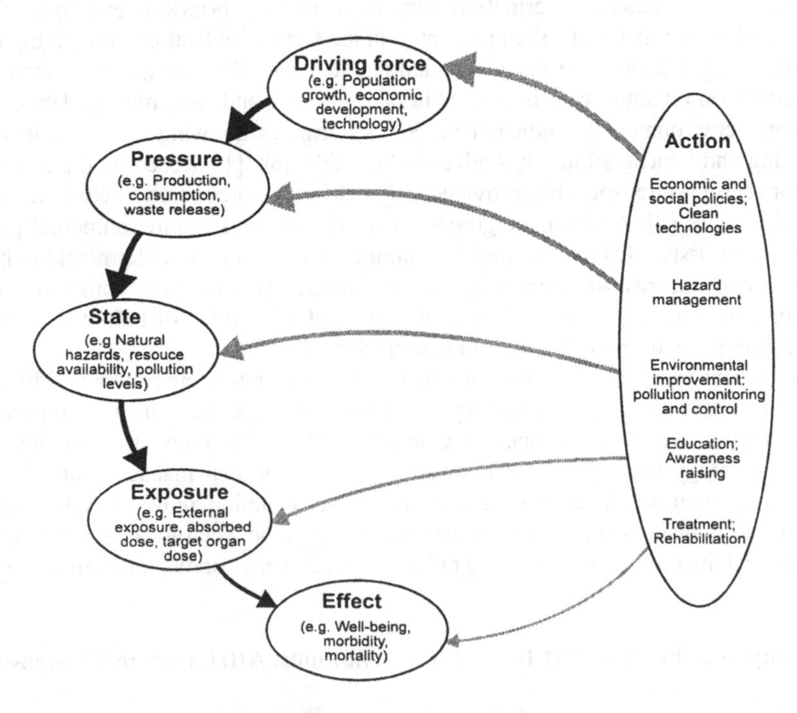

Figure 1. The DPSEEA framework [3].

TABLE 1. List of the environmental health indicators core set.

EH issue	Indicator title	Code	DPSExEA
Air quality	Kilometers driven per transport mode per person	Air_D1	D
	Consumption of fuel type from road transport	Air_D2	D
	Consumption of leaded gasoline	Air_P1	P
	Emissions of air pollutants	Air_P2	P
	Population-weighted exceedance of pollutants referent concentrations	Air_Ex1	Ex
	Infant mortality due to respiratory diseases	Air_E1	E
	Mortality due to respiratory diseases all ages	Air_E2	E
	Mortality due to diseases of circulatory system, all ages	Air_E3	E
	Participation in international agreements and environmental initiatives	Air_A1	A
	Policies to reduce environmental tobacco smoke exposure	Air_A2	A
Housing and settlements	Living floor area per person	Hous_S1	S
	Population living in substandard housing	Hous_Ex1	Ex
	Mortality due to external causes in children under 5 years of age	Hous_E1	E
	Scope and application of building regulations for housing	Hous_A1	A
	Land use and urban planning regulations	Hous_A2	A
Traffic	Mortality from traffic accidents	Traf_E1	E
	Rate of injuries from traffic accidents	Traf_E2	E
Noise	Population annoyance by certain sources of noise	Noise_E1	E
	Sleep disturbance by noise	Noise_E2	E
	Application of regulations, restrictions and noise abatement measures	Noise_A1	A
Waste	Hazardous waste generation	Waste_P1	P
	Contaminated land area	Waste_S1	S
	Hazardous waste policies	Waste_A1	A
	Municipal waste collection	Waste_A2	A
Radiation	Incidence of skin cancer	Rad_E1	E
	Topicality of permits on the use of radioactive substances	Rad_A1	A
	Efficient environmental monitoring of radiation activity	Rad_A2	A
Water and sanitation	Waste water treatment coverage	WatSan_P1	P
	Exceedance of recreational water limits for microbiological parameters	WatSan_S1	S
	Exceedance of WHO guideline values for microbial. Parameters	WatSan_S2	S
	Exceedance of WHO guideline values for chemical parameters	WatSan_S3	S
	Access to drinking water complying with WHO guideline values	WatSan_Ex1	Ex
	Access to safe drinking water	WatSan_Ex2	Ex
	Supply from public water supplies	WatSan_Ex3	Ex
	Access to adequate sanitation	WatSan_Ex4	Ex
	Outbreaks of water-borne diseases	WatSan_E1	E
	Diarrhoea morbidity in children	WatSan_E2	E
Food safety	Monitoring chemical hazards in food: potential exposure	Food_Ex1	Ex
	Food-borne illness	Food_E1	E
Chemical emergency	Sites containing large quantities of hazardous chemicals	Chem_P1	P
	Mortality from chemical incidents	Chem_E1	E
	Chemical incidents registers	Chem_A1	A
	Poison centre service according to quality standards	Chem_A2	A
	Medical treatment guidelines	Chem_A3	A
	Governmental preparedness	Chem_A4	A
Workplace	Occupational fatality rate	Work_E1	E
	Rates of injuries	Work_E2	E
	Statutory reports of occupational diseases	Work_E3	E

To provide a clear and detailed formulation, the EH indicators' operational forms or the "methodology sheets" were constructed according to the WHO guidelines on indicator profiles [3]. The following elements were included:

- Indicator definition
- Underlying definitions and concepts
- Specifications of data needed and data sources
- Computation procedures and units of measurement
- Geographical scale of application
- Interpretation remarks: limitations and complexities
- Linkage with other indicators in the core set
- Related indicators sets and databases

This approach allows precise identification of the measurable indicators and the comparability of the data. It also enables one to take a closer account of earlier international and national work and related indicator sets, thus preventing unnecessary duplication of efforts. Two illustrative examples on air quality indicators are given below. More details can be found in [5].

Example 1: Exposure to ambient air pollutants (urban)

DEFINITION: Population-weighted exceedance of reference concentrations for NO_2, SO2, PM_{10}, TSP, BS (daily), O_3 (8-hours)

UNDERLYING DEFINITIONS:
- ✓ Mean annual concentrations and reference values (RV) for the pollutants
- ✓ Population-weighting: fraction of population living in exceedance area
- ✓ Urban area

COMPUTATION: Exposure to a pollutant y with RV_Y, concentration Cy_i in subpopulation P_i and population P:

$$Ex_Y = \sum_i (P_i / P) * (Cy_i - RV_Y)$$

where $P = \sum_i (P_i)$

UNITS OF MEASUREMENT: µg / m³

A smaller value of this indicator shows that a smaller proportion of the population is exposed to higher, more harmful concentrations of air pollutant; a zero value indicates no population exposure above the reference.

The second example refers to an action indicator. It should be noted that, in order to enable between/within country comparisons, scoring is attributed to the action indicators; and three levels are considered ("not", "partly", "fully") to report the degree of policy implementation.

Example 2: Policies to reduce environmental tobacco smoke exposure

DEFINITION: Capability for implementing policies to reduce environmental tobacco smoke (ETS) and to promote smoke-free zones

UNDERLYING DEFINITIONS: Ten components on existence, implementation and enforcement of instruments and measures to prohibit/ restrict smoking

COMPUTATION:

$$I = \sum_{i} (C_i) \qquad i = 1, \ldots, 10$$

where C_i scoring 0 (not existing, not clearly stated)
1 (yes, partly implemented or enforced)
2 (yes, implemented and enforced)

C_i smoking restricted/ prohibited in: .... Schools
.... Day-care centres
.... Governmental offices and buildings
.... Public traffic vehicles (urban)
.... Public traffic vehicles (long distance)
.... Hospitals
.... Workplaces
.... Cinemas, theatres, museums
.... Bars, restaurants
.... Advertisement of cigarettes prohibited

UNITS OF MEASUREMENT: **score**

The maximum value of 20 shows full implementation of all policy elements specified in the indicator's definition.

2.1. THE USE OF THE EH INDICATORS

The EH indicators were designed to facilitate and promote information exchange on the EH issue they address at national as well as at the WHO European regional level. They serve as the guiding structure to the reports and also provide the contents structure for the establishment of an electronic network for data access/exchange. The most straightforward use of the indicators is descriptive: to show time-trends and spatial patterns on public health status and the environment. The comparability of the underlined datasets is of utmost importance. Such reports reflect the situation without reference to how it should be. Nevertheless, relevant comparisons, e.g., versus standards, between countries, can exert pressure on decision makers to undertake action. When the distance between a specific set of reference conditions (inter- and/or national policy targets) and the current situation is presented, the report can serve as a basis for monitoring progress towards officially adopted policy objectives. The reports based on performance indicators are currently in use by different environmental bodies, e.g., the European Environment Agency (EEA) [1]. They have rarely been used by the health sector because, with a few exceptions, setting national public health targets is not a common practice.

The added value of using the EH indicators to address policy questions is in the integrated assessment of the environmental health impacts. The disease burden caused by environmental exposure, and the preventable part of it, are major elements, which can guide decision making, the setting of priorities, and public health targets. Two basic approaches are used to assess disease burden from environmental risk factors: the exposure-based and the outcome-based approach [6].

Indicators on Air Quality are among those designed to support health impact assessment [7]. They are based on observations from numerous epidemiological studies, summarized, for example in the WHO Air Quality Guidelines [8], indicating that the incidence of several health outcomes increases linearly with exposure level. Therefore, the value of the exposure indicator Air_Ex1 for a pollutant is proportional to the increase in health risk (with the proportionality coefficient determined by epidemiological studies). Consider the following illustrative example.

Example 3: Use of indicator Air_Ex1 for health impact assessment

The indicators are used to compare health relevance of pollution with particulate matter in cities of two regions of a country C. In region A, there are two cities with air quality monitoring data: Aa (500,000 residents) and Ab (200,000 residents). The annual average PM_{10} level in city Aa is 55 $\mu g/m^3$, and in Ab 45 $\mu g/m^3$. In region B, the three cities are Ba (700,000 residents, mean pollution level 44 $\mu g/m^3$), Bb (300,000 residents and 46 $\mu g/m^3$) and Bc (200,000 residents and 46 $\mu g/m^3$). The reference value set in the definition of the indicator Air_Ex1 for PM_{10} is 40 $\mu g/m^3$. The following values for the indicator Air_Ex1 can be calculated:

Region A: 12.1 $\mu g/m^3$
Region B: 4.8 $\mu g/m^3$.

This means that the increase in risk due to pollution exceeding the reference level is more than double in A than in B. Assuming that the relative risk for natural mortality estimated from a cohort study is 1.10 per 10 $\mu g/m^3$ of long-term average PM_{10} [8], the exposure leads to 12.1% in mortality due to all causes (except accidents) in region A, and 4.8% in region B. With the background mortality assumed to be 800 per 100,000 (taken from national statistics), this leads to:

In Region A: 12.1% x 800 x 7 = 678 deaths associated with pollution
In Region B: 4.8% x 800 x 12 = 461 deaths associated with pollution.

The absolute magnitude of the estimated impacts does not differ as much as the relative risks due to the greater urban population of Region B.

The estimated numbers must be seen as a rough approximation of the burden of pollution (exceeding the reference value) on health. A discussion of the methodology, its applicability and limitations can be found in the relevant literature, e.g., in [9].

In order to increase the reliability of the estimates similar to those presented in Example 3, one should carefully evaluate the quality and the relevance of the data collected from environmental monitoring for health impact assessment purposes.

3. Testing Phase: Procedures and First Results

The first part of the project is a pilot evaluation of the relevance of EH indicators for wide-scale implementation in the European region of the WHO. A feasibility study was initiated in October 2000 in several countries to test, according to a standardized protocol, the availability and quality of the necessary data, their accessibility and exchange mechanisms, use and usefulness of the proposed indicators in (sub)-national context, and the capacity for multi-agency networking on the information. The following main issues were addressed:

- *Data quality and reliability:* use of standardized methodology for data collection and quality control/quality assurance, completeness of the data over time, statutory requirements for data collection, possibility to derive population-based estimates from routine monitoring (spatial coverage and spatial resolution of sources, pollutants and population concerned), sensitivity of the health surveillance system, and the existence of regular population-based surveys;
- *Data flow and accessibility:* existence of institutional framework for data access/exchange, accessibility through electronic networks in common simple formats, accessibility of the data at central level, and timeliness versus accessibility.

  Fourteen Member States (MS) are participating in the study: Bulgaria, The Czech Republic, Estonia, Finland, Hungary, Latvia, Lithuania, The Netherlands, Poland, Romania, Russian Federation (one region), Slovakia, Spain, and Switzerland. The progress of the study was reviewed at a meeting, recently convened by the WHO – ECEH.

Feedback from the countries shows the following points as very useful outcomes of the study:

- The study enables the creation of standardized inventories of data availability and quality on a wide range of environmental health issues. The inventories include information on the data-flows and holders as well as on the data quality and availability. Therefore, the study enables the establishment of a multi-agency network for streamlining the information and stimulates working relationships with the numerous data-providers. The standardized assessment from the study should serve as a basis for a meta-information system.
- The results of the feasibility study also indicate which data collection techniques require further harmonization or methodological developments.
- The knowledge gained on the existing information systems related to EH creates synergies with other ongoing indicator projects, thus preventing unnecessary duplication of the work.

The main challenges and difficulties are seen in:

- The necessity to co-operate with numerous institutions, including those responsible for data collection and processing, emerges as a common problem due most probably to the lack of tradition in health and environment reporting. The system requires the establishment of links between more than ten different agencies and the ability to cope with the reluctance and insufficient understanding reflected by the data-providers.
- The organizational changes, the transitional period in changing legislation and data collection, and the ongoing reform of the health sector in many countries further reinforced the difficulties in establishing working contacts with the numerous data-providers.
- Human resources were also seen as an important factor, in particular under conditions of limited funding.
- There is a large variety of projects and parallel initiatives on indicators and indicator-based reporting at national and international scale, several of the core indicators being already reported to the EEA, EUROSTAT, OECD, etc.

The first round of evaluation of the EH indicators set, considering the criteria of availability, quality, and usefulness/interpretability, was performed by the countries participating in the study. The average scores for the indicators on air quality, noise, housing and water and sanitation are plotted in Fig. 2.

The results show that the majority of the indicators are likely to be feasible and relevant for use in national reports; a few indicators were identified for further refinement. The indicator Air_P1 related to lead exposure was rejected in view of the decreased policy relevance. As it can be seen, the indicators on noise, drinking water quality, and associated health effects were evaluated to be very useful; the lesser quality and availability of data necessary for those indicators require development of methods and documented guidance for data collection using household surveys.

The usefulness of some indicators related to "infrastructure", like habitation facilities, drinking water supply, sanitation and wastewater treatment facilities, is under debate due mainly to the mandatory requirements in some countries, e.g., The Netherlands and Switzerland, where very high standards have already been achieved. This indicates the need to increase the indicators' relevance for highly developed countries. Possible solutions may include differential setting of "reference values" while maintaining comparability across Europe. Establishing a close coordination with the European sustainability indicators project would be very useful in view of the changes in benchmarking in these countries.

The limited usefulness reported on the health effect indicators of air quality, despite good availability and quality, points out the necessity for integration of environment and health linkage analysis and impact assessment methodology in the reporting mechanisms. Such a product-oriented approach, starting already in the first phases of the pilot process, is beneficial as it reinforces the feedback loop: raises interest, gets comments, and increases quality. It also enables the formulation of policy messages according to the users' needs to identify priority data needs and areas of research. In this respect, a better inter-agency coordination of WHO, EEA, OECD, EUROSTAT, etc., is also very important.

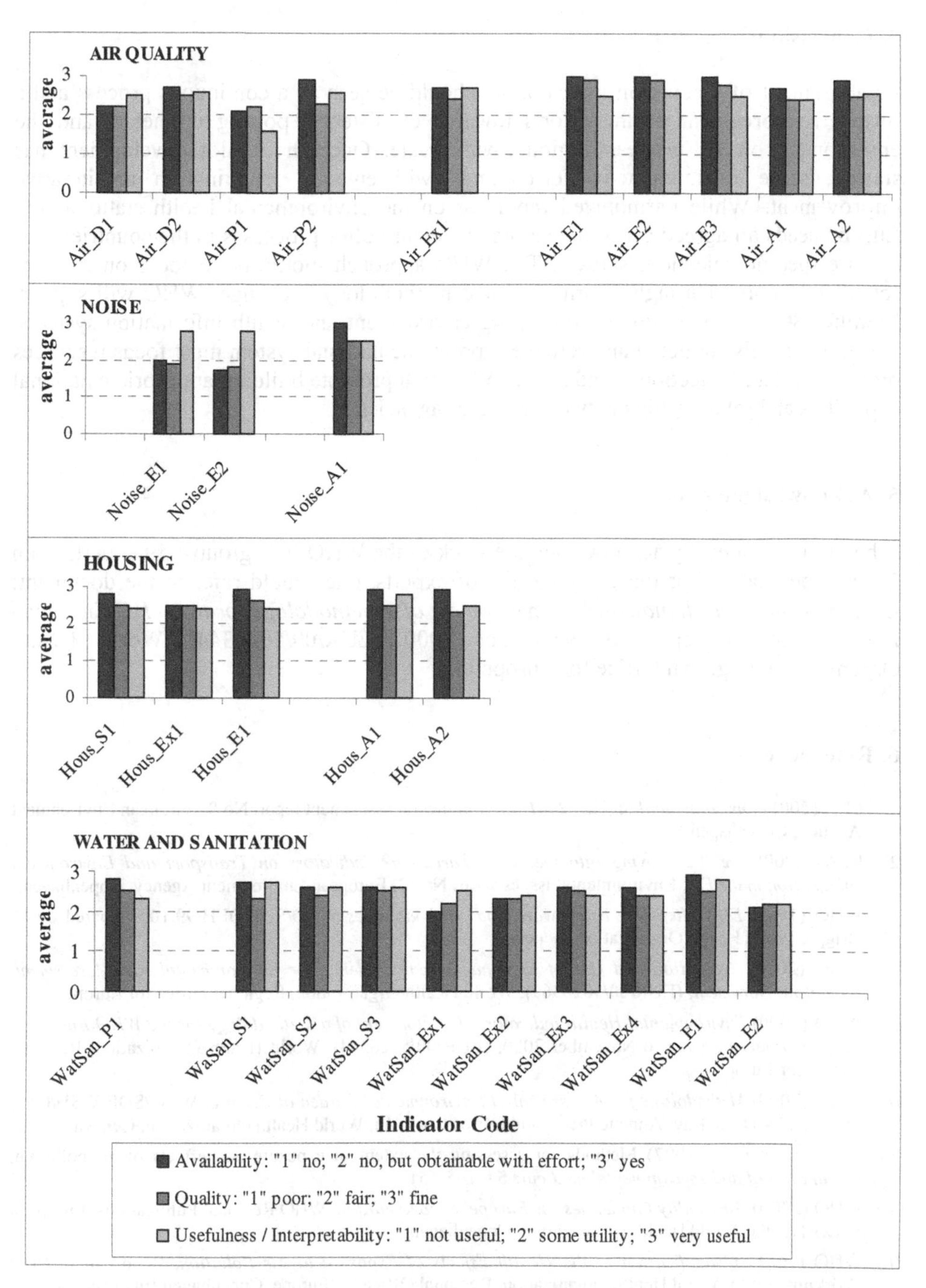

Figure 2. Average scores of the countries' ratings of the availability, quality and usefulness/interpretability for the air quality, housing, noise, and water and sanitation indicators (see Table 1 for the indicator names).

4. Conclusions

Establishment of a common indicator set should be seen as a continuous process and a further reinforcement of the efforts towards consistent reporting on health and the environment on a European regional-wide scale. Once their pilot development has started, some unanticipated shortcomings will emerge, resulting in an iterative improvement. While harmonized reporting on the environmental health status across Europe needs an agreed set of indicators, different policy processes in the countries may require specific selections/subsets. The WHO approach should be to focus on the core set of indicators. Through information and methodology exchange, WHO will support Member States in upgrading the existing environment and health information systems. Aiming at timely, targeted and reliable reports, the national system must focus resources on priority data collection. In addition, WHO will promote building appropriate national capacities and relationships between relevant agencies.

5. Acknowledgements

The authors gratefully acknowledge the work of the WHO WG groups: 59 experts from 27 member states. For the complete list of experts, one should refer to the document: *Environmental health indicators: development of a methodology for the WHO European Region.* Interim report 6 November 2000 (EUR/00/5026344) World Health Organization, Regional Office for Europe [5].

6. References

1. EEA (2001) *Environmental Signals 2001*, Environmental assessment report No 8, European Environment Agency, Copenhagen.
2. EEA (2000) *Are We Moving into the Right Direction? Indicators on Transport and Environment Integration in the EU*, Environmental issues series No 12, European Environment Agency, Copenhagen.
3. WHO (1999) *EH Indicators: Framework and Methodologies* WHO/SDE/OEH/99.10, prepared by D. Briggs, World Health Organization, Geneva.
4. WHO (2000) *Evaluation and Use of Epidemiological Evidence for Environmental Risk Assessment, Guideline Document*, (EUR/00/5020369), World Health Organization, Regional Office for Europe.
5. WHO (2000) *Environmental Health Indicators: Development of a Methodology for the WHO European Region, Interim Report,* 6 November 2000, (EUR/00/5026344), World Health Organization, Regional Office for Europe.
6. WHO (2000) *Methodology for Assessment of Environmental Burden of Disease,* WHO/SDE/WSH/00.7, prepared by David Kay, Annette Prüss, and Carlos Corvalán, World Health Organization, Geneva.
7. Krzyzanowski, M. (1997) Methods for assessing the extent of exposure and effects of air pollution, *Occupational and environmental medicine* **54**, 145-151.
8. WHO (2000) *Air Quality Guidelines for Europe, Second edition,* WHO Regional Publications, European Series No. 91, World Health Organization, Copenhagen.
9. WHO (2001) *Quantification of the Health Effects of Exposure to Air Pollution*, Report of a WHO Working Group, World Health Organization, Regional Office for Europe, Copenhagen (in press).

Part VII

DATA PROCESSING, ANALYSIS AND MODELING

DOWNSCALING OF CONTINENTAL-SCALE ATMOSPHERIC FORECASTS TO THE SCALE OF A WATERSHED FOR HYDROLOGIC FORECASTING

M. L. KAVVAS, Z. CHEN, and M. ANDERSON
Hydrologic Research Laboratory
Department of Civil and Environmental Engineering
University of California
Davis, CA95616, U.S.A.

Abstract. A study, utilizing atmospheric forecasts for runoff prediction, has been performed at Calaveras River watershed in Northern California. Eta atmospheric model forecasts were used to provide input to a mesoscale atmospheric model, MM5. The MM5 model was then run to provide refined precipitation forecasts over the Calaveras River watershed. Using these precipitation forecasts as input, the HEC-HMS watershed hydrology model was then used to obtain 48-hour-ahead runoff forecasts at Calaveras watershed.

1. Introduction

In the past, accurate forecasts of the reservoir inflows resulting from watershed runoff have been made, but only after the water had entered the main channel. During flooding events, this limits the amount of time available for the implementation of emergency management procedures. An example of this limitation and its consequences occurred during the December 1996/ January 1997 flooding events over Northern California [1].

Recent efforts in river forecasting have focused on quantifying rainfall amounts from radar images [see 2, 3, 4, 5]. Research has also progressed in the continued development of models capable of predicting the spatial and temporal evolution of the flood wave as it moves down the channel [see 6, 7, 8]. However, these reported studies are limited to the time frame of radar images and water that is already in the main channel. The only way to gain additional lead-time in runoff forecasting is to gain precipitation information ahead of its occurrence.

One way in which this can be accomplished is by translating precipitation forecasts into runoff forecasts. The National Center for Environmental Prediction (NCEP) Eta model [9, 10] provides 48 hour-ahead forecasting of precipitation in 6-hour intervals in a 40 square km gridded form over the entire United States. Through the use of a mesoscale model, MM5, and a rainfall-runoff model, HEC-HMS, this information can be translated into runoff forecasts with a 48-hour lead-time. Accurate runoff forecasts of this nature would greatly improve the lead-time necessary for emergency management procedures such as evacuations to be carried out.

As a means of evaluating this approach, a feasibility study has been completed on the Calaveras River watershed in Northern California. Eta model forecasts were obtained to provide input and boundary conditions for the mesoscale model, MM5. The MM5 model was then run to provide a refined precipitation forecast with spatial and temporal scales suitable for use in the United States Army Corps of Engineers (USACE) watershed model HEC-HMS. HEC-HMS was run with the precipitation forecast data in order to provide a 48-hour forecast of runoff entering a reservoir on the Calaveras River.

The results of this study are presented in this paper, which starts with an overview of the methodology used to translate the Eta-model precipitation forecasts into HEC-HMS runoff forecasts. After describing the computer models MM5 and HEC-HMS that are used in the study, a description of the Calaveras River watershed, used in the case study, is presented including its representation in HEC-HMS. Results of the study are then presented and discussed. The paper concludes with an assessment of the methodology including future directions for its further development.

2. Methodology and Model Descriptions

A schematic of the methodology for translating precipitation forecasts into runoff forecasts is shown in Fig. 1. Eta-model products are first downloaded from NCEP's Internet web site. This information provides the input and boundary conditions for the mesoscale model, MM5, which can be nested into the Eta model. The mesoscale model, MM5, is run to produce 48 one-hour precipitation depth forecasts on a grid covering the Calaveras River basin (see Fig. 2). That same grid is used by the watershed model, HEC-HMS, to compute runoff.

However, before HEC-HMS can be run, the results of the MM5 simulation must be put into the proper format for use by HEC-HMS. This is done by two utility programs that put the precipitation data from MM5 into the database HEC-DSS in a series of 48 records of gridded one-hour precipitation depths. Once the HEC-HMS model of the study area is created, calibrated and verified, it can then be run with the gridded precipitation to generate the runoff forecast. In the following subsections, the Eta model forecast products, and the MM5 and HEC-HMS models are described, including the input that is necessary for this simulation methodology.

2.1. ETA MODEL FORECAST PRODUCTS

A series of operational Eta models for atmospheric numerical forecasts have been developed for use in the United States [11, 12, 13]. NCEP releases forecasts at 00Z and 12Z UTC (Coordinated Universal Time) each day. The letter Z means that the time is the local time at the Zero degree longitude (Greenwich meridian). It may be necessary to convert the UTC time to a local time (minus 8 hours for Pacific Standard Time where the Calaveras Basin is located). Each forecast has a lead-time of 48 hours with time intervals of 6-hours, 3-hours, or 1-hour. Pressure, temperature, three dimensional wind fields, moisture fields, and precipitation depths per time interval are provided over a given forecast region (the continental United States, for example) on a variety of AWIPS (Advanced Weather Interactive Processing System) grids.

The gridded Eta model results can be downloaded from an ftp site in GRIB format. The 6-hour interval data is available to the public from the NCEP Data Repository Site: ftp.ncep.noaa.gov. GRIB (GRIdded Binary) is a general purpose, bit-oriented data exchange format that is an efficient vehicle for transmitting large volumes of gridded data to automated centers over high-speed telecommunication lines using modern protocols. A GRIB decoder is required to read and to process the forecast data in their raw formats.

The current operational Eta model has 32 km horizontal resolution and 45 vertical layers and runs over a domain which encompasses nearly all of North and Central America, including surrounding oceans, Alaska and Hawaii. Initial conditions are provided by the Eta Data Assimilation System (EDAS), which runs on a 3-hour forecast/analysis/update cycle for 12 hours prior to the start time of a model run. Boundary conditions are provided by the previous cycle's Aviation model (AVN) run of the NCEP Global Spectral Model.

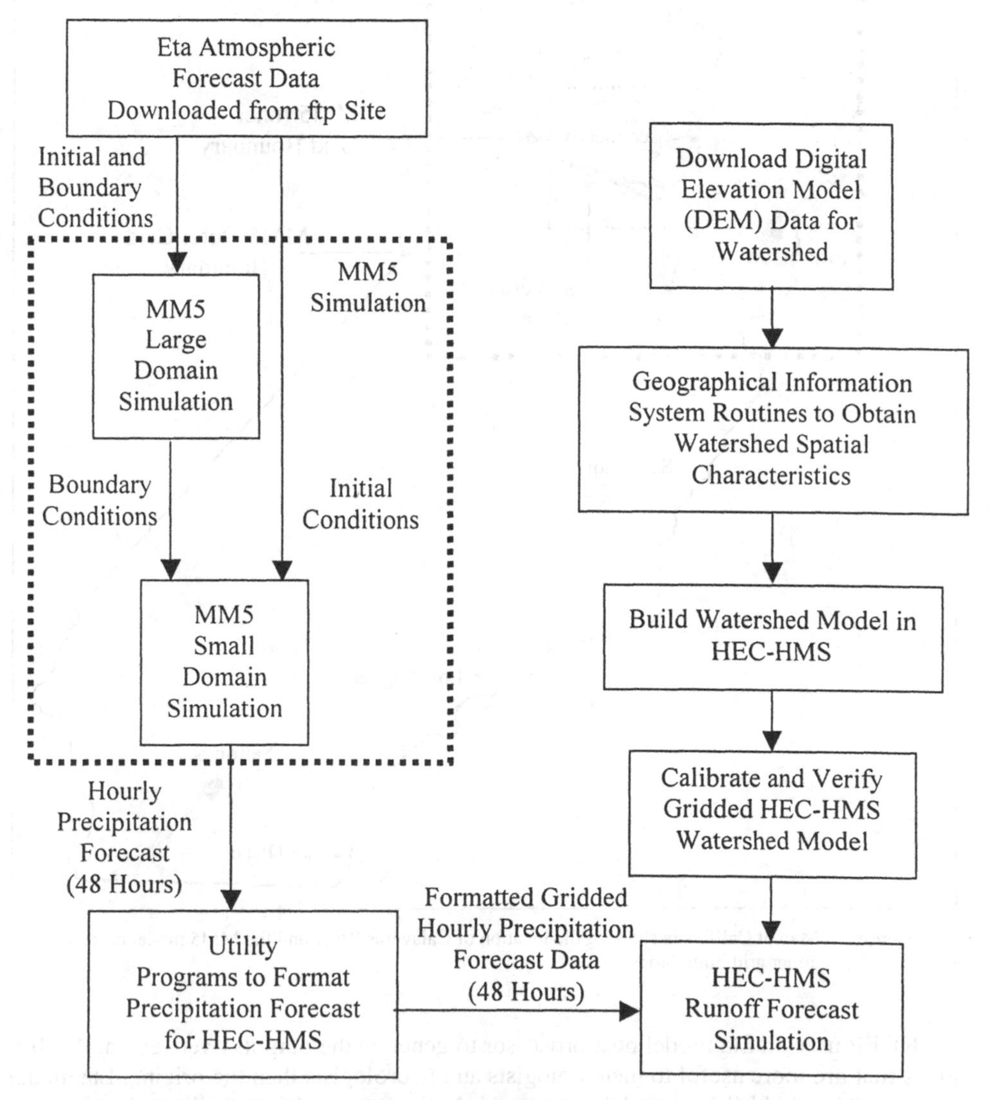

Figure 1. Schematic diagram of the process for obtaining runoff forecasts from HEC-HMS.

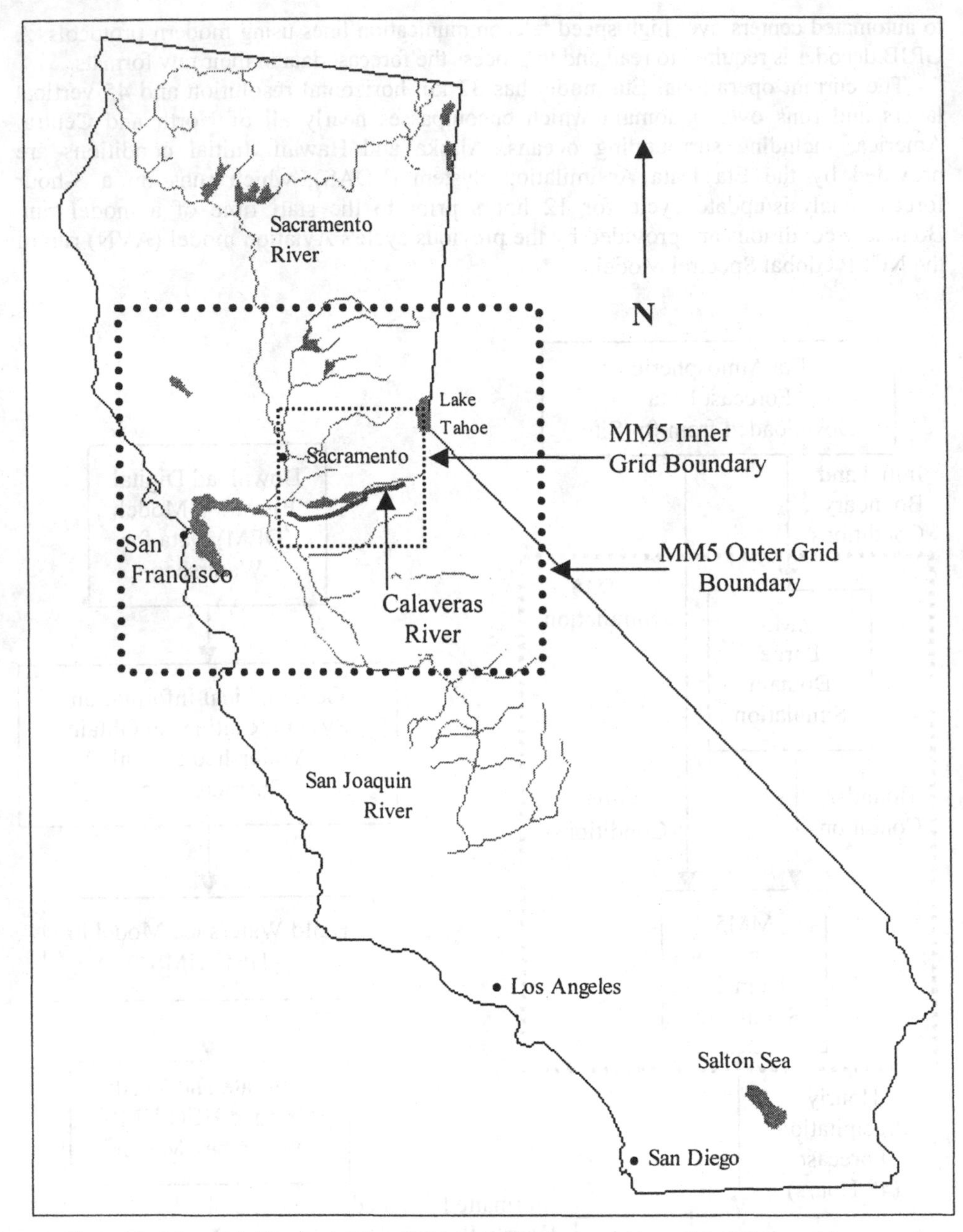

Figure 2. Map of California showing the location of Calaveras River and the MM5 model outer and inner grid boundaries.

NCEP uses an Eta model post-processor to generate the outputs over several AWIPS grids that are more useful to meteorologists and hydrologists than the original Eta model outputs. The gridded Eta model outputs are further grouped into 3-dimensional output files and surface output files. AWIPS grid #212 was used for this study. This grid is defined on a Lambert Conformal projection with a nominal horizontal grid resolution of 40-km and grid dimensions of 185 x 129. The latitude/longitude locations of the corner points of this grid are given in Table 1. This domain covers most of North America and the nearby oceans, including the 48 contiguous United States, the southern half of Canada, and most of Mexico.

TABLE 1. Grid Points of the corners of AWIPS Grid #212

(I, J)	Latitude / Longitude
(1, 1)	12.190 N / 133.459 W
(1, 129)	54.536 N / 152.856 W
(185, 129)	57.290 N / 49.385 W
(185, 1)	14.335 N / 65.091 W

2.2. THE MESOSCALE MODEL MM5

The MM5 modeling system is the fifth generation mesoscale model developed by the National Center for Atmospheric Research (NCAR) and the Pennsylvania State University. It is a globally re-locatable model with three map projections: polar stereographic, Lambert conformal and Mercator, that support different true latitudes. The MM5 model uses terrain-following coordinates in the vertical and can be run under either hydrostatic or nonhydrostatic dynamic frameworks. The hydrostatic framework is used for large-domain simulations and approximates the vertical momentum equation with the hydrostatic relation where pressure varies linearly with altitude. The Eta model uses this approach in its simulations. The nonhydrostatic framework, used in this study, simulates the full vertical momentum equation and allows the model to be used at a few-kilometer scale necessary for watershed simulations.

The state variables of the MM5 modeling system with the nonhydrostatic framework include pressure, temperature, density, and wind velocities. Parameterized processes include advection, diffusion, radiation, boundary-layer processes, surface-layer processes, cumulus convection, and routines for all three phases of water in the atmosphere. Several options exist for the parameterization of moist convection and boundary layer processes for the simulation of atmospheric phenomena at different scales and different characteristics [14, 15].

The MM5 modeling system provides variable resolution terrain elevation and land use data sets. The model grid sizes are flexible, and multiple nesting capabilities exist. The model can be configured to run from the global scale down to the cloud scale in one model with both 2-way and 1-way nesting modes. The 2-way nesting mode can handle multiple nests and moving nests. The MM5 modeling system can also be coupled with global models and other regional models. It uses the other model's output either as a first guess for objective analysis, or as lateral boundary conditions as was done in this study with the NCEP Eta model [14, 15].

The MM5 model for the Calaveras River basin is a 32-layer model with two nested grids identified as the outer grid and inner grid. The 31 x 31 node outer grid has a spatial resolution of 12 km and spans central California, part of Nevada near Lake Tahoe, and part of the Pacific Ocean by San Francisco and Half Moon Bays (see Fig. 2). A comparison of topography over this region at 40-km resolution (used by the Eta model) and at 12-km resolution (used by the MM5 model) is shown in Fig. 3. The refined topography combined with the nonhydrostatic representation of vertical motions used in the MM5 model provides more detail for orographic influences on precipitation.

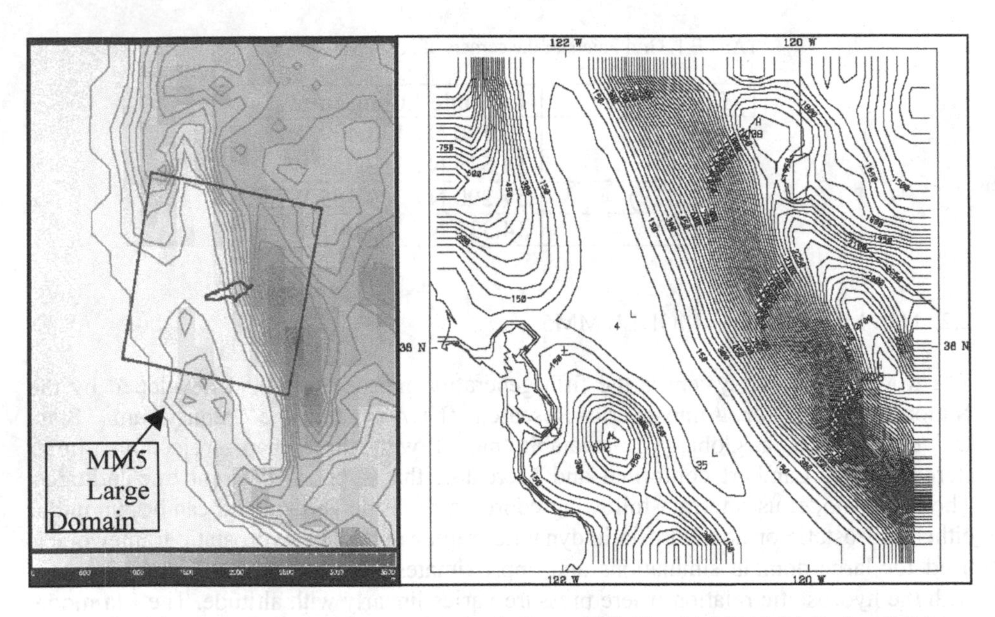

Figure 3. Comparison of topographic representation in Eta model (left) and MM5 (right) for the MM5 large domain spanning Central California and the Calaveras Basin.

The MM5 model receives its time-dependent boundary conditions from the NCEP Eta model forecast on AWIPS grid #212, including three-dimensional forecasts of temperatures, wind vectors, relative humidity, geopotential heights, and surface pressures every 6 hours for a total of 48 hours. Nested within the outer grid is a 34 x 34 cell inner grid with a resolution of 4 km. The inner grid spans the Calaveras River basin and its immediate surrounding areas (see Fig. 2) and uses relaxation boundary conditions. The computational time interval is twenty seconds and the time interval of the MM5 precipitation output is set to one hour to coincide with the HEC-HMS computational time interval for the Calaveras River basin.

The following steps were used to nest the MM5 model within the NCEP Eta model forecast. First, the Eta data files were obtained from the ftp site and decoded. The Eta data were then projected onto the Calaveras MM5 domain, using a utility program in the MM5 modeling system called DATAGRID. The initial and boundary conditions for the Calaveras MM5 simulation were obtained, using the utility program INTERP. The MM5 simulation was then run, producing hourly precipitation depth data over the inner grid covering the Calaveras basin. The parameterizations used in the MM5 simulation of the Calaveras basin play an important role in the refinement of the spatial and temporal scales of the precipitation forecast and are described below.

A complete description of the parameterizations available in the MM5 modeling system can be found in Grell *et al.* [14]. The parameterizations that were used in the Calaveras basin simulation are described here. Atmospheric radiation was computed using Dudhia's radiation scheme every one-half hour without considering the effects due to snow cover. Shortwave radiation is computed, using transmissivity functions for clear and cloudy sky absorption, and longwave radiation is computed, using an emissivity approach with the Stefan-Boltzmann law [14].

The planetary boundary layer processes are modeled using the Blackadar force-restore method. This method uses a thermal inertia parameterization to obtain the thermal capacity per unit area parameter, Cg, which is used in the surface temperature equation. A multiple-layer soil model was used to compute the heat flow into and out of the ground [14]. The planetary boundary layer is heated from solar radiation, longwave radiation, and sensible and latent heat fluxes. The sensible heat fluxes are computed, using a bulk aerodynamic formulation which uses exchange coefficients multiplied by the temperature difference between the ground surface and the first atmospheric layer. The exchange coefficients are computed based upon stability conditions of the boundary layer. The exchange coefficients are also used to compute the surface moisture flux, which also uses a moisture availability parameter.

Precipitation parameterizations used in the MM5 model of the Calaveras basin include one for nonconvective precipitation and one for convective precipitation. The saturation threshold for producing nonconvective precipitation was set to 100%. Both the outer and inner domains used an explicit treatment of cloud water, rainwater, snow and ice for the moisture parameterization. In this scheme, three equations are used to simulate the time evolution of the mixing ratios for water vapor, cloud/ice water and rain/snow water. Computations are made to determine the amount of water that changes from one type of atmospheric moisture to another for all possible paths, and adjustments are made to each of the mixing ratio equations. For example, computations are made to track the changes from water vapor to cloud, from water vapor to ice, from water vapor to rain, and from water vapor to snow. Calculations are also made to track the changes from cloud to rain or rain to cloud, rain to snow or snow to rain, from ice to snow, and from cloud to ice. Ice-phase processes are assumed to exist when the temperature is less than zero degrees Celsius. Details of this parameterization can be found in Grell *et al.* [14].

For convective precipitation computations, the Grell cumulus parameterization was chosen [see 14]. In this scheme, there is an updraft and a downdraft circulation that are considered to be in a steady state. The only mixing between cloudy air and environmental air occurs at the top and bottom of the cloud, which yields a mass flux that is constant with height. This also assumes that there is no mixing or entrainment of air along the boundaries of the updraft or downdraft. Thermodynamic variables are computed for any updraft or downdraft, starting at any level and allowing for maximum buoyancy. With this scheme, the cloud base for the updraft can be anywhere in the troposphere, not just in the boundary layer. Condensation and evaporation are computed by tracking a moisture budget in the updraft and downdraft. All condensation is converted to rain and no cloud water is assumed to exist from the convective process. These parameterizations provide the basis for the refinement of the precipitation forecast, which is used by HEC-HMS to obtain a runoff forecast for the Calaveras basin. The HEC-HMS model is described next.

2.3. HEC-HMS

HEC-HMS is the updated and expanded version of the United States Army Corps of Engineers (USACE) HEC-1 rainfall-runoff model [16]. It utilizes a graphical user interface to build a watershed model and to set up the precipitation and control variables for simulation. For this project, the Calaveras River watershed in California was modeled with a focus on the upper watershed that provides inflow into New Hogan Reservoir. A map of the location of the watershed is shown in Fig. 2.

The watershed model created in HEC-HMS follows the form of the Sacramento District Corps office HEC-1 forecast model of the basin [17]. This model utilizes one subbasin above New Hogan Reservoir, that provides runoff into the reservoir. Below the reservoir, there is a river channel routing reach and another subbasin that provides local runoff to a sink point at Bellota. Bellota is a point on the lower Calaveras River where a flow and stage gaging station is located. A schematic description of the basin is shown in Fig. 4.

In Fig. 4, the outline of the watershed and branches of the upper Calaveras River are shown underneath the schematic components of the watershed model. The upper Calaveras watershed is represented by the subwatershed component Upper Basin. This subwatershed component provides inflows into New Hogan Reservoir, which is represented with a triangle. A second subwatershed is made up of the watershed below the reservoir and is titled Lower Basin. The gaging station at Bellota is represented in the watershed model as a sink, and the river reach between New Hogan Reservoir and Bellota is represented by a channel reach which routes streamflow, using the Muskingum method. For this study, the inflows into New Hogan Reservoir from the upper watershed were the only components examined.

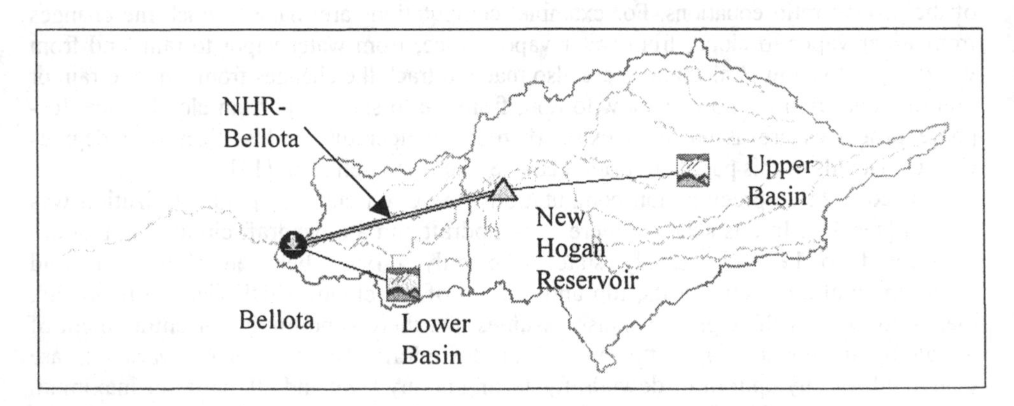

Figure 4. HEC-HMS schematic description of Calaveras River basin.

In order to use the gridded precipitation component of HEC-HMS, a GIS-based spatial representation of the watershed is required. This spatial representation can be created using a set of UNIX-based ARC/INFO routines called GridParm [18]. The software is currently being replaced by new ArcView extensions called HEC-GeoHMS [19]. The GridParm routines use data from digital elevation maps (DEMs) available from the USGS to determine watershed boundaries, subbasin areas, and channel reaches. This information is stored in two files (one for the map background and one for grid definition) for use by HEC-HMS. The spatial representation of the basin is treated as a collection of cells, each of which has a response time that is based upon its relative location to the subwatershed outlet. The precipitation on each grid cell is subject to infiltration. The excess water is routed to the subbasin outlet, using a cell-to-cell flow distance and attenuation of the ModClark method [20]. For this study, one set of infiltration parameters is used for all the cells in the watershed. With appropriate land cover and soil data, infiltration parameters could be computed for each cell. With this representation, HEC-HMS can transform spatial representations of rainfall data into runoff at the subwatershed outlet.

The calibration period for the HEC-HMS model of the Calaveras basin was a 48-hour period from February 8, 1999 through February 9, 1999. Rainfall data were obtained from 5 raingages in the Calaveras Basin listed in Table 2. Calibration and verification were performed, using rainfall data from these 5 ground-based raingages. The point-gauge-based model uses a Green-Ampt infiltration/loss parameterization, the Clark hydrograph transformation routine, and a recession base flow component. The gridded precipitation model uses the ModClark hydrograph transformation routine, with time of concentration and storage coefficient values taken from the point-gauge-based model. The initial loss and initial flow are treated as initial conditions and vary from simulation to simulation.

TABLE 2. Raingages used in the calibration of Calaveras Watershed HEC-HMS Model

Raingage	**Latitude (degrees)**	**Longitude (degrees)**
Esparanza	38.242	-120.497
Railroad Flat	38.314	-120.543
Sheep Ranch	38.210	-120.462
New Hogan Reservoir	38.155	-120.814
Robidart Ranch	38.137	-121.030

Final calibration parameters for the HEC-HMS model are shown in Table 3. Parameter values were chosen, based upon their ability to fit the observed runoff. The unusually high value of percent imperviousness is likely the result of the need to compensate for variations in rainfall within the basin, that are not represented by the raingage data. A discussion of the calibration process for this methodology is presented in a later section.

A plot of the observed inflow into New Hogan Reservoir versus model simulated inflow, using the calibration parameters of Table 3 is shown in Fig. 5. Note that the observed and model simulated flows match well, except that the peak of the simulated flow is slightly later than the observed flow. The opposite occurs for the verification run.

TABLE 3. Calibration coefficients for the Calaveras Watershed HEC-HMS Model

Model Parameter	Calibrated Value	Model Parameter	Calibrated Value
Vol. Moist. Deficit	0.1 inch	Time of Concentration	5.5 hour
Wet. Front Suct.	0.1 inch	Storage Coefficient	0.5
Hydraulic Conductivity	0.125 in./hr.	Recession Constant	0.25
% Impervious	22%	Threshold Flow	3000 cfs

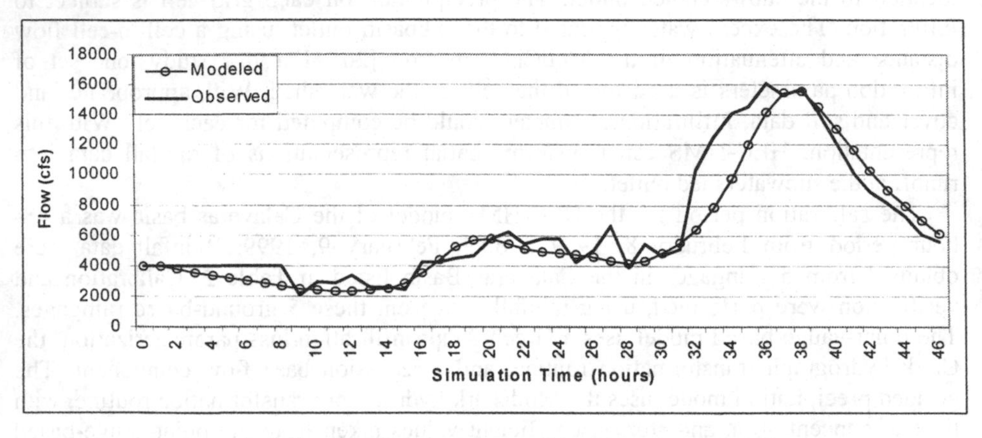

Figure 5. Calibrated modeled vs. observed New Hogan Reservoir inflows.

For verification purposes, a second 48-hour period from February 16, 1999 to February 18, 1999 was used. A plot of the observed versus predicted inflow into New Hogan Reservoir is shown in Fig. 6. Note that the predicted peak is early compared to the observed flow. The time of concentration value in the ModClark hydrograph transformation routine was set to match the timing of the calibration and verification peaks as close as possible. Using this set of parameters for the Calaveras River watershed model, shown in Table 3, an application of the runoff forecasting process was conducted.

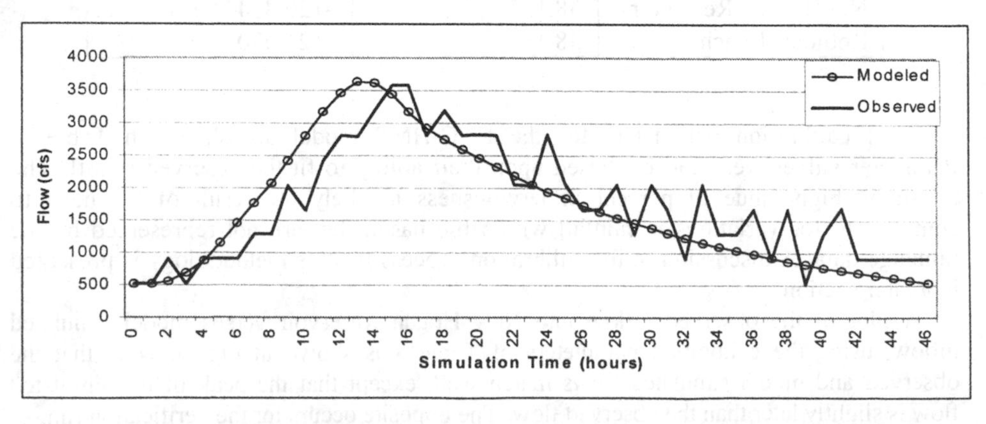

Figure 6. Verification results on predicted vs. observed New Hogan Reservoir inflows.

3. Application

In order to demonstrate the use of forecast data for the prediction of watershed runoff, two sets of simulations are presented in this section. The first set of simulations utilizes precipitation data directly from the Eta forecast data. The data are interpolated linearly in space, and the six-hour precipitation depths are divided uniformly into one-hour depths for use in HEC-HMS. In the second set of simulations, precipitation data are used from the MM5 simulations, where no interpolation is necessary. Results of both simulations are compared to observed runoff and results obtained from raingage data. A 48-hour forecast period from January 19, 1999 through January 21, 1999 was selected for simulations. During this time period, raingages measured between two and three inches of rainfall in the upper part of the Calaveras basin.

Figure 7 shows the 48-hour Eta forecast of precipitation in the Calaveras River basin and its surrounding region for the above-mentioned time period of the study. There are 8 plots in Fig. 7. Each plot represents the spatial distribution of the accumulated precipitation depth during a 6-hour time period. Each square in the plots represents one grid square for the AWIPS #212 grid. These AWIPS grids have a size of roughly 40 km x 40 km. In these plots, the heavy black outline indicates the boundary of Calaveras River basin. The gray shades indicate the 6-hour accumulated precipitation in mm in each grid where representative grid lines are shown in the upper left plot. Six AWIPS grids cover the Calaveras River watershed as is shown in Fig. 7.

The 48 hour-ahead forecast of precipitation by the MM5 model covered the time period of 12:00 GMT January 19, 1999 to 12:00 GMT January 21, 1999. A sample of six hourly plots of the spatial distribution of the hourly precipitation depths, which were generated by MM5, is shown in Fig. 8. In these figures, the heavy black outline indicates the boundary of the Calaveras River basin. The gray shades indicate the precipitation in mm/hr in each grid where a representation of the grid is shown in the upper left plot. The spatial and temporal evolution of the precipitation fields in Calaveras River basin and its surrounding region are shown clearly in these figures.

Using the precipitation inputs represented in Fig. 7 and Fig. 8, HEC-HMS was run with the gridded precipitation routine, using the calibrated values shown in Table 3. For the Eta model forecast precipitation, the precipitation values were interpolated bilinearly in space and divided evenly in time in order to obtain hourly precipitation data on the HEC-HMS precipitation grid. A plot of observed inflows into New Hogan Reservoir compared to the HEC-HMS model forecasts using the Eta model precipitation, the MM5 model precipitation, and point gage precipitation is shown in Fig. 9. The 40-km resolution of the Eta model forecast precipitation is too coarse to capture the orographic precipitation effects of the modeled region. The resulting 6-hour precipitation values are too small to lead to any runoff, especially when divided uniformly into one-hour values for use in the point-gage calibrated HEC-HMS model. This can be seen in Fig. 9, where the Eta model runoff prediction is a recession curve only. Using the refined precipitation data from the MM5 simulation, the predicted runoff shows a response to the precipitation, but it is significantly less than the observed values. The underprediction may be due in part to the fact that the calibration and verification of HEC-HMS were performed with point raingage data which may not correspond as well with the gridded precipitation formulation of the MM5 and HEC-HMS models. Another possible reason for the underprediction is that the default parameter settings for MM5 were used.

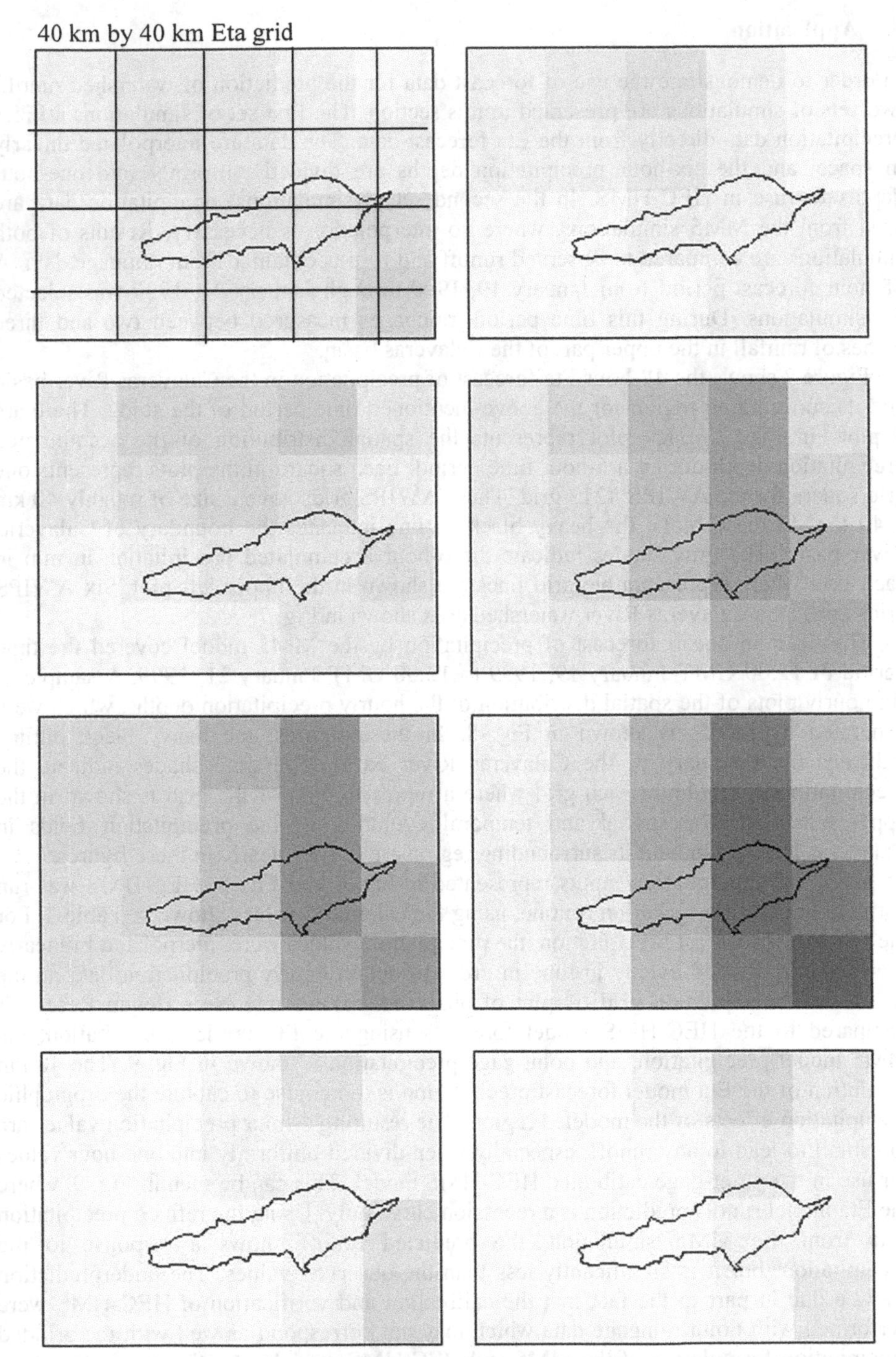

Figure 7. Eta model forecast: 6-hour accumulated precipitation depths over Calaveras River Basin from 1999-01-19_12:00Z to 1999-01-21_12:00Z.

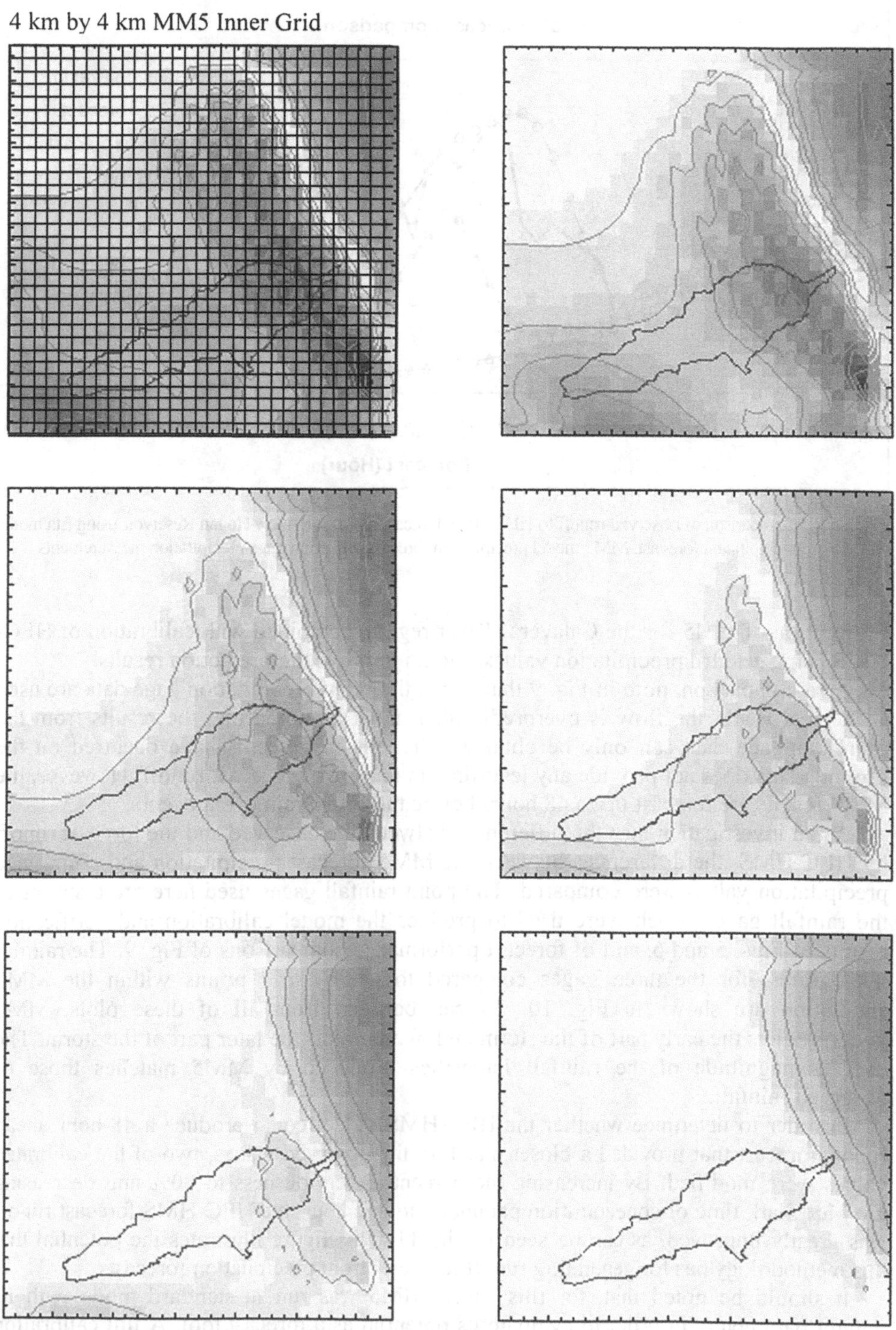

Figure 8. Sample of MM5 model forecasts of 1-hour accumulated precipitation over Calaveras River basin.

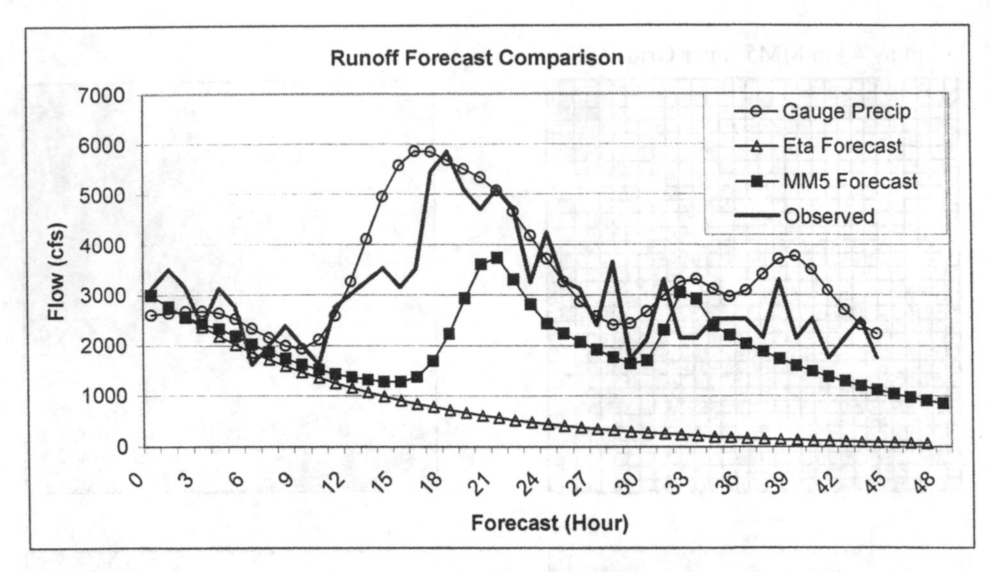

Figure 9. Comparison of observed runoff to HEC-HMS forecast runoff into New Hogan Reservoir using Eta model precipitation forecast, MM5 model precipitation forecast, and point gage precipitation measurements.

Calibration of MM5 for the Calaveras River region, combined with calibration of HEC-HMS using gridded precipitation values, should improve the prediction results.

As a comparison, note in Fig. 9 that, when the point precipitation gage data are used with HEC-HMS, the flow is overpredicted. It must be noted that the results from the point raingage data can only be obtained after the rain events have occurred on the ground. This does not provide any lead-time in terms of a forecast runoff. However, the MM5 results are forecast up to 48 hours before the actual rain event occurs.

As an investigation into the differences between the observed and the forecast runoff by HEC-HMS, the differences between the MM5 forecast precipitation and point gage precipitation values were compared. The point rainfall gages used here are a subset of the rainfall gages which were used to produce the model calibration and verification results of Figs. 5 and 6, and of forecast performance comparisons of Fig. 9. The rainfall hyetographs for the three gages compared to nearby grid points within the MM5 simulation are shown in Fig. 10. As can be seen from all of these plots, MM5 underpredicts the early part of the storm and overpredicts the later part of the storm. The overall magnitude of the rainfall intensities produced by MM5 matches those of observed rainfall.

In order to determine whether the HEC-HMS model could produce a 48-hour ahead runoff forecast that provided a closer match to the observed values, two of the calibrated values were modified. By increasing the percent imperviousness to 80% and decreasing the ModClark time of concentration parameter to one hour, the HEC-HMS forecast runoff was greatly improved, as can be seen in Fig. 11. This figure illustrates the potential that this methodology has for generating runoff forecasts from precipitation forecasts.

It should be noted that, for this study, MM5 was run in standard mode with its default parameters in order to evaluate its potential as a forecast tool. A full calibration and verification exercise of both MM5 and HEC-HMS, using Eta model data, would likely improve both the precipitation and the runoff prediction results as Fig. 11 indicates. Such a calibration/verification exercise would also lead to more realistic values for parameters such as percent imperviousness, which are unusually high in this study. However, the current results illustrate the promise that this methodology has for obtaining runoff predictions with a lead-time much greater than is currently available.

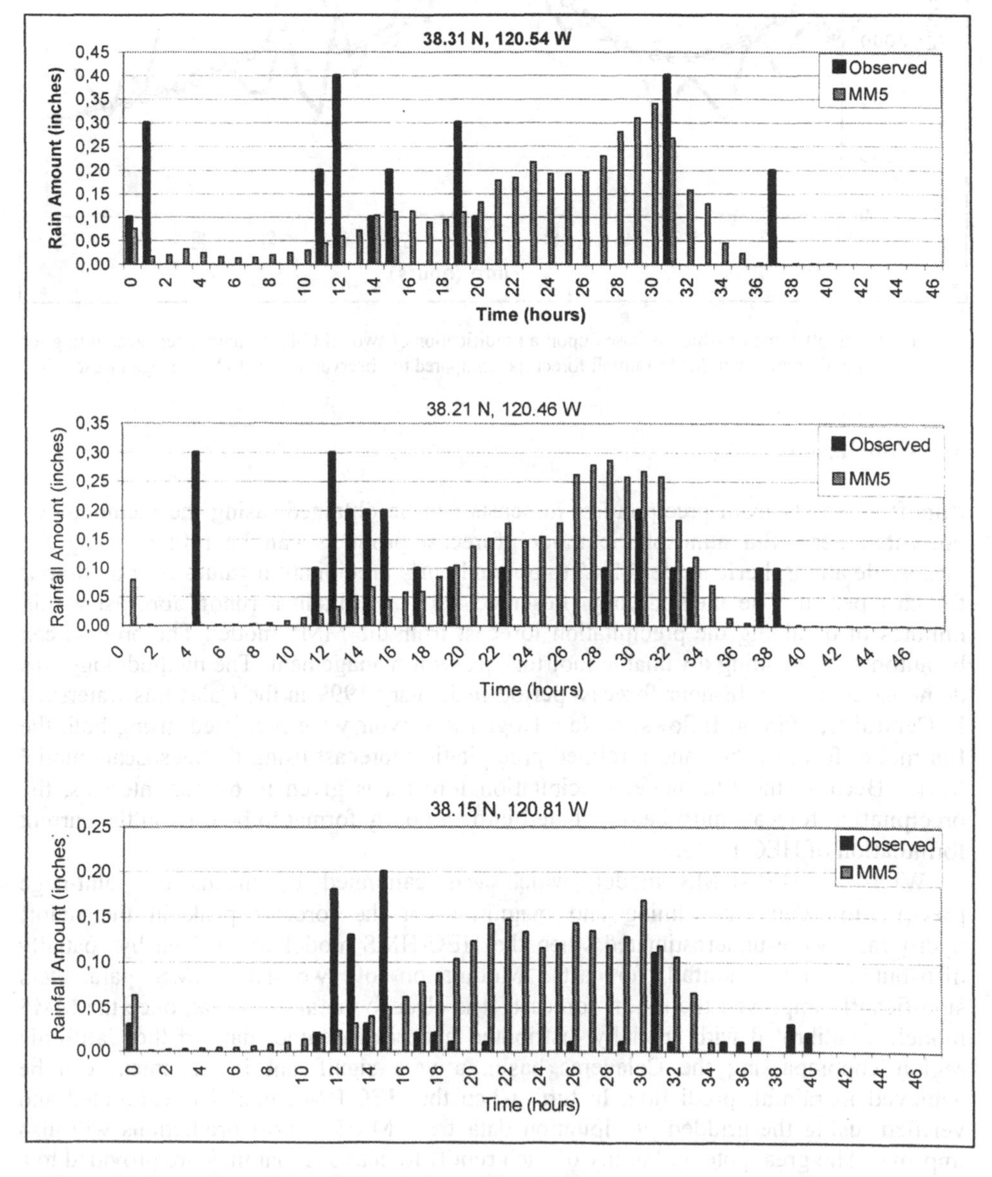

Figure 10. Comparison of observed versus forecasted precipitation for three sites in the Calaveras Basin.

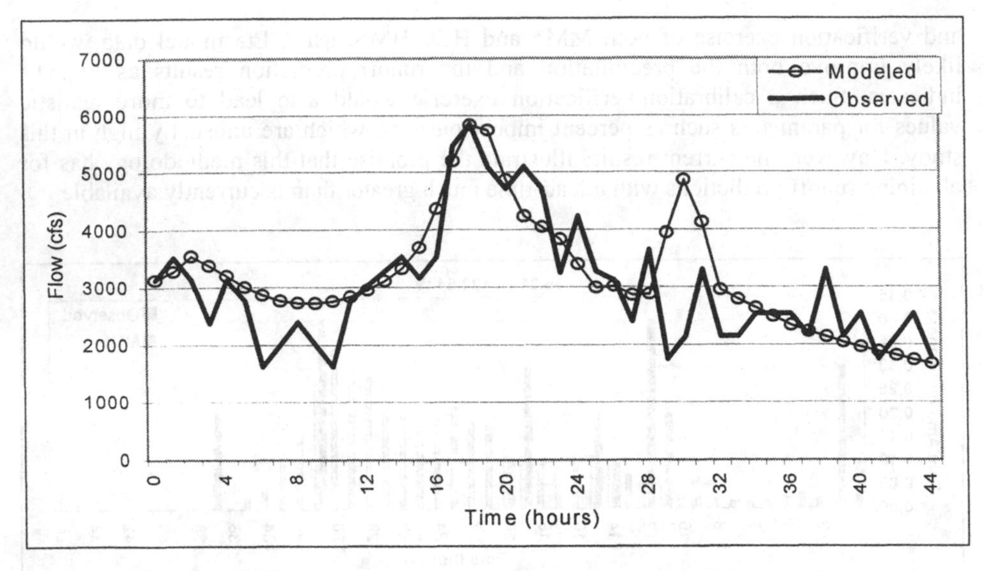

Figure 11. Runoff forecast which is based upon a modification of two HEC-HMS parameters accounting for the discrepancy in MM5 rainfall forecasts, compared to observed inflow to New Hogan Reservoir.

4. Conclusions

Runoff forecasts from precipitation forecasts can be obtained, using the methodology presented here. Eta atmospheric model forecast products can be refined using the mesoscale atmospheric model MM5 to obtain hourly precipitation values over a 48-hour forecast period. The methodology presented here can obtain a runoff forecast within minutes of obtaining the precipitation forecast from the MM5 model. The process can be automated, yielding a valuable tool for reservoir management. The methodology was demonstrated for a 48-hour forecast period in January 1999 in the Calaveras watershed in Central California. Inflows to New Hogan Reservoir were predicted, using both the Eta model forecast data and a refined precipitation forecast using the mesoscale model MM5. Because the Eta model precipitation forecast is given in 6-hour intervals, the precipitation forecast must be distributed into an hourly format to be used in the current formulation of HEC-HMS.

With the HEC-HMS model, which was calibrated by means of point-gage precipitation data, the timing and magnitude of the forecast peak in the runoff hydrograph were underestimated when the HEC-HMS model was driven by spatially distributed MM5 rainfall forecasts. Modulation of two HEC-HMS parameters significantly improved the runoff forecast. It is clear to the authors that, once the MM5 model is calibrated with spatially distributed historical weather data of the California region encompassing the Calaveras basin, a very significant improvement can be achieved in rainfall prediction. In turn, when the HEC-HMS model is calibrated and verified, using the gridded precipitation data from MM5, runoff predictions will also improve. The great potential utility of such runoff forecasts is that they are provided to a river basin before the actual precipitation falls on the basin, and 48 hours ahead of the occurrence of the actual critical runoff event. The whole procedure can be fully automated in order to provide routine runoff forecasts over a specified river basin.

5. References

1. National Weather Service (1997) Disastrous floods from the severe winter storms in California, Nevada, Washington, Oregon, and Idaho: December 1996 - January 1997", *Natural Disaster Survey Report*, U. S. Department of Commerce National Oceanographic and Atmospheric Administration National Weather Service, Silver Spring Maryland.

2. Charley, W. (1986) Weather Radar as an Aid to Real-Time Water Control, Master's Thesis, University of California, Davis.

3. Coontz, R. (1994) Digital Rivers - Flood forecasting is about to get a badly needed overhaul, *Science* 34 **(4)**, 21.

4. James, W., Robinson, C., and Bell, J. (1993) Radar-assisted real-time flood forecasting, *Journal of Water Resources Planning and Management - ASCE* **119 (1)**, 32-44.

5. Mimikou, M. (1996) Flood forecasting based on radar rainfall measurements, *Journal of Water Resources Planning and Management - ASCE* **122 (3)**, 151-156.

6. Yapo, P., Sorooshian, S., and Gupta, V. (1993) A Markov chain flow model for flood forecasting, *Water Resources Research* **29 (7)**, 2427-2436.

7. Franchini, M. and Lamberti, P. (1994) A flood routing Muskingum type simulation model based on level data alone, *Water Resources Research* **30 (7)**, 2183-2196.

8. Lamberti, P. and Pilati, S. (1996) Flood propagation models for real-time forecasting, *Journal of Hydrology* **175 (1-4)**, 239-265.

9. Staudenmaier, M.J. (1996a) A Description of the Meso Eta Model, NWS Western Region Technical Attachment No. 96-06.

10. Staudenmaier, M.J. (1996b) The Initialization Procedure in the Meso Eta Model, NWS Western Region Technical Attachment No. 96-30.

11. Burks, J.E. and Staudenmaier, M.J. (1996) A Comparison of the Eta and the Meso Eta Models During the 11-12 December 1995 Storm of the Decade, WR-Technical Attachment 96-21.

12. Janish, P.R. and Weiss, S.J. (1996) Evaluation of various mesoscale phenomena associated with severe convection during VORTEX-95 using the Meso Eta model, Preprints, 15th Conf. on Weather Analysis and Forecasting, AMS, Norfolk, VA, August 1996.

13. Schneider, R.S., Junker, N.W., Eckert, M.T., and Considine, T.M. (1996) The performance of the 29 km Meso Eta model in support of forecasting at the Hydrometeorological Prediction Center, Preprints, 15th Conf. on Weather Analysis and Forecasting, AMS, Norfolk, VA, August 1996.

14. Grell, G., Dudhia, J., and Stauffer, D. (1994) A Description of the Fifth-Generation Penn State/NCAR Mesoscale Model (MM5), NCAR Technical Note NCAR/TN-398+STR, National Center for Atmospheric Research, Boulder CO.

15. Kavvas, M. and Chen, Z. (1998) Meteorologic Model Interface for HEC-HMS: NCEP Eta Atmospheric Model and HEC Hydrologic Modeling System, Report to the United States Army Corps of Engineers Hydrologic Engineering Center.

16. USACE-HEC (1998) HEC-HMS Hydrologic Modeling System User's Manual.

17. USACE (1987) Real-Time Flood Forecasting and Reservoir Regulation in the Calaveras River Basin, Project Report No. 87-2.

18. USACE-HEC (1996) GridParm Procedures for Deriving Grid Cell Parameters for the ModClark Rainfall-Runoff Model User's Manual.

19. USACE-HEC (2000) GEO-HMS User's Manual.

20. Kull, D. and Feldman, A. (1998) Evolution of Clark's unit graph method for spatially distributed runoff, *Journal of Hydrologic Engineering – ASCE* **3 (1)**, 9-19.

21. Dudhia, J. (1993) A nonhydrostatic version of the Penn State/NCAR mesoscale model: Validation tests and simulation of an Atlantic cyclone and cold front, *Mon. Wea. Rev.* **121**, 1493-1515.

UPSCALING SURFACE FLOW EQUATIONS DEPENDING UPON DATA AVAILABILITY AT DIFFERENT SCALES

G. TAYFUR
Izmir Institute of Technology, Department of Civil Engineering
Urla, Izmir 35437 Turkey

Abstract. St.Venant equations, which are used to model sheet flows, are point-scale, depth-averaged equations, requiring data on model parameters at a very fine scale. When data are available at the scale of a hillslope transect, the point equations need to be upscaled to conserve the mass and momentum at that scale. Hillslope-scale upscaled model must be developed if data are available at that scale. The performance of the three models applied to simulate flows from non-rilled surfaces revealed that the hillslope-scale upscaled model performs as good as the point-scale model though it uses far less data. The transectionally-upscaled model slightly underestimates the observed data.

1. Introduction

Overland flow is an important part of flow dynamics in a watershed. It is mostly responsible for sheet erosion, sediment transport, and the flashy response in the stream hydrograph. There have been many experimental studies [1, 2, 3] and modeling efforts [4, 5] on overland flows. Most of the modeling studies replaced the actual varying microtopographic surfaces by smooth surfaces. This is because it is too difficult to obtain data on model parameters at a very fine scale. Only had Tayfur *et al.* [5] tried to model overland flows over varying microtopographic surfaces. However, they pointed out that, in order to obtain stable solutions, the actual varying microtopographic surfaces have to be smoothed out to satisfy the gradually varied assumption embedded in derivation of flow equations.

When one replaces the actual varying microtopographic surfaces by smooth surfaces, he/she, in a way, treats the depth-averaged equations as large-scale-averaged equations. Such a treatment, in fact, does not sound mathematically right. The proper way is to upscale the depth-averaged equations to conserve the mass and momentum at a larger scale. Tayfur and Kavvas [6] were the first to develop transectionally averaged flow equations. They properly averaged the depth-averaged point scale flow equations over a hillslope transect. In their model, the effects of rills on flow dynamics were considered. The flow on interrill areas was treated in two dimensions, having the profile of a sine function, and the flow in rills was analyzed in one dimension. When finding the expectation of the terms containing more than one variable, Tayfur and Kavvas [6] assumed that the randomness in the state variable is due to the randomness in flow dynamics. Hence, they had to develop equations for the fluctuating part of the state variables. They solved the resulting transectionally averaged quasi-two dimensional flow equations by the implicit-centered finite difference method, routing the flow from transect to transect.

Following the transectionally-upscaled model, Tayfur and Kavvas [7] averaged the depth-averaged flow equations over a hillslope surface to obtain areal-averaged equations. They again treated the flow at interrill area sections in two dimensions having the profile of a sine function while the flow in rills was studied in one-dimension. They first mathematically averaged flow over an individual interrill area and in an individual rill section. Then, they statistically averaged the resulting equations over a hillslope surface. In the case of areal-averaging, it was assumed that the randomness in the state variables is due to the randomness in the parameters of the model. The resulting areal-averaged flow equations were only time dependent, for which the solution required a very simple numerical procedure.

In this study, three models were employed to simulate flows over non-rilled surfaces. The performance of the three models was investigated quantitatively.

2. Mathematical Description

2.1. POINT-SCALE MODEL

Overland flows are modeled by St.Venant equations. These equations are point-scale depth averaged equations, which require data at a very fine scale (e.g., 0.2 m by 0.2 m section). However, even when data are available at this fine scale, numerical solutions of these highly nonlinear equations fail to converge. While the numerical solution of these equations requires a fine mesh dictated by computational accuracy, data measurements need to be made at a larger scale (e.g., 2.0 m by 2.0 m section), yielding smoother surfaces to satisfy the gradually varying assumption embedded in the derivation of the equations.

The two-dimensional St.Venant equations can be expressed as:

$$\frac{\partial h}{\partial t}+\frac{\partial hu}{\partial x}+\frac{\partial hv}{\partial y}=q_l \tag{1}$$

$$\frac{\partial u}{\partial t}+u\frac{\partial u}{\partial x}+v\frac{\partial u}{\partial y}+g\frac{\partial h}{\partial x}=g\left(S_{ox}-S_{fx}\right) \tag{2}$$

$$\frac{\partial v}{\partial t}+u\frac{\partial v}{\partial x}+v\frac{\partial v}{\partial y}+g\frac{\partial h}{\partial y}=g\left(S_{oy}-S_{fy}\right) \tag{3}$$

where,

$$S_{fx}=\frac{n^2 u\sqrt{u^2+v^2}}{h^{1.33}} \tag{4}$$

$$S_{fy}=\frac{n^2 v\sqrt{u^2+v^2}}{h^{1.33}} \tag{5}$$

where h is the flow depth; u and v, the flow velocities in x- and y-directions, respectively; g, the gravitational acceleration; q_l, the lateral flow (rainfall-infiltration); S_{ox} and S_{oy}, bed-slopes in x- and y-directions, respectively; S_{fx} and S_{fy}, friction slopes in x-and y-directions, respectively; and n, the Manning's roughness coefficient (Fig.1).

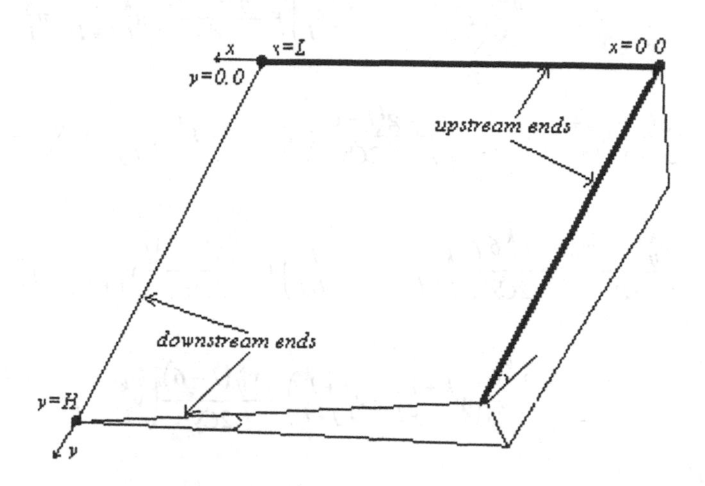

Figure 1. Schematic representation of a hillslope.

Equations (1) through (3) are solved numerically. In this study, the implicit-centered finite difference method is employed. Figure 2 shows a typical 'computational cell', where i and j are spatial node numbers in x- and y-directions, respectively, and k is the time-step number. The finite difference form of the equations may be written as:

$$
\begin{aligned}
C_{ij} = {} & \frac{h_{ij}^{k+1} - h_{ij}^{k}}{\Delta t} + \frac{u_{ij}^{k} 6}{2\acute{C}x}\left(h_{i+1,j}^{k+1} - h_{i-1,j}^{k+1}\right) + \frac{u_{ij}^{k}(1-6)}{2\acute{C}x}\left(h_{i+1,j}^{k} - h_{i-1,j}^{k}\right) + \\
& \frac{h_{ij}^{k} 6}{2\acute{C}x}\left(u_{i+1,j}^{k+1} - u_{i-1,j}^{k+1}\right) + \frac{h_{ij}^{k}(1-6)}{2\acute{C}x}\left(u_{i+1,j}^{k} - u_{i-1,j}^{k}\right) + \\
& \frac{v_{ij}^{k} 6}{2\acute{C}y}\left(h_{i,j+1}^{k+1} - h_{i,j-1}^{k+1}\right) + \frac{v_{ij}^{k}(1-6)}{2\acute{C}y}\left(h_{i,j+1}^{k} - h_{i,j-1}^{k}\right) + \\
& \frac{h_{ij}^{k} 6}{2\acute{C}y}\left(v_{i,j+1}^{k+1} - v_{i,j-1}^{k+1}\right) + \frac{h_{ij}^{k}(1-6)}{2\acute{C}y}\left(v_{i,j+1}^{k} - v_{i,j-1}^{k}\right) - q_l
\end{aligned}
\tag{6}
$$

$$M_{x_{ij}} = \frac{u_{ij}^{k+1} - u_{ij}^{k}}{\Delta t} + \frac{u_{ij}^{k}\theta}{2\Delta x}\left(u_{i+1,j}^{k+1} - u_{i-1,j}^{k+1}\right) + \frac{u_{ij}^{k}(1-\theta)}{2\Delta x}\left(u_{i+1,j}^{k} - u_{i-1,j}^{k}\right) +$$

$$\frac{v_{ij}^{k}\theta}{2\Delta y}\left(u_{i,j+1}^{k+1} - u_{i,j-1}^{k+1}\right) + \frac{v_{ij}^{k}(1-\theta)}{2\Delta y}\left(u_{i,j+1}^{k} - u_{i,j-1}^{k}\right) + \qquad (7)$$

$$\frac{g\theta}{2\Delta x}\left(h_{i+1,j}^{k+1} - h_{i-1,j}^{k+1}\right) + \frac{g(1-\theta)}{2\Delta x}\left(h_{i+1,j}^{k} - h_{i-1,j}^{k}\right) - g(S_{ox} - S_{fx})$$

$$M_{y_{ij}} = \frac{v_{ij}^{k+1} - v_{ij}^{k}}{\Delta t} + \frac{u_{ij}^{k}\theta}{2\Delta x}\left(v_{i+1,j}^{k+1} - v_{i-1,j}^{k+1}\right) + \frac{u_{ij}^{k}(1-\theta)}{2\Delta x}\left(v_{i+1,j}^{k} - v_{i-1,j}^{k}\right) +$$

$$\frac{v_{ij}^{k}\theta}{2\Delta y}\left(v_{i,j+1}^{k+1} - v_{i,j-1}^{k+1}\right) + \frac{v_{ij}^{k}(1-\theta)}{2\Delta y}\left(v_{i,j+1}^{k} - v_{i,j-1}^{k}\right) + \qquad (8)$$

$$\frac{g\theta}{2\Delta y}\left(h_{i,j+1}^{k+1} - h_{i,j-1}^{k+1}\right) + \frac{g(1-\theta)}{2\Delta y}\left(h_{i,j+1}^{k} - h_{i,j-1}^{k}\right) - g(S_{oy} - S_{fy})$$

where C_{ij} is the finite difference approximation of the continuity equation (1); M_{xij} and M_{yij}, the finite difference approximations of the momentum equations of (2) and (3), respectively; and θ, a weighting parameter which takes values from 0.5 to 1.0 to give stable solutions.

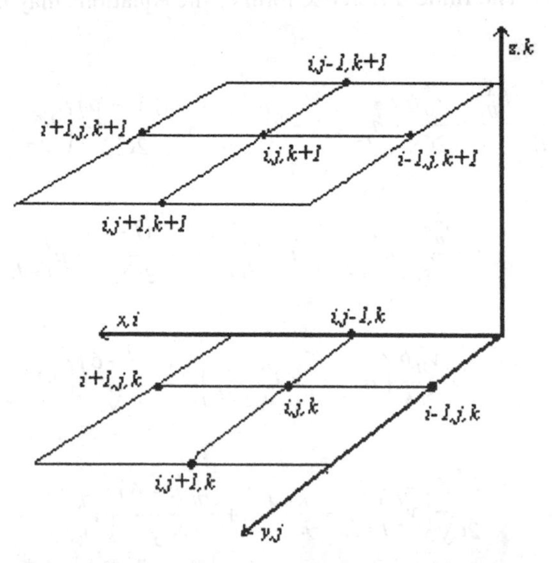

Figure 2. Computational cell for the numerical scheme.

The Newton-Raphson iterative technique is used to solve the set of nonlinear equations resulting from the implicit procedure. The iterative technique requires the evaluation of the derivative terms so as to form the coefficient matrix. The derivatives of equations (6), (7), and (8) with respect to flow depth *(h)*, and flow velocities *(u, and v)* at time step *(k+1)* are required.

Since the two-dimensional flow is an initial-boundary value problem, the initial and boundary conditions need to be specified. As the flow starts on a dry surface, the flow depth and velocities initially must be specified as zero. However, such a specification causes a singularity problem. In order to avoid such a problem, a very thin layer of flow depth of 0.0001 mm and very low flow velocities of 0.0001 m/s are initially assumed.

As upstream boundary conditions (Fig.1), zero flow depth and zero-flow velocities are assumed. Zero-depth gradient and zero-flow velocity gradient boundary conditions are assumed as downstream boundary conditions (Fig.1). These conditions can be expressed as:

$$\frac{\partial h(L,y,t)}{\partial x} = 0.0\,, \qquad t>0.0 \tag{9}$$

$$\frac{\partial u(L,y,t)}{\partial x} = 0.0\,, \qquad t>0.0 \tag{10}$$

$$\frac{\partial v(L,y,t)}{\partial x} = 0.0\,, \qquad t>0.0 \tag{11}$$

$$\frac{\partial h(x,H,t)}{\partial y} = 0.0\,, \qquad t>0.0 \tag{12}$$

$$\frac{\partial u(x,H,t)}{\partial y} = 0.0\,, \qquad t>0.0 \tag{13}$$

$$\frac{\partial v(x,H,t)}{\partial y} = 0.0\,, \qquad t>0.0 \tag{14}$$

The numerical method requires the finite difference of the derivative boundary conditions. The derivatives of the related difference equations of the boundary conditions with respect to flow depth and velocities at time step *(k+1)* are also required by the coefficient matrix of the numerical scheme.

2.2. TRANSECT-SCALE MODEL

When data are available at the scale of a hillslope transect (Fig.3), the point-scale flow equations need to be upscaled to conserve the mass and momentum at that scale (e.g., 2.0 m by 20.0 m section). Tayfur and Kavvas [6] averaged the point-scale flow equations, based on the kinematic wave approximation to obtain transectionally-averaged model. They considered surfaces having rill and interrill area sections. They treated flow on interrill areas in two dimensions, having the profile of a sine function, and considered flow in rill sections in one-dimension, having the dynamics of channel flow. First, they mathematically averaged the two-dimensional point-equation to obtain the quasi-two dimensional locally averaged interrill area flow equation. Then, the locally averaged interrill and rill flows were statistically averaged to obtain transectionally-averaged model. Tayfur and Kavvas [6], in statistical averaging, assumed that all the randomness in the state variables is due to the randomness in flow dynamics. Hence, they had to develop an equation for the variance of the state variables. The details of the averaging can be obtained from Tayfur and Kavvas [6]. In this study, non-rilled surfaces are considered. Following the work of Tayfur and Kavvas [6], the transectionally-upscaled equations for a non-rilled surface can be expressed as:

$$\frac{\partial <h>}{\partial t}+\frac{\partial}{\partial x}\left[0.44<C_x><h>^{1.67}+0.56<C_x><h>^{-0.33}<h^2>\right]=$$
$$<q_l>-2.13\left[0.44\frac{<C_y>}{l}<h>^{1.67}+0.56\frac{<C_y>}{l}<h>^{-0.33}<h^2>\right] \tag{15}$$

$$\frac{\partial <h^2>}{\partial t}+\frac{\partial}{\partial x}\left[-2.44<C_x><h>^{2.67}+4.44<C_x><h>^{0.67}<h^2>\right]-$$
$$\left[0.33<C_x>\frac{\partial}{\partial x}<h>^{2.67}+1.67<C_x><h^2>\frac{\partial}{\partial x}<h>^{0.67}\right]= \tag{16}$$
$$2<h><q_l>-2.13\left[-2.44\frac{<C_y>}{l}<h>^{2.67}+4.44\frac{<C_y>}{l}<h>^{0.67}<h^2>\right]$$

Equation (15) is for the transectionally averaged flow depth, $<h>$, and Eq. (16) is for the transectionally averaged deviatic part of flow depth, $<h^2>$. l is the sub-section area width (Fig.3). Note that $<.>$ stands for the expectation of a variable at a hillslope transect. C_x and C_y are defined as:

$$C_x=\frac{S_{ox}^{0.5}}{n\left[1+\left(\frac{S_{oy}}{S_{ox}}\right)^2\right]^{0.25}} \tag{17}$$

$$C_y = \frac{S_{oy}^{0.5}}{n\left[1+\left(\frac{S_{ox}}{S_{oy}}\right)^2\right]^{0.25}} \tag{18}$$

Equations (15) and (16) are solved simultaneously by the implicit-centered finite difference method. The resulting nonlinear difference equations are solved by the Newton-Raphson method. Initially, a very thin layer of water is assumed over the surface. Zero flow depth at the upstream boundary and zero-flow depth gradient at the downstream boundary are assumed. Note that the average values of the parameters, i.e., bed slopes, are used at each transect and the flow is routed from transect to transect.

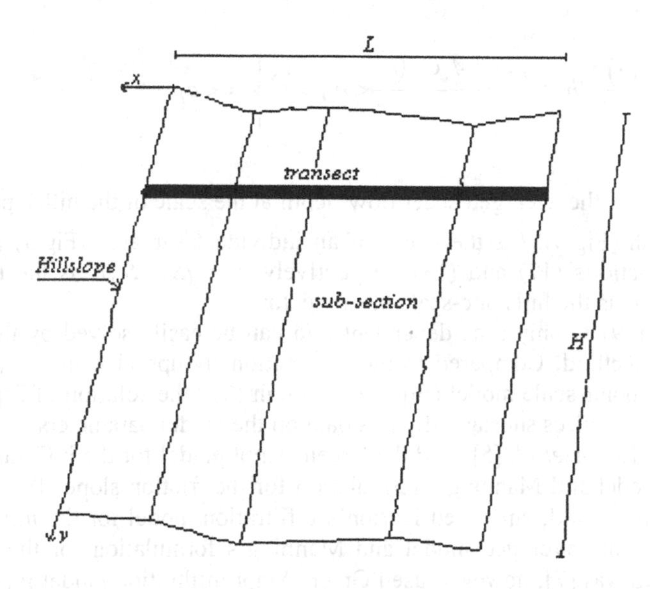

Figure 3. Schematic representation of upscaling sections.

2.3. HILLSLOPE-SCALE MODEL

When data are available at a hillslope scale (Fig.3) (e.g., 30 m by 30 m), the point-scale flow equations need to be upscaled to conserve the mass and momentum at that scale. Tayfur and Kavvas [7] developed areal-averaged flow equations by averaging the point-scale flow equations, based on the kinematic wave approximation over a hillslope section. They considered surfaces consisting of rill and interrill area sections. First, they mathematically averaged two dimensional sheet flow having the profile of a sine function over an individual interrill area section. Then, they averaged rill flow having the dynamics of channel flow over an individual rill section. Next, they statistically averaged locally averaged flow equations over a whole hillslope section. In statistical averaging, they assumed that the randomness in the flow variables is due to the randomness in the parameters of the model. This assumption resulted in time-dependent areal averaged equations. The details of the averaging can be obtained from Tayfur and Kavvas [7]. In this study, non-rilled surfaces are considered. Following the work of Tayfur and Kavvas [7], the hillslope-scale upscaled equation for a non-rilled surface can be expressed as:

$$\frac{\partial < h_o >}{\partial t} + \sum_{i=1}^{n} \sum_{j=1}^{n} Cov(r_i, r_j) \left\{ \frac{\partial^2 \left[\frac{C_x(\bar{r})}{L} < h_o >^{1.5} \right]}{\partial r_i \, \partial r_j} + 2 \frac{\partial^2 \left[\frac{C_y(\bar{r})}{l} < h_o >^{1.5} \right]}{\partial r_i \, \partial r_j} \right\}$$

$$+ \left\{ \frac{2\, C_x(\bar{r})}{L} < h_o >^{1.5} + \frac{4\, C_y(\bar{r})}{l} < h_o >^{1.5} \right\} = < q_l > \qquad (19)$$

where $< h_o >$ is the averaged sheet flow depth at the scale of the hillslope, and L is the hillslope width (Fig.3). l is the width of an individual sub-area (Fig.3); C_x and C_y are given by Equations (17) and (18), respectively. $r = (S_{ox}, S_{oy})$ is the random vector variable, and $\bar{r}$ is the hillslope-scale mean vector.

Equation (19) is only time-dependent and can be easily solved by the fourth order Runga-Kutta method. Compared to the transectionally-upscaled model (Eqs. (15) and (16)) and the point-scale model (Eqs. (1) through (3)), the solution of Eq. (19) is much simpler, and it requires substantially less data on the model parameters.

Note that Tayfur *et al.* [5] used the Green-Ampt model for the infiltration part of the point-scale model and Manning's formulation for the friction slope. Tayfur and Kavvas [6], on the other hand, employed Horton's infiltration model for the infiltration part of the transectionally averaged model and Manning's formulation for the friction slope. Tayfur and Kavvas [7], however, used Green-Ampt infiltration model for the infiltration component of the areal-averaged model but employed Chezy formulation for the friction slope. Hence, there was no healthy way of comparing the performances of these three models. In this study, for purposes of consistency, Horton's formulation for the infiltration component and Manning's formulation for the friction slope are employed in the three models. Furthermore, non-rilled surfaces are considered.

3. Application of the Models

The three models were applied to simulate observed flow discharges from three different experimental plots, which were constructed by the researchers in the Department of Agricultural Engineering of University of Kentucky [8]. The plots were subjected to different rainfall applications. The data related to the experimental rainfall simulations and plots are summarized in Table 1. From the microtopographic data of the plots, local slopes at 0.6 m by 0.6 m grids were obtained. From the computed local slopes, the average slope values were determined at each transect (0.6 m by 4.5 m) of the total 36 transects of each experimental plot. The ranges for the local slopes and the average slopes at each transect of the experimental plots are summarized in Table 2. The statistical information on model parameters is summarized in Table 3. Note that the point-scale model utilizes local slopes, the transect-scale model utilizes average slopes at transects, and finally the hillslope-scale model utilizes the average main slopes.

TABLE 1. Data on rainfall experiments and experimental plots

Plots	Dimensions L/H (m)	Average Slopes S_{ox}/S_{oy}	Rainfall Duration (min)	Ponding Time (min)	Rainfall Intensity (mm/h)
S1R2	4.5/22	0.021/0.093	103	14	62
S2R2	4.5/22	0.052/0.092	90	18	97
S3R2	4.5/22	0.036/0.086	90	26	78

TABLE 2. Data on the range of slopes at local and transect scales

Plots	Range for Local Slopes S_{ox}/S_{oy}	Range for Transect Slopes S_{ox}/S_{oy}
S1R2	0.0004 — 0.037 / 0.05 — 0.11	0.012 — 0.103 / 0.022 —0.092
S2R2	0.0003 — 0.041 / 0.04 — 0.13	0.010 — 0.089 / 0.031 —0.132
S3R2	0.0001 — 0.057 / 0.008 — 0.14	0.025 — 0.078 / 0.043 —0.118

TABLE 3. Statistical information on model parameters

Plots	Var (S_{ox})	Var (S_{oy})	Cov (S_{ox}, S_{oy})
S1R2	1.98E-03	0.00162	3.04E-04
S2R2	3.24E-04	0.00112	1.02E-03
S3R2	1.38E-05	0.00186	2.78E-04

Figures 4, 5, and 6 show the simulations of the observed flow discharge data by the three models. The calibrated Horton's infiltration parameters are: f_c = *24.4 mm/h (constant infiltration capacity)*, and K = *0.25 (proportionality factor)*. The calibrated *roughness coefficient* is: n = *0.012* [6]. As can be seen from Figs. 4, 5, and 6, all three models simulate the observed runoff data satisfactorily. The transect-scale model slightly underestimates the observed data. However, the simulation of runoff data of the experiments S2R2 and S3R2 by the point-scale and the hillslope-scale models are quite satisfactory (Figs. 5 and 6). The rising and recession limbs of the hydrographs are well replicated the by the hillslope-scale model. The three models slightly underestimate the runoff data of the experiment S1R2 (Fig.1). However, as seen in the figure, the observed data sharply increase after 60 minutes. This is due to the fact that a constant rainfall rate application may not be maintained throughout the experiment.

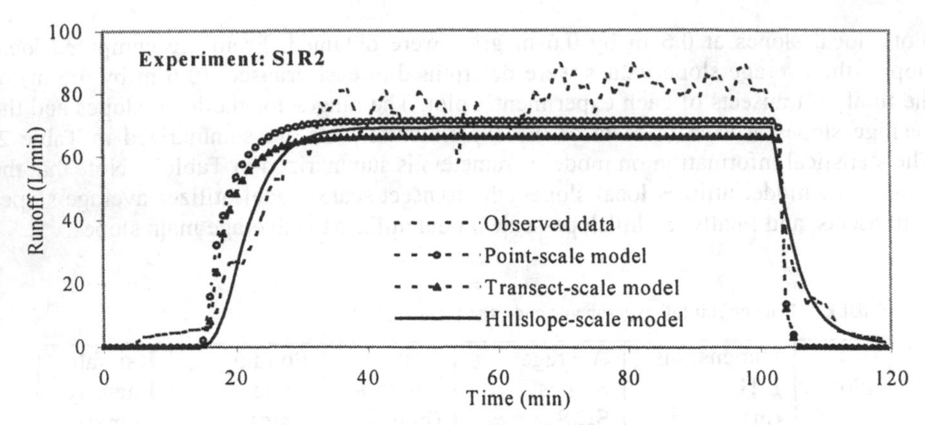

Figure 4. Simulation of the observed data by the three models.

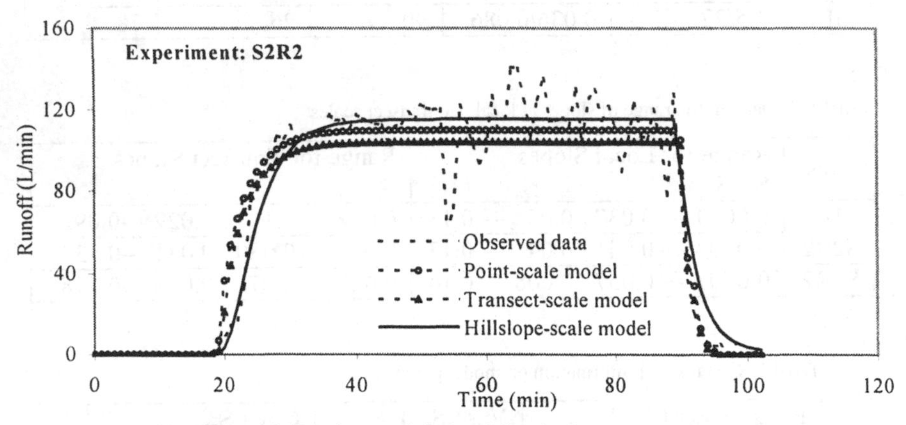

Figure 5. Simulation of observed data by the three models.

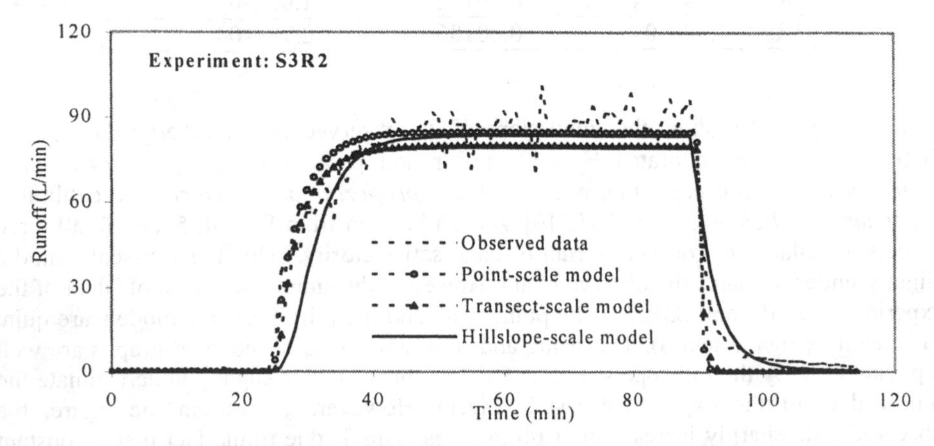

Figure 6. Simulation of observed data by the three models.

4. Concluding Remarks

In this study, three different models, i.e., point-scale, transect-scale, and hillslope-scale, simulating flows over non-rilled surfaces are presented. Each model utilizes data at different scales related to the same land-surface. The point-scale model utilizes microtopographic data at a local grid scale, while the transect-scale model uses data at a transect scale. On the other hand, hillslope-scale model utilizes data at a land-surface scale.

The point-scale model requires local x- and y-direction slope values at every nodal point of a computational network mesh over a land surface. This results in a very substantial parameter-estimation problem. Meanwhile, the transect-scale model requires only the transectionally averaged values of one x-direction slope and one y-direction slope. Thus, compared to the point-scale model, the transect-scale model significantly reduces the parameter estimation. On the other hand, the hillslope-scale model requires average values of one x-direction slope and one y-direction slope over a whole land surface. Hence, compared to the transect-scale model, the hillslope-scale model further reduces the parameter estimation problem.

In real world problems, one has neither the detailed microtopographic data nor the data at the scale of a hillslope transect. At best, by a digital elevation map with a standard resolution of 30 m, one can obtain data on the mean and covariances of the model parameters at 30 m by 30 m grid-scale required by the hillslope-scale model. A typical hillslope has dimensions of 300 m by 600 m. If 30 m by 30 m resolution data were provided, there would be 200 resolution-scale surface flow parameter sample values, which could be utilized to estimate the mean and the covariance of the model parameters for the hillslope-scale model.

The simulation results show that the hillslope-scale model performs as good as, in some cases better than, the other two models though it utilizes far less data on the model parameters. Since the performance of the hillslope-scale model is better than that of the transect-scale model, the assumption that the randomness in the state variable is due to the randomness in the parameters of the process is more plausible than associating it with the randomness in the flow dynamics.

Because of its good performance and less data requirements, it can be concluded that the hillslope-scale model is superior to the other two models. Hence the hillslope-scale model has important ramifications in terms of practical applications for modeling flows at larger scales.

5. Acknowledgements

The author is grateful to Prof. M. Levent Kavvas of the Dept. of Civil and Environmental Engineering, University of California, Davis, USA, for co-forming the foundation for this study through the already published journal papers as given in the reference list. The author is also grateful to B.J. Barfield and D.E. Storm of the Department of Agricultural Engineering, University of Kentucky, for the experimental data presented herein.

6. References

1. Kilinc, M. and Richardson, E.V. (1973) Mechanics of soil erosion from overland flow generated by simulated rainfall, Hydrology Papers, Colorado State University, Fort Collins, Colorado, Paper 63.

2. Abrahams, A.D., Parsons, A.J., and Luk, S.H. (1989) Distribution of depth of overland flow on desert hillslopes and its implication for modeling soil erosion, *J. of Hydrology* **106(1)**, 177-184.

3. Abrahams, A.D. and Parsons, A.J. (1990) Determining the mean depth of overland flow in field studies of flow hydraulics, *Water Resources Research* **26**, 501-503.

4. Zhang, W. and Cundy, T.W. (1989) Modeling of two-dimensional overland flows,. *Water Resources Research* **25(9)**, 2019-2035.

5. Tayfur, G., Kavvas, M.L., Govindaraju, R.S., and Storm, D.E. (1993) Applicability of St.Venant equations for two dimensional overland flows over rough infiltrating surfaces,. *J. of Hydraulic Engineering, ASCE* **119(1)**, 51-63.

6. Tayfur, G. and Kavvas, M.L. (1994) Spatially averaged conservation equations for interacting rill-interrill area overland flows, *J. of Hydraulic Engineering, ASCE* **120(12)**, 1426-1448.

7. Tayfur, G. and Kavvas, M.L. (1998) Areal-averaged overland flow equations at hillslope scale, *Hydrologic Sciences Journal, IAHS* **43(3)**, 361-378.

8. Barfield, B.J., Barnhisel, R.I., Powell, J.L., Hirschi, M.C., and Moore, I.D. (1983) Erodibilities and eroded size distribution of Western Kentucky mine spoil and reconstructed topsoil, Ins. for Min. and Minerals Res., Final Report, University of Kentucky, Lexington, Kentucky.

INTEGRATION OF INTELLIGENT TECHNIQUES FOR ENVIRONMENTAL DATA PROCESSING

E. CHAROU[1], N. VASSILAS[2], S. PERANTONIS[1]
and S. VAROUFAKIS[1]
[1]*National Research Center "Demokritos"*
Institute of Informatics and Telecommunications
153 10 Agia Paraskevi, Greece
[2]*Technical Educational Institution of Athens*
Department of Computer Science
Ag. Spyridonos St., 122 10 Egaleo, Greece

Abstract. In this paper, some aspects of the usefulness of intelligent techniques in environmental data processing are discussed. The capabilities of neural networks in improving memory requirements for storage of environmental data and the increase in processing speed are analyzed. Finally, a software package for processing multi-source (geophysical, geochemical, satellite, etc.) data using various neural, fuzzy, multimodular, pattern-recognition and image processing algorithms is presented.

1. Introduction

The processing and analysis of environmental data are of great importance for effective modeling and understanding of complex environmental processes and, eventually, for the solution of urgent environmental problems. Usually, environmental data are more complex than in other fields as they may come from a variety of different, rather non-homogeneous information sources (i.e., measurement data from monitoring networks, satellite data, geophysical grid data, structural data on chemical substances, etc.). Furthermore, they possess spatial and temporal dimensions and are produced in large amounts. On the other hand, for environmental information to be useful, it must be *timely;* that is, it must be available when needed [1]. In a number of recent studies, artificial intelligence techniques have proven to be effective in various environmental data processing procedures [2]. However, the integration of these techniques with the conventional ones could lead to more powerful and useful technologies and tools.

This paper deals with an integrated software which incorporates various classification algorithms and other pattern recognition techniques in a more powerful and useful way. The key features of this tool are the multimodular classifier, which integrates the decision made by multiple classifiers, and a methodology for time and memory efficient clustering and classification of images or grid spatial data.

2. The Multimodular Classifier

Data classification is an essential step in many environmental applications, such as land-use or land-cover identification and change detection, geological evaluation of land surface, soil contamination, sea ice monitoring, and oil spill detection, to name a few. Several techniques have been proposed for classification. Such techniques include traditional statistics, neural networks, and fuzzy logic and can be distinguished in the following two general categories: a) supervised techniques in which labeled training samples are used for parameter optimization, and b) unsupervised techniques (automatic classification), using a data clustering algorithm. Neural Networks (NNs) [3] are computational systems whose architecture and operation are based on our present-day knowledge about biological nervous systems. They are very well suited for classification, mainly due to their processing speed, robustness, generalization capabilities and easiness to deal with high-dimensional spaces. Fuzzy classification techniques are capable to solve problems that frequently arise due to vague, uncertain or incomplete data or knowledge, as they were developed to deal with such uncertainties. However, classification results obtained by a single classifier may be absolutely dependent on the particular design and properties of the classifier. Such dependence may have serious effects on the final performance, especially when there is a significant overlap of the categories and when the optimum (in the Bayesian sense) boundaries are nonlinear. For further improvement of final classification results, an integration of the power of classifiers and a combination of several classification models with independent decisions are needed. This integration could be realized through a multimodular decision making architecture. The multimodular architecture considered here (Fig. 1) incorporates neural, fuzzy, and statistical classifiers and utilizes the simple voting schemes suggested by [4]. Accordingly, each classifier (module) is allocated a vote, and the final decision is made by following a relative or absolute majority rule. Absolute majority rules are applied for a *primal* classification. Independence of individual decisions is guaranteed by using different classification models. Depending on the majority rule, input patterns may be rejected from classification, such as the *don't know* cases resulting from classifier disagreement. To resolve the problem of *don't know* pixels in the primal classification, one may exploit the spatial property of the image data. To this end, a *don't know* pixel may be given the label of the majority of its local neighbors found in a window centered around the *don't know* pixel. Such a technique can be viewed as spatial *noise filtering*. This cleans the final image and homogenizes its classification regions. With respect to individual classifiers, the results with different combinations of modules show an overall superiority of the multimodular system in terms of classification accuracy.

3. Memory Savings Using Kohonen's Neural Network

Kohonen's self-organizing maps (SOM) [5, 6, 7] are among the most popular neural networks. The Kohonen network consists of nodes or *neurons* arranged on 1-D or usually 2-D lattices. It is based on neurobiological establishments that the brain uses for spatial mapping to model complex data structures internally.

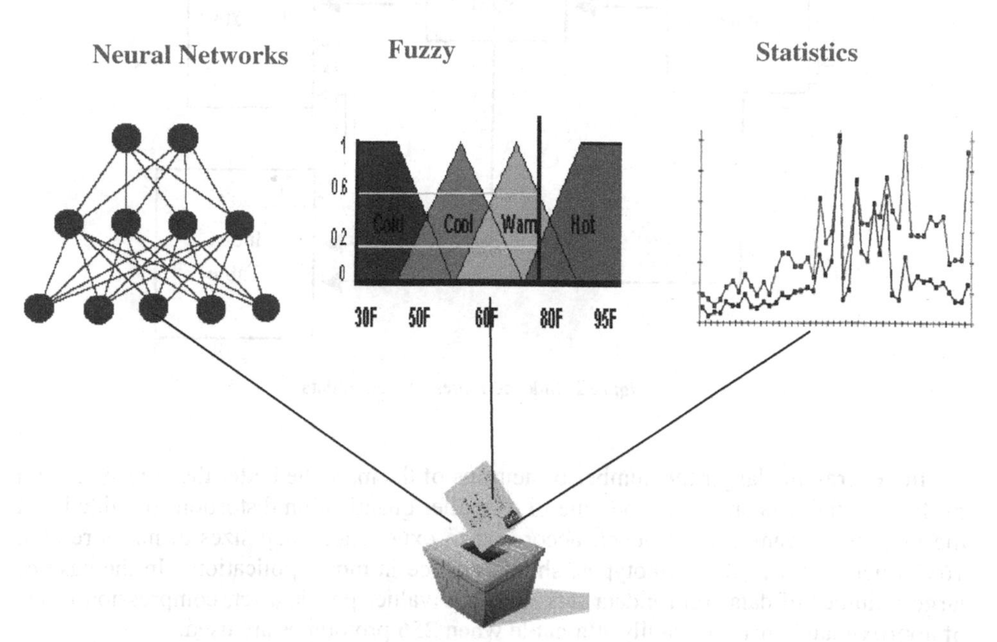

Figure 1. The multimodular classifier.

The SOM competitive algorithm, which is used for the unsupervised training of this network, is summarized as follows: A sequence of input patterns (vectors) is randomly presented to the network (neuronal map) to compared to weights (vectors) "stored" at each node. Where inputs match closest to the node weights, that area of the map is selectively optimized, and its weights are updated so as to reproduce the input probability distribution as closely as possible. The weights self-organize in the sense that neighboring neurons respond to neighboring inputs (topology which preserves mapping of the input space to the neurons of the map) and the tendency towards asymptotic values that quantize the input space in an optimal way. Using the Euclidean distance metric, the SOM algorithm performs a Voronoi tessellation of the input space [5, 6], and the asymptotic weight vectors can then be considered as a catalogue of *prototypes*, with each prototype representing all data from its corresponding Voronoi cell.

This catalogue of prototypes along with indexing techniques can be used for a compressed representation of original data [8, 9]. For the presentation, MxN images obtained from n grided data sets are assumed. The input space $\mathbf{R}^n$ is used to represent the image as a set of MxN points, whose coordinates are the corresponding values of each data set. An MxN *index table* is constructed to store pointers from pixels of the original image to their closest prototypes. The replacement of the original data with the catalogue of prototypes and the index table (Fig. 2) constitutes the indexed representation of the image and results in compression of the original image.

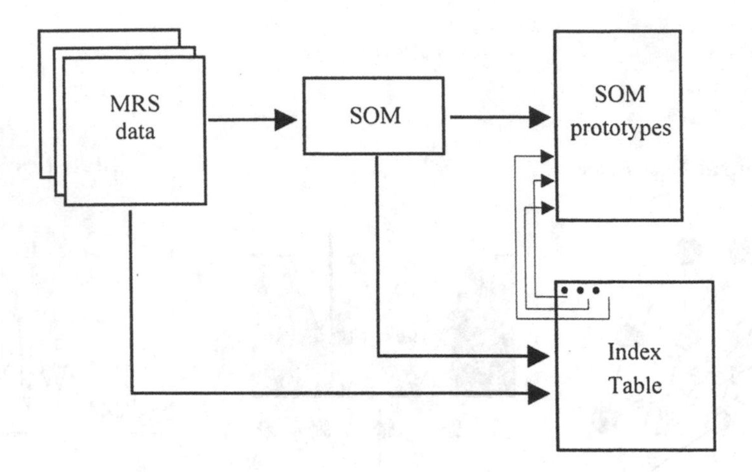

Figure 2. Indexed representation of data.

In general, the larger the number of neurons of the map, the better the approximation of the original data space will be, due to a smaller quantization distortion (provided that the map self-organizes). However, according to experience, map sizes of no more than 16x16 neurons, i.e., 256 prototypes, should suffice in most applications. In the case of large volumes of data from *n* data sets with 256 values per data set, compression ratios of approximately n:1 are readily attainable when 256 prototypes are used.

4. Acceleration of Classification Methods

The indexed representation of original data results not only in data compression but also in a significant speedup of training, data clustering, and final classification.

a) Fast Training. In supervised classification applications, the training phase involves the use of appropriately selected training and, possibly, validation samples of known classification. These sets are usually composed of several thousands of pixels and, along with the complexity of the classification task (i.e., the number of categories as well as the optimal shapes of class boundaries), they are responsible for the long training times observed. On the other hand, using the SOM prototypes, one can quantize the training set and reduce its size by deleting duplicate (prototype) samples [8]. To preserve the between-- and within--class relative frequencies needed to specify optimal boundary placement in overlapping regions, the new (compressed) training set, as well as the supervised training algorithms, are modified to include the multiplicities of the deleted samples. The result is the reduction of redundancy in the training data and, thus, a significant training speedup, approximately proportional to the ratio between the original training set size and the number of prototypes, provided that most of the prototypes exist in the compressed training set.

b) Fast Clustering. Typically, automatic classification involves clustering of the data space, followed by label assignment. However, clustering performed on the original data is inefficient due to the large number of data points. On the other hand, in the proposed methodology, clustering is performed on the SOM prototypes, thus achieving a speedup, which allows one to use even the computationally complex hierarchical algorithms [10].

c) Efficient Indexed Classification. At the final stage of the proposed methodology, only the SOM prototypes need to be classified instead of the traditional pixel-by-pixel classification that requires a computational time proportional to the original image dimensions. The result is a catalogue of labels (e.g., gray levels or colors) in complete correspondence with the SOM prototypes (Fig. 3). For the supervised techniques, ANN is used to label the SOM prototypes; whereas, for the unsupervised techniques, labels are given directly after the clustering procedure. This, in turn, allows for fast indexed classification, thus avoiding expensive computations, since the result is now obtained by following the pointers of the index table and accessing the corresponding labels as shown in Fig. 4.

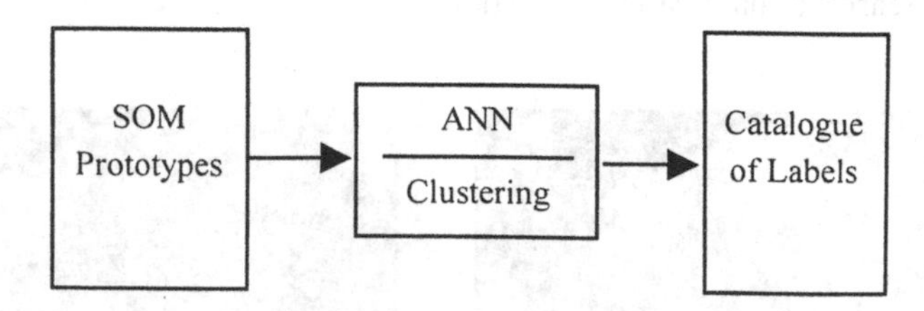

Figure 3. Creation of the catalogue of labels.

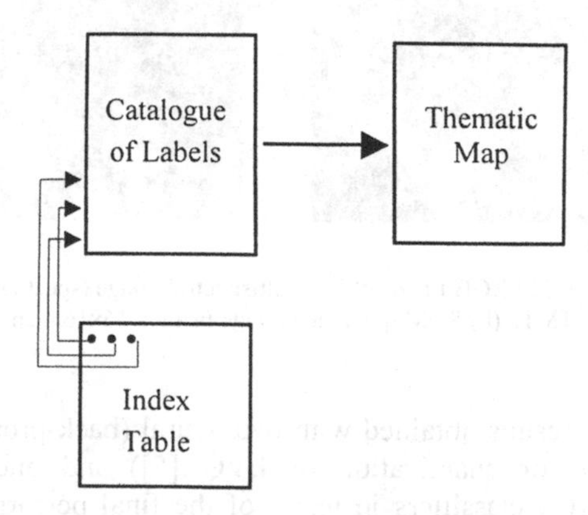

Figure 4. Fast indexed classification.

The above methodology is applied [8, 9] for supervised and unsupervised classification of multispectral data consisting of the three bands TM3, TM2 and TM1 (256 gray levels each) of a Landsat TM 512x512 image over the Lesvos island in

Greece. The original RGB image is shown in Fig. 5a. The goal here was to classify the original image into the following 4 land-cover categories: *a) forest; b) sea; c) agricultural;* and *d) inhabited areas-bare rock-quarries-land with less than 10% vegetation.* Two labeled sets of 6011 and 3324 samples from the above four categories were selected respectively for ANN training and testing their classification performance. To assess the generalization capabilities of the supervised algorithms during training, the first of these sets was further randomly split to generate a training set of 4209 samples and a validation set consisting of 1802 samples.

All programs were run on a SUN ULTRA II Enterprise workstation. A 16x16 map was used for quantization of Fig. 5a. SOM training (100000 iterations) and index table construction required 23.12 sec and 54.30 sec, respectively, leading to a compression ratio of 2.96. The resulting quantized image is shown in Fig. 5b. At this point, it should be noted that both SOM training and index table construction usually involve off-line computations with larger compression ratios for more spectral bands or smaller maps. However, reduction of map sizes should be applied with caution in order not to avoid a significant increase of quantization distortion.

Figure 5. (a) Original 512x512 RGB Landsat TM multispectral image (spectral bands used: TM3, TM2, TM1), (b) SOM quantized image using a 16x16 map.

Table 1 shows the results obtained with two neural (back-propagation or simply BP [11] and learning vector quantization or LVQ [5]) and one statistical (k-nearest neighbors or k-NN [10]) classifiers in terms of the final performance on the training, validation and test sets. Table 2 shows the respective increases in training speeds, achieved by the proposed methodology.

Experiments performed with Isodata [10, 12] Fuzzy Isodata [13], and the hierarchical clustering algorithms show significant increases in clustering speed (Table 3).

TABLE 1. Performance of the three supervised classifiers for the classical and proposed methodologies (F-*forest*, S-*sea*, A-*agricultural*, R-*rock*).

		Classical Method			Proposed Method		
Model	Category	Training	Valid.	Test	Training	Valid.	Test
BP	F	98.25%	98.64%	94.72%	96.51%	97.62%	91.80%
	S	100.00%	100.00%	100.00%	99.91%	100.00%	98.27%
	A	96.44%	94.79%	93.22%	96.58%	95.11%	94.92%
	R	89.60%	91.22%	95.19%	93.64%	92.57%	97.36%
	Total	***97.36%***	***97.06%***	***95.40%***	***97.15%***	***96.95%***	***95.01%***
LVQ	F	97.16%	98.47%	93.90%	97.53%	98.64%	94.26%
	S	100.00%	100.00%	100.00%	99.91%	99.78%	98.12%
	A	97.07%	95.77%	94.70%	96.37%	95.60%	94.70%
	R	90.46%	91.22%	96.12%	89.31%	89.86%	94.26%
	Total	***97.29%***	***97.34%***	***95.73%***	***97.05%***	***97.17%***	***95.13%***
k-NN	F	97.75%	97.96%	92.35%	97.82%	98.13%	92.53%
	S	100.00%	100.00%	100.00%	99.91%	99.78%	98.12%
	A	97.84%	94.79%	92.58%	97.42%	95.60%	93.64%
	R	91.62%	92.57%	96.12%	87.57%	87.16%	93.49%
	Total	***97.84%***	***96.95%***	***94.61%***	***97.36%***	***96.78%***	***94.10%***

TABLE 2. Increases in training speed for the three classifiers.

	Classical Method		Proposed Method		
Classifier	Number of Presentations	Time	Number of Presentations	Time	Speedup
BP	6313500	2847.47sec	62480	5.31sec	536.25
LVQ	1000	0.58sec	2200	0.13sec	4.51
k-NN	-	14.57sec	-	0.26sec	56.04

TABLE 3. Computational times and clustering gain (per iteration) for a map of 16x16 neurons. The symbol ∞ means extremely large clustering time.

Clustering Algorithm	Classical Method	Proposed Method	Speedup
Isodata	2.74sec	2.67msec	1026
Fuzzy Isodata	10.84sec	9.67msec	1121
Hierarchical	∞	11.90sec	∞

5. The ANNGIE Software

The Artificial Neural Networks Graphical Integrated Environment (ANNGIE) is a prototype software tool, which was originally developed in the frame of *GeoNickel*

BRITE / EURAM project (BE 1117) funded by the European Community [14]. *ANNGIE* has the capacity to analyze a range of different types of remotely sensed and/or geophysical data. ANNGIE integrates various neural, fuzzy, statistical and multimodular classifiers, pattern-recognition and image-processing algorithms within a user friendly GUI environment developed in Tcl/Tk. Currently, it is also running in MatLab.

The aim of ANNGIE is to have modern information processing technologies assist in tasks related to classification of multisource images or grid data. It has the capability of selecting training sets through polygonal region definition and labeling by the user. ANNGIE also provides techniques for image processing and pattern recognition and supports TIFF, PGM, GRD, ASCII and MapInfo-GEOTIFF formats. The package contains the following groups of algorithms:

UNSUPERVISED CLASSIFICATION METHODS
Unsupervised classification methods are very useful especially when a training set does not exist. The algorithms included in this group are Kohonen's Self-Organizing Maps neural network algorithm, the Batch-Map neuro-statistical clustering algorithm, the hierarchical min-max statistical clustering algorithm, the ISODATA statistical clustering algorithm, and the Fuzzy ISODATA clustering algorithm.

SUPERVISED CLASSIFICATION METHODS
The algorithms included in this group are the multi-layer feed-forward neural network classifiers: Back Propagation, ALECO-1 [15] and ALECO-2 [16], with the preprocessing option of forming high-order correlations of the input data; the reputedly fast single-layer neural network classifiers LVQ and *k*-LVQ; the fuzzy Pal-Majumder [17] classifier; and the multimodular classifier described in paragraph 2 above.

A NEURAL NETWORK BASED METHODOLOGY FOR EFFICIENT CLASSIFICATION (Described in paragraphs 3 and 4 above).

PATTERN RECOGNITION AND IMAGE PROCESSING METHODS
A group of pattern recognition and image processing algorithms is provided to perform automatic lineament extraction. This group includes an algorithm to introduce a threshold into the gray level classification where the result is produced by any clustering or classification (gray-level, percentage, foreground or background threshold options are provided); an algorithm for determining the connected regions in binary image, using four- or eight- neighbor connectivity; extraction of various region shape descriptors (area, elongation, orientation); and a novel weighted Hough transformation algorithm to locate the most prominent lines.

6. Conclusions and Future Work

Various intelligent and conventional classification and other processing methods were integrated in a powerful and useful software package. There is still room for future refinement of this prototype software. Significant improvements can be made by adding the true georeferencing of data, extending the allowed formats to those normally used in environmental applications, and by adding also point data to the suite of data types allowed in the classification. The task of integration with other systems so that results can be directly and automatically communicated to them is also a welcome challenge. Application of these novel methods to real environmental problems also calls for further familiarization of the environmentalists with this newly developed tool. Co-operation between environmentalists and information processing scientists towards this goal is a very appealing perspective, and it is believed that this could lead to mutual benefits and could help in developing a deeper insight into environmental phenomena and successful environmental monitoring.

7. References

1. Harmancioglu, N.B. *et al.* (1997) Conclusions and recommendations, in Harmancioglu *et al.* (eds.), *Integrated Approach to Environmental Data Management Systems*, Kluwer Academic Publishers, The Netherlands, pp. 423-434.

2. Avouris, N. and Vassilas, N. (1999) Introduction in ACAI'99, Workshop 07, *Intelligent Techniques for Spatiotemporal Data Analysis in Environmental Applications*, Chania, Greece, pp. 6-9.

3. Haykin, S. (1994) *Neural Networks: A Comprehensive Foundation*, MacMillan, Englewood Cliffs, New Jersey.

4. Battiti, R. and Colla, A.M. (1994) Democracy in neural nets: Voting schemes for classification, *Neural Networks* 7, 691-707.

5. Kohonen, T. (1989) *Self-Organization and Associative Memory*, 3rd ed., Springer, Berlin-Heidelberg-New York.

6. Kohonen, T. (1995) *Self-Organizing Maps*, Springer-Verlag, Berlin-Heidelberg-New York-Tokyo.

7. Ienne, P., Thiran, P., and Vassilas, N. (1997) Modified Self-Organizing Feature Map Algorithms for Efficient Digital Hardware Implementation, *IEEE Trans. Neural Networks* **8(2)**, 315-330.

8. Vassilas, N. and Charou, E. (1999) A new methodology for efficient classification of multispectral satellite images using neural network techniques, *Neural Processing Letters* **9(1)**, 35-43.

9. Vassilas, N., Perantonis, S., Charou, E., and Varoufakis, S. (1998) Intelligent techniques for efficient generation of ground-cover maps, in *Proc. 20th Canadaian Symposium on Remote Sensing*, Calgary, Canada, pp. 255-258.

10. Duda, R.D. and Hart, P.E. (1973) *Pattern Classification and Scene Analysis*, Wiley, New York.

11. Rumelhart, D.E., Hinton, G.E., and Williams, R.J. (1986) Learning Representations by Back Propagating Errors, *Nature* **323,** 533-536.

12. Ball, G.H. and Hall, D.J. (1967) A Clustering Technique for Summarizing Multivariate Data, *Behavioral Science* **12**, 153-155.

13. Bezdeck, J.C. (1976) A Physical Interpretation of Fuzzy ISODATA, *IEEE Trans. Systems, Man and Cybernetics* **6**, 387-389.

14. Aarnisalo, J., Makela, K., Bourgeois, B., Spyropoulos, C., Varoufakis, S., Morten, E., Maglaras, K., Soininen, H., Angelopoulos, A., Eliopoulos, D., Lamberg, P., Papunen, H., Hattula, A., Pietila, R., Elo, S., Ripis, C., Dimou, E., Apostolikas, A., Melakis, M., Economou-Eliopoulos, M., Vassilas, N., Perantonis, S., Ampazis, N., Charou, E., Gustavsson, N., Tiainen, M., Stefouli, M., Delfini, D, Venturini, P., Vouros, G., Stavrakas, Y., and Katsoulas, D. (1999) Integrated technologies for minerals exploration; pilot project for nickel ore deposits, *Transactions of the Institution of Mining and Metallurgy, Section B Applied Earth Science* **108**, 151-163.

15. Karras, D.A. and Perantonis, S.J. (1995) An efficient training algorithm for feedforward networks, *IEEE Tr. Neural Networks* **6**, 1420-1434.

16. Perantonis, S.J. and Karras, D.A. (1995) An efficient learning algorithm with momentum acceleration, *Neural Networks* **8**, 237-249.

17. Pal, S.K. and Majumder, D.D. (1977) Fuzzy sets and decision making-approaches in vowel and speaker recognition, *IEEE Transactions on Systems, Man and Cybernetics* **7**, 625-629.

INTEGRATED USE OF MONITORING AND MODELING IN WATER RESOURCES RESEARCH

K. HAVNØ, H. MADSEN and V. BABOVIC
DHI Water & Environment
Agern Allé 11
DK-2970 Hørsholm, Denmark

Abstract. Despite the many advances in sensors and recording techniques, monitoring programs can still be relatively expensive. In practice, this often limits the density of monitoring programs. Yet, large amounts of data are monitored and filed without proper analysis of their information contents. The combined use of monitoring and simulation models can reduce the costs and facilitate rigorous analyses of monitored data. Physically based simulation models provide the best means of interpolating between measurement points (in space and time). The models can also aid the effective design of monitoring programs. Field data can be used to improve the quality of simulation models. For real time monitoring, information can be fed back into the simulation models through automatic update routines. These combined techniques, long used for hydraulic data, are now also developed for water quality data. Whenever possible, the integration between monitoring and modeling should be designed from the outset to obtain full benefit. New techniques are developed for linking the two methodologies including data mining, data validation, and data assimilation techniques. The paper describes some of the recent developments in this field, giving examples of practical applications.

1. Data, A Rapidly Increasing Resource

In 1991, the American Computer Scientist, David Gelertner, wrote a visionary book called "Mirror World", which advanced the idea of a virtual copy of the real world: a Mirror World could be made by a combination of mathematical modeling and a wide-range data collection, streaming data into the Mirror at high bandwidths. At the time, the technical limitations for this vision were obvious; however, within the last 10 years, the appropriate technologies have advanced and are still advancing rapidly. World Mirrors are now being constructed and applied within many disciplines; in the water sector, they are used for e.g., water supply management, drainage, flood forecasting, tidal and storm surge forecasting, navigation, environmental monitoring, and many other fields.

The technologies involved combine numerical modeling software and on-line monitoring relying heavily on automatic and continuous measurements. All the information is stored in a Water Information System (WIS), containing observations and simulations in hindcast, nowcast and forecast mode, being viewed through GIS & Web interfaces.

The tools are used for numerous purposes: they make it possible to analyze situations from past and their cause-effect relationships and to make scenario analysis and forecasts by parallel simulation. They are also used as interpolation tools (in space and time) between monitoring points and are indispensable tools for planning and optimizing monitoring programs. An important element, which links models and measurements, is the data assimilation. A brief description of this method is given in the following.

2. Data Assimilation in Hydro Modeling

In general, data assimilation or model updating is a method to combine a model of a system with measurements in order to obtain a better-combined knowledge of the system. This is motivated by the fact that a model will never be able to exactly describe the state of the system due to a number of different error sources, e.g., the use of non-optimal model parameters, uncertain model forcing, errors in boundary and initial conditions, and neglected or badly described physical processes. On the other hand, measurements are expensive and will therefore always be sparse in both time and space. Thus, they will not be able to fully resolve the dynamics of the system at all spatial and temporal scales of interest.

Application of data assimilation in a model enables to spread the information obtained from point measurements to the entire domain in a consistent manner and according to the model dynamics (interpolation in space). Furthermore, by updating the state of the system whenever measurements are available, it is ensured that the model will not drift away from the measurements (interpolation in time).

2.1. DATA ASSIMILATION TECHNIQUES

Data assimilation (DA) is a feedback process where the model prediction is conditioned to the observations of system variables. Data assimilation techniques can be classified according to the variables that are modified in the feedback process, i.e., input variables, model states, model parameters, and output variables [1]. In hydrodynamic and hydrologic modeling, updating of state variables or output variables (also known as error correction) is usually adopted. Updating of input variables (model forcing) can be included as an integrated part of state updating (see [2]). Model parameters are usually considered as being fixed, and DA techniques are applied off-line for automatic estimation.

2.2. STATE UPDATING

State updating is usually formulated as a filtering problem, where model and observations are melded by distributing the model residuals (differences between point measurements and model forecasts) to the entire grid and state variables. The formulation of the weighting matrix to distribute the model residuals is the most essential part of the data assimilation scheme, and the different schemes mainly differ from each other in the way this matrix is calculated.

The most comprehensive melding scheme is the Kalman filter (KF), where the weighting matrix (denoted the Kalman gain) is determined, based on the minimization of the expected error of the updated state in terms of the errors of both the model dynamics and the data. The KF was originally designed for linear model dynamics, but a nonlinear extension (the extended KF) has been developed. The main strength of the KF is that it explicitly takes model and data uncertainties into account in the updating process and provides an estimate of the uncertainty of the system. The propagation of errors in the KF, however, usually imposes unacceptable computational burdens and storage requirements. In recent years, several KF schemes have been formulated, that use different approximations of the error modeling to reduce the computational costs.

State-of-art procedures which have been implemented in hydrodynamic and hydrologic modeling systems include the reduced rank square root filter and the ensemble Kalman filter [2, 3, 4]. The computational costs of these filters are in the order of 100 model simulations, which may still be too expensive in operational systems.

A cost-effective approximation of the KF is the steady state filter, in which the Kalman gain is predefined and assumed constant in the entire simulation period. In this case, error propagation is not part of the DA scheme, implying that the computational costs are only slightly larger than a normal model run. The steady state Kalman gain can be determined from an off-line simulation, using a time varying Kalman filter [5]. The steady state filter prescribes time invariant error statistics and fixed measurement positions, which may cause restrictions on the applicability of the method. A variant of the steady state filter is optimal interpolation, where the weighting matrix is based on a statistical description of the correlation between the state of the system and the model residuals in measurement points.

Another class of state updating DA schemes is nudging or relaxation. In these methods, observable state variables are relaxed towards the observations by smoothing the observed point residuals in space and time. Other state variables are then modified via the dynamical evolution of these corrections.

2.3. ERROR CORRECTION

Data assimilation using error correction is based on updating the modeled output with a forecast of the model error. In this case, an error correction forecast model is built, based on the observed model residuals, and this model is then superimposed on the simulation model. Error correction is used for updating single points in the modeled system, such as catchment runoff in hydrologic modeling [1, 6] and water levels in specific locations as part of a hydrodynamic modeling system [7, 8, 9]. Error correction techniques are generally very cost-effective and, hence, efficient for DA in operational systems.

Error correction models have generally been developed, using linear statistical models or the autoregressive moving average type of models with or without inclusion of exogenous input variables. Recently, studies have been conducted, that use a general nonlinear model formulation [6, 8]. These models are based on state-of-art data mining techniques, including artificial neural networks, genetic programming, local linear models, and support vector machines. Compared to the traditional linear models, these techniques have shown to offer a great potential in improving forecast skills [8]. Selected error correction models are presented and compared in the following.

2.4. LOCAL LINEAR MODEL

The AR(p) model approximates the observed points in the p-dimensional space of model residuals $x_t = (\varepsilon_t, \varepsilon_{t-1}, ..., \varepsilon_{t+1-p})$, using a global linear function. Local linear modeling is a method inspired from chaos theory [10], where the points in the p-dimensional space are approximated by local linear functions. In this case, only the nearest neighbors of x_t in the p-dimensional state-space are used to estimate the local autoregressive parameters. Since individual linear models are used for each prediction, the resulting model is globally nonlinear.

In this study, the local linear model algorithm by Babovic and Keijzer [11] is adopted. The calibration data are split into a training and a testing set. The training data constitute the database for which the nearest neighbors of a new prediction point x_t are searched. In this case, a fixed neighborhood size is used for estimation of each local linear model. The optimal neighborhood size and the optimal order of the autoregressive model (embedding dimension) are evaluated, based on the testing data.

In Fig. 1, the RMSE of the one-step ahead prediction based on the testing data for the calibrated Orgeval catchment is shown for varying neighborhood sizes and model orders. The local AR(1) model has a minimum RMSE for a neighborhood size of about 10. For larger neighborhoods, the RMSE increases and converges to the RMSE of the global AR(1) model. Thus, an optimal neighborhood size of 10 should be used in the case of a local AR(1) model. The AR(2) model has no well-defined RMSE-mimimum, whereas the AR(3) model has a very flat RMSE-minimum for a neighborhood size of about 100. The higher order AR models have RMSE curves virtually identical to the AR(3) model. A local AR(3) model with a neighborhood size of 100 has been used in the validation test presented below.

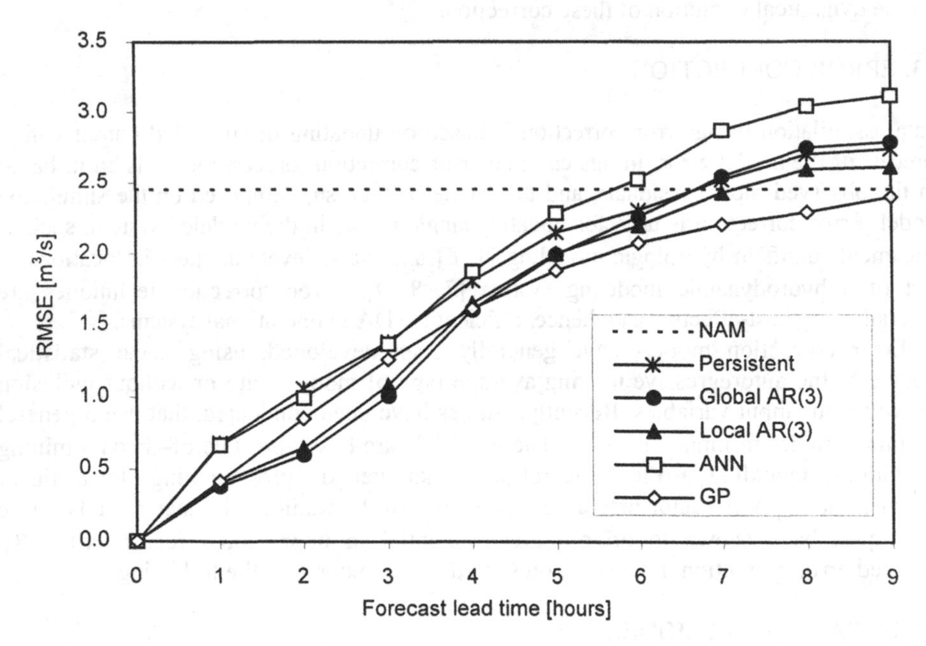

Figure 1. Average RMSE of the six validation events for the different error correction procedures based on the calibrated Orgeval model.

2.5. ARTIFICIAL NEURAL NETWORKS

In recent years, the use of artificial neural networks (ANN) has gained increasing interest in hydrologic modeling. In many of these applications, ANN is used for prediction of the rainfall-runoff process directly [12]. For time series prediction and modeling, dynamic neural networks form an important and powerful class of network topologies. They can be viewed as a nonlinear extension of linear filters such as AR models or as a variant of static networks with local recurrent structures providing short-term memory.

In this paper, a time-lagged recurrent network (TLRN) structure is used [13]. The main advantage of TLRN's is that smaller networks can be used for forecasting. The correlation analysis above suggests that a third order AR structure is appropriate so that the depth of memory (taps) was fixed at 3. A three-layer network topology was used, consisting of one input layer, one hidden layer containing 6 processing elements, and a linear output layer to improve the extrapolation properties. The network was used as a one-step ahead error predictor, i.e.:

$$\hat{\varepsilon}_{t+1} = N(\varepsilon_t) \tag{1}$$

where N represents the time-lagged recurrent network. The network was developed using the NeuroSolutions^{TM} package and trained on data from the calibration period, with a representative subsample for cross-validation.

2.6. GENETIC PROGRAMMING

All the methods described above belong to the class of numerical regression techniques, where the parameters of a fixed model structure are optimized from the available data. A relatively new, more versatile technique is the symbolic regression [14], where the model structure is also an explicit part of the model optimization. In this regard, genetic programming (GP) has shown to be an efficient optimization algorithm.

In this study, the GP algorithm by Babovic and Keijzer [15] was applied. Ten storm events from the calibration period were selected for training the GP. As independent variables, $\{Q_{sim,t}, Q_{sim,t-1}, .., Q_{sim,t-5}\}$ and $\{\varepsilon_t, \varepsilon_{t-1}, .., \varepsilon_{t-5}\}$ were used, and the functional set consisted of the basic algebraic operators $\{+,-,*,/\}$. To circumvent the problem of dimension inconsistency, the independent variables were normalized with respect to their maximum values. For optimization, the RMSE of the one-step ahead prediction was used as the objective function.

For the calibrated Orgeval model, the best functional form for the GP error correction model was found to be:

$$\begin{aligned}\hat{\varepsilon}_{t+1} = {} & 0.009 + 1.611\,\varepsilon_t - 0.644\,\varepsilon_{t-1} \\ & + 0.087\,\varepsilon_t\,(Q_{sim,t-2} - Q_{sim,t-1})\end{aligned} \tag{2}$$

The equation consists of two parts: a linear part with an autoregressive structure and a nonlinear part that describes an interaction between the model error and the simulated discharge. The nonlinear part prescribes that different weights should be given to the previous model error, depending on the gradient of the simulated hydrograph. On a rising limb ($Q_{sim,t-2} < Q_{sim,t-1}$), less weight is given to ε_t than on a falling limb.

3. Forecasting Results

The updating procedures were applied for simulated real-time forecasting of the Orgeval catchment in France, which was also used in the WMO intercomparison project [16]. The Orgeval catchment is located about 80 km east of Paris and drains to the Marne River. The catchment has an area of 104 km^2 with elevation ranging between 70 – 180 m.

The test data consist of hourly data of average catchment rainfall and runoff at the catchment outlet and daily data of potential evapotranspiration. For calibration of the rainfall-runoff model and estimation of the different error correction models, data from the period October 1972 to December 1974 were used. For testing the updating procedures, six events in the period December 1978 – July 1980 were adopted. The peak runoff for the 6 events ranges between 10.2 – 28.8 m^3/s. It should be noted that all of these events have higher peak runoff than the maximum observed peak in the calibration period.

The simulation model used in this study is the NAM rainfall-runoff model, which forms part of the MIKE 11 river modeling system [17]. It is a multi-storage model, which continuously accounts for the water content in four storages representing: (1) snow storage; (2) surface storage; (3) lower zone storage; and (4) groundwater storage.

The different error correction models described above were applied to the six validation events. For each event, 7 forecasts were carried out every three hours with lead times from 1 to 9 hours. In the forecast period, observed rainfall was used as forecasted rainfall.

To assess the performance of the forecasting procedures, the RMSE between observed and forecasted runoff was calculated for each lead time. The average RMSE of the six events for the different forecasting procedures based on the calibrated model are shown in Fig. 1. For all models, the RMSE increases with forecast lead time. For short lead times, error correction significantly improves the forecasting accuracy as compared to model forecasting without error correction. For longer lead times, the RMSE of the different procedures approach the RMSE of the direct model forecast. All forecasting procedures, except the ANN model, perform significantly better than the persistent error correction model for lead times smaller than about 6 hours, which indicates the positive impact of using more sophisticated error correction models for small lead times.

The global AR(3) model has the best performance for lead times smaller than 4 hours. The performance of the local AR(3) model is slightly worse for small lead times, but it has a smaller RMSE for lead times larger than 6 hours. Thus, in this case, as also illustrated in Fig. 1, the expected improved forecast accuracy, when using a local linear model as compared to a global AR model, is not pronounced. The GP model performs worse than the global and the local AR models for small lead times; whereas for lead times larger than 4 hours, it produces the most accurate forecasts of the different models considered. In this respect, mainly two aspects should be noted for the performance of the GP model as compared to the other models. First, the GP model is the only model that also uses the simulated runoff for predicting the error corrections; and, secondly, only the peak runoff events were used for training the GP. The ANN model has the worst performance for all lead times. For small lead times, it is only slightly better than the persistent error correction model. For lead times larger than 4 hours, the ANN model produces less accurate forecasts than the persistent error correction model, indicating that the added model complexity and flexibility introduced in the ANN model has a negative impact in this case.

4. The Use of Genetic Programming for Data Mining

Genetic programming algorithms have a huge potential apart from error correction modeling. Due to their evolutionary nature, they are also effective for data-driven discovery applications. Especially, dimensionally aware genetic programming [18] have shown to be effective, and an application to derive equations from available data is presented below for the case of faecal pellet settling.

The settling velocity of faecal pellets produced by marine organisms contributes to different oceanic processes including sedimentation rates, geochemical cycles, and nutrient availability. Because faecal pellets are aggregates of smaller particles, the pellet sinking rates can be much larger than the rates of the individual particles, increasing the sedimentation flux and possibly the rate of particle deposition. Faecal pellets influence sediment transport processes in the benthic boundary layer, and an evaluation of faecal pellet settling rates contributes to the study of sediment mobility on the sea floor.

Faecal pellet settling velocity equations have been presented in the literature for both pelagic organisms and benthic organisms. Significantly larger pellets with higher settling velocities are produced by these benthic organisms. This case study concentrates on these faecal pellets of benthic origin. The data themselves originate from Taghon *et al.* [19], where the faecal pellets were produced by the benthic feeder *Amphicteis scaphobranchiata*. The faecal pellet data consist of length (L), width at widest point (W), density (ρ), and measured settling velocity (v_s) for each individual pellet. Settling velocities were measured in seawater. The data were organized in two separate groups: group 1 (where $37<R_e<178$) consisted of pellets produced by feeding on < 61μm sediment fraction, and group 2 (where $45<R_e<117$) consisted of pellets produced by feeding on 61- 250 μm sediment fraction. Faecal pellet settling velocity equations, which have been fitted to these data, as well as settling velocity equations which have been commonly applied to high-density sedimentary particles are examined.

4.1. SETTLING VELOCITY EQUATIONS FITTED TO THE FAECAL PELLET DATA

A number of equations presented in the literature have been fitted to the data. They assumed the pellet to be cylindrical and calculated the nominal diameter (d_n), based on the equal volume sphere. Taghon *et al.* [19] used a regression analysis to yield (using the cgs unit system):

$$v_s = 1.30\, d_n + 9\rho - 9.08 \tag{3}$$

Komar & Taghon [20] used the pellets nominal diameter (d_n) to produce the following:

$$v_s = 0.275 \left(\frac{\left((\rho - \rho_f) g / \rho_f \right)}{\nu} \right)^{0.2} \tag{4}$$

where ρ_f denotes the density of freshwater.

They also found a relationship between the pellet settling velocity (v_s) and the settling velocity (v_t) of the 'equivalent' sphere, the sphere with the pellets nominal diameter. The calculation of v_t was carried out by using either [21] or [22]. They presented the following [cm/sec]:

$$v_s = 0.824\, v_t^{\,0.767} \qquad \text{based on [21]} \tag{5}$$

$$v_s = 1.08\, v_t^{\,0.686} \qquad \text{based on [22]} \tag{6}$$

Although the original settling velocity data were measured in seawater, it appears that Komar & Taghon [20] used freshwater conditions in developing the above equations.

4.2. NATURAL SEDIMENTARY PARTICLE SETTLING VELOCITY EQUATIONS

A number of equations have been presented in the literature for natural sedimentary high-density particles (with densities higher than for faecal pellets). These equations include those presented by Rubey [23], Gibbs *et al.* [21], Hallermeier [24], Dietrich [25], and Van Rijn [26]. The equations of Rubey [23], Gibbs *et al.* [21], and Hallermeier [24] were fitted to the faecal pellet data by Taghon *et al.* [19]. They found the fit of all equations, except that by Rubey [23], to be unsatisfactory and concluded that the goodness of fit of the Rubey [23] equation was coincidental. Of the remaining equations, only the Dietrich [25] equation has potential as it includes particle shape and roundness terms. These equations are not analyzed here in further detail, and the interested reader is referred to Babovic *et al.* [27] for a thorough survey and discussion.

4.3. RESULTS CREATED USING GENETIC PROGRAMMING

By way of comparison, a dimensionally aware genetic programming environment was set-up in such a way as to comprehend all measured data (that is the data from both groups 1 and 2). Since pre-processing of raw observations was not employed here, it can be argued that GP was confronted with a problem of trying to formulate a solution from first principles. The evolutionary processes resulted in a number of expressions, of which only the most interesting one is presented:

$$v_s = 0.4322\sqrt{g \cdot d_n \sqrt{\frac{\rho_s - \rho_{sw}}{\rho_{sw}}}} \tag{7}$$

The equation shows that settling velocity increases with increasing differential density ($\rho_s - \rho_{sw}$) (Fig. 2). Figure 3 shows that the fit of the equation to the measured data reduces as the settling velocities increase. The expression is generally quite similar to the form of the semi-empirical Eq.(4). Both equations indicate an increasing settling velocity with relative floating density and d_n. This information does not only reinforce the validity of GP as a way of knowledge induction technique, but it can also be of help in the design of future data collection efforts. Table 1 presents an intercomparison of these equations. It can be seen that Eq.(7) provides accuracy within 10%, based on the average variation from measured data approach.

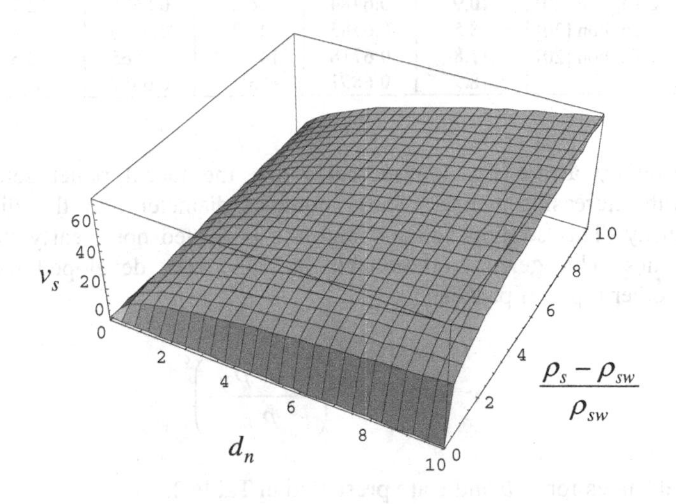

Figure 2. Settling velocity as a function of differential density and geometrical properties

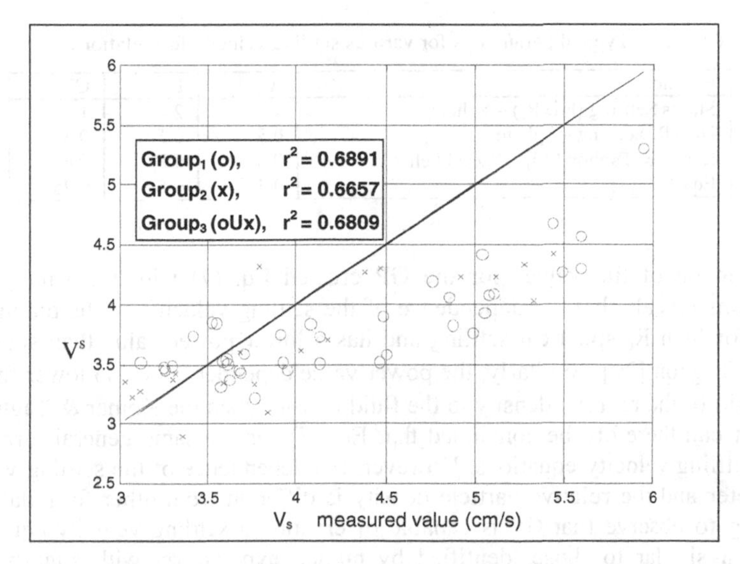

Figure 3. Scatter plot. GP vs. Measured Seawater Data.

TABLE 1. Settling velocity equation fit to measured seawater data

Equation	Group 1		Group 2		Combined Groups 1 & 2	
	Variation (%)	r^2	Variation (%)	r^2	Variation (%)	r^2
Eq (3) Taghon *et al.* [19]	0.6	0.6614	3.1	0.6526	1.4	0.6575
Eq (4) Komar & Taghon [20]	10.9	0.6484	16.0	0.6675	12.7	0.6572
Eq (5) Komar & Taghon [20]	8.5	0.6783	10.3	0.6690	9.2	0.6768
Eq (6) Komar & Taghon [20]	11.8	0.6718	14	0.6665	12.5	0.6713
Eq (7) GP	8.9	0.6891	6.7	0.6657	8.2	0.6809

The dimensionally aware Eq. (7) indicates that the faecal pellet settling velocity increases with increasing values of the nominal diameter and the differential (or floating) density. This settling velocity increase is related nonlinearly to these faecal pellet properties. This general relationship has also been developed for the settling velocities of other types of particles:

$$v_s \; \alpha \; g^a \; d^b \left(\frac{\rho - \rho_f}{\rho_f} \right)^c \tag{8}$$

where typical values for *a*, *b* and *c* are presented in Table 2.

The faecal pellets examined lie in the intermediate Reynolds Number range ($37 < R_e < 178$). This explains why Eq. (4) has values for *b* and *c* which lie between the range of values for Stokes and the high R_e settling presented in Table 2.

TABLE 2. Typical parameters for various settling velocity formulations.

Equation	A	B	C
Stokes Settling (low R_e) – Sphere	1	2	1
High R_e Settling – Sphere	0.5	0.5	0.5
Komar & Taghon [20] – Faecal Pellet	0.6	0.8	0.6
Equation (7)	0.5	0.5	0.25

Comparison of the values for the GP created Eq. (7) with the values by other formulations reveals that the dependence of the settling velocity on the diameter is the same as for high R_e spherical settling and has a lower power value than suggested by Komar & Taghon [20]. Similarly, the power value dependence is also lower for Eq. (7), for the ratio of the relative density to the fluid density, than the Komar & Taghon's [20] Eq. (4). It can therefore be concluded that Eq. (7) has the same general form as other particle settling velocity equations. However, the dependence of the settling velocity on the diameter and the relative particle density is different from other formulations. It is fascinating to observe that GP is capable of creating a settling velocity equation with phenomena similar to those identified by human experts, yet with slightly different nonlinear components.

5. Conclusions

The means of data collection, transmission, and storage are advancing at an extraordinary rate. The same cannot be asserted about advances in information and knowledge extraction from data. In the field of water resources, the combination of data, numerical simulation models, and data assimilation techniques provide the most effective means of transforming measured data into concrete decision support tools with predictive capabilities.

In fields where basic knowledge and understanding are still lacking to formulate reliable simulation models, tools are evolving, which can extract relations and equations from available data. Hybrid approaches, where the different methodologies are used in combination, have great potential in enhancing the information and understanding, which can be achieved from measured data.

6. References

1. Refsgaard, J.C. (1997) Validation and intercomparison of different updating procedures for real-time forecasting, *Nordic Hydrology* **28**, 65-84.

2. Madsen, H. and Cañizares, R. (1999) Comparison of extended and ensemble Kalman filter for data assimilation in coastal area modeling, *International Journal for Numerical Methods in Fluids* **31**, 961-981.

3. Verlaan, M. and Heemink, A.W. (1997) Tidal flow forecasting using reduced rank square root filters, *Stochastic Hydrol. Hydraul.* **11**, 349-368.

4. Hartnack, J. and Madsen, H. (2001) Data assimilation in river flow modeling, 4th DHI Software Conference, 6-8 June, 2001, Scanticon Conference Centre, Helsingør, Denmark.

5. Cañizares, R., Madsen, H., Jensen, H.R., and Vested, H.J. (2001) Developments in operational shelf sea modeling in Danish waters, Estuarine and Coastal Shelf Science (in press).

6. Madsen, H., Butts, M.B., Khu, S.T., and Liong, S.Y. (2000) Data assimilation in rainfall-runoff forecasting, Proceedings of Hydroinformatics 2000, 4th Conference on Hydroinformatics, Cedar Rapids, Iowa, USA, 23-27 July 2000.

7. Rungø, M., Refsgaard, J.C., and Havnø, K. (1991) The updating procedure in the MIKE 11 modeling system for real-time forecasting, in: Proceedings of the International Symposium on Hydrological Applications on Weather Radar, University of Salford, 14-17 August, Ellis Horword, 497-508.

8. Babovic, V., Keijzer, M., and Bundzel, M. (2000) From global to local modeling: A case study in error correction of deterministic models, Proceedings of Hydroinformatics 2000, 4th Conference on Hydroinformatics, Cedar Rapids, Iowa, USA, 23-27 July 2000.

9. Babovic, V., Cañizares, R., Jensen, H.R., and Klinting, A. (2001) Neural networks as routine for error updating of numerical models, *J. Hydraul. Eng., ASCE* **127(3)**, 181-193.

10. Farmer, J.D. and Sidorowich, J.J. (1987) Predicting chaotic time series, *Physical Review Letters* **59(8)**, 62-65.

11. Babovic, V. and Keijzer, M. (1999) Forecasting river discharges in the presence of chaos and noise, in J. Marsalek (ed.), *Coping with Floods: Lessons Learned from Recent Experiences*, Kluwer, Dordrecht.

12. Minns, A.W. and Hall, M.J. (1996) Artificial neural networks as rainfall-runoff models, *Journal of Hydrological Sciences* **41(3)**, 399-417.

13. Wan E. (1993) Time Series Prediction by using a connectionist network with internal delays, in A.S. Weigend and N.A. Gershenfeld (eds), *Time Series Prediction: Forecasting the Future and Understanding the Past*, SFI Studies in the Sciences of Complexity, Proc. Vol XV, Addison-Wesley.

14. Koza, J.R. (1992) *Genetic Programming: On the Programming of Computers by Natural Selection*, MIT Press, Cambridge.

15. Babovic, V. and Keijzer, M. (2000) Genetic programming as a model induction engine, *Journal of Hydroinformatics* **2(1)**, 35-60.

16. WMO (1992) *Simulated Real-Time Intercomparison of Hydrological Models*, Operational Hydrology Report no. 38, World Meteorological Organization, Geneva.

17. Havnø, K., Madsen, M.N., and Dørge, J. (1995) MIKE 11 – a generalized river modelling package, in V.P. Singh (ed.), *Computer Models of Watershed Hydrology*, Water Resources Publications, Colorado, pp. 733-782.

18. Keijzer, M., and Babovic, V. (1999) Dimensionally aware genetic programming, in W. Banzhaf, J. Daida, A. E. Eiben, M. H. Garzon, V. Honavar, M. Jakiela, and R. E. Smith, (eds.), *GECCO-99: Proceedings of the Genetic and Evolutionary Computation Conference*, July 13-17, 1999, Orlando, Florida USA. San Francisco, CA: Morgan Kaufmann.

19. Taghon, G.L, Nowell, A.R.M., and Jumars, P.A. (1984) Transport and breakdown of fecal pellets: biological and sedimentological consequences, *Limnol. Oceanogr.* **29(1)**, 64-72.

20. Komar, P.D. and Taghon, G.L. (1985) Analysis of the settling velocities of fecal pellets from the subtidal polychaete amphicteis scaphobronchiate, *Jnl. of Marine Research* **43(3)**, 605-614.

21. Gibbs, R.J., Matthews, M.D., and Link, D.A. (1971) The relationship between sphere size and settling velocity, *Jnl. of Sed. Petrology* **41(1)**, 7-18.

22. Davies, C.N. (1945) Definitive equations for the fluid resistance of spheres, *Proc. of the Physical Society* **57(4)**, No. 322.

23. Rubey, W.W. (1933) Settling velocities of gravel, sand, and silt particles, *American Journal of Science* **25**, 325-338.

24. Hallermeier, R.J. (1981) Terminal settling velocity of commonly occuring sands, *Sedimentology* **28**, 859-865.

25. Dietrich, W.E. (1982) Settling velocity of natural particles, *Water Resources Research* **18(6)**, 1615-1626.

26. Van Rijn, L.C. (1989) *Handbook of Sediment Transport By Currents And Waves*, Delft Hydraulics, Report H 461.

27. Babovic, V., Keijzer, M., Rodríguez, A.D., and Harrington, J. (2001) *Automated discovery of settling velocity equations*, D2K Technical Report, D2K-0201-1, http://d2k.dk/Publications.

28. Babovic, V. (1996) *Emergence, Evolution, Intelligence; Hydroinformatics*, Balkema, Rotterdam.

29. Babovic, V. and Abbott, M.B. (1997) The evolution of equations from hydraulic data, Part I: Theory, *Journal of Hydraulic Research* **35(3)**, 1-14.

30. Heemink, A., Verlaan, M., and Segers, A.J. (2000) Variance reduced ensemble Kalman filtering, Report 00-03, Department of Applied Mathematical Analysis, Delft University of Technology, The Netherlands.

31. Kompare, B. (1995) The use of artificial intelligence in ecological modeling, Ph.D. Thesis, University of Copenhagen, Denmark.

32. Madsen, H. (2000) Automatic calibration of a conceptual rainfall-runoff model using multiple objectives, *Journal of Hydrology* (accepted).

SPATIALLY DISTRIBUTED PREDICTION OF WATER DEFICIT PERIODS

G. MENDICINO
Department of Soil Defence, University of Calabria
Ponte Pietro Bucci
87036 Arcavacata di Rende (CS), Italy

Abstract: In this paper, the main characteristics of an integrated system to predict space-time variations of water deficit in southern Italy are presented. The system, which has already been in use for a year, has been developed within the European INTERREG IIC Programme. Using GIS technologies integrated with a telemetering network, hydrological databases and distributed water balance models, the system allows the analysis of different aspects of the drought phenomenon. Specifically, the system has been realized according to the features required by GIS-Web applications, which foresee the use of Internet as a tool of diffusion of the spatially distributed information managed by the same system. The integrated system proposed herein represents both an original methodological approach and a suitable tool to determine the critical areas characterized by water deficit which occurs widely and frequently in the analyzed region.

1. Introduction

Water entering the hydrologic cycle is a small percentage of the global amount present on Earth. However, this volume could adequately satisfy the world population if it were equally distributed. Unfortunately, this doesn't happen, and more and more regions are forced to face the degradation of ecosystems due to water deficit. As a result, crucial problems arise in different areas of the world, and many of these have become by now extremely intense, leading to situations of dramatic crisis. The problems in some countries of the 'third world' are so serious that, for centuries, agriculturists have fought for developing few hectares of cultivable ground in the desert; and frequently, conflicts originate between countries over the control of the few water resources available.

Furthermore, in the more developed countries, the situation has become worrisome because of the well-known climatic changes that have interested the whole planet for some decades. Italian watersheds are not free of such an unbalanced alternation of intense precipitation and drought periods. This trend amplifies the problems of flood control and hydrogeological risk. It also affects the availability of water resources, leading to greater restrictions in the southern regions and on the major islands of Italy.

This situation seems to be worsened by the progressive increase of droughts observed in recent years, not only in Italy, but also in southern Europe [1]. In this context, numerous initiatives have been promoted by the European Community to mitigate the effects induced by droughts on the Member States. Among these, the main initiative is represented by the INTERREG IIC Programme, which has permitted the realization of several transnational planning programmes and which includes, among its objectives, a strategy of prevention and co-operation in controlling the reduction of water resources due to droughts. Specifically, within the sub-programme entitled

"Analysis of the Hydrological Cycle", the southern Italian regions have been analyzed in detail because of their highly unstable climatic regime, characterized by strong intermittences of the water balance.

In this context, an integrated monitoring system for the analysis and the forecast of the effects of prolonged water deficit periods has been realized for the whole of southern Italy [2, 3]. The system comprises a commercial GIS (Arc-View), connected with a Data Acquisition System to store real time data recorded by the National telemetering hydro-meteorological network of the Servizio Idrografico e Mareografico Nazionale (SIMN). The system has been in operation since the beginning of the year 2000, and the telemetering data are constantly updated and managed according to the input required by the simulation models. These models allow the estimation of spatially distributed hydrologic quantities, such as solar radiation [4], potential evapotranspiration [5, 6], water deficit [7], surface and subsurface runoff, etc. Further analyses have been carried out, using indices based on rainfall observations. Among these, the Standardized Precipitation Index (SPI), suggested by McKee *et al.* [8], was used as a drought predictor. The SPI is estimated monthly on the whole territory for different accumulated precipitation values, and it is spatially represented on maps characterized by the same spatial resolution of the other hydrologic quantities.

The next section describes the main criteria used in developing the integrated monitoring system. Section 3 outlines the models utilized within the spatially distributed water balance procedure. The embedded coupling of the water balance procedure with the GIS is shown in section 4. Section 5 presents the analyses carried out for parameter estimation and discusses the first results obtained by the integrated system during the years 1999 and 2000. Finally, the output data produced by the system and their diffusion into Internet is described in section 6.

2. Integrated Monitoring System

The integrated system developed in southern Italy for the analysis and forecast of the prolonged water deficit effects [2, 3] is based on a GIS-embedded water balance modeling, directly linked with a Data Acquisition System to store real time data recorded by the National telemetering hydrometeorological network of the Servizio Idrografico e Mareografico Nazionale (SIMN). This network has been strengthened with a great number of stations, most of which are aimed to measure specific quantities to be directly used for water balance estimates over the southern Italian regions (Fig. 1). Specifically, measurements of rainfall, water level, groundwater level, temperature, solar radiation, evaporation, wind speed and direction, soil moisture, and soil water matric potential are available in real-time.

For the whole regions of southern Italy, all the geographic information has been stored within the Arc-View GIS according to two data structures, respectively vector and raster. Administrative boundaries, watershed boundaries, river networks, contours, reservoirs, towns and villages, highways and railways, geo-lithological and soil use coverages are available in vector format with different levels of detail. In the case of raster data, the system allows the use of different Digital Terrain Models (DTM), with spatial resolution varying from the regional scale (250 m square grid) to the basin scale (20 m square grid). Furthermore, all the spatially distributed hydrologic quantities are elaborated, managed and stored using a raster data structure characterized by a spatial resolution equal to the one utilized for the DTM.

Hydrologic coverages are directly obtained by linking the GIS with the hydrologic databases. Specifically, hydrometeorological information is managed, considering two different data types. The first is based on daily hydrometeorological values recorded by all gauging stations (*historic DB*) since 1925. The latter is based on real time data recorded by telemetering stations with a 20-minute time step (*real time DB*). The structure of databases has been designed, considering both the great amount of data recorded by historic and real time gauging stations during the period 1925-2000, and the easiness of management and query of the same information through the Internet. The hydrologic databases have been developed, using Microsoft® SQL Server™ software to facilitate data management and updating.

Hydrometeorological data are organized within the database in logical components, entirely transparent to the user. The interaction with the database is mainly provided through the analysis of tables, which represent the physical container of the data themselves. The information flux from and toward the databases can be managed both locally and as a remote server, considering also the access through local networks or Internet.

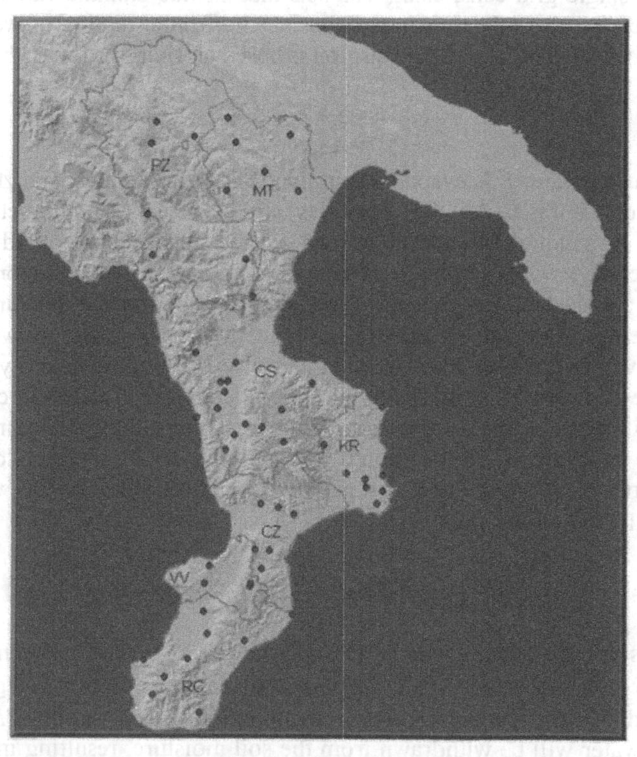

Figure 1. Telemetering hydrometeorological stations in southern Italy, managed by the integrated system.

The interaction with databases has been improved by some Windows-based software applications, ensuring a simpler and more flexible data query and allowing the user to extract synthetic graphs and reports. The user-DB interfaces can work both autonomously and directly inside the GIS by simple graphical selections on the gauging stations (Fig. 2). These additional functionalities are directly embedded within the GIS through a set of Avenue scripts. Specifically, they allow to dynamically query the different station types available in the databases and show the results obtained inside the GIS. For example, the GIS is able to singularly show rainfall or temperature stations; it is also capable of representing the gauging stations with a fixed number of observations or of recording data within a fixed time interval. Furthermore, the addition or the elimination of a station within the hydrologic databases is dynamically underlined to the user by the GIS.

3. Simulation Models and Forecasting

3.1. GENERAL STRUCTURE OF THE SIMULATION MODEL

In the region analyzed, water deficit is evaluated monthly, using a spatially distributed water balance model. This model follows the original approach suggested by Thornthwaite and Mather [7] and simulates soil moisture variations, evapotranspiration, and runoff on single grid cells, using data sets that include climatic variables, vegetation, and soil properties. The model does not consider horizontal motion of water on the land surface or in the soil. The governing equation is based on simplified mass balance:

$$P = S + E + \Delta W \tag{1}$$

where P is precipitation; E is evapotranspiration; S is water surplus; and ΔW is the change in soil moisture storage. All the quantities are evaluated in millimeters per month. Equation (1) does not differentiate between surface runoff and groundwater runoff; it allows to determine water surplus S as the water which does not evaporate or does not remain in soil storage but is available to generate surface and subsurface runoff.

The procedure schematizes the soil column through a reservoir whose maximum capacity is given by the soil-water holding capacity WHC. This quantity is obtained for given soil use and type by multiplying the difference between the field capacity and the permanent wilting point by the root zone thickness. The state variable representing the soil moisture at the end of the month i is defined as W_i, which depends on the difference between precipitation P_i and potential evapotranspiration PE_i values.

Specifically, if $P_i \geq PE_i$ then:

$$W_i = min\left[W_{i-1} + (P_i - PE_i)\,,\, WHC\right] \tag{2}$$

and soil moisture is recharged up to the maximum value WHC. When W_i is equal to WHC, further positive values $(P_i - PE_i)$ are considered as surplus. Negative values of $(P_i - PE_i)$ indicate the amount by which precipitation fails to supply PE_i requirements. In this case, water will be withdrawn from the soil moisture, resulting in an exponential soil moisture depletion, and the actual evapotranspiration E_i is less than PE_i. The soil moisture depletion curve is given by the following equation:

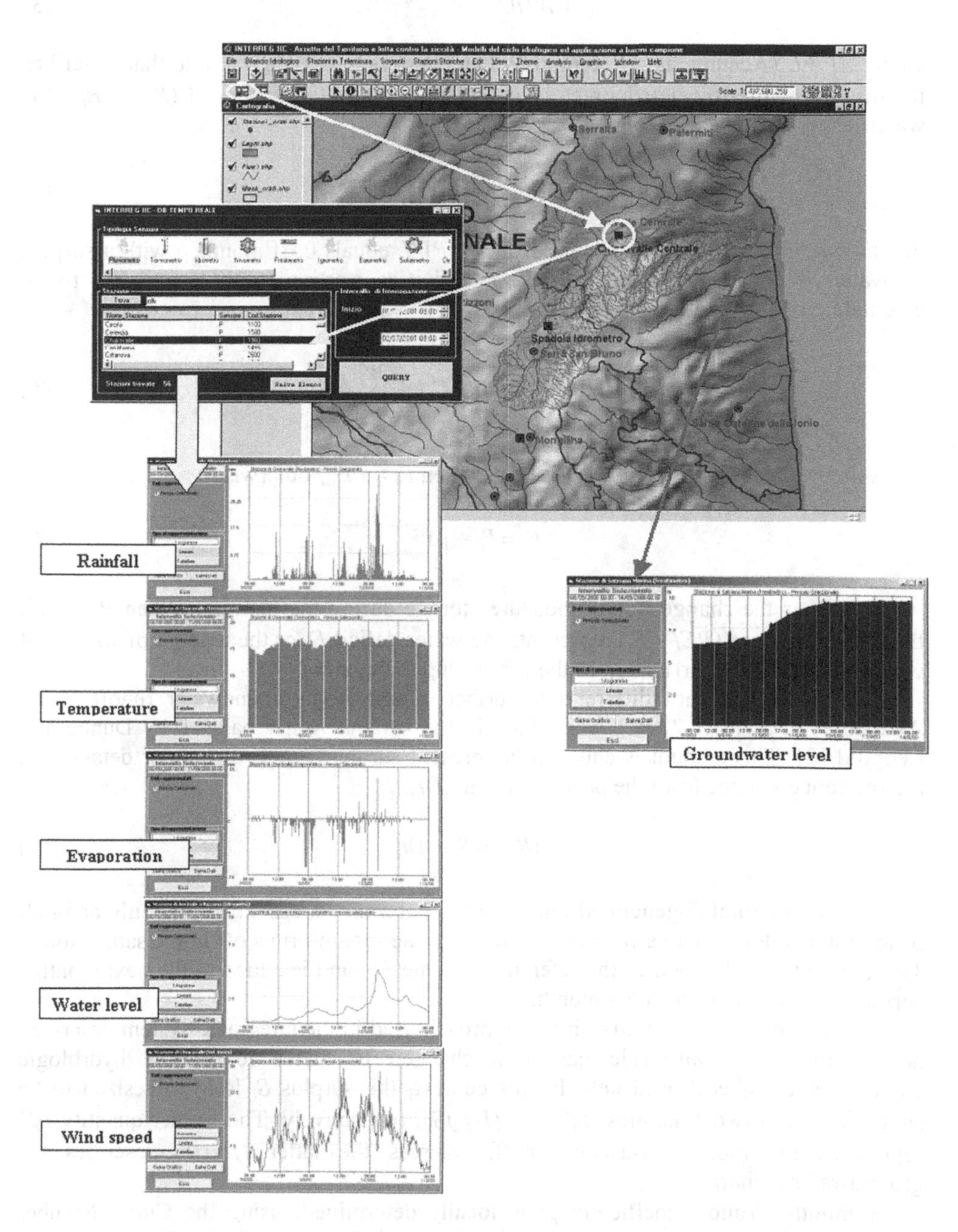

Figure 2. Embedded coupling of GIS-DB for real time analysis of hydrologic geo-referenced data (the figure shows the possibility of querying all sensors of a selected telemetering station or the query of more stations simultaneously).

$$W_i = WHC\ e^{(-APWL_i / WHC)} \tag{3}$$

where $APWL$ (Accumulated Potential Water Loss) represents a variable that describes the dryness of the soil. For months characterized by negative values of $(P_i - PE_i)$, this water loss is calculated as follows:

$$APWL_i = APWL_{i-1} + (P_i - PE_i) \tag{4}$$

For months characterized by water surplus, $APWL$ equals 0. If month$_{i-1}$, with a surplus of water, is followed by month$_i$ with a deficit, a starting $APWL$ value has to be calculated using the following equation:

$$APWL_{i-1} = -WHC\ ln\left(\frac{W_{i-1}}{WHC}\right) \tag{5}$$

Actual evapotranspiration E_i equals PE_i when $P_i > PE_i$; otherwise:

$$E_i = P_i + |\Delta W_i| \tag{6}$$

where ΔW_i is the change in soil moisture storage during the month i. When $PE_i > E_i$, then the difference $(PE_i - E_i)$ represents the water deficit D_i or the amount of water that would be supplied by irrigation to the soil during the month i.

The model does not differentiate surface runoff from groundwater runoff. Total Available for RunOff $TARO_i$ was added in the original water balance by Dunne and Leopold [9]. This amount is equal to the present surplus S_i plus any water detained in the reference volume from the previous month Dr_{i-1}:

$$TARO_i = S_i + Dr_{i-1} \tag{7}$$

Finally, the rainfall-generated runoff RO_i is determined, assuming that only one-half of the total water available for runoff $TARO_i$, in the month i runs off in the same month. The other half is detained in the reference volume Dr_i and is added to the next month's surplus S_{i+1}, to run off the next month.

Because of the spatially distributed approach used in the integrated system described herein, the "one half" rule has been changed to better reflect the hydrologic characteristics of each grid cell. In this context, the surplus S_i is hypothesized to be subdivided into two quantities, $\mu_i S_i$ and $(1-\mu_i)S_i$, respectively. The former quantity $\mu_i S_i$ represents the monthly surface runoff, whereas the latter $(1-\mu_i)S_i$ describes the groundwater recharge.

A monthly runoff coefficient μ_i is locally determined, using the Curve Number method suggested by the Soil Conservation Service [10] and modifying it to account for antecedent soil moisture conditions [11]. Specifically, coefficient μ_i is determined for each month according to the following equation:

$$\mu_i = \left\{ \frac{\left[P_i - 0.2\left(\frac{25400}{CN_I} - 254 \right)\left(\frac{W_{sat} - W_i}{W_{sat} - W_{wp}} \right) \right]^2}{P_i + 0.8\left(\frac{25400}{CN_I} - 254 \right)\left(\frac{W_{sat} - W_i}{W_{sat} - W_{wp}} \right)} \right\} / P_i \tag{8}$$

where CN_I is the curve number for dry antecedent moisture conditions; W_i is the actual soil moisture; W_{sat} is the soil moisture at saturation; and W_{wp} is the soil moisture at wilting point.

Groundwater recharge $(1-\mu_i)S_i$ is used to determine the runoff delay caused by water transport through groundwater before it enters river channels. If a linear groundwater reservoir is assumed, then its monthly runoff detention Dr_i (in millimeters) can be expressed as follows:

$$\frac{Dr_i - Dr_{i-1}}{\Delta t} = S_i(1-\mu_i) - \frac{Dr_{i-1}}{\beta'} \tag{9}$$

where β is the linear reservoir constant and Dr_{i-1}/β is the groundwater runoff. Finally, monthly river runoff Q_i is obtained as follows:

$$Q_i = S_i \mu_i + \frac{Dr_{i-1}}{\beta'} \tag{10}$$

The computation of soil-moisture surplus appears to be an iterative procedure where an initial soil moisture storage W_0 is needed before the computation can start. Since the initial soil moisture storage is typically unknown, the following water balancing procedure is applied to force the net change in soil moisture from the beginning to the end of a specified balancing period to zero, i.e., $W_0 - W_{n+1} < \xi$, where n is the number of time steps of the computational period, and ξ is a user-specified tolerance.

3.2. PRECIPITATION

In southern Italy, a network of uniformly distributed high-resolution precipitation stations exists. Therefore, spatially distributed rainfall estimates are obtained by using only interpolation techniques based on bi-dimensional splines.

GIS directly manages rainfall data, automatically querying the hydrologic database, which carries both the historic monthly means and the actual monthly-accumulated values for the entire network. These values are interpolated on a 250 m square grid and stored into the GIS, which allows mapping over the whole of southern Italy and the loading of data into the water balance model (Fig. 3).

Until present, analyses carried out by the integrated system have produced 36 spatial distributions of precipitation, consisting of 12 historic maps and 24 actual maps based on monthly-accumulated values observed during the years 1999 and 2000. Each month, rainfall deficit is determined by comparing the actual and the historic spatial distributions and by observing the spatial deviation of the former with respect to the latter.

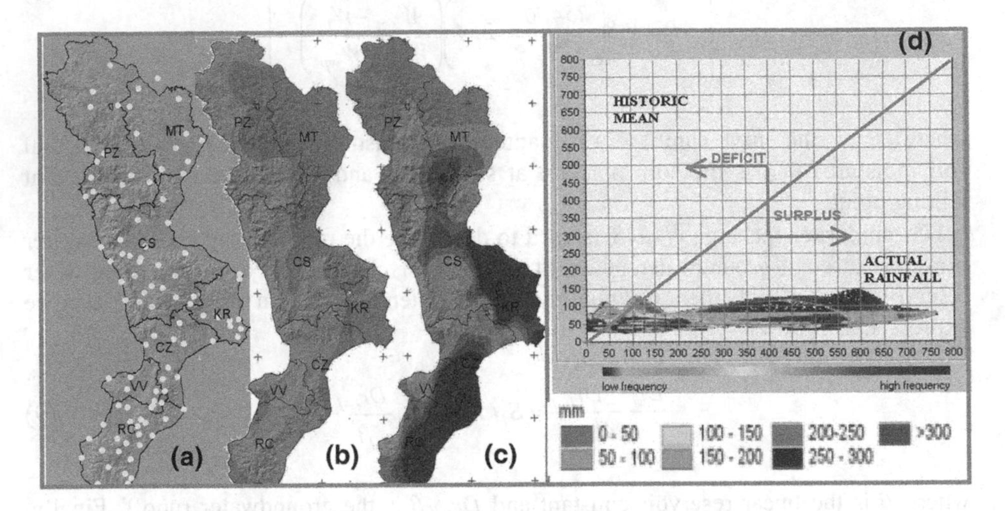

Figure 3. Example of monthly precipitation analysis: a) spatial distribution of the precipitation network; b) historic map - September; c) actual map - September 2000; d) spatial comparison between historic and actual maps.

Prediction of rainfall deficit is also obtained, by using the Standardized Precipitation Index (SPI), suggested by McKee *et al.* [8]. This index is based on precipitation alone and can be determined for a variety of time scales (1, 3, 6, 12, 24 and 48 months). At an assigned location and for a specific time period, this index is calculated by determining the probability distribution function of long-term precipitation records. This distribution function is transformed, using probabilities equal to a normal distribution with a mean of zero and standard deviation of one. Such a transformation allows the SPI values to be represented in standard deviations, so that positive index values indicate greater than median precipitation; on the other hand, negative index values show less than median precipitation. Using the same SPI classification scale suggested by McKee *et al.* [8], each long-term precipitation series has been analyzed, determining a set of rainfall threshold curves for every month of the year, which represent drought severity indicators for different time scales (Fig. 4a). Threshold curves are available for every telemetering precipitation station, and they are stored into the GIS. The GIS compares for each month the threshold values with the observed accumulated rainfall and determines SPI maps for different time scales (Fig. 4b).

3.3. POTENTIAL EVAPOTRANSPIRATION

Potential evapotranspiration *PE* (millimeters per day) in the soil-water balance model is estimated by the Penman equation [5], modified by Monteith [6].

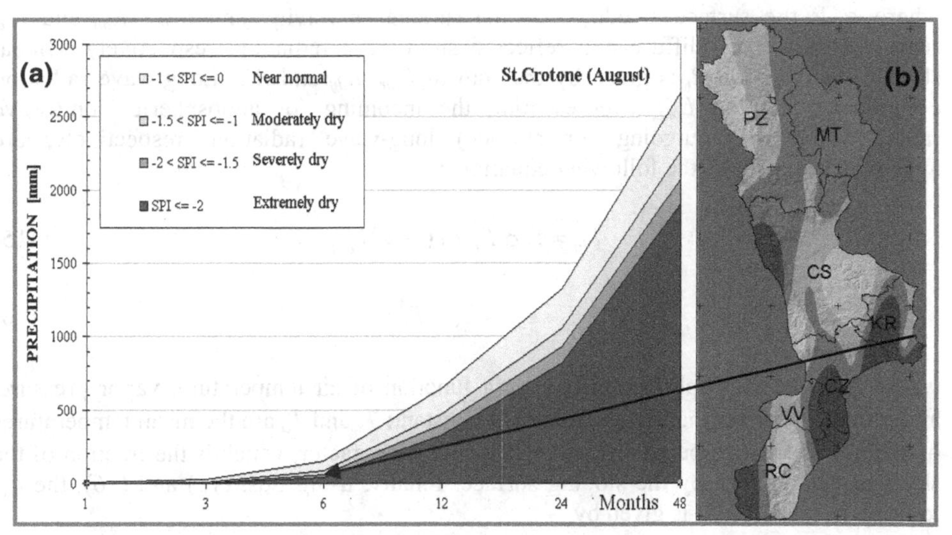

Figure 4. Rainfall deficit prediction based on SPI: a) example of threshold curves for a given precipitation station (Crotone) during August; b) a 6-month SPI map for August 2000.

This equation is expressed as follows:

$$PE = PE_{rad} + PE_{aero} \tag{11}$$

with:

$$PE_{rad} = \frac{1}{\lambda}\frac{\Delta}{\Delta+\gamma^*}(R_n - G) \tag{12}$$

and:

$$EP_{aero} = \frac{86.4}{\lambda}\frac{1}{\Delta+\gamma^*}\frac{\rho\, c_p}{r_a}(e_a - e_d) \tag{13}$$

where γ^* is the modified psychometric constant [kPa °C^{-1}]; λ, the latent heat of vaporization [MJ kg^{-1}]; Δ, the vapor pressure gradient with temperature [kPa °C^{-1}]; R_n, the net radiation [MJ m^{-2} d^{-1}]; G, the soil heat flux [MJ m^{-2} d^{-1}]; 86.4, a conversion factor;0 ρ, the atmospheric density [kg/m^{-3}]; e_a, the saturated vapor pressure at air temperature [kPa]; e_d, the actual vapor pressure [kPa]; c_p, the specific heat moist air [kJ kg^{-1} °C^{-1}]; and r_a, the aerodynamic resistance [s m^{-1}].

The evapotranspiration equation requires an estimate of the net radiation flux R_n received by an inclined surface. The spatial distribution of this quantity is determined, using a modified version of the model originally suggested by Moore *et al.* [4].

According to the model, net radiation flux R_n received by an inclined surface is given by the equation:

$$R_n = (1-\alpha_s)(R_{dir} + R_{dif} + R_{ref}) + \varepsilon_s L_{in} - L_{out} = (1-\alpha_s)R_t + L_n \tag{14}$$

where α_s is the surface albedo; ε_s is the surface emissivity; and R_{dir} , R_{dif} and R_{ref} represent the direct, diffuse and reflected short-wave radiation, respectively. Global short-wave radiation R_t is given by the sum of R_{dir}, R_{dif} and R_{ref}. Long-wave radiation components L_{in} and L_{out} , representing the incoming (or atmospheric) long-wave radiation and the outgoing (or surface) long-wave radiation, respectively, are approximated through the following equations:

$$L_{in} = \varepsilon_a \, \sigma T_a^4 \nu + (1 - \nu) L_{out} \tag{15}$$

$$L_{out} = \varepsilon_s \, \sigma T_s^4 \tag{16}$$

where ε_a is the atmospheric emissivity (a function of air temperature, vapor pressure, and cloudiness); σ is the Stefan-Boltzman constant; T_s and T_a are the mean temperatures of surface and air, respectively; and ν is the skyview factor, which is the fraction of the sky that can be seen by the sloping surface. Finally, using Eqs. (15) and (16), the net long-wave radiation L_n is given by:

$$L_n = \varepsilon_s \, L_{in} - L_{out} \tag{17}$$

All these quantities are spatially estimated by analyzing the topographically heterogeneous landscape information represented by the DTM. The variables of the model, such as albedo, emissivity, sunshine fraction, mean air and surface temperatures, and clear sky transmittance, can be varied on a monthly time scale. The solar radiation model has been developed to include also algorithms for the determination of short-wave radiation components, taking into account the effects of shading from direct sunlight by the surrounding terrain at enclaved sites.

Net radiation and potential evapotranspiration are spatially estimated in southern Italy on the same 250 m square grid used for precipitation. In addition, historic (monthly mean) and actual (monthly values observed during the years 1999 and 2000) spatial distributions of these quantities are available to the integrated system.

4. Integration of Arc View GIS and Models

The water balance model described in the previous section has been entirely developed by a GIS-embedded modeling procedure. The model takes advantage of the Avenue scripting language, which is part of the Arc View GIS software, transforming the simulation phases in a set of GIS functionalities.

The first of these functionalities aggregates the 250 m square grid data of climatic characteristics and soil properties on 25 km^2 square areas, as shown in Fig. 5a. In this way, spatially distributed monthly values of precipitation, potential evapotranspiration, and soil properties (soil-water holding capacity and curve number) can be stored in apposite GIS-tables which represent the water balance input. After the selection of a simulation time period and the input of some parameters, hypothesized as constant for the whole region, the water balance model is activated on user-selected areas (Fig. 5b).

For each of these areas, monthly values of actual evapotranspiration, accumulated potential water loss, soil moisture, water deficit, surface runoff, groundwater runoff, and river runoff are obtained.

Finally, output data are stored into GIS, that allows representing them schematically as a table and a graph, or spatially distributed as a map (Fig. 5c).

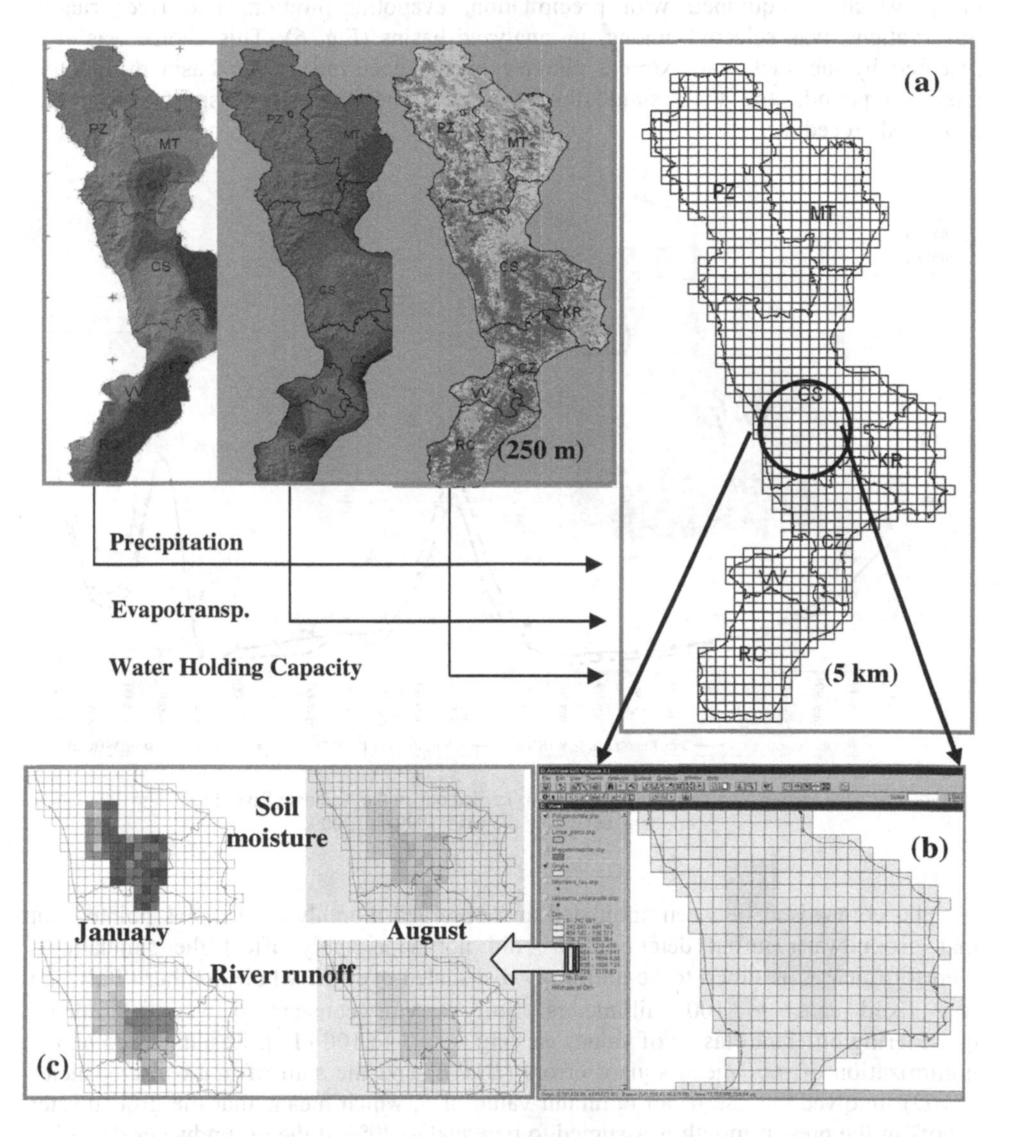

Figure 5. a) Southern Italy grid (5 x 5 km) used for water balance simulations; b) example of user-selected areas where monthly GIS-water balance model is activated; c) output data mapped by GIS.

5. Simulations

Model performance has been verified on few experimental basins in southern Italy before proceeding with the analysis of the regional water balance. Further analyses have been carried out by varying some characteristic parameters (μ, β, WHC, Dr_0) and observing model sensitivity. For the years 1999 and 2000, Ancinale River basin (116 km^2), which is equipped with precipitation, evapotranspiration, and river runoff observations, was selected among the analyzed basins (Fig. 6). This choice was also justified by the fact that extreme climatic events occurred in the basin during the analyzed period, and their simulation revealed some interesting aspects about the proposed procedure.

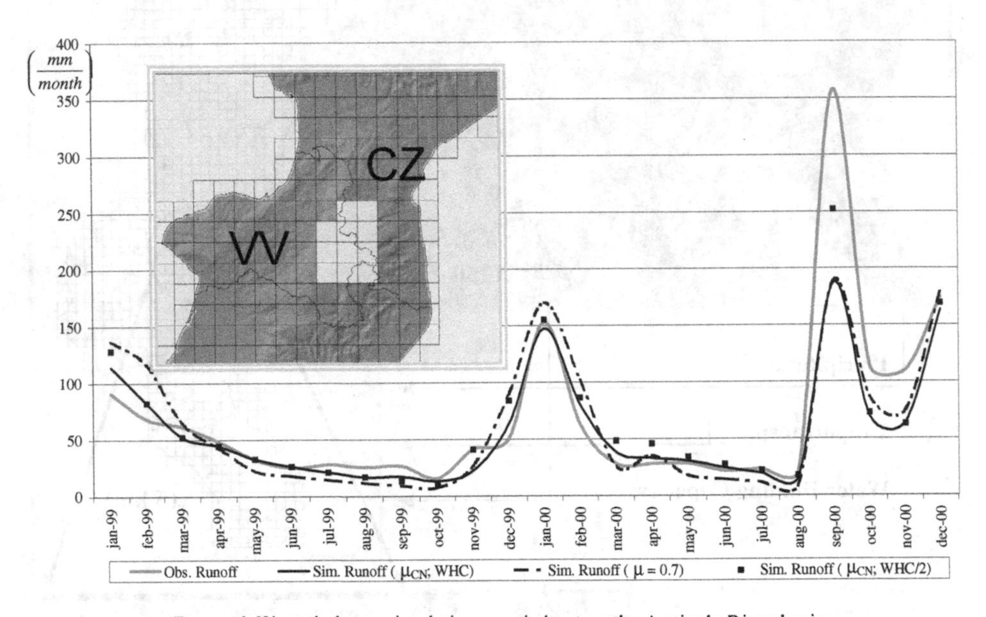

Figure 6. Water balance simulations carried out on the Ancinale River basin (bright areas) during the years 1999 and 2000.

The comparison between simulated and observed monthly river flows pointed out that groundwater initial detention Dr_0 did not appreciably affect the simulations; therefore, it was assumed to be uniformly distributed over the whole southern Italy and was fixed equal to 100 millimeters. Soil-reservoir constant β was determined, considering different classes of values varying from 1 to 100 [T^{-1}]. Results based on the minimization of root mean square errors (*RMSE*) and the sum of mass discrepancies (*SMD*) involved the use of an optimum value of 5, which means that the groundwater runoff in the present month is assumed to be equal to 20% of the groundwater detention of the previous month. Changes in soil-water holding capacity did not produce appreciable differences in simulated river runoff, with the exception of months characterized by very extreme precipitation events. For example, it was observed that halving the *WHC* values for the Ancinale basin, the simulated river runoff remained unchanged. Just for the very extreme September 2000, characterized by a 600-millimeter monthly rainfall, a substantial increase in simulated river runoff was achieved (Fig. 6). Finally, the use of a spatially distributed *CN*-based runoff coefficient showed a greater reliability in simulating river runoff than that obtained by using a uniformly distributed μ value (whose optimum value was determined to be equal to 0.7). This aspect is pointed out in Fig. 6, where the use of a uniformly distributed optimum μ value shows a systematic under-prediction of low flows during the dry season and an over-prediction of high flows during the wet season.

Results obtained at basin scale have been extended to the whole region of southern Italy, without changing the original soil-water holding capacity spatial distribution and assuming uniform distributions for groundwater initial detention Dr_0 and soil-reservoir constant β.

Water balance has been developed by considering historic monthly mean data first (for the period 1925-1998) and then the monthly values observed during the years 1999 and 2000. Simulations carried out on actual data, which provide useful information about the hydrologic characteristics of the analyzed region, may be criticized because of the simplifications introduced into the water balance model. On the other hand, their comparison with the historic mean quantities univocally defines the critical areas subject to water deficit.

Figures 7a, b and c show the comparison between the actual and the historic monthly values obtained by averaging the spatially distributed water balance quantities over the whole region of southern Italy. The same figures point out a net reduction in actual precipitation data, that involves a more evident reduction in river runoff. Specifically, October 1999 and August 2000 appear to be mostly struck by water deficit (Fig. 7c). For these months, GIS-embedded modeling allows to automatically determine the critical areas where water deficit values appear to be appreciably far from the corresponding historic mean (Fig. 7d).

6. GIS-Web Interface

GIS-based applications are considered to be high-technology products, and they appear to be surrounded by the aegis of modernity and scientific rigor with which developers, advisors and experts in the sector endow them. As a consequence, there are many intermediaries between the users and the suppliers of the system (technicians, experts, analysts, advisors, programmers, and so on). This leads to a lack of interactivity between the users and GIS, an interactivity that represents the base of an efficient, effective and productive use of the system. A further obstacle to the use of GIS is the one associated with the institutional and organizational problems prevailing in the transfer of technologies. While it is true that computer technologies help the democratization process of information by making it transparent, always usable, and more fairly accessible, it is also true that the opposite risk exists, that is, making a powerful information and control tool available only to an exclusive set of experts capable of using it.

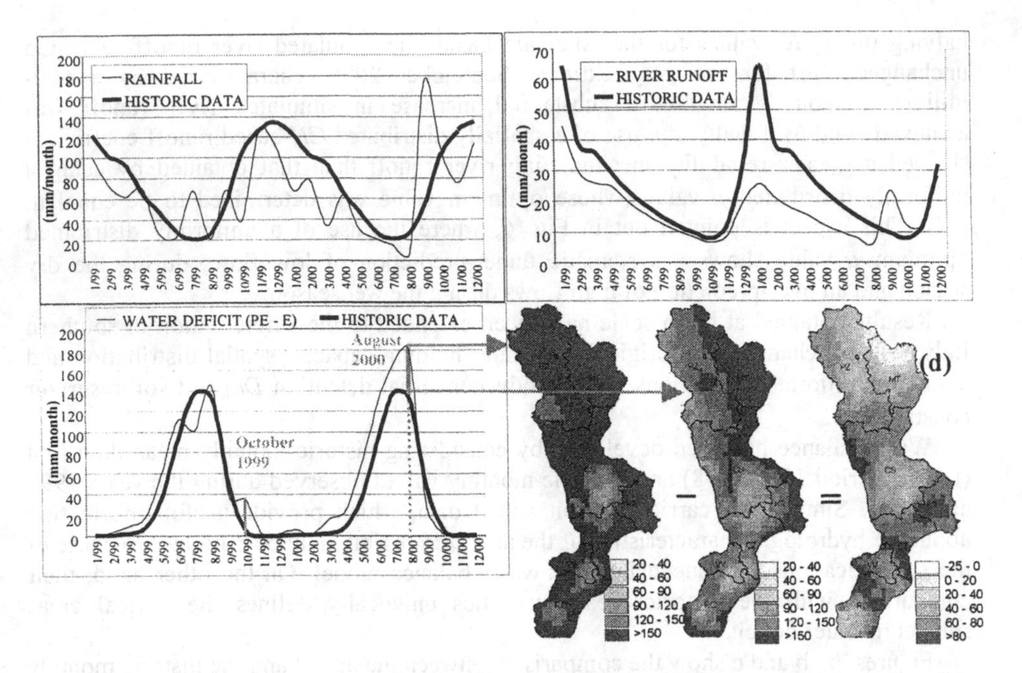

Figure 7. Comparison between the actual and the historic monthly water balance quantities in southern Italy: a) spatially-averaged rainfall values; b) spatially-averaged river runoff values; c) spatially-averaged water deficit values; d) example of spatially distributed water deficit differences for the critical month of August 2000.

In this context, a Geographic Information System has to consider a wider number of users represented by the Internet network; and, as a consequence, it has to be developed according to a GIS-Web environment whose architecture is based on the logic of client/server systems, or on an approach characterized by competing accesses to the same application by many users.

The integrated system described herein is equipped with automatisms allowing the publication and dissemination of real time data via World Wide Web at http://www.camilab.unical.it. Specifically, the web interface allows the user to find useful information about water deficit or surplus at regional scale. At present, the monthly hydrometeorological data of all the telemetering stations are available on the web. Starting from 1925, long-term series can be queried according to different selection criteria (basin, altitude, and country). For each precipitation station, a user can obtain historic information regarding chronological state of drought periods as well as monthly estimates of drought indices.

All hydrologic spatial distributions are available via web according to two sets of monthly maps. The former regards historic data based on monthly mean values recorded during the period 1925 – 1998; the latter comprises actual data observed each month during the years 1999 and 2000. For these periods, rainfall, temperature, net solar radiation, wind speed, potential evapotranspiration and water deficit maps are available. Finally, actual SPI maps can also be obtained for different time scales (1, 3, 6, 12, 24 and 48 months); in addition, for a given precipitation station, monthly SPI threshold curves can be observed or downloaded.

7. Conclusion

In this paper, an integrated GIS-Web monitoring system to analyze, control and map water deficit periods occurring in southern Italy is described.

Real time hydrometeorological data are stored in databases, which are directly linked to a GIS-embedded water balance model. GIS is able to spatially determine critical areas where water deficit values appear to be far from the corresponding historic means. This is realized monthly by comparing historic mean quantities with actual observed values. Analyses carried out during the years 1999 and 2000 have shown a diffuse drought event characterized by an increasing water deficit trend. Specifically, this trend is more evident along the Ionian zones.

The original aspect of the proposed system is represented by its GIS-Web architecture, through which all telemetering recorded data and spatially distributed water balance analyses are available, both as tables and maps, via World Wide Web at http://www.camilab.unical.it.

Future developments will be in two main directions: the first one will be to improve the water balance model, especially for what concerns soil properties and the correct definition of groundwater volume; the other one will be to introduce quantities related to irrigation and water supply into the water balance model.

8. References

1. Bordi, I. and Sutera, A. (1999) Identification of drought: Meteorological aspects, *Workshop INTERREG IIC "Assetto del Territorio e Lotta Contro la Siccità"*, Taormina, Italy.
2. Mendicino, G. and Versace, P. (1999) Water balance model based on GIS procedures, *Workshop INTERREG IIC "Assetto del Territorio e Lotta Contro la Siccità"*, Taormina, Italy.
3. Mendicino, G. and Versace, P. (2000) Siccità: Monitoraggio, analisi e previsione, *Workshop INTERREG IIC "Assetto del Territorio e Lotta Contro la Siccità"*, Villasimius, Cagliari, Italy.
4. Moore, I.D., Norton, T.W., and Williams, J.E. (1993) Modelling environmental heterogeneity in forested landscapes, *Journal of Hydrology* **150**, 717-747.
5. Penman, H.L. (1948) Natural evaporation from open water bare soil and grass, *Proc. R. Soc. London, Ser. A* **193**, 120-145.
6. Monteith, J.L. (1965) Evaporation and the environment, in *The State and Movement of Water in Living Organisms, XIXth Symp. Soc. Exp. Biol., Swansea*, Cambridge University Press, pp. 205-234.
7. Thornthwaite, C. W. and Mather, J.R. (1955) *The Water Balance, Climatology,* Drexel Inst. Of Technology, Centeron, New Jersey.
8. McKee, T.B., Doesken, N.J., and Kleist, J. (1993) The relationship of drought frequency and duration to time scales, *Eighth Conf. On Applied Climatology, Amer. Meteor. Soc.*, Anaheim, CA, pp. 179-184.
9. Dunne, T. and Leopold, L.B. (1978) *Water in Environmental Planning*, W.H. Freeman Company, New York.

10. Soil Conservation Service (1968) Hydrology, *Supplement A to Section 4, National Engineering Handbook*, Washington, D.C.: U.S. Department of Agriculture.

11. Heatwole, C.D., Campbell, K.L., and Bottcher, A.B. (1987) Modified CREAMS hydrology model for coastal plain flatwoods, *Trans. ASAE* **30(4)**, 1014-1022.

Part VIII

REMOTE SENSING AND GIS

DBMS/GIS APPLICATIONS IN INTEGRATED MARINE DATA MANAGEMENT

N. N. MIKHAILOV and A. A. VORONTSOV
Russian Research Institute of Hydrometeorological Information
World Data Center (RIHMI-WDC)
National Oceanographic Data Center of Russia
6, Korolev St., Obninsk, Kaluga Region
Russian Federation, 249035

Abstract: The exploitation of marine resources depends considerably on the level of information support available for this activity. This requires a full technological cycle of data management, ranging from the acquisition of observational data up to provision of data to end-users with complex information on all environmental aspects. The presented paper discusses the features of an integrated information technology and its use for marine data management. This methodology is based on a two-level presentation of its architecture. The upper level visible to a user is occupied by three functional subsystems: subsystems of archive banks, an integrated bank, and problem-oriented applications. The subsystems comprise a set of components (methodical, program-technological, language, etc.) which form a basic low level of technology and which provide a multi-faceted environment of the developed technology.

At present, the approach considered is being implemented in definite developments under various research programs.

1. Basic Aspects of Information Technology Development

Due to requirements imposed on the informational support for world ocean investigations and exploration, it is necessary to consider a complete technological data management cycle. This cycle covers all activities starting from the collection of data and extending to provision of an end-user with complex and validated information required for proper understanding of natural processes and for making correct decisions. Such a scheme of data management is now usually named as an integrated "end-to-end" information technology (hereafter referred to as IIT) [1, 2].

The fundamental features of IIT, as well as its importance in investigation and exploitation of the marine environment, may be briefly illustrated by the following simple example. Speaking of information technologies, three areas of the representation of natural processes and phenomena under study should be mentioned.

The first area comprises the real world or part of it (application domain, AD), where objects exist and have specific properties. The second area comprises the properties of an object, which are preserved and used in the information technology. The third area includes ideas and information on people (scientists, designers, customers, information system developers and programmers) who are in one way on another connected with the objects of the application domain.

One can imagine an abstract situation where the following are available:

- a system of AD objects;
- an expert observer who is able, on one hand, to perceive a state of the system through the system's data and fix them in his/her memory in a specific form and, on the other hand, to analyze and interpret the state as being fixed for some scientific or applied aims.

In this case, the "data" on the system of "objects" are accumulated in the memory of the expert observer. Using the data, the expert observer creates "new" information to understand processes and phenomena occurring in the system of objects. The function of these expert observers, that relates all the three concepts of the application domain in the general case, is performed by information technologies or by their visible representation, i.e., information systems. However, the integrated technology realizes this relation, irrespective of specific and changing concepts of analysts and programmers.

The application domain of the information technology is defined by the requirements imposed on the information support for the investigation and exploitation of the marine environment. Generally, these requirements may be divided into several classes of tasks:

- provision of requested data and information as hard copies or in electronic format;
- obtaining climatic characteristics of the marine environment in the region investigated;
- assessment of human impacts upon the environment;
- monitoring of the current state of the marine environment and of possible natural dangerous conditions.

To perform the aforementioned and other tasks, IIT should handle data and information on a broad spectrum of AD objects:

- environmental conditions of the region under study;
- legal and regulatory acts related to marine environment;
- methods, algorithms and models to calculate various characteristics of the marine environment.

The scope of both the tasks and the AD generates the need to create fairly complicated sets of tools to realize IIT. These tools may be divided into components and functional subsystems. The components (methods, general design decisions, data description languages, codes and codifiers, instrumental software systems, etc.) form the basic level of IIT tools and comprise the environment for the development of functional elements. The functional subsystems are generated from the components and are intended to perform appropriate information activities.

The methodology adopted for development of the information technology has made it possible to clearly represent its architecture as a set of functional subsystems, consisting of archived data banks, integrated data bank, and problem-oriented applications.

The first (input) block of the technology, i.e., the archived data bank subsystem (ADB), accumulates data on the marine environment, systematizes and converts them into internal information standards. The outcome of the ADB consists of documented data files on different aspects of the marine environment (hydrometeorology, geocreology, geomorphology, biota, and others). The model of the ADB information content is constructed within the context of the AD model.

Data sets formed in ADB arrive at the next block or the integrated data bank (IDB). The major function of IDB is to provide integration on the basis of a more complicated and unified data model, which considers both the AD and the functional requirements. The outcome of the IDB consists of the complex database (results of observations and calculations, textual data, topographical and thematic maps, and others), maintained in actual conditions to "feed" the next block, which is the subsystem of problem - oriented applications (POA).

In the broad sense, POA may be described as a set of specially selected subject-oriented data, knowledge obtained earlier, and applied programs which carry methods and models to compute environmental characteristics. POA is oriented to obtain new information for the selection of reasonable and economically feasible design decisions related to the exploitation of the marine environment.

The aforementioned subsystems are interconnected as they are developed on the basis of unified components. Whereas the ADB subsystem interacts with other subsystems only on the basis of information standards (formats, metadata structure and others), IDB and POA subsystems have a higher level of interconnection based on the client - server architecture, using GIS, DBMS and WWW technologies.

In both cases, the optimum organization of data management reflects the greatest complexity. In this regard, when designing the IIT, various aspects of the representation and connections of various data types were considered, including:

- metadata - «data about data»;
- factographical data - observed, derived, calculated and modeled data,
- spatial data - electronic marine navigation and thematic maps;
- image data (graphics, figures, photos);
- textual data (documents, description, papers);
- software modules realizing particular algorithms of data processing or modeling to produce information.

Data management in the IDB and the POA subsystems is based on exchange procedures of elementary data units in a client-server architecture. Thus, the POA subsystem, including GIS applications and a certain part of the data, is on the client side; and the IDB subsystem supporting the basic database is placed on the server side. The data are placed both on the client (POA) side and on the server (IDB) side, depending on their type. The server side supports factographical, image and textual data as tables of DBMS or through the references to data files. The spatial data, including semantic layers of thematic maps and a certain part of the specialized textual and image data, are placed on the client side as an internal database of GIS applications. As an example, the internal attributive GIS table, where each string corresponds to the information about one coastal station, is shown in Fig.1.

If necessary, the GIS application may contact the basic database, which, in turn, is under the management of a IDB subsystem. For this purpose, the GIS application addresses the server side of the IIT, where the basic database is placed under DBMS management. To use data from the basic database, the GIS application provides ODBC links (path, names of the users, passwords, etc.) with DBMS, which are preserved as long as a session of the GIS application runs.

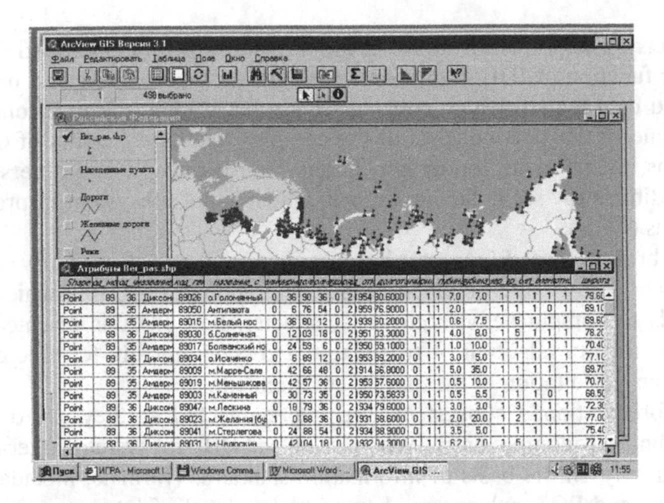

Figure 1. The view of an attributive table and a map in the GIS application.

Data management through the GIS application is carried out by using a special link table. This table represents the nucleus of the standard module, which is called the Navigator and which is used in nearly all GIS applications. The Navigator module and the link table are constructed on basis of the unified classification of marine objects and are applied in all IIT subsystems. An example of such a link table is shown in Table 1.

TABLE 1. An example of a link table in GIS applications.

Name	Level	Ref	Description	Type	Kind	Code
Td	4	td	Cotidal maps	Book	Open	none
Td	4	cotdl200.doc	Method of calculation	Doc	Doc	200
Td	4	cotidal\cotidal.dll	Executable module	Exe	Exe	100
Td	4	dstdc001.bmp	Isoamplitudes of wave M2	Bmp	Bmp	108
Td	4	Tidal	Tidal data	Table	Tbl	109

In this case, spatial data management is provided by internal GIS application tools, using attributive tables. As a result, the data which are available in different forms (factographical, textual, graphic, spatial) and which are placed in the external and internal (relative to GIS) database are jointly used by applying a Navigator link table and an internal attributive table of the GIS application. An example of the interactive tables mentioned is shown in Fig. 2.

A key feature of the POA is the attachment of user programs, which carry out computations of various characteristics and modeling of natural marine processes. A special input and output parameter definition language is developed for the interaction between th e POA shell and the user software. The language includes a number of constructs, allowing parameter specification required for the software to be set up and run. At the same time, the language permits one to describe the user interface of an application (buttons, menu bar, sliders, etc.).

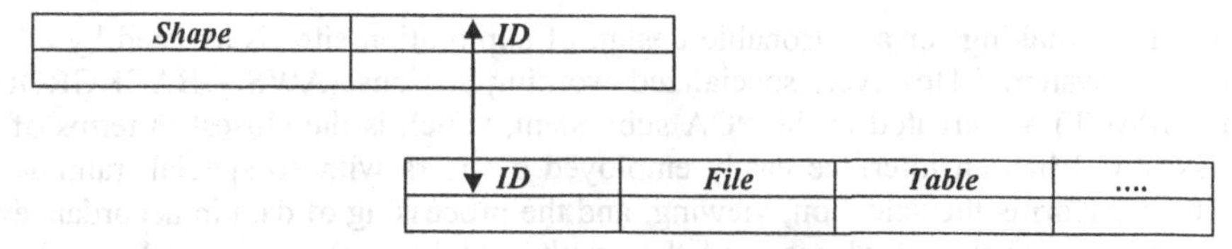

Figure 2. An example of interacting Navigator link table and an internal attributive table of the GIS application.

2. Examples on Realization of the Information Technology

In the last few years, the described approach used in the construction of the integrated information technology has been employed to develop several specialized information systems to support the exploration of petroleum and gas deposits in the Arctic seas [3, 4]. In 1999, the study on the IIT development acquired a large-scale character due to the development of a unified system of information about the World's Oceans. The essential development of IIT components on the basis of Web-technology was also significant. Some examples of the IIT subsystem are described below.

Specialized Information System (SIS-Yamal) is developed in the framework of «Yamal», the project of RISC «Gasprom». It is intended for the collection, accumulation, processing, and the use of data and information on Yamal Peninsula and the adjacent area of the Kara Sea, required for the design, construction, and exploitation of Bovanenkovo and Kharasavei gas condensation fields (GCF).

By now, the operating model of SIS-Yamal has been developed, which can fulfill three tasks:

- formation of a multi-purpose initial information data bank (ADB subsystem);
- operation in request-reply mode (REQUESTS task), providing data and information (IDB subsystem);
- preparation of materials for the design (POA subsystem) of a number of sites on the basis of the «natural limitations» principle by fulfilling BACKGROUND and IMPACT tasks.

The initial information bank of the system contains databases on the following:

- hydrometeorology, geocreology, hydrology of rivers, pollution, and other environmental aspects (the data are held in the form of factographical, textual and spatial data sets);
- technical-technological characteristics of sites: location plans of Bovanenkovo GCF exploration and Baydaratskaya Bay pipeline construction;
- methods and models of computation: description and software modules to compute environmental characteristics;
- legal and regulatory information: description of documents related to the application domain of the system.

Maps at scales of 1:1000000 (small scale), 1:100000 (large scale) and 1:2500 (detailed scale) serve as a topographical base of the geoinformation block. Subject-oriented maps reflecting environmental conditions are related to the unified topographical base.

Decision making for a reasonable design of exploration sites is assisted by all SIS-Yamal subsystems. However, specialized working stations (AWS - BACKGROUND and IMPACT) are created in the POA subsystem, which is the closest in terms of user access. AWS has an interface easily employed by users with no special training. The interface performs the selection, viewing, and the processing of data in accordance with prearranged scenarios, analyses, and the results obtained. Some examples on system operation are given below.

When design decisions are made, it is necessary to make use of the most detailed possible information on current environmental conditions and on characteristics of technical objects in the region studied. The system allows a review of the automated catalogues of the ADB subsystem and a selection of required materials in the form of data files and their descriptions.

IDB and POA subsystems provide improved tools for the user. Figure 3 shows user interfaces of several AWS in these subsystems, providing access to observations and the generalized data, subject-oriented maps describing environmental conditions, and characteristics of engineering decisions related to the GCF exploration and pipeline construction.

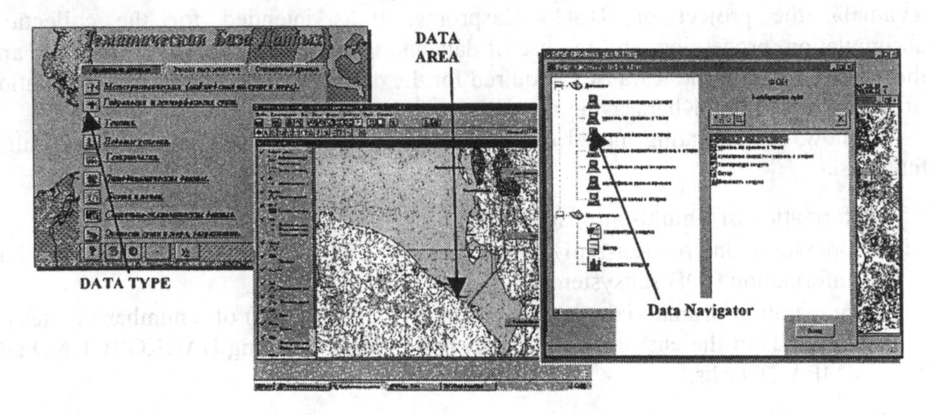

Figure 3. Access to data on environmental conditions.

To estimate an impact of an underwater pipeline on the environment, it is required that various hydrometeorological characteristics (currents, sea level, waves, etc) are known. For these purposes, POA contains a detailed block (arbitrarily called "Hydrodynamics") which combines several analytical modules.

The coastal area of Baydaratskaya Bay is an area of organic matter accumulation. The impact of the pipeline construction on the environment is primarily connected with the probable pollution of Baydaratskaya Bay by technogenic suspended matter, which may occur in the course of construction works and hydrotests of pipelines. The impact of the ground dumping is considered here in more detail. The interface of the corresponding analytical module contains "Windows" to be filled with the following information: fractions of ground composition, type of dredge, amount of dumping, location of intake and dumping point, and other parameters.

After computations, a dynamic picture of the pollution in the bay by ground suspended matter is displayed, and a warning is given when the permissible level of pollution is exceeded. Computations may be repeated for any point of the bay with a change of parameters.

The examples described above cover at the moment only a part of the applied tasks which may be solved by users of the SIS-Yamal.

IIT methodology is also tested in the Black Sea region through the construction of IAS "Blue Stream". The model of the POA subsystem of IAS, related to access and visualization of marine environmental data, has been created in GIS media, using high-level algorithmic languages. It includes a database, a special software environment developed for the model, and a standard software GIS.

The test database consists of spatial and factographical data sets. Space-oriented data are represented by topographical and bathymetry maps (hydrometeorology, geocreology and geomorphology, biota, and others) of the Black Sea and adjacent land areas. Factographical data include temperature and salinity climate sets, as well as three components of current vectors for each month of the year.

To provide a quick access to data, data visualization, processing and analysis, as well as the computation and visualization of additional environmental characteristics, several GIS applications are realized within the IAS model.

These applications contain two groups of modules:

1) modules extending standard GIS possibilities, as far as map visualization and attributive information on the maps are concerned;
2) modules for the acquisition of new information in GIS media via factographical data processing in a mode set up by the user and the subsequent visualization of results.

The following modules are included in the first group:

- module for quick positioning of a map;
- module for construction of explanatory legend for a map or a diagram;
- module for construction of a scale rule in a map window;
- module for construction of a grid for selected area of a map;
- module for search and representation of factographical data.

The second group includes the following modules:

- module for computation and visualization of horizontal water temperature and salinity fields;
- module for computation and visualization of vertical profiles and time series of water temperature and salinity;
- module for computation and visualization of horizontal current vectors;
- module for computation and visualization of trajectories of tracers released from pressing grid points at standard horizons;
- module for computation of spatial and temporal statistical characteristics by factographical data sets.

Figure 4 shows a fragment of the IAS model study. The user can select one of the existing map views, form a new view, or delete any view. He/she can see the subject-oriented maps of water temperature and current vectors at the Black Sea surface, constructed by program modules on the basis of three-dimensional sets of corresponding factographical and modeling data. Sets of three components of current vectors are obtained with the help of a nonlinear barocline model of temperature, salinity and hydrodynamic representation of current fields [5].

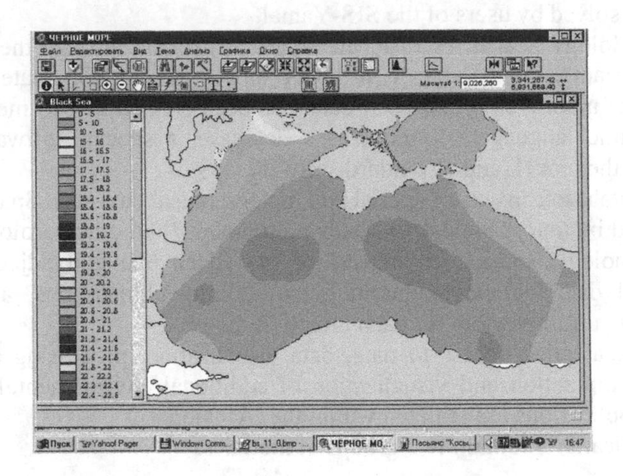

Figure 4. The view of the thematic map of the climatic temperature field for Black Sea in the GIS application.

To assess the distribution of pollutants in the Black Sea, the module for the computation and visualization of Lagrangian tracer trajectories is created. This module has an interface where the following parameters are specified: initial position of a tracer (selected right on the map), current field used, horizon, duration of movement, and discreteness of fixing successive positions of a tracer. These trajectories illustrate the climatic regime of the distribution of conservative pollutants in the Black Sea.

The presented fragments of IAS "Blue Stream" model reflect the opportunities provided for the acquisition of background characteristics of the marine environment and characteristics of processes, defining the technogenic impact on the environment.

The architecture of another version of the application is created on the basis of the local network Windows NT, which also uses IIT methodology. This version of the application is realized as a client application IDB, which is created in the media of DBMS Oracle for access, visualization, and miscellaneous processing of marine data. It is designed in the Delphi 3/4 media with the application of the cartographical module MapObjects and modules written in other object oriented programming languages. The general performances are the multi-component independent software product with a remote main database under the control of DBMS Oracle8i and its independent replicate fragment in Oracle Lite.

MapObjects represents a set of mapping tools and the technology of geoinformation systems (GIS). It includes control elements for linking and implantation of objects (OLE Control) into user programs. MapObjects is based on standard OLE 2.0 (Microsoft's Object Linking and Embedding). It includes more than 30 programmable

OLE-objects (programmable OLE Automation objects). It provides the developers of the applications with modern resources of mapping and GIS. MapObjects allows one to use shapefiles, layers of spatial data SDE, ARC/INFO, and snapshots, including coordinate linking in formats of tif, bmp, and many others. The link from the DBMS is carried out through standard drivers of Microsoft's ODBC or on the basis of BDE (Borland Data Engine).

In applications created with MapObjects, it is possible to fulfill a set of operations, i.e., the display of spatial data stored in formats of shapefiles or in various raster formats, and so on. The operation of the application is carried out under the control of the system shell, which provides the operation of separate models called from the multiwindow menu and/or based on internal program links. The data search is made on the basis of sampling criteria, preset by the user, and is presented in a tabulated form with boundaries of research area shown on a map.

For example, for the data on HMS, it is possible to set the following list of sampling criteria:

1) station number (list denoted with a title, with the choice of one or several strings(lines));
2) title (list with a choice of one or several strings);
3) min. latitude (degrees, fraction of degree);
4) max. latitude (degrees, fraction of degree);
5) min. longitude (degrees, fraction of degree);
6) max. longitude (degrees, fraction of degree);
7) year, initial;
8) year, end;
9) month, initial;
10) month, end;
11) parameter or a group of parameters (list with the choice of one or several strings).

All data from DBMS are transferred in the application as fixed tabulated View-performances. For example, for coastal data, the following table is used:

prop. COAST_OBS_V
STATION NUMBER (5) - STATION NUMBER
TIME NUMBER (2) - HOUR
START_TIME DATE - DATE
VISIBLE NUMBER (5) - VISIBILITY
WIND_DIRECTION NUMBER (3) - DIRECTION of a WIND
WIND_SPEED NUMBER (2) - WIND SPEED
WIND_SPEED_MAX NUMBER (2) - MAX OF WIND SPEED
WATER_TEMPERATURE NUMBER (3,1) - TEMPERATURE of WATER
AIR_TEMPERATURE NUMBER (3,1) - TEMPERATURE of AIR
PRESSURE NUMBER (6,1) - PRESSURE

After obtaining data from View-performances, further management is carried out by the application. This management is based on a table of correspondence, which uses codes applied to DBMS.

To gain the necessary experience for the development of the POA subsystem, design and implementation of the test variant of the GIS application are accomplished. This application allows an access to and a visualization of the marine environmental data (Fig.5).

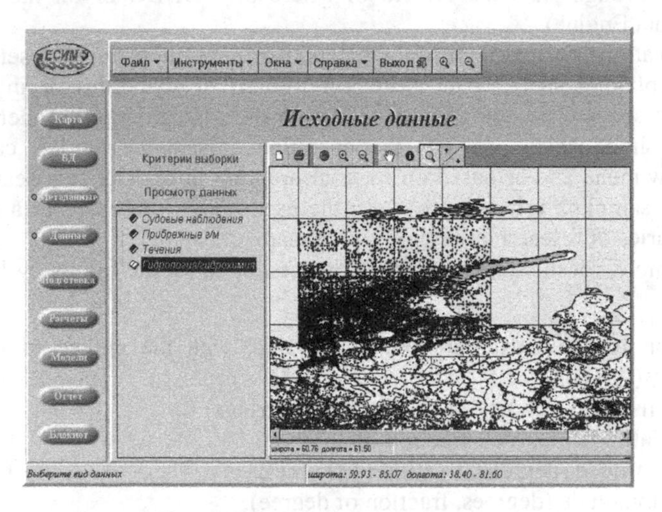

Figure 5. The use of the POA for processing of marine data.

The practical realization of the described technology in Russian NODC will make it possible to meet requirements of modern users, concerning the data content, speed of access to data, data processing and analysis, and the form of presentation of output products.

3. References

1. Fedra K. (1997) Integrated environmental information systems: from data to information in (N. Harmancioglu et.al. (eds.), *Integrated Approach to Environmental Data Management Systems*, Kluwer Academic Publisher, Dordrecht, 367-378.
2. Pospelov G.S. (1983) Artificial intellect: modern information environment -Vestnik AN USSR, N6: 31-42.
3. Odisharia G.E., Tsvetsinsky A.S., Mikhailov N.N., and Dubikov G.I. (1997) Specialized Information System on Environment of Yamal Peninsula and Baydaratskaya Bay, *Proc. Int. Offshore and Polar Eng. Conf. (ISOPE-97)*, Honolulu, USA, Vol. 1, pp 574-581.
4. Odisharia G.E., Shershneva L.V., Tsvetsinsky A.S., Mikhailov N.N., Arkhipov B.V., Vorontsov A.A., Batalkina S.A., and Tuzhilkin V.S. (1998) On application of GIS technologies for design of Jamal gas production and transportation sites, *ArcReview* **4(7)**, 12-13.
5. Trukchev, D, Kosarev, A, and Tuzhilkin, V. (1995) Specific features of the Black Sea seasonal climatic circulation: Part I. Variability of the upper layer circulation, *Comtes rendus de l'Academie Bulgare des Sciences* **48(8)**, 21-24.

THE USE OF SATELLITE REMOTE SENSING DATA IN NUMERICAL MODELING OF THE NORTH PACIFIC CIRCULATION

V. KUZIN, V. MOISEEV and A. MARTYNOV
Institute of Computational Mathematics and Mathematical Geophysics
Siberian Division of Russian Academy of Sciences
Novosibirsk, Russian Federation

Abstract. A numerical model[1] constructed at the Institute of Computational Mathematics and Mathematical Geophysics (the former Novosibirsk Computing Center), based on the finite element method, has been used for investigation of the sensitivity of the Pacific Ocean circulation to a climatic and satellite derived wind forcing. The model grid covers the region between 30°S and 60°N at a space resolution of 2° in longitude and 1° in latitude with 18 levels in depth. The model includes a block of vertical mixture in the upper layer. Diagnostic experiments were carried out for April and October, 1994. These two months were selected for modeling different states of ocean circulation with weak wind in April (switch from winter monsoon to summer monsoon) and strong wind in October (switch from summer monsoon to winter monsoon). In addition, the short-range prognostic experiments for both climatic and satellite data were carried out for a 3-year period with a seasonal cycle.

Analysis of the results showed that the general circulation of the Pacific Ocean did not essentially change, depending on the different data used. Nevertheless, a more complicated current structure was obtained with the use of satellite data. At the same time, the main boundary currents were more intensive, and the transport volume estimates indicated higher values when climatological data were employed.

1. Introduction

In order to understand and describe the interannual variability of ocean circulation processes by the use of numerical models, one has to consider the quality of various sources of surface wind as one of the important forcing parameters, as well as the sensitivity of the model to the variation of this forcing. A comparison between different surface wind stress products shows that the application of modern satellite data may correct or improve climatologies traditionally used by modelers and may produce more realistic ocean circulation estimates [4, 5].

There are several surface wind products which may be used to force the ocean circulation models:

- Hellerman-Rosenstein climatology [1];
- Comprehensive Ocean-Atmosphere Data Set (COADS);
- European Centre for Medium-Range Weather Forecast (ECMWF);
- Goddard Earth Observing System (GEOS);
- An atlas of monthly mean distribution of SSMI surface wind speed; AVHRR/2 sea surface temperature; AMI surface wind velocity, and, TOPEX/POSEIDON sea surface height during 1987 - 1997, etc.

[1] Supported by the Russian Foundation for Basic Research.

In this paper, the Finite Element Ocean Circulation Model (FEOCM) of the Novosibirsk Computing Center was applied for the comparison of two sources:

- monthly averaged wind stress data of Hellerman and Rosenstein [1]; and,
- monthly mean AMI surface wind speed [2].

In previous studies, this model was used for the reconstruction of the 3D velocity fields in the North Pacific, employing the climatic data [6, 7, 8]. In the present study, the first stage of numerical experiments includes the diagnostic calculations for two months (April and October, 1994) with the use of climatic and satellite wind stress. At the next stage of the study, short-range prognostic experiments were carried out on the seasonal spin-up of the model with climatic forcing for a 3-year period. Next, the calculation of 2.5-year seasonal cycles was realized with satellite derived wind forcing.

2. Ocean Circulation Model

The system of equations for calculation of the 3-D fields of velocity, temperature and salinity can be written in the form of:

$$\frac{dU}{dt}+\left(f-\delta\right)\vec{k}\times U=-\frac{1}{\rho_0}\nabla P+\frac{\partial}{\partial z}v\frac{\partial U}{dz}+\vec{F} \tag{1}$$

$$divU+\frac{\partial w}{\partial z}=0,\quad \frac{\partial P}{\partial z}=g\,\rho,\quad \rho=\rho\left(T,S\right) \tag{2}$$

$$\frac{d\left(T,S\right)}{dt}=\frac{\partial}{\partial z}\kappa\frac{\partial\left(T,S\right)}{\partial z}+\nabla\mu\nabla\left(T,S\right) \tag{3}$$

Equation (1) through Eq. (3) are given in the coordinates (λ, θ, z) on a sphere of radius α, where λ is longitude; $\theta = \varphi + \pi / 2$; φ, latitude; z, the vertical coordinate with the positive direction from the surface toward the center of the Earth; $U = (u, v)$, the vector of horizontal velocity components; w, the vertical velocity component; m = $1/\alpha \sin\theta$; n = $1/\alpha$; $\alpha = 6.38 \times 10^8$ cm; f = $-2\omega\cos\theta$, the Coriolis parameter; $\omega = 0.73 \times 10^{-4}$, the angular speed of the Earth's rotation; ρ_0 = const, the standard density; ρ, density; P, pressure; ν and κ, the vertical eddy viscosity and diffusivity coefficients, respectively; μ, the horizontal diffusivity coefficient; ∇, the spherical horizontal gradient operator; T, temperature (°C); S, salinity (°/₀₀); k, a unit vector along z-direction; and $\delta = m\cos\theta\, u$, with:

$$\frac{d\varphi}{dt}=\frac{\partial\varphi}{\partial t}+mu\frac{\partial\varphi}{\partial\lambda}+nv\frac{\partial\varphi}{\partial\theta}+\frac{\partial\varphi}{\partial z} \tag{4}$$

$$\nabla P=\left(m\frac{\partial P}{\partial\lambda},n\frac{\partial P}{\partial\theta}\right),\quad divU=m\left(\frac{\partial u}{\partial\lambda}+\frac{\partial}{\partial\theta}\frac{n}{m}v\right) \tag{5}$$

$$F=A_1\left(m\Delta U+\left(n^2-m^2\cos^2\theta\right)U-2m^2\cos\theta\cdot\vec{k}\times\frac{\partial U}{\partial\lambda}\right) \tag{6}$$

where A_l is the horizontal eddy viscosity coefficient, and:

$$\Delta\varphi = \frac{\partial}{\partial\lambda} m \frac{\partial}{\partial\lambda} + \frac{\partial}{\partial\theta} \frac{n^2}{m} \frac{\partial\varphi}{\partial\theta}, \quad \frac{\partial\varphi}{\partial t} = \frac{\partial\varphi}{\partial t} + mu\frac{\partial\varphi}{\partial\lambda} + nv\frac{\partial\varphi}{\partial\theta} + w\frac{\partial\varphi}{\partial z} \tag{7}$$

The boundary conditions for Eq. (1) through Eq. (3) are as follows:

at the surface:

$$z = 0: \quad w = 0, \quad \mathrm{v}\frac{\partial U}{\partial z} = -\frac{\vec{\tau}}{\rho_0}, \quad (\mathrm{T},\mathrm{S})) = \left(T^0, S^0\right) \tag{8}$$

at the bottom:

$$z = H(\lambda,\theta): \quad w = U \cdot \nabla H, \quad \nu\frac{\partial U}{\partial z} = -R\overline{U}$$

$$\overline{U} = \frac{1}{H}\int_0^H U dz, \quad \frac{\partial(T,S)}{\partial z} = 0 \tag{9}$$

at the cylindrical lateral boundaries $\Gamma = \Gamma_0 \cup \Gamma_1$:

a) "solid" boundary:

$$\Gamma_0: \quad \frac{\partial U \cdot \vec{l}}{\partial n} = 0, \quad \vec{U} \cdot \vec{n} = 0, \quad \frac{\partial(T,S)}{\partial n} = 0 \tag{10}$$

b) "liquid" boundary:

$$\Gamma_1: \ U = U^0, \ \ (T,S) = \left(T^0, S^0\right) \tag{11}$$

In Eq. (8) through Eq. (11), $\vec{\tau}$ is a wind-stress vector; R, a bottom drag coefficient; and $\vec{l}, \vec{n}$ are the tangent and normal unit vectors to the lateral boundary Γ, respectively. The index (^{o}) marks the values that are specified. At the initial time step, the values, u^o, v^o, T^o, S^o are prescribed. The system of equations (1) - (3) is solved by the finite element technique on the staggered E-grid [9]. Equation (3) is solved by splitting [10] along the directions λ_1, θ_1, where $\left(\lambda,\hat{\lambda}_l\right) = \left(\theta,\hat{\theta}_l\right) = \pi/4$.

The principal features of the model are as follows:

- separation of the external and internal modes;
- transformation of advective terms from the gradient form to some special divergent form;
- splitting of equations with respect to the physical processes;
- finite element method (FEM) discretization with respect to space;
- splitting of the multi-point FEM grid operators into a series of three-point operators;
- using implicit and semi-implicit schemes with respect to time.

3. Data Sources and Analysis

The present study uses the climatological monthly mean temperature and salinity fields of the "World Ocean Atlas 1994" [11] (climatological or the Levitus data), the climatological monthly mean wind stress by Hellerman - Rosenstein [1] (climatological or the Hellerman data), and satellite derived data from "Atlas of monthly mean distribution of SSMI surface wind speed, AVHRR/2 sea surface temperature, AMI surface wind velocity, and TOPEX/POSEIDON sea surface height during 1994" [2] (satellite or the Halpern data). The satellite data were obtained from the FTP site ftp://poddac.jpl.nasa.gov/pub.

To use satellite data in numerical modeling of the general circulation of the Pacific Ocean, the wind satellite data of Halpern for April and October 1994 were selected for the diagnostic experiment. A weaker wind distribution is typical in April (switch from the winter monsoon to the summer monsoon), and a stronger wind is observed in October (switch from the summer monsoon to the winter monsoon).

The short-range prognostic computations were successively realized for the 5.5-year period with seasonal cycles, using the Hellerman and the Halpern wind products to force the model. The seasonal average of the satellite data was obtained, using an approach analogous to that applied by Levitus in his Atlas [11] for: Winter (February, March, April); Spring (May, June, July); Summer (August, September, October); and Autumn (November, December, January).

The following calculation of the wind stress was realized, using the ERS-1 wind vectors on the basis of the method presented in [1]:

$$\tau^x = \rho\, C\, D^u\, (u^2 + v^2)^{1/2} \quad , \tau^y = \rho\, C\, D^v\, (u^2 + v^2)^{1/2}$$

$$CD = 0.934 \cdot 10^{-3} + 0.788 \cdot 10^{-4}\, M + 0.868 \cdot 10^{-4}\, \Delta T - 0.616 \cdot 10^{-6}\, M^2 - 0.120 \cdot 10^{-5}\, (\Delta T)^2 - 0.214 \cdot 10^{-5}\, M \Delta T$$

where $M = (u^2 + v^2)^{1/2}$; $\Delta T = T_a - T_s$; u and v, the components of the ERS-1 surface wind vectors in cm/s; T_a , the air temperature near the sea surface; and T_s is the sea surface temperature. As there were no data for T_a , it was assumed to be equal to T_s.

The climatological subsurface temperature and salinity data were used as initial values for all the experiments. Both climatological and satellite data were employed on the surface in different simulations.

A comparison of climatological data with the satellite derived data results in the following:

Wind-stress data. The study of the wind-stress data showed an essential structural difference, as well as a difference in the maximum and the mean values. A significant difference in the whole general wind field was observed at 30° N and south of Japan in April (Figs. 1a and 1b) with the maximum: 0.164 N (Hellerman) and 0.103 N (Halpern); and the mean values: 0.062 N (Hellerman) and 0.041 N (Halpern).

The satellite wind-stress data were rather different from the climatological data in the sea of the Kuril Islands and the sea of Kamchatka peninsula in October. Two large cyclones with centers at about 37° N, 175° E and 35° N, 135° W were also observed in these data (Figs. 1c and 1d) with the maximum: 0.185 N (Hellerman) and 0.092 N (Halpern) and the mean values: 0.063 N (Hellerman) and 0.038 N (Halpern).

It should be noted that climatological wind-stress data usually have higher values in comparison with satellite data. At the same time, the satellite derived data are more complicated and have finer structural features.

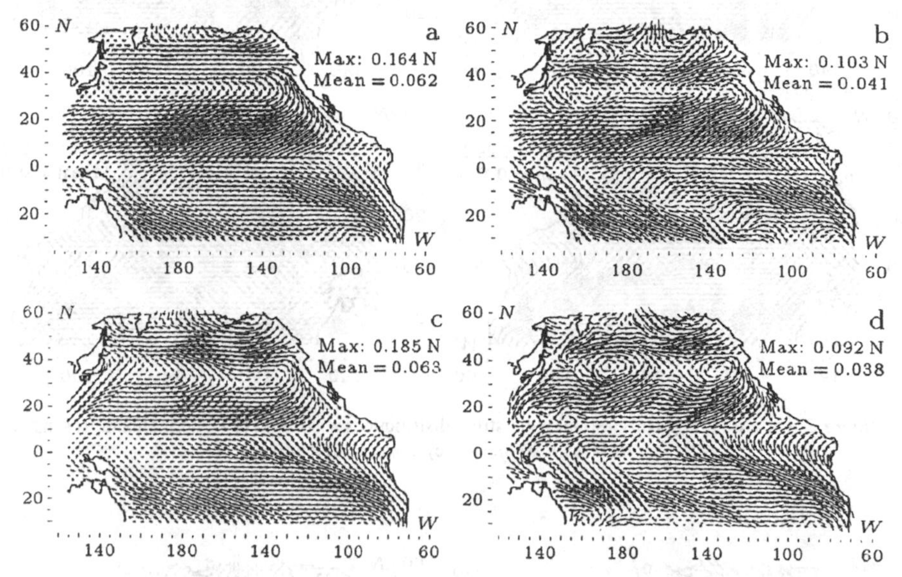

Figure 1. The monthly mean wind stress distributions in N for April (upper figures) and October (lower figures) according to Hellerman and Rosenstein (a and c) and Halpern (b and d).

The seasonal wind-stress fields were calculated for the prognostic experiments, using both the climatological (Fig. 2) and satellite derived data (Fig. 3). These fields are rather different in structure. As a rule, they are spatially smoother for climatological distributions. Both stresses are more intensive during *winter* and *autumn* and weaker during *spring* and *summer*.

The discrepancy between climatological and satellite derived data is shown in the seasonal zonally averaged curl distribution in Figure 4. The satellite zonally averaged curl is weaker for all the seasons except for spring at high latitudes (50 - 60° N). It is comparable to climatology for *spring*, *summer* and *autumn* in the westerlies. The summer curl is stronger in tropics, whereas the winter one is weaker. There is a fluctuation in the tropical extremum position relative to the equator. It shifts northward from winter to summer. The same feature is observed for the other extremum located to the south of the previous one in the curve. It is better developed in *summer* with the location to the north of the equator. Both satellite and climatological curls show similarity at the south of the equator, where the curl maximum is observed in *spring*.

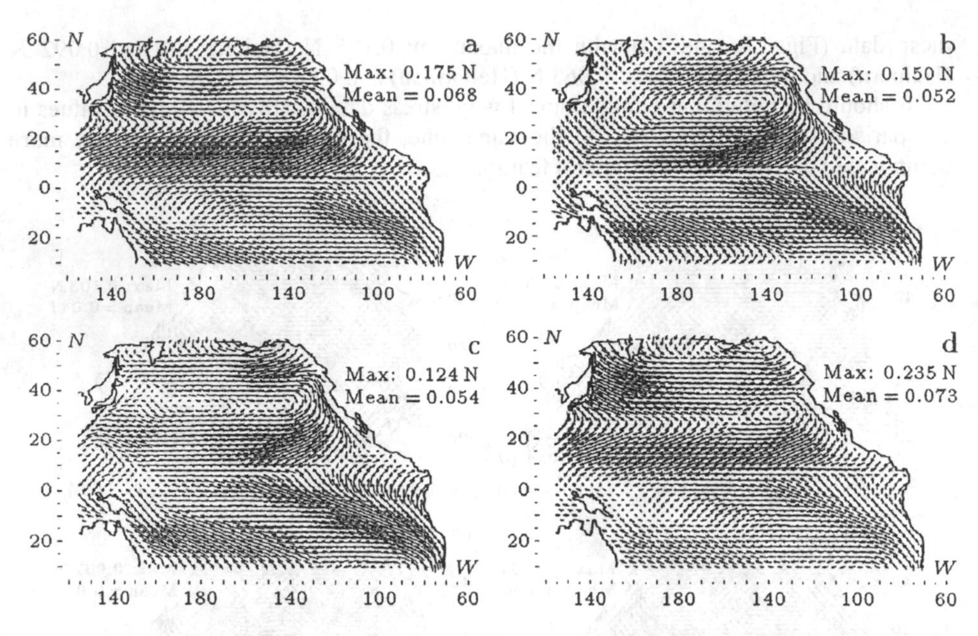

Figure 2. The climatological seasonal wind stress distributions according to Hellerman and Rosenstein: a) *Winter*; b) *Spring*; c) *Summer*; d) *Autumn*.

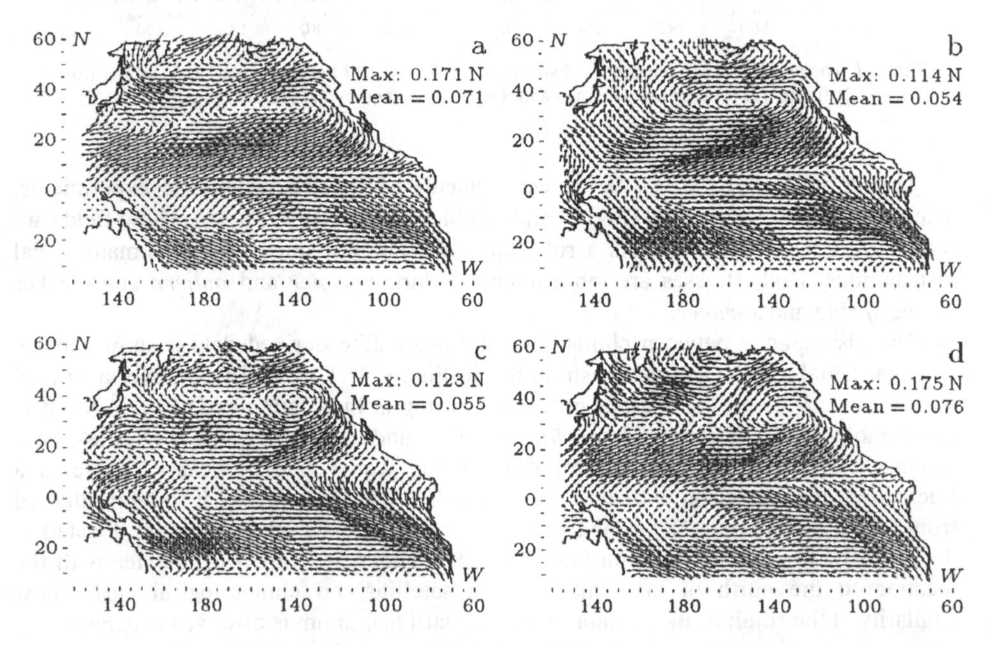

Figure 3. The satellite seasonal wind stress distributions according to Halpern: a) *Winter*; b) *Spring*; c) *Summer*; d) *Autumn*.

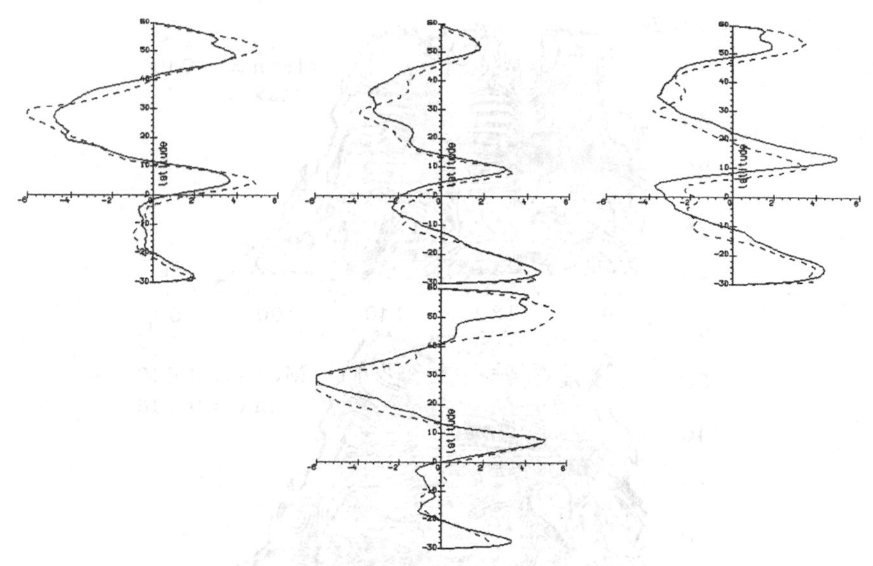

Figure 4. Seasonal zonally averaged curl distribution derived from Hellerman and Rosenstein climatology (the dashed line), and from Halpern (solid line): *Winter, Spring, Summer, Autumn.*

4. Numerical Experiments

To test the satellite data and compare the results obtained with the climatological data and the satellite data, the following experiments were carried out:

- Ia. diagnostic experiment with climatological data for April, 1994;
- Ib. the same with satellite data for April, 1994;
- IIa. diagnostic experiment with climatological data for October;
- IIb. the same with satellite data for October, 1994;
- IIIa. short-range prognostic experiment with climatological data (3 years);
- IIIb. short-range prognostic experiment with satellite data for the seasonal cycle of 1994 (2.5 years).

The diffusivity and viscosity coefficients in the experiments are as follows:

the horizontal viscosity:	$A_l = 2 \times 10^6$ cm^2/s;
the vertical viscosity:	$\nu = 10$ cm^2/s;
the horizontal diffusivity:	$\mu = 2 \times 10^6$ cm^2/s;
the vertical diffusivity:	$\kappa = 1$ cm^2/s.
the bottom drag coefficient:	$R = 10^{-7} s^{-1}$.

Experiments Ia and Ib. All diagnostic experiments started with the state of rest and were run for a period of 60 days. According to the kinetic energy distribution, the solution attained the equilibrium state both in the whole domain and in the selected regions.

The simulated current fields for April and October, both for climatological and satellite wind forcing, are shown in Fig. 5. The southern part of the model domain is deleted in these figures because the diagnostic calculations do not adequately reproduce the circulation in the vicinity of the equator.

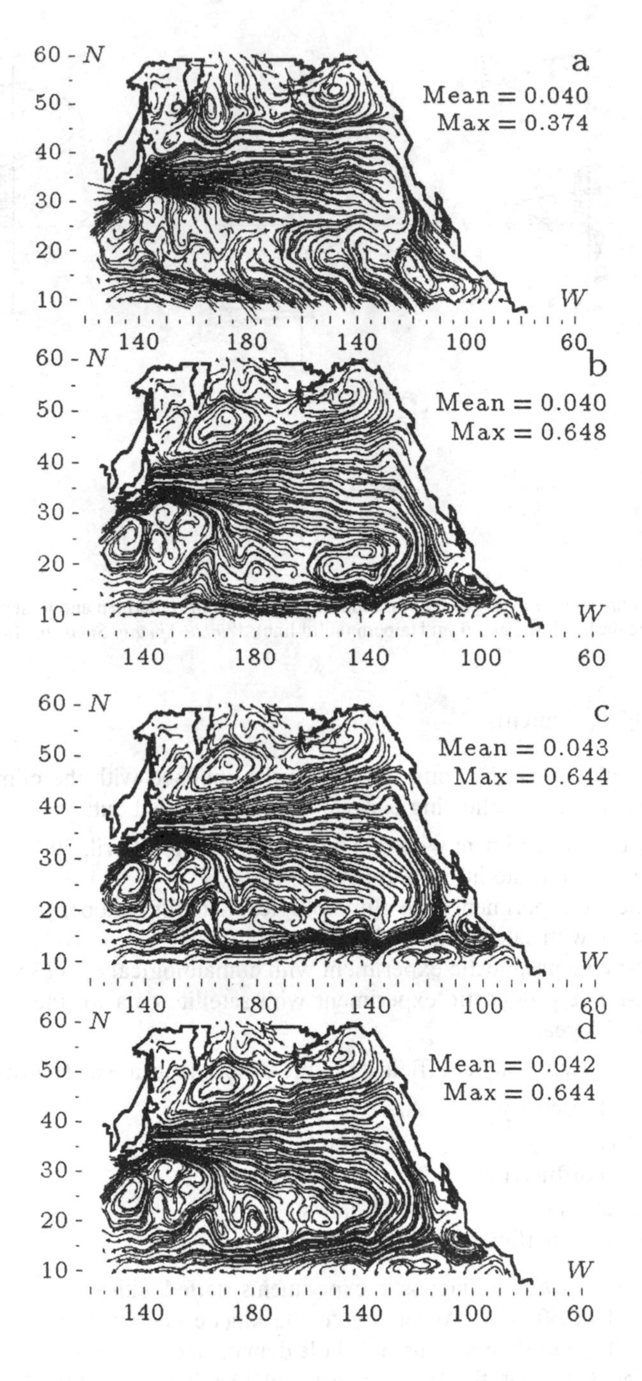

Figure 5. The simulated current velocities at the depth Z=50 m for April (upper figures) and October (lower figures) according to Hellerman and Rosenstein (a and c) and Halpern (b and d).

The simulated currents in Figs. 5a and 5b have reproduced all the main currents of the North Pacific circulation. The Kuroshio and the Oyashio, as well as the West Wind Drift, are more compact in Experiment Ib where satellite data are used. Two anticyclonic eddies to the right side of the Kuroshio stream correspond to a well-developed wind vorticity to south of Japan in Fig. 5b. Such eddies are observed in about the same positions as those in the observations. The Alaska current and the California current seem to be stronger and are better recognized in Experiment Ia which uses climatological data. It is well known that the eastern boundary currents are tightly connected with the dominant winds in this region.

The stream function distribution (Figs. 6a and 6b) shows a general similarity although a subpolar gyre is much better developed in Experiment Ia; and the difference in the maximum values of the integrated mass transport is fairly large: Max = 85 Sv (Hellerman) and 48 Sv (Halpern); Min = -90 Sv (Hellerman) and -70 Sv (Halpern). These differences in maxima are caused by a more intensive wind-stress field in the climatological data.

Figures 6c and 6d show the distribution of the vertical velocity sign at a depth of 50 m. A solid region of upwelling is observed along the western coast of North America; it is better developed in Experiment Ia in comparison with Experiment Ib. However, in the northeastern part of the Pacific Ocean and south of Japan (nearby the Idzu-Ogasavara Ridge), the region covered with upwelling is bigger in Experiment Ib. All these features are derived by the wind structure above the North Pacific.

Experiments IIa and IIb. Figures 5c and 5d show the horizontal velocity field at the level of 50 m. Differences in results obtained with the climatological and the satellite data are less than those in Experiments Ia and Ib.

Figures 7a and 7b for the stream function show the same tendency as in Experiments Ia and Ib with the maximum values of volume transport: Max = 78 Sv (Hellerman) and 48 Sv (Halpern); Min = -50 Sv (Hellerman) and -50 Sv (Halpern).

The structure of the integral circulation in the medium and high latitudes for different kinds of data is changed insignificantly. Implementation of satellite data has led to a less regular circulation nearby the equator.

It should be noted, nevertheless, that the obtained results for October show larger regions covered with upwelling (Figs. 7c and 7d) in the subtropical region in comparison with similar results for April (for both kinds of data).

Experiments IIIa and IIIb. The next stage of the simulation was devoted to the study of the model response to two kinds of boundary conditions varying with respect to time. The monthly averaged climatological wind-stress data were averaged to seasonal ones, and a linear interpolation was made between them. At first, a 3-year spin-up was carried out on the basis of the climatological data (Experiment IIIa). The quasi-flux conditions calculated on the basis of the Levitus data were set up for temperature and salinity on the surface. Then, satellite seasonal wind-stresses were used for the wind forcing. The calculations were continued for the 2.5-year period (Experiment IIIb).

The results of simulation for Experiments IIIa and IIIb are presented in Figs. 8 - 12. The integral stream function has qualitatively the same structure as in diagnostic Experiments Ia,b and IIa,b. However, as it is mentioned above, the differences in wind-stress amplitudes lead to essential deviations in the values of the mass transport in some zones. The deviations of mass transport values calculated with the use of the satellite data after the 2-year cycle from the initial diagnostic values are presented in Fig. 8.

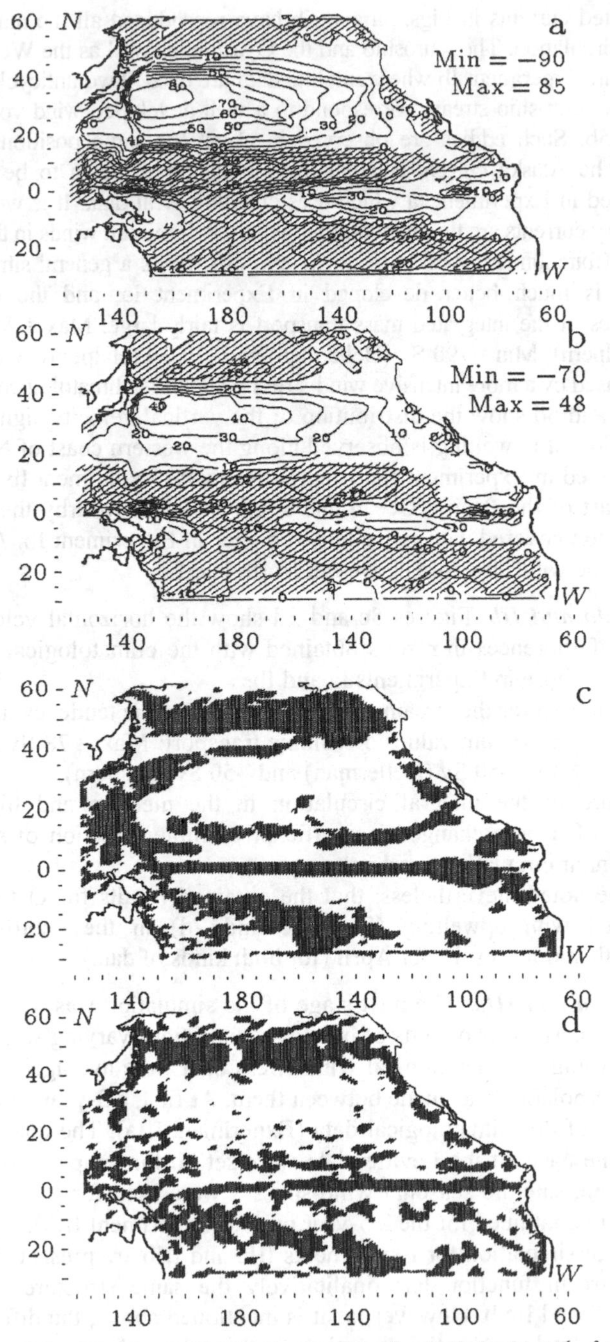

Figure 6. The simulated integral stream function in Sv (1 Sv = 10^6 m^3s^{-1}) (upper figures) and W - component at the depth Z=50 m (lower figures, zones of upwelling are shaded) for April according to Hellerman and Rosenstein (a and c) and Halpern (b and d).

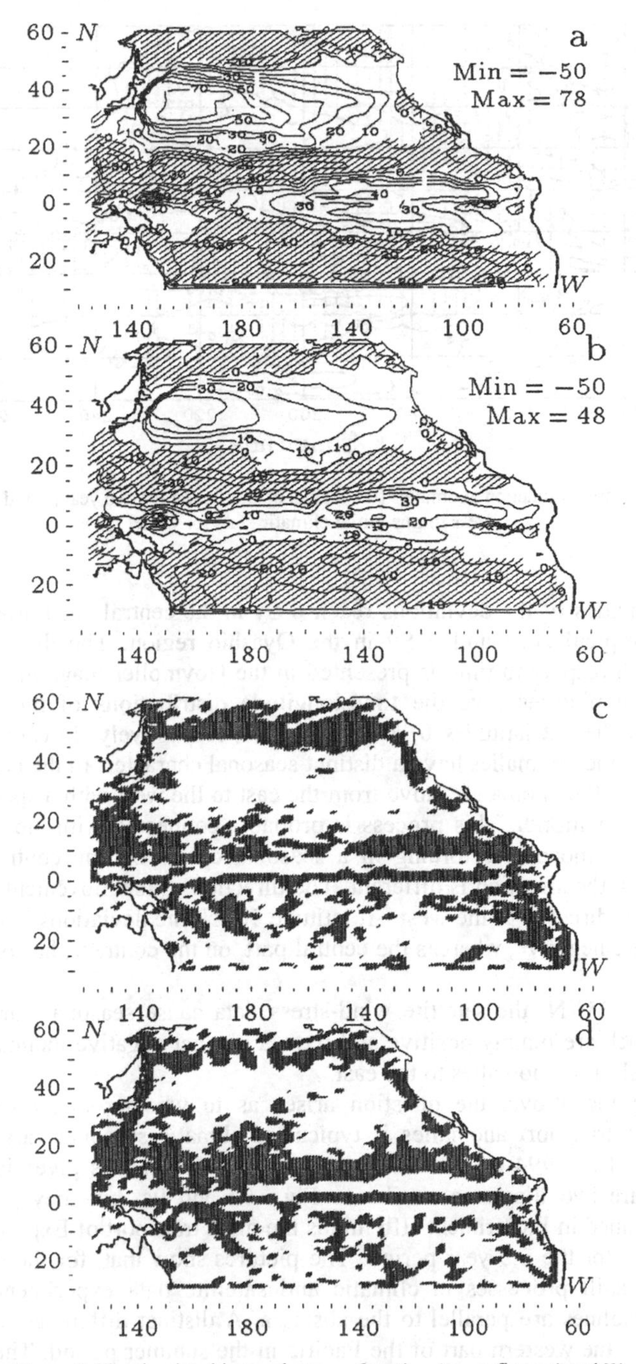

Figure 7. **The simulated integral stream function (upper figures) and W - component at the depth Z=50 m (lower figures, zones of upwelling are shaded) for October according to Hellerman (a and c) and Rosenstein and Halpern (b and d).**

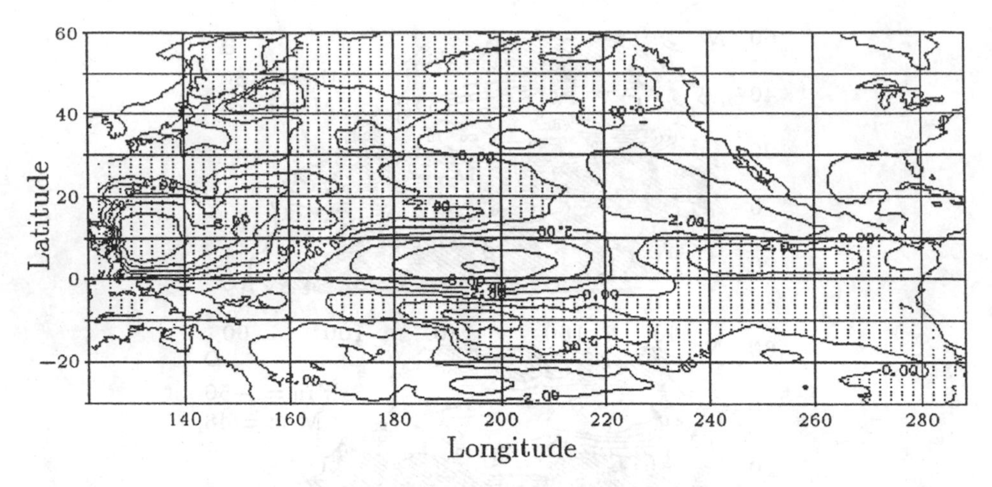

Figure 8. Integral stream function anomalies (the results of simulation for two year period with Halpern wind stress minus initial climatic state for winter).

The amplitudes of the deviations reach 6 Sv in the central equatorial zone, 8 Sv in the eastern tropical zone and 6 Sv in the Oyashio region. The distribution of these anomalies with respect to time is presented in the Hovmoller diagrams in Figs. 9a, 9b, and 9c. These pictures give the time-longitude distributions of the integral stream function anomalies at latitudes 0, 10, and 35° N, respectively. In each zone, the time dependence of the anomalies have a distinct seasonal character. In the central part of the equatorial zone, the anomalies move from the east to the west with a speed of about two thousand km per month. This process is probably connected with the changing of the centers of the atmospheric forcing in a seasonal cycle. On the contrary, at latitudes 10 ° and 35° N, the situation is different. The direction of the movement of anomalies is reversed and is directed to the west. At latitude 10° N, the deviations in the western part of the basin are negative, whereas the central part, on the contrary, has positively signed deviations.

At latitude 35° N, the satellite wind-stress data cause seasonal variations in mass transport, which are mainly positive and have a distinct negative inclination of isolines to testify the shift of anomalies to the east.

Following the above, the question arises as to whether this distribution of the seasonal mass transport anomalies is typical of climatic seasonal variations as in the concrete period of 1994 or not. The answer to this question is given in Figs. 10a and 10b. These are two diagrams similar to Figs. 9a and 9c, but they present the mass transport obtained in Experiment IIIb minus the mass transport of Experiment IIIa in the seasonal cycle for the 2.5-year period. The pictures show that, for the equatorial zone, the anomaly shift processes in climatic and satellite data experiments are identical because the isolines are parallel to the abscissa. A distinct difference in the amplitude can be seen in the western part of the Pacific in the summer period. The reason is that, for this period, there is an essential sharpening of the wind-stress curl in the equatorial region in the climatic data. In 1994, the atmospheric circulation for this period in the equatorial region is much weaker. At latitude 10 ° N, the situation is similar.

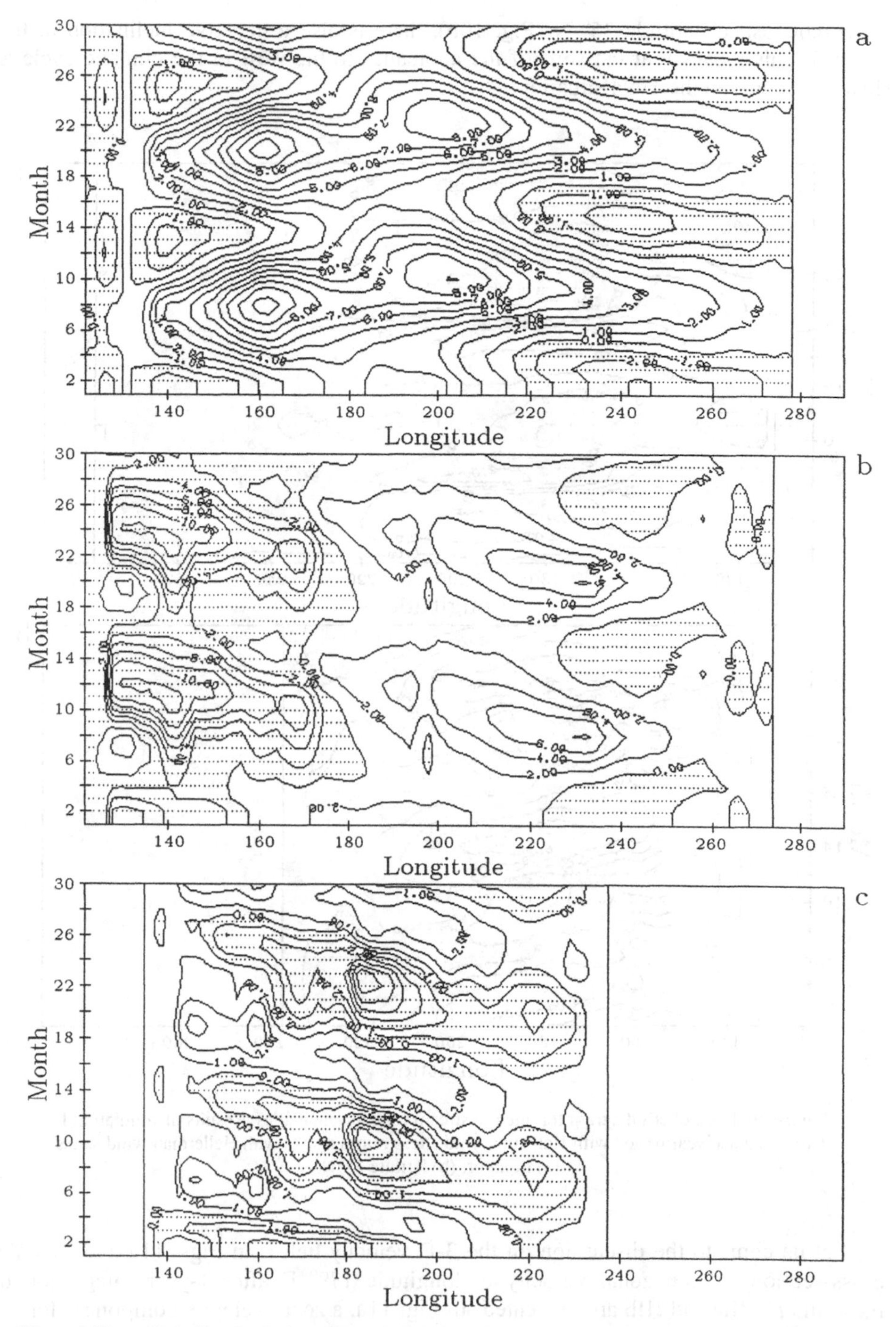

Figure 9. Hovmoller's diagram for integral stream function anomalies: results of simulation for two and a half year period with Halpern's wind stress minus initial climatic state: (a) Equator; (b) latitute 10° N; (c) latitude 35° N.

In contrast, at latitude 35° N (Fig. 10b), there is also a negative inclination of the isolines. This means that the shift of the anomalies to the west in the climatic cycle is slower than that in the 1994 situation.

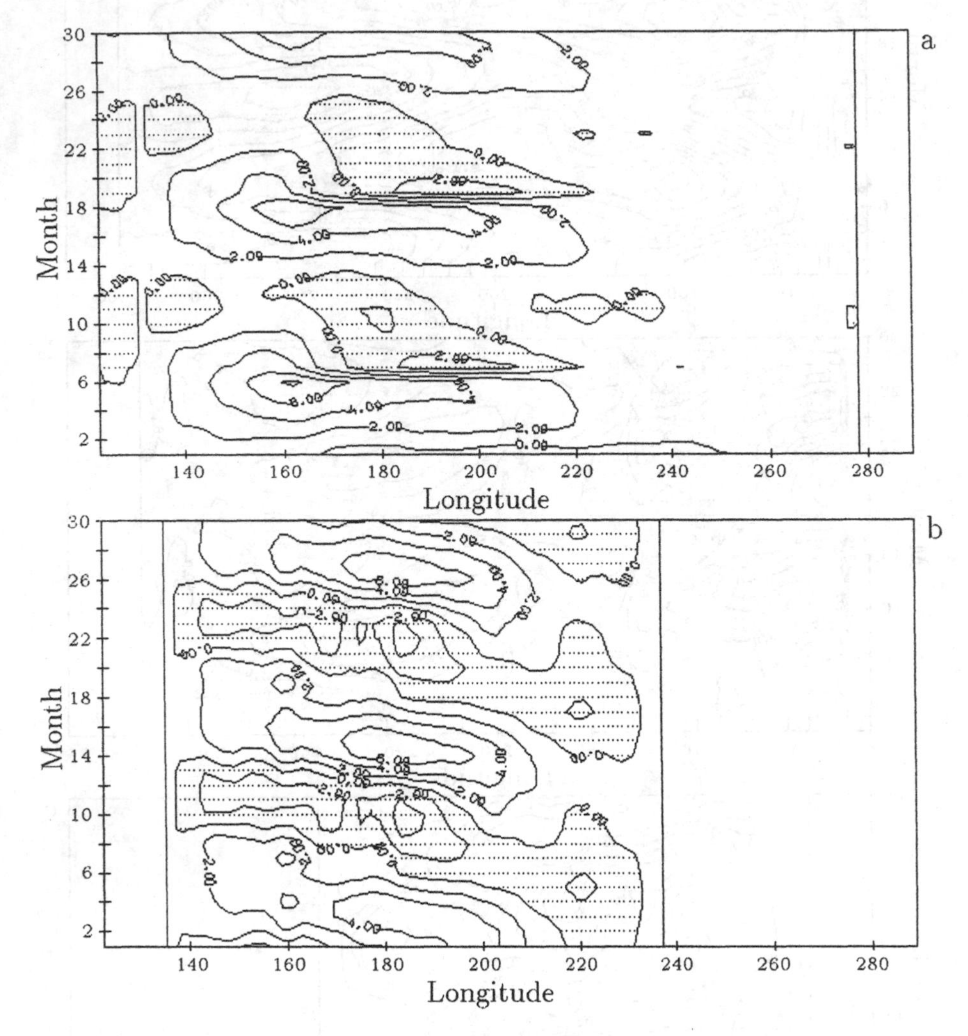

Figure 10. Hovmoller's diagram for integral stream function anomalies. Results of simulation for two and a half year period with Halpern's wind stress minus results with Hellerman wind stress: (a) Equator; (b) latitute 35° N.

Let us come to the discussion on the 3-D velocity fields. In Figs. 11a and 11b, the cross-sections of the zonal velocity at longitude 179° E after 1-year integration of Experiments IIIa and IIIb are presented. In Fig. 11a, a zonal velocity component for the winter season in Experiment IIIa is plotted. The main specific features are observed in the equatorial zone. There is a westward jet near the surface, which reaches a depth of 100 m. Below 150 m, there is an undercurrent which was formed during the prognostic calculation with a climatic seasonal cycle. The center of this jet is located at a depth of 300 m, which is in agreement with observations. However, the value of the undercurrent is not as strong as that in nature and reaches only 25 cm/s. This is typical of numerical models with coarse grids. Below the undercurrent at a depth of about 1000 m, one more less intensive stream exists, which is directed to the west.

Figure 11b presents the same characteristic for numerical Experiment IIIb with satellite wind-stress for 1994. The main features are identical; but the intensity of the equatorial jets is somewhat weaker, which is connected with the less intensive wind-stress in this case. In the tropical, subtropical, and subpolar zones, the main currents are concentrated near the surface, and their signs correspond to the pictures of the horizontal velocity fields.

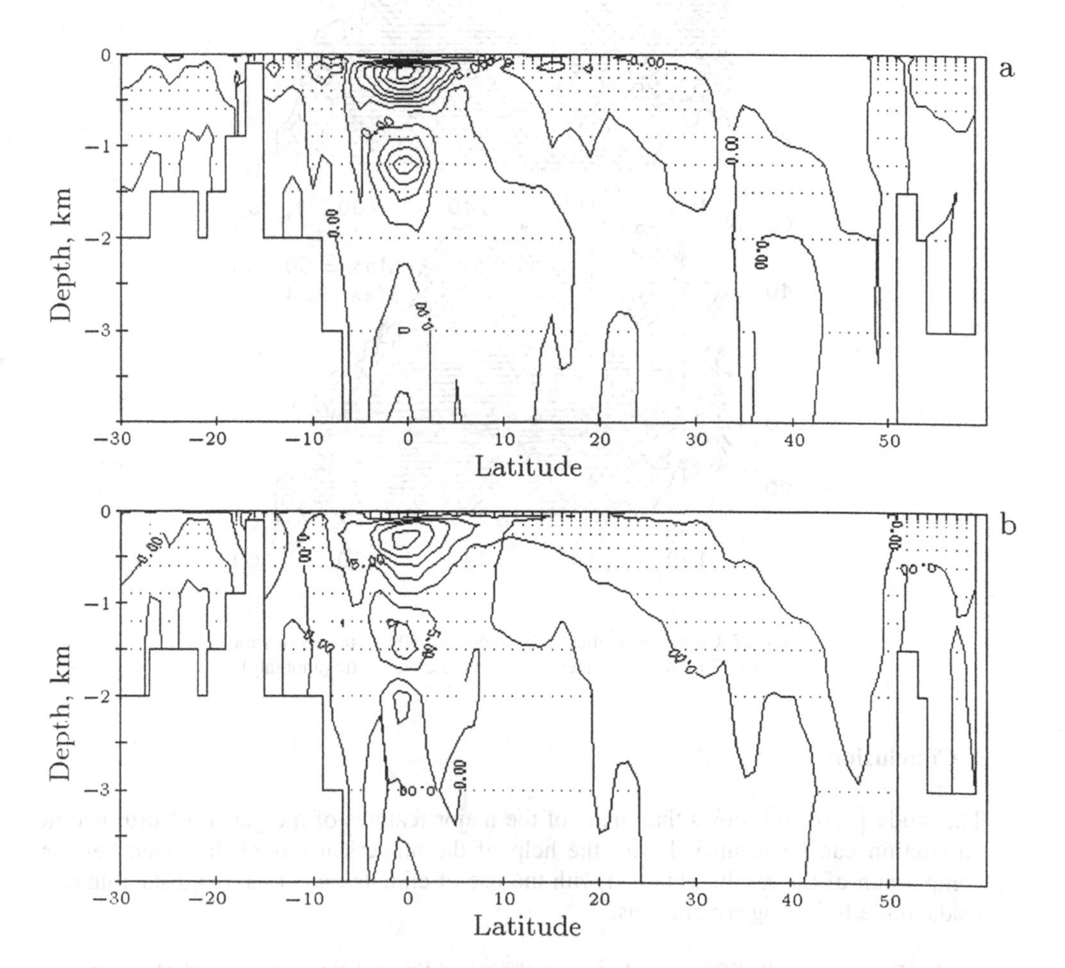

Figure 11. Cross section of the calculated currents U – component: results of simulation for 13 months with (a) Hellerman and (b) Halpern wind stress.

These distributions are presented in Figs. 12a and 12b. Figure 12a shows the horizontal velocity field for Experiment IIIb after the 2-year period of integration at a depth of 100 m. One can see that there also exist the main Pacific Ocean circulation gyres but with weak boundary currents. The maximum velocity values are observed near the equator area. Figure 12b shows the difference of the velocity fields at a depth of 100 m between Experiments IIIb and IIIa. The picture also testifies that the climatic data give a more intensive forcing for the currents than the satellite data.

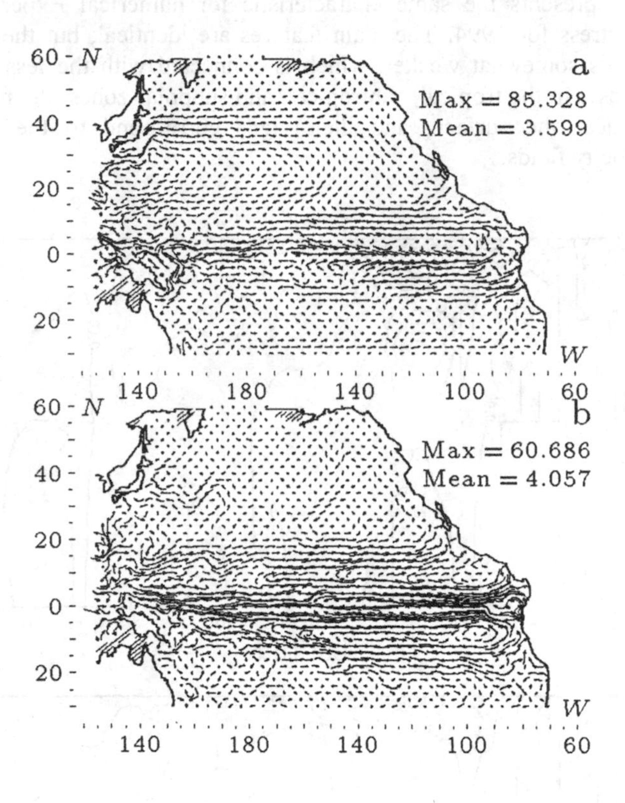

Figure 12. Current velocities at the depth Z = 100 m: results of simulation for 13 months: (a) satellite; (b) satellite – climatic (anomaly).

5. Conclusion

The study presented shows that many of the major features of the general North Pacific circulation can be identified with the help of the numerical model discussed here. A comparison of the results obtained with the use of climatological data and satellite data leads to the following conclusions:

1. Even a small difference between the climatic and the satellite wind-stress data can cause substantial differences in the simulated currents.

2. The use of satellite data has resulted in decreased values of the volume transport although the general circulation has retained its main structural features in all the experiments.
3. Vorticity of the horizontal circulation becomes more intensive with the use of satellite data.
4. Distinct changes are observed in diagnostic results in the vertical structure of the boundary currents, depending on the data used (climatological or satellite); new eddy formations appear in regions with intensive wind vorticity (satellite data).
5. Comparison of the integral stream function distributions in the prognostic experiments shows essential variability of the centers of circulation, both for the climatic and the satellite surface forcing. Seasonal anomalies in mass transport are caused by the atmospheric influence and move to the east or to the west according to the variation of wind anomalies.
6. The 3-D velocity fields obtained in prognostic experiments from the satellite data are much weaker with respect to the horizontal and the vertical directions in comparison with those derived from climatological data.

Accordingly, the use of satellite data in numerical modeling has shown both favorable and unfavorable results. Nevertheless, it is obvious that this new source of data with high resolution in space and time develops new possibilities for research on ocean circulation. The general ocean circulation can be investigated not only in terms of climatic states, but one can also study the interannual variability of the general circulation, as well as the tendency of a current to change in a particular region.

6. Acknowledgements

The authors gratefully acknowledge Dr. David Halpern's useful comments.

7. References

1. Hellerman S. and Rosenstein M. (1983) Normal monthly wind-stress over the World Ocean with error estimates, *J. Phys. Oceanogr.* **13**, 1093 - 1104.
2. Halpern D., Zlotnicki V., Brown O., Freilich M., and Wentz F. (1997) *An Atlas of Monthly Mean Distribution of SSMI Surface Wind Speed, AVHRR/2 Sea Surface Temperature, AMI Surface Wind Velocity, and TOPEX/POSEIDON Sea Surface Height During 1994*, Jet Propulsion Laboratory, JPL Publ. 97-1, Pasadena.
3. Platov G.A. (1989) Modeling of interacted boundary layers of atmosphere and ocean in the Kuroshio region, (Preprint / AN USSR. Sib. Branch. Computing Center; 850), Novosibirsk, (in Russian).
4. Liu W.T. and Tang W. (1993) Atlas R. Sea surface temperature exhibited by an ocean general circulation model in response to wind forcing derived from satellite data, in I.S.F. Jones, Y. Sugimori, and R.W. Stewart (eds.), *Remote Sensing of the Oceanic Environment*, Seibutsu Kenkyusha, Tokyo, pp. 350 - 355.
5. Rienecker M.M., Atlas R.M., Schubert S.D., and Willet C.S. (1996) A comparison of surface wind products over the North Pacific ocean, *J. Geoph. Res.* **101**, 1011 - 1023.

6. Kuzin V.I. and Moiseev V.M. (1993) Model of the North Pacific circulation, Proc. of Computing Center SD RAS, Vol. 1, pp. 19 - 46 (in Russian).

7. Kuzin V.I. and Moiseev V.M. (1996) Diagnostic and adjustment current calculations in the North Pacific, *Izvestia AN. Physics of Atmosphere and Ocean*, **32(5)**, 680 - 689.

8. Kuzin V.I., Moiseev V.M., and Martynov A.V. (1998) The response of the North Pacific circulation model to various surface boundary conditions, *Proceedings of 7th Annual PICES Meeting*, Fairbanks, USA.

9. Marchuk G.I. (1972) *Numerical Solution of the Problems of the Atmosphere and Ocean Dynamics*, Leningrad, Gidrometeoizdat, (in Russian).

10. Kuzin V.I. (1985) *The Finite Element Method in the Ocean Processes Modeling*, Computing Center, Novosibirsk, (in Russian).

11. Levitus S. (1994) *World Ocean Atlas 1994*, National Oceanographic Data Center, Ocean Climate Laboratory, Washington, D.C.

APPLICATION OF GIS TECHNOLOGY IN HYDROMETEOROLOGICAL MODELING

A. VORONTSOV[1], V. PLOTNIKOV[1], E. FEDOROVA[2] and Y. ZHUKOV[3]
[1]*National Oceanographic Data Centre, All Russia Research Institute of Hydrometeorological Information World Data Centre*
6, Korolyov St., Obninsk, Kaluga reg., 249035 Russia

[2]*State Oceanographic Institute*
6, Kropotkinsky per., Moscow 119838 Russia

[3]*State Research Institute Navigation and Cartography*
41, Kozhevenaya line, St.-Petersburg, 199106 Russia

Abstract. The efficiency in application of GIS for tasks of illustrative referencing, hydrometeorological supporting, ecological monitoring etc., is explained through mapping of objects in various scales based on properties of GIS. The paper analyzes the concept and the methodology of a GIS-technology for hydrometeorological analyses. Examples of practical tasks are presented.

1. Introduction

An effective support provided by data and information on the environment is one of the major requirements for environmental planning and decision making. It requires the integration of large volumes of data, data analysis with the use of various methods and models, and effective means of representing the results.

Modern computer technology allows the development of databases, computational methods and models, standards and manuals, supervisory programs, and application software to be combined at a functional level as an integrated information system comprising commercial geographical information systems (GIS) and database management systems (DBMS). Thus, it allows the creation of a certain toolkit for processing of data and obtaining new information on the process and model characteristics. This toolkit should include not only an informational base (initial, calculated, model and reference data), but also programs for hydrodynamic and probability simulations, along with a software for output presentation in cartographical, text, tabular, or graphical forms.

Such a toolkit in the form of an interconnected information technology [1, 2] based on GIS is developed and applied to different tasks at the National Oceanographic Data Centre of Russia at present. The presented paper discusses some examples on the application of this toolkit.

2. Practical Implementation of the GIS Technology

Modern information systems comprise the following functional subsystems (Fig. 1): the subsystem of a data archive, the subsystem of an integrated data bank, and the subsystem of problem-oriented applications.

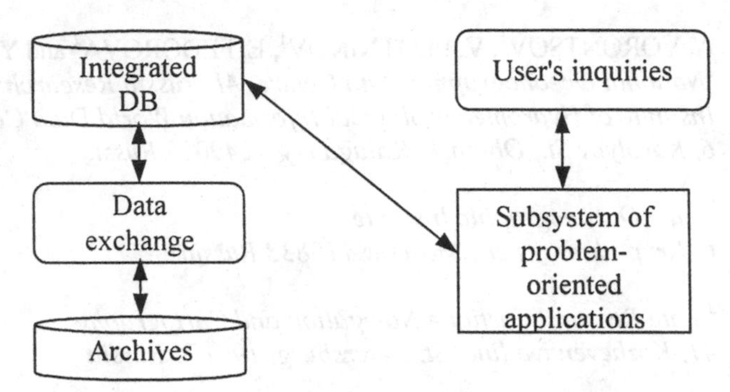

Figure 1. The general scheme of an information system.

The subsystem of problem-oriented applications (Fig. 2) is intended for the solution of different tasks and deals with information processing and dissemination to end users. It is based on the DBMS and GIS technology, with GIS playing a predominant role in integration and joint analysis of the information on marine environments.

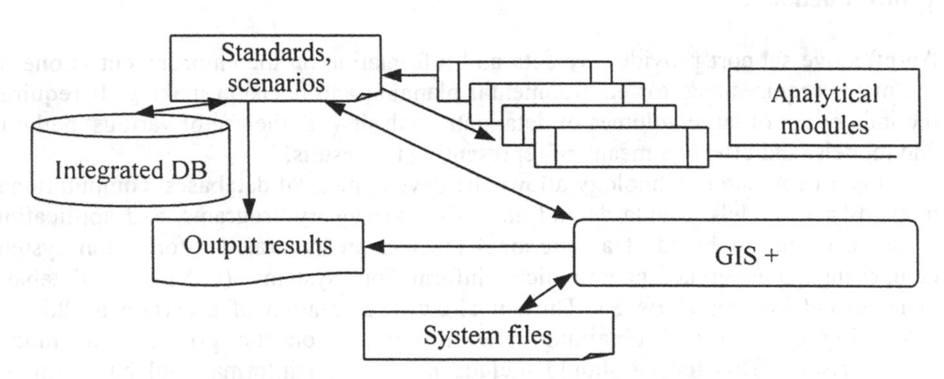

Figure 2. The general scheme of problem-oriented applications.

The main functions of GIS are: visualization of data, information, and model results; definition of input data for models; and definition and change of output parameters for models.

Let's consider an example on the application of the common GIS technology for the particular problem of "constructing hydrophysical fields of time and space averages". The scheme of implementation of this example is shown in Fig. 3.

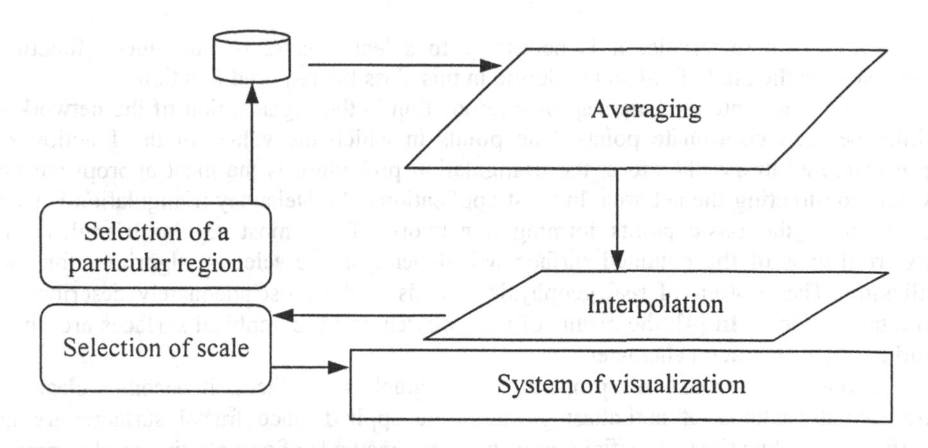

Figure 3. The scheme for application of the common GIS-technology.

In this scheme, there are some basic procedures. The first is to obtain averages. The second is the interpolation procedure. The third is a system of visualization.

To estimate the arithmetic average of a certain scalar oceanographic characteristic, the following algorithm is suggested. For a relatively large water area, there is a data array of N measurements for vertical allocation of the parameter. Let these measurements be defined in multi time points $\{x_i, y_i, \tau_i\}_{i=1}^{N}$, where x_i, y_i are the space coordinates, and τ_i is the time coordinate of the ith measurement. The task is to estimate the average arithmetic value of an oceanographic characteristic at a fixed depth for both the whole area and its any part.

The average value of the function $\overline{T}$ based on n measurements is equal to:

$$\overline{T}_n = \frac{1}{n} \sum T\left(x_{i_k}, y_{i_k}, \tau_{i_k}\right) \tag{1}$$

The error of $\overline{T}$ is determined by the formula:

$$\delta\left(\overline{T}_n\right) = \left|\, \overline{T}_n - M\left[T(x, y, \tau)\right] \right| \tag{2}$$

where $M[T(x, y, \tau)]$ is the true value of the arithmetic average $\overline{T}(x, y, \tau)$.

The task here is to define a subset $n \subseteq N$, which gives the minimum value for Eq. (2). The solution for this task can be based on the algorithm described in [3].

The second step in preprocessing the initial information is to select the interpolation procedure. The initial task of interpolation is usually formulated as follows: on a plane tangential to the earth's surface, there are points x_i, y_i, being members of the sets S, $i = \overline{1, N}$, where the values of a certain geophysical characteristic f_i are preset. In the general case, these points are located arbitrarily. For visualization of the field of this geophysical characteristic, it is necessary to select a class of continuous functions representing the study field and to define in this class the required function.

A necessary intermediate step of interpolation is the organization of the network of links between coordinate points. The points in which the values of the function are preset are arbitrary; therefore, the triangulation procedure is the most appropriate one when constructing the net area. In most applications, the Delauney triangulation is used. It connects the basic points forming a network of the most regular triangles. The external view of the obtained surface will depend on the selected algorithm for anti-aliasing. The contour of real geophysical fields is the most adequately described by fractal functions. In [4], the results of the research on topographical surfaces are given, indicating their fractal character.

Taking into account this property of topographical surfaces, it becomes clear that traditional methods of anti-aliasing cannot be applied since fractal surfaces are not differentiated functions by definition; whereas the methods of anti-aliasing used at present result in differentiated surfaces. Therefore, for construction of maps in traditional GIS applications, it is possible to consider only the method of triangulation; and for interpolation, new methods are required, such as the method of fractal interpolation [5]. This method is based on the design of a self-affined fractal surface by means of a system of iterative functions. In the theory of fractals, this task is called the inverse fractal problem. From our point of view, it is most expedient to use the wavelet conversion [6], as its use in GIS applications for generalization of geophysical fields does not theoretically result in distortions. Traditionally, hydrometeorological parameters are described by continuous smooth functions. The latest research on hydrometeorological processes has shown that their variability is more adequately described by fractals. The adequacy of fractal functions is not unexpected; it follows from the laws of nonlinear dynamics. Thus, it is necessary that GIS supports the operation with such functions. It should be noted that the set of fractal functions has, as a subset, the set of all continuous smooth functions.

The main statistical property of the fractal representation of the variability of hydrometeorological processes is written for the one-dimensional case as:

$$\Delta f(\lambda\tau) \equiv \lambda^{H} \Delta f(\tau) \qquad (3)$$

where, $\Delta f(\tau) = f(t+\tau) - f(t)$ and H is the numerical parameter ($0<H<1$).

This property determines the variability of the hydrometeorological process for their rescaling. With the help of fractals, the variability of a lot of hydrometeorological parameters can be described.

The procedure for generalization of hydrometeorological fields in cartography can be illustrated on the following example. Let's assume that it is necessary to map a cloudness cover of the Earth. The picture of cloudness will vary with the distance from the Earth because the fine details of cloudness derivations cease to differ. On the contrary, we shall see an opposite picture closer to the Earth. The degree of visible cloudness will be defined by the distance and the resolution of an eye. In cartography, the distance from which the Earth is watched is an analogue of the scale of a map; the resolution of an eye is the resolution of a map, and the change in visible cloudness structure with a change in distance is a clone of the procedure for generalization.

Let's imagine that we consider the cloudness of the Earth in a telescope from the Moon. The change in telescope magnification is similar to the change in distance from the Earth, and the resolution of the telescope is similar to the resolution of an eye. We need to find formal tools for GIS, which would fulfill the functions of such a telescope. As such formal tools, the wavelet theory can be used [6]. In wavelet conversion, functions like the following are used:

$$g_{a,b}(x)=\frac{1}{\sqrt{a}}g\left(\frac{x-b}{a}\right) \tag{4}$$

The wavelet function $g_{a,b}(x)$ is also the "mathematical telescope". The graph of the function $g_{a,b}(x)$ characterizes an optical behavior of the "telescope"; the parameter b defines the point to which it is directed; and the value $1/a$ defines its increase. The parameter a defines the scale of the study. For practical applications, a specific kind of $g_{a,b}(x)$ is defined for purposes of practical use.

Wavelet conversion of the one-dimensional hydrometeorological signal $f(x)$ is written as:

$$f(x)\rightarrow T_g(a,b)=\frac{1}{\sqrt{a}}\int_{-\infty}^{\infty}f(x)\cdot g^*\left(\frac{x-b}{a}\right)dx\,,\ a>0 \tag{5}$$

where the asterisk denotes a complex interface. The set of parsing wavelet functions $a^{1/2}g((x-b)/a)$ is derived from the function $g(x)$ by means of the shift b and the expansion a. For this conversion, there is an inversion formula:

$$f(x)=\frac{1}{c_g}\int_{-\infty}^{\infty}db\int_{0}^{\infty}\frac{da}{a^2}g_{a,b}(x)T_g(a,b) \tag{6}$$

The wavelet conversion gives more detailed information about a signal than that produced by the standard Fourier analysis, as the integral wavelet conversion gives the local information on the signal and its Fourier transform. As wavelet conversion realizes a hierarchical multiscale representation of the signal, it has many favorable features. Thus, the use of the wavelet conversion in GIS for generalization does not produce distortions in the visualization of hydrometeorological fields.

3. Conclusions

It is clear that modern integrated information technologies make it possible to receive a rather complete set of characteristics on the sea environment, based on modeling of environmental processes. The technical implementation is a powerful program, comprising the production of output production with a modern DB application to include sets of input data, regime materials, metadata, and cartographic information for each data type. However, any technical and technological toolkit should work on the basis of verified techniques and algorithms with adequate mathematical tools. Otherwise, the obtained results will be rather far from practice and the nature of the phenomena studied.

4. References

1. Vorontsov A., Plotnikov V. (2001) Application of GIS/RDBMS-technology for creation of electronic environmental references, in *Proc of the 4th Russian Scientific Conf. NO-2001,* St.-Petersburg, 6-9 June 2001, 1, 235-237.
2. Karageorgis A., Mikhailov N., Drakopoulou P., Vorontsov A., Anagnosou Ch., Lykiardopoulos A. (2000) GIS applications for the investigation of hydrometeorological and biochemical conditions of coastal areas, *Mediterranean Marine Science* **1,** 157-164.
3. Sobol I.M. (1969) *The Many-Dimensional Quadrature Formulas and Functions of Haar,* Nauka, Moscow.
4. Pheder E. (1991) *Fractals,* Mir, Moscow.
5. Cronover R.M. (2000) *Fractals and Chaos in Dynamic Systems,* Moscow.
6. Struzik Z.R. (1995) The wavelet transform in the solution to the Inverse Fractal Problem, *Fractals* **3(2)**, 329-350.

SATELLITE OBSERVATION OF ARAL SEA

S.V. STANICHNY[1], D.M. SOLOVIEV[1], V.M. BURDUGOV[1]
YU.B. RATNER[1], R.R. STANICHNAYA[1] and U. HORSTMAN[2]
[1]*Marine Hydrophysical Institute of National Academy of Science of Ukraine, 2 Kapitanskaya st., 99000, Sevastopol, Crimea, Ukraine*

[2]*Institute of Marine Research at the University of Kiel Fosternbrooker Weg 20, Kiel, Germany, 24105*

Abstract. Aral Sea is one of the best examples of an anthropogenic catastrophe caused by improper water resource management. Uncontrolled water withdrawal has led the Aral Sea level drop by 21 meters for the last 20 years. The main aim of this paper is to demonstrate the use of the AVHRR NOAA satellite data for investigation of the environmental processes in the Aral Sea. Direct and indirect parameters of the studied area can be estimated by using satellite data and GIS elements.

1. Introduction

Aral Sea is one of the best examples of an anthropogenic catastrophe caused by improper water resource management. Uncontrolled water withdrawal has led the Aral Sea level drop as much as 21 meters for the last 20 years.

In its stable state, Aral Sea had an area of 66 000 km^2 and a volume of 1050 km^3 [1]. Salinity of the sea was 9-10‰. It played a significant role in freshwater fishery in the Soviet Union and produced up to 10% of all catches.

Water losses due to evaporation were compensated for by precipitation and discharges of two rivers (Amu Darya and Syr Darya). Intensive irrigation in this region practically made full use of the waters of the Amu Darya River for agriculture. A significant drop in the Aral Sea level began at the end of the 70's. Regular investigations of the Aral Sea were stopped after the collapse of the Soviet Union, and satellite data became the main source of information. The availability of the high quality and user-friendly satellite data archives provides the unique ability for multidisciplinary investigation of the Aral Sea.

The main aim of this paper is to demonstrate the use of the AVHRR NOAA satellite data for investigation of the environmental processes in the Aral Sea.

2. Sea Level

Sea level estimation is based on the intercomparison between the real (obtained from satellite images) shape of the coastline and model bathymetry. A digital map of the sea bottom topography, based on the high-resolution navigation map, was prepared for these purposes. Raw AVHRR Satellite Data from SAA NOAA (www.saa.noaa.gov) were downloaded for the period between 1989-1998. These data were added to the AVHRR images received at the Marine Hydrophysical Institute (MHI) station for 1998-2001. Both data sets were processed by MHI software. AVHRR channel 2 maps were used for identification of the coastline. A two-step geographical correction was applied for development of a map with a spatial grid of 1.1x1.1 km.

The procedure for sea level estimation consists of selecting the model sea depth isoline with the best correspondence to the coastline shape on satellite images. The East part of the sea with a bottom slope of ~0.00015 -0.0002 is used for intercomparison. The estimated accuracy of the method for sea level restoration is 0.2-0.3m (taking into account the spatial satellite resolution). Figure 1 demonstrates the state of the Aral Sea for November 2000.

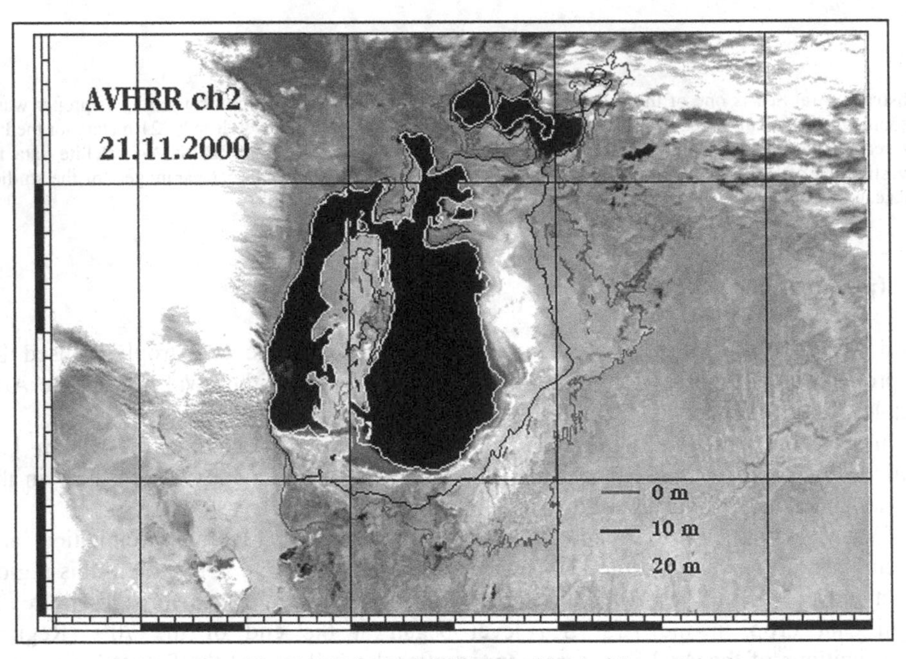

Figure 1. Satellite (AVHRR, NOAA-16) map of the Aral Sea and depth curves.

Note that the north part of the Sea, separated from the south part, has had a stable level since 1989 (approximately a drop of 12 m). The south part lost waters mainly due to evaporation, and the mean yearly level drop was 0.6 m for the time period between 1989-2001 (Fig. 2).

Level drops exceeding 1m in 1990-1991 and 1999-2000 cannot be explained by evaporation alone and may be resulting from other water loss factors like bottom intrusion. Annual level variations are shown in Fig. 3 for 1989, 1993 and 1998. Maximum level drops were observed during the summer seasons for all years. Note that, after 1998, significant level rise in the spring was not observed.

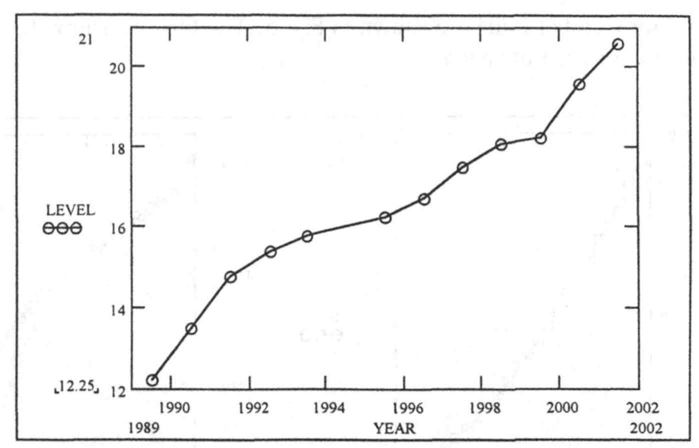

Figure 2. Sea level drop from the stable state (in meters) for June 1989-2001 (retrieved from satellite images).

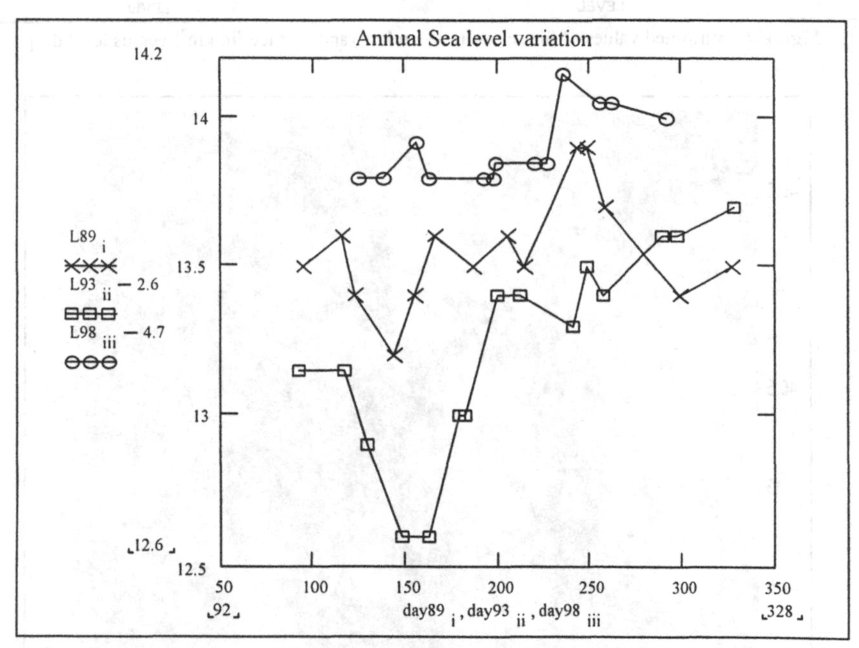

Figure 3. Annual variation of the sea level drop for 1989, 1993 and 1998 (the data for 1993 and 1998 are shifted on 2.6 and 4.7 meters for a better presentation in the graph).

3. Surface and Volume

The integration of GIS technology (digital map) and satellite data made it possible to estimate the variations in the sea surface and volume. Figure 4 shows the dependence of the sea surface and volume upon the level drop. At present, the surface has decreased 3 times and the volume 9 times. The current state of the sea (for October 11, 2001) and the prediction for October 2004 are shown in Fig. 5. The forecast shows that the sea will be divided into three separate parts.

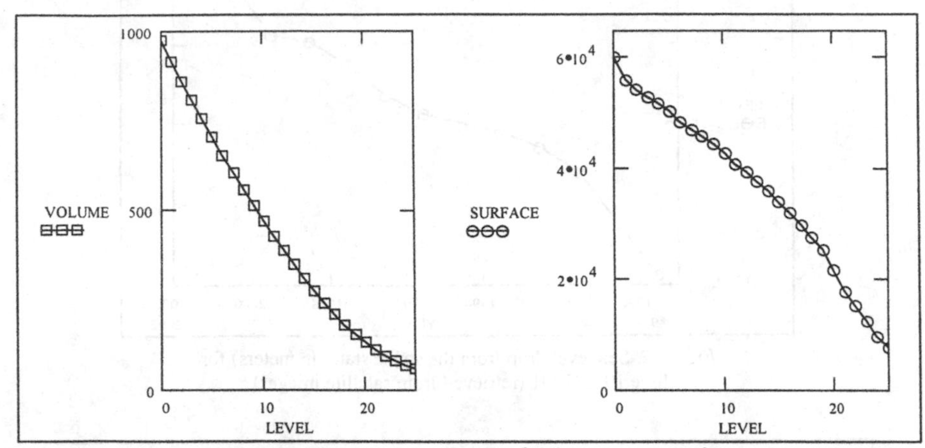

Figure 4. Computed values of the sea volume (in km^3) and surface (in km^2) versus level drop.

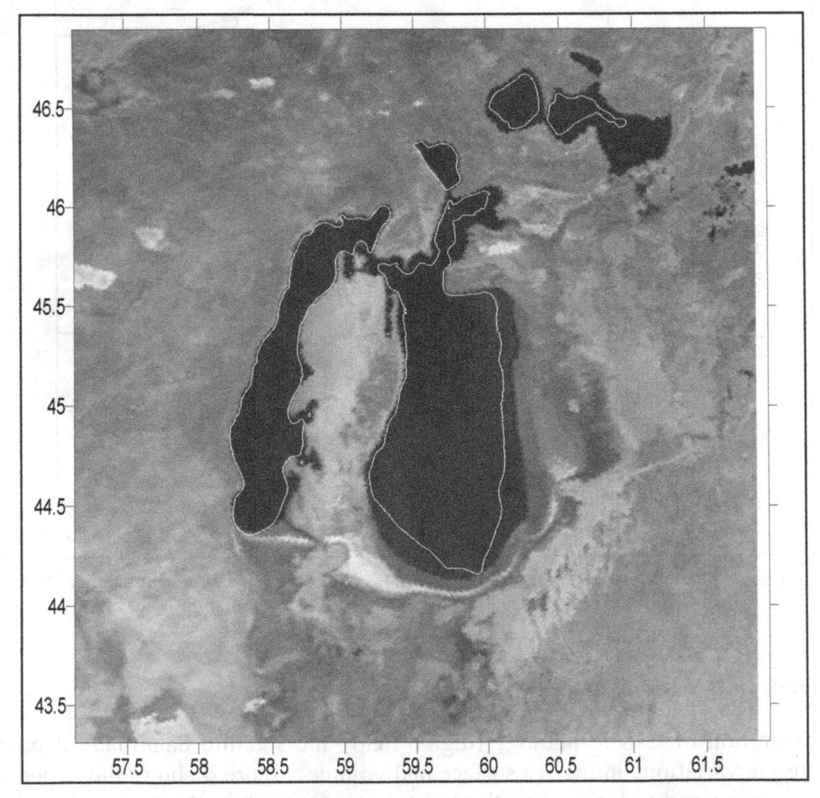

Figure 5. AVHRR map of the Aral Sea for 11 October 2001(21 m level drop), and the forecast for October 2004 (22.4m level drop) - white line.

4. Salinity and Ice Coverage

Dramatic changes in sea volume caused an increase in water salinity. By means of satellite data, an indirect estimation of this parameter can be realized by measuring the freezing temperature, which depends on salinity. The minimum sea surface temperature measured by AVHRR radiometer in February 2001 was -3.9^{0}C and the corresponding salinity ~50‰. Increased salinity leads to changes in water density, convection processes and heat fluxes. Within the years, full ice coverage was not observed. Figure 6 demonstrates a typical situation with ice coverage only in the East (shallow) part.

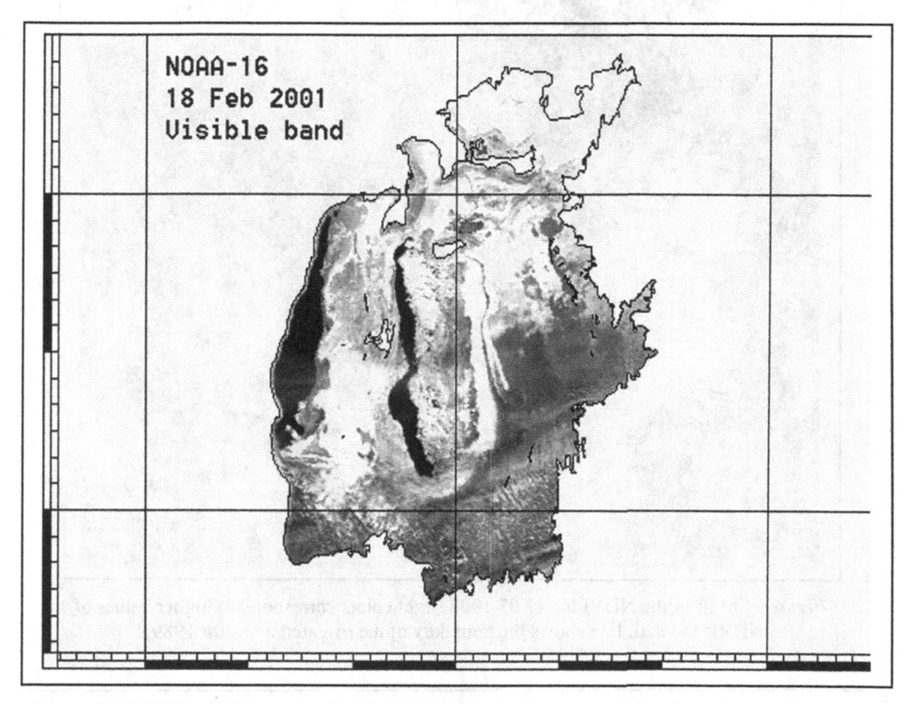

Figure 6. AVHRR Aral Sea map for 18.02.2001 (the dark area corresponds to the open sea; contour line represents the stable state of Aral Sea).

5. Vegetation

Changes in heat fluxes due to changing bottom and surface thermal regimes caused variations in the local weather and affected the vegetation cycle in the surrounding lands. One of the satellite-derived parameters describing surface greenness is Normalized Difference Vegetation Index (NDVI). A sample of the AVHRR NDVI map of the Aral Sea region for July 11 1998 is shown in Fig. 7, where the dark line corresponds to the boundary of irrigated areas for 1989. Monthly mean NDVI data from DAAC GSFC (http://daac.gsfc.nasa.gov) are used for analyses of the vegetation for the period 1989 -2000. Figure 8 demonstrates the variation of the NDVI in agricultural regions in the low part of the Amu Darya River for 1981-1986 and 1995-2000.

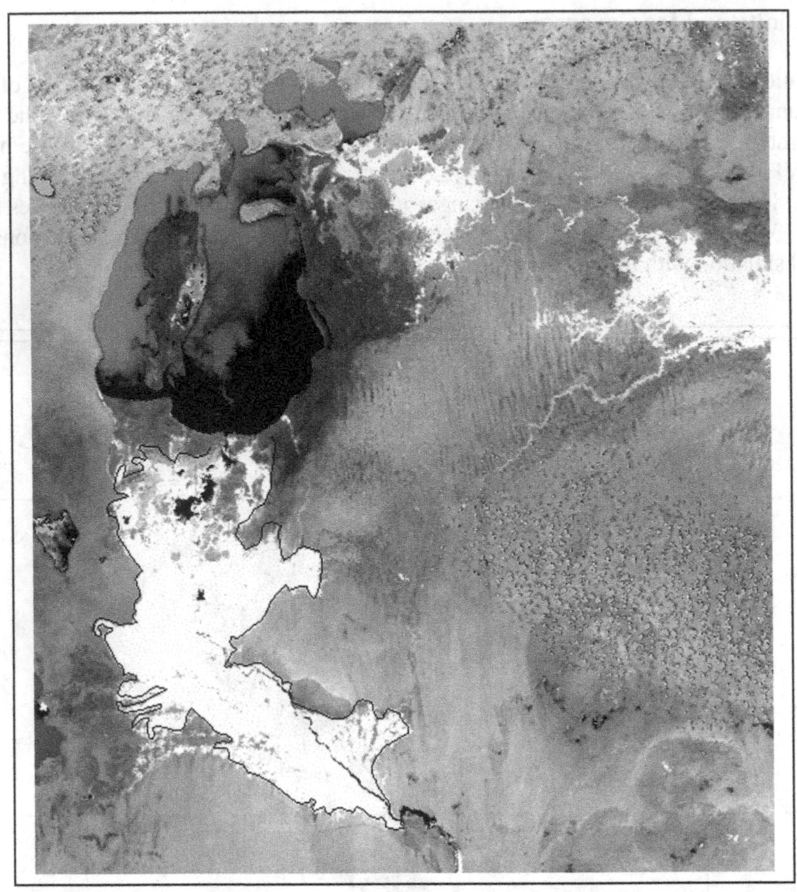

Figure 7. Map of the NDVI for 11.07.1998 (light colors correspond to higher values of NDVI; the dark line shows the boundary of the irrigated areas for 1989).

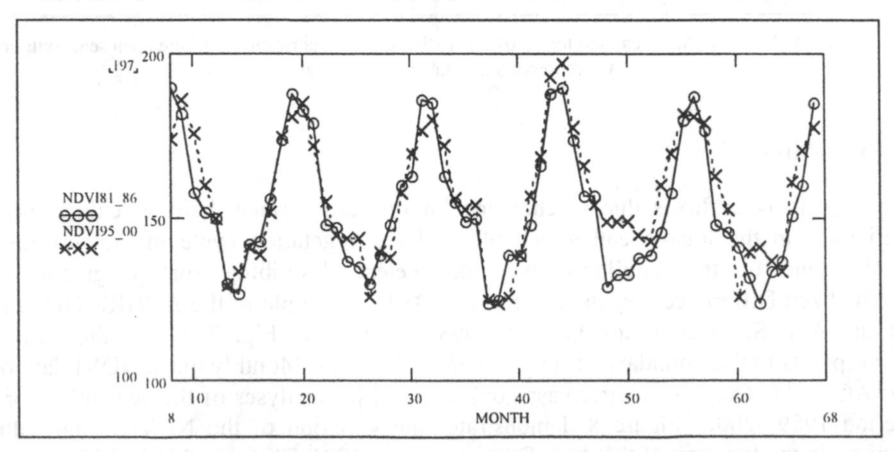

Figure 8. Variation of the NDVI for 1981-1986 and 1995-2000.

6. Conclusions

This short paper demonstrates, in case of the Aral Sea, the use of AVHRR data in investigating the sea and land environmental changes. Direct and indirect parameters of the studied area can be estimated by using satellite data and GIS elements. Note that SAA NOAA and DAAC GSFC are free of charge archives and have user-friendly Internet interfaces for data access.

7. Acknowledgements

This study was supported by the European Community grant INCO-Copernicus ICA-CT-2000-10023.

8. References

1 Kuksa, V.I (1994) *South Seas under the anthropogenic stress*, Gidrometizdat, Sankt -Peterburg (in Russian).

REMOTE SENSING OF THE LACUSTRINE ENVIRONMENT: DATA SOURCES AND ANALYSIS

S.V. SEMOVSKI
Limnological Institute SB RAS
P.O.Box 4199 Irkutsk 664033 Russia

Abstract. Recently, there occurred a marked shift in the Earth Observation policies of practically all governments and space agencies of the world, namely the trend changing from large observation platforms, with virtually all-round capabilities, to smaller spacecrafts with very specific objectives. The benefit of smaller projects lies in their far greater flexibility, accountability, and responsiveness in serving specific needs of the problems to be solved. At the same time, the present and the future manned orbital stations (MIR and International Space Station) and big automatic platforms (Envisat) provide unique capabilities for controlled experiments to detect and monitor special features of the environment.

1. Introduction

Earth observation by remote sensing changes our view and perception of the world. Remote sensing permits, for the first time in history, a total system view of the Earth. The view of the Earth from space has brought about sweeping revisions in the Earth sciences, in particular in such fields as meteorology, oceanography, hydrology, geology, geography, forestry, agriculture, geodynamics, solar-terrestrial interactions, and many others.

For highly variable water environments, a systematic remote sensing survey is the only way to estimate statistical characteristics of variability and to infer on trends. Investigations of lakes, however, need satellite-based instruments having some specific characteristics. First, the frequency band and the spectral resolution of passive or active sensors should be applicable for studies of the water surface, subsurface features or the coast. Second, the spatial (aerial) resolution should be suitable for detection of details on the surface of the lake studied, and the time coverage should be enough for investigations of mesoscale, annual or interannual variability. The cost of satellite surveys is also an essential factor for planning remote sensing surveys. A typology of present difficulties in applying remote sensing to limnology was outlined in [1], covering intrinsic (complexity of inland waters), technological (land resource satellite bands not designed for water sensing; revisit time too long) and institutional difficulties (cost of data and equipment, and the scarcity of limnologists proficient in remote sensing techniques).

The synergetic use of remote sensing data with available observations and models of lake environments is considered as a new perspective, including numerical methods of data assimilation in models. Some examples of the synthesis of satellite images, field observations and models in studies of mesoscale features in the phytoplankton field can be seen in [2].

Recently, there occurred a marked shift in the Earth Observation policies of practically all governments and space agencies of the world, namely the trend changing from large observation platforms, with virtually all-round capabilities, to smaller spacecrafts with very specific objectives. The benefit of smaller projects lies in their far greater flexibility, accountability, and responsiveness in serving specific needs of the problems to be solved. At the same time, the present and the future manned orbital stations (MIR and International Space Station) and big automatic platforms (Envisat) provide unique capabilities for controlled experiments to detect and monitor special features of the environment. Note that the resolutions of modern airborne sensors are at least an order of magnitude better than those of spaceborne data due to the proximity of the sensors to the target (earth surface). Technology will eventually come to grips with these problems to handle immense data volumes (along with onboard processing and data compression techniques) and very high resolution imaging data so that spaceborne surveys of the future will offer data of almost the same resolution qualities that are nowadays achieved in airborne applications.

2. High Resolution Imagery

Among sources of remote sensing data, the most widely known are the satellite photos of high quality, made by automatic instruments or by astronauts on manned spacecrafts. These images have a high spatial resolution (Landsat-5 MSS, 80 m; Landsat-5 TM, 30m; Spot HRV, 20-10 m; ADEOS, 20 m; planned Resurs MK-4, 5 m; new commercial US and Russian platforms with 1-2 m resolution) and high prices. These images have very limited applications in studies of lacustrine environments. Spectral bands used are usually not suitable for resolving structures on the water surface. Coverage is very limited in time and space, and many images are unique. The black and white version of a color image of Lake Biwa, obtained by the Japanese instrument AVNIR with 16 m resolution, is presented in Fig.1, where light grey areas on the lake surface correspond to more shallow areas and higher concentrations of suspended matter (see section 4).

Lake shoreline data derived from high quality imagery can present significant features because, for many lakes, the position of the water edge is not stable and depends on the water level (see case study of the Indian lagoon lake in [3]). Studies of lake areas are also interesting (see [4] for studies of Lough Neagh (area 390 km^2) using the Advanced Very High Resolution Radiometer (AVHRR) and [5] for Chinese paleolakes investigation).

3. Monitoring of Surface Temperature, Ice and Snow Cover

Passive electromagnetic remote sensing in the infrared band is a basis for the *estimation of surface temperature* for the water environment. The signal detected by the satellite is formed by the thin "skin-layer" of water. The temperature of this layer can differ from that of the upper mixed layer for sea or lake. Due to this feature and to the effect of the atmosphere, the accuracy of surface temperature estimation is comparably low (up to 0.5 ^{0}C; see, for example, [6]). However, the data on surface temperature distribution are used intensively in oceanography and limnology for climatic variability studies and energy

Figure 1. Lake Biwa image obtained by AVNIR scanner (satellite ADEOS-Minori, Japan), courtesy of NASDA, Japan.

balance calculations ([7]), etc. Zones of high temperature gradients can be detected with high accuracy on satellite images, and this fact is a basis for temperature front studies using infrared imagery. The thermal bar, that is typical for temperate lakes during spring and autumn, can be clearly seen on thermal images. The variability of the thermal bar and its movement can be a good indicator of climatic changes. Remote sensing of Ladoga Lake thermal bar was presented in [8]. The distribution and the variability of Lake Baikal thermal fronts, including the thermal bar, are investigated in [9].

Sequences of temperature fields derived from satellite imagery can be used in numerical procedures for the estimation of velocity fields (the so called "net advection of heat"), diffusion, and heat flux [9].

One of the best sources of data for surface temperature studies is the imagery of AVHRR (Advanced Very High Resolution Radiometer) based on NOAA polar-orbit meteorological satellites (National Ocean and Atmospheric Administration, USA). These instruments produce daily images of every region on the Earth's surface. The price of images depends only on the availability of a receiving station because NOAA carries the policy of "open sky", and data are free for every receiver. However, the spatial resolution of AVHRR images (1.1 km or worse) makes them suitable only for big lake studies.

Figure 2 presents temperature field changes of Lake Ladoga, derived from AVHRR NOAA imagery for June 1998 (images, courtesy of N.Yu.Mogilev). The northward propagation of the thermal bar front can be clearly seen from the sequence of images.

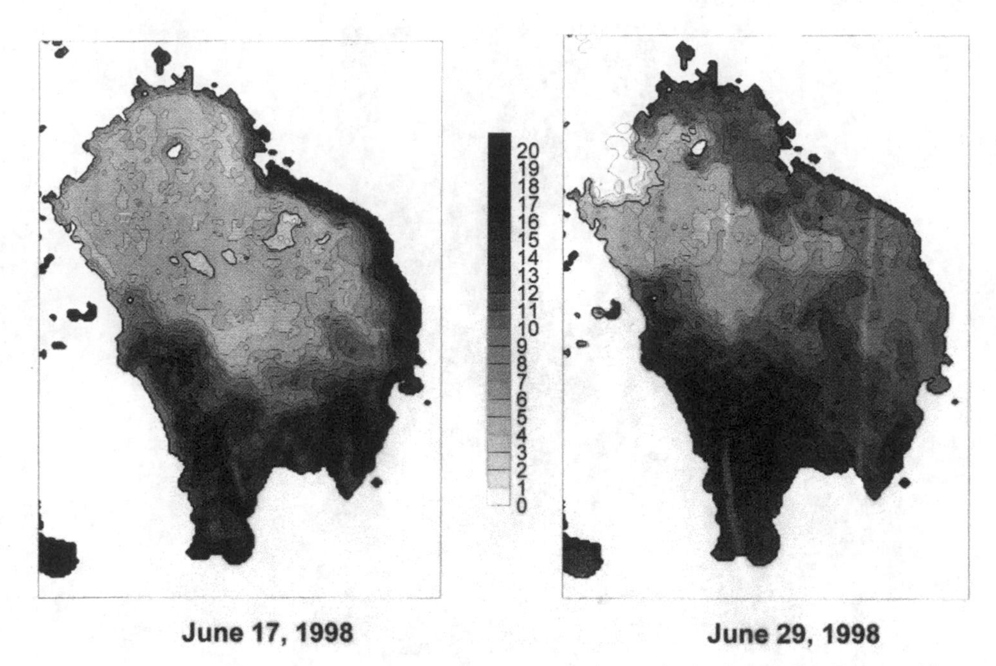

Figure 2. Surface temperature of the Lake Ladoga fields derived on the basis of AVHRR multispectral image (NOAA satellite, USA).

Infrared imagery coupled with data in visual spectral bands can be used effectively for studies of *ice and snow cover variability*. It is well known that freezing and the break-up of lakes are important indicators of climatic variability. For many temperate and polar lakes, ice and snow cover conditions play an important role in the intensity of under-ice phytoplankton bloom. For Lake Baikal, the first attempt to realize the multivariate analysis of the AVHRR data for studies of ice and snow cover variability was made in [10, 11]. It should be noted that the use of the satellite-based Synthetic Aperture Radar with high spatial resolution allows for detailed ice studies even for small lakes. Data on the Alaska receiving station were used in [12] for studies on the freezing of small Alaska lakes. Availability of ERS images depends on the position of the receiving station (see section 5). High resolution images in the optical band are also used for verification, based on moderate-resolution AVHRR images. In Fig. 3, the image of the melting ice structure is presented for North Baikal (Resurs satellite).

4. Remote Sensing in Optical Band: Detection of Suspended and Dissolved Substances

Water color is a well-known indicator of the trophic state of a lake. Optical properties of the lake surface are formed by attenuation and scattering of optically active substances of water, both dissolved and suspended. This fact is a basis for satellite and airborn remote sensing in the visual (optical) band.

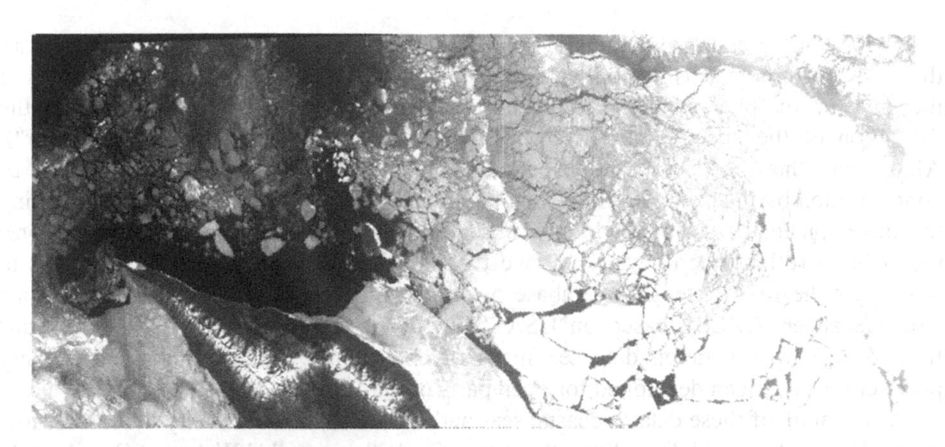

Figure 3. Ice break-up in North Baikal on the Resurs image (23.05.1999).

Nowadays, multispectral satellite-based scanners, specially designed for oceanographic applications, and high resolution optical scanners of Landsat type produce great amounts of data. This information can be used for studies of the lacustrine environment. Data of oceanographic instruments are often distributed free for scientific purposes (CZCS, MOS, OCTS, SeaWiFS, MODIS, etc.). These data have a good spectral resolution, and algorithms are developed and verified for the calculation of optically active substance concentrations in coastal waters. Part of these methods can be used for the study of lakes as well. On other hand, global data of these scanners have a moderate spatial resolution (usually about 1 km) and can be used regularly only for big lakes. High spatial resolution data of Landsat, Spot, Resurs, AVNIR and other platforms have high prices. The spectral resolution of these instruments does not permit the development of universal algorithms suitable for different water environments. For few lakes, the possibility was investigated for the use of these data in estimation of suspended matter concentration, chlorophyll content, and other environmental indicators.

The use of AVHRR (NOAA satellites) data in the visual band for analysis of suspended matter and phytoplankton distribution was investigated in a number of works. Methods for estimation of suspended matter concentration were developed in [13] for regions with high turbidity. AVHRR data in the visual and infrared band were used for Lake Baikal phytoplankton studies [10].

Experimental results of the Lake Biwa and Lake Kasimigaura (Japan) environment studies using multispectral MOS-1 satellite-based scanner data were presented in [14, 15]. Multispectral satellite imagery in 700 to 840 nm spectral band was used in [16] for the classification of the tropical status of Karelian Isthmus lakes (at least 200 m long and wide).

Airplane-borne color scanners (CASI, etc.) with high spatial (down to 5 m) and spectral (up to 200 spectral bands) resolution can be used with success in monitoring of the lacustrine environment [17]. These instruments are based on small and low cost airplanes; however, the price of instruments is usually very high.

For a number of years, detailed investigations have been realized on Lake Chicot (Arkansas, USA) to assess the use of Landsat high resolution imagery in the visual band for the estimation of chlorophyll and concentrations of other substances [18].

Color signatures in reflections of solar radiation due to phytoplankton pigments are the basis for remote sensing applications in phytoplankton studies. The peculiar form of the spectra of photosynthetic pigments in living phytoplankton cells affects the formation of the spectral coefficient of diffusive reflection (albedo, "water color"). Algorithms that serve for the estimation of C_a (chlorophyll *a* + pheophytin *a*) concentration by the spectra of upwelled radiation are known since the 1960s. Following the development of the methodology, observational tools were constructed to analyze the spectral reflectance of water surface from airplane or satellite and from tracers to scanners. The first systematic database of ocean color was produced by Coastal Zone Color Scanner (CZCS), based on USA "Nimbus-7" satellite (1978-1984). Using this limited spectral resolution data set, maps of chlorophyll concentrations and primary production have been developed for open parts of the World Ocean [19, 20].

Application of these data to coastal seas and lakes was, however, very limited because the CZCS spectral resolution allows the study of only the so-called Waters of Case 1, in A-Morel terminology. The alternative class, namely, the Waters of Case 2, comprise coastal seas and lakes. For these water basins, the impact of other optically active components of the ecosystem (dissolved organic suspended matter) upon spectral reflectance is significant. New observational tools should be used for studies of productive coastal areas and lakes of different types, and new algorithms should be developed for data analysis.

A new stage in water ecosystem studies using remote sensing methods began after the launching of several new instruments (OCTS, Japan, 1996; MOS-IRS, Germany-India, 1996; SeaWiFS, USA, 1997; recently launched ROCSAT, Taiwan, 1999). These satellite-based instruments, as well as their near future successors (MODIS, MERIS, etc.), can measure reflected radiance in few spectral bands. Algorithms are developed in the USA and Europe for the estimation of chlorophyll and other substances with use of modern data. As an example, studies of [21, 22] should be noted, where, for a lake system in the Netherlands, Secchi disk transparency, seston dry weight, vertical attenuation coefficients, chlorophyll-a and phaeopigments were determined by estimating inherent optical properties from apparent optical properties measured on the basis of high spectral and spatial resolution remote sensing data. Numerical assimilation of multispectral satellite data in models of the lake environment can be realized by using bio-optical ecodynamics model [23] to simulate the dynamics of optically active components in the lake ecosystem.

However, it should be noted that modern multispectral satellite-based scanner data are applicable only for big lakes due to their spatial resolution (usually about 1.1 km). Even for such big lakes as Baikal, this resolution is not enough for the investigation of correlations between local hydrodynamics and productivity, studies of frontal structures in phytoplankton concentrations, "patching", etc. Chlorophyll concentration for Lake Biwa (Japan) is presented in Fig. 4, derived from OCTS scanner data (ADEOS-Minori satellite, Japan).

Another disadvantage of the data is spectral resolution, which is still not good enough for more detailed studies of lake ecosystems. Note, as an example, the practically important problem of discrimination between different taxonomical groups of phytoplankton, such as diatoms, dinoflagellats, coccolites, picophytoplankton, etc. For many water basins, picophytoplankton consists mainly of blue-green algae (*Cyanobacteria*); often, the variability of its concentration is correlated with eutrophication processes. It is well known that the specific attenuation spectra of blue-green algae has a peculiar maximum in the area of 600-640 nm. Generally speaking, multispectral observations of spectral reflection can serve as a basis for discrimination between different taxonomic groups of lake phytoplankton. Unfortunately, this spectral band is not presented in modern satellite-based scanner data. Results of the last discussion (November-December 1998) among the Internet mailing group of specialists in remote sensing of inland water media show that general requirements should be formulated for a satellite-based scanner of new generations, specially constructed for studies of freshwater basins with extended SeaWiFS spectral bands and spatial resolution of 50-300 m. Now, possibilities are investigated to design low-cost specialized instruments with spectral and spatial resolutions good enough for studies of inland water bodies. One of the Internet discussions was organized in the framework of the International Alliance of Marine Remote Sensing for preparation of a proposal in the framework of the NASA small satellites program. Another discussion on delineation of requirements for free AVHRR data with 50 m resolution is to be organized in the framework of a program supported by the European Union.

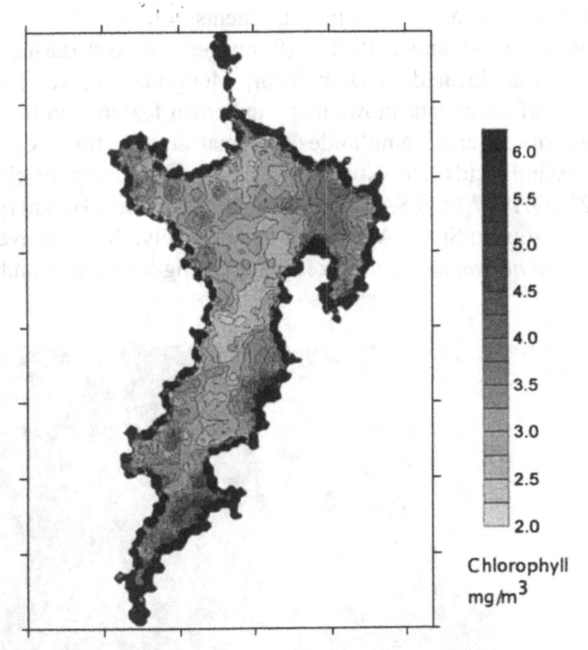

Figure 4. Chlorophyll concentrations on the surface of Lake Biwa (25.04.1996), calculated from mulstispectral information of OCTS scanner (satellite ADEOS-Minori, Japan), courtesy to NASDA, Japan.

5. Active Methods of Remote Sensing (SAR Imagery, Altimetry, Lidars)

The *synthetic aperture radar* (SAR), which can map roughness of the water surface and detect some features, has many potential applications in hydrology. High spatial resolution (about 25 m for ERS satellites) and independence of atmospheric conditions make these instruments applicable for studies of small inland water bodies. Recently, satellite-borne SAR imaging radar has demonstrated the significant capability to image a large variety of phenomena on the water surface (surface waves, thermal fronts, upwellings, internal waves, oil slicks, etc.) as well as phenomena associated with the atmospheric boundary layer. Studies of oil slicks and films of biologic origin, using SAR imagery, were also demonstrated in numerous papers. It is well known that SAR imagery has high capabilities to detect differences in ice and snow coverage; this possibility makes SAR data very useful for studies of temperate and polar lakes. However, there are no universal algorithms for interpretation of SAR imagery.

There are only few examples of analyses on lakes, using satellite-borne SAR observations. In [24], an attempt was made to describe patterns of high algae accumulation in correlation with peculiar signatures on ERS images. The development of internal waves in Ladoga Lake (see [4]) was studied, and mechanisms of their generation were discussed. Studies of the small, semi-arid and climatically sensitive lakes, based on SAR data, satellite altimetry and AVHRR imagery, was presented in [25]. In [24], the hydrobiological situation in Lake Onega was studied, using joint spaceborne, airborne radar, and *in situ* measurements.

For Lake Baikal, ERS-1 and ERS-2 SAR images received during 1997-98 by ESA mobile receiving station located in Ulan-Bator, Mongolia, are currently being studied. Preliminary analysis of these data shows many unknown features on the lake surface, such as internal waves of different amplitudes, thermal fronts, films of biological origin, inhomogenities in wind fields, etc. Numerous slicks of biological origin (black lines) are presented in the 27.09.1997 ERS SAR image fragment of North Baikal (presented in Fig.5, image courtesy to European Space Agency). Most probably, the intensive surface bloom of the green-blue *Anabaena lemmermanii* takes place during low-wind conditions.

Figure 5. The image of North Baikal (27.09.1997) made by Synthetic Aperture Radar (ERS satellite) demonstrates numerous patterns of slicks of biological origin (black lines).

The high cost of SAR imagery is a limiting factor for a broad use of this source of data, as well as the necessity to have the receiving station in close neighborhood of the area under study. Such future platforms as ESA (the Euopean Space Agency) Envisat (planned launch time was 2001) and the International Space Station will have recording facilities onboard and special host satellite systems for data transmission.

Satellite altimetry is a high-accuracy method of measuring the satellite height over the Earth's surface; this method is suitable for water level studies. Water level data convey important information on mesoscale (internal seishes, floods, etc.), annual and climatic variability. TOPEX/POSEIDON and ERS altimetry data are available free of charge for scientific purposes. Some preliminary results of satellite altimetry data applications for climatic variability studies of big lakes are presented in [26].

Active *laser Lidar systems* to be launched in the future seem to be powerful tools for studies of optically active substances in the surface layer of lakes (chlorophyll, dissolved organic matter, oil spills, etc.). Modern Lidar instruments are suitable for investigating vertical profiles of backscattering; however, they are now based on onboard research vessels or aeroplanes.

The development of software support systems has not been discussed here for the use of databases of infrared and other multispectral imagery. This problem in beyond the framework of this article as well as other important questions such as:

- *georeferencing*, making available comparisons between images produced at different times and by different instruments (specially aspects of automatic georeferencing);
- *correct computation* of such physical parameters as surface temperature and other parameters;
- *estimation of the vertical structure* of hydrophysical or hydrobiological parameters, using surface measurements by space-borne instruments;
- software support for *mutual analysis of satellite imagery obtained by different instruments* and having different spatial and spectral resolution, which is one of the interesting and the most complicated issues in modern remote sensing.

6. Acknowledgement

I am grateful to Dr. William Silvert for his help in improving my English. Long time collaboration with Nikolay Yu. Mogilev was very fruitful for understanding Baikal processes, using satellite images. Discussions with Prof. Werner Alpers were very important for studying the role of SAR imagery in Lake Baikal studies.

7. References

1. Jaquet J.M. (1989) Limnology and remote sensing: Present situation and future developments, *Rev.Sci.Eau* **2**, 457–481.
2. Semovski S.V., Dowell M., Beszczynska-Moller A., Darecki M., and Szczucka J. (1999) The integration of remotely sensed, seatruth and modeled data in the investigation of mesoscale features in the Baltic coastal phytoplankton field, *Int. J. of Remote Sensing* **20**, 1265-1287.
3. Mahapatra K., Sudarshana R., and Das, N.C. (1994) Remote sensing based study on spatio-temporal variabilities of the coastal lagoonal features — a case study at Chilka Lake, India, coastal zone, in P.G. Wells and P.J. Ricketts (eds.), *Canada-94, Cooperation in the Coastal Zone*, Dartmouth, **3**, 936–949.

4. Harris A.R. and Mason I.M. (1989) Lake area measurement using AVHRR - A case study, *Int. J. of Remote Sens.* **10**, 885–895.
5. Zou S. (1987) Study of modern vicissitudes of the Jianghan Lake group by using remote sensing techniques, *Oceanol. Limnol. Sin.* **18**, 469–476.
6. Xin J.N. and Shih S.F. (1993) Lake surface temperature estimation using NOAA satellite APT data, *Int. J. of Remote Sensing* **14**, 1325–1337.
7. Schneider K. and Mauser W. (1992) *Utilization of NOAA/AVHRR data for energy balance of Lake Constance (Switzerland)*, Freiburg Univ., Inst. fuer Physische Geographie (Germany).
8. Malm J. and Jonsson L. (1993) A study of the thermal bar in Lake Ladoga using water surface temperature data from satellite images, *Remote Sensing of Envir.* **44**, 35–46.
9. Semovski S.V., Shimaraev M.N., Minko N.P., and Gnatovsky R.Yu. (1998) Satellite observations using for the lake Baikal thermal front studies, *Issledovanija Zemli iz Kosmosa (Earth Studies from Space)* **5**, 65–75 (in Russian).
10. Semovski S.V., Bondarenko N.A., Sherstyankin P.P., Minko N.P., and Mogilov N.Yu. (1998) Lake Baikal phytoplankton annual cycle studies using AVHRR imagery collection, in *Proceedings of 27th Int.Symposium on Remote Sensing of Environment, 7-12 June 1998*, Tromso, Norway, 320–323.
11. Semovski S.V., Mogilev N.Yu., and Sherstyankin P.P. (2000) Lake Baikal ice: Analysis of AVHRR imagery and simulation of under-ice phytoplankton bloom, *J. of Marine Systems* **27**, 117-130.
12. Jeffries M.O., Morris K., Weeks W.F., and Wakabayashi, H. (1994) Structural and stratigraphic features and ERS 1 synthetic aperture radar backscatter characteristics of ice growing on shallow lakes in NW Alaska, winter 1991-1992, *J. of Geophys. Res.* **99**, 22,459–22,471.
13. Stumpf R.P. (1987) *Application of AVHRR Satellite to the Study of Sediment and Chlorophyll in Turbid Coastal Water*, Tech. Memo. NESDIS AISC 7, Natl. Environ. Satell. Data and Inf. Serv., NOAA, Washington D.C.
14. Kumagai M., Maeda H., and Onishi Y. (1990) *Monitoring of Lake Environments by MOS-1*, Lake Biwa Research Inst., Otsu (Japan).
15. Sugihara S. and Kishino M. (1990) *Retrieval of Water Quality Parameters from Radiance Detected by MESSR*, Institute of Physical and Chemical Research, Saitama (Japan).
16. Kondratyev K.Ya., Lvov V.A., and Shumakov F.D. (1992) Use of data from multiband satellite photographs to evaluate the trophic status of lake systems, *Dokl. Earth Sci. Sect.* **313**, 36–39.
17. Theimann S. and Kaufmann H. (1998) Secchi depth and chlorophyll-a determination using field spectrometer and airborn hyperspectral data in the Mecklenburg lake district, Germany, in *Proceedings of 27th Int. Symposium on Remote Sensing of Environment, 7-12 June 1998,* Tromso, Norway, 325–328.
18. Schiebe F.R. and Harrington J.A. (1992) Remote sensing of suspended sediments: the Lake Chicot, Arcansas project, *Int. J. of Remote Sensing* **13**, 1487–1509.
19. Esaias W. E., Fridman G. G., McClain C. R., and Elrod J. A. (1986) Monthly satellite-derived phytoplankton pigment distribution for the North Atlantic Ocean basin, *EOS* **67**, 835–837.
20. Vinogradov M.N., Shushkina E.A., Vedernikov V.I., Gagarin V.I., Nezlin N.P., and Sheberstov S.V. (1995) Characteristics of epipelagic ecosystem of Pacific Ocean based on satellite and field data - Abiotic parameters and production indicators of phytoplankton, *Okeanologia* **35**, 226–236.
21. Dekker A.G., Malthus T.J., and Seyhan E. (1991) Quantitative modeling of inland water quality for high-resolution MSS systems, *IEEE Trans. Geosci. Remote Sens.* **29**, 89–95.
22. Dekker A.G. and Hoogenboom H.J. (1996) Predictive modeling of AVRIS performance over inland waters, in R.O. Green (ed.), *Summaries of the 6th Annual JPL Airborn Earth Science Workshop*, **1**.
23. Semovski S.V. (1999) The Baltic Sea and Lake Baikal underwater bio-optical Fields Simulation using Ecodynamical Model, *Ecological Modelling* **116**, 149–163.
24. Naumenko M.A., Beletsky D.V., and Rumyantsev V.B. (1994) Investigations of the hydrobiological situation in Lake Onega using joint spaceborn, airborn radars and *in situ* measurements, *Int. J. of Remote Sensing* **15**, 2039–2049.
25. Harris A.R. (1994) Time series remote sensing of a climatically sensitive lake, *Remote Sens. of Environ.* **50**, 83–94.
26. Birkett C.M. (1995) The contribution of TOPEX/POSEIDON to the global monitoring of climatically sensitive lakes, *J. of Geophys. Res.* **100**, 25,179–25,204.

Part IX

TRANSFER OF DATA INTO INFORMATION

FROM DATA MANAGEMENT TO DECISION SUPPORT

K. FEDRA
Environmental Software & Services GmbH
A-2351 Gumpoldskirchen AUSTRIA

Abstract. Water resources management requires potentially large volumes of data of different nature, ranging from long term historical time series to large-scale spatially distributed data and to real time monitoring and telemetry. The acquisition and processing of these data and turning them into useful information for policy and decision making pose a number of challenges. Information technology and, in particular, the rapid growth of the Internet promise new approaches and solutions that can help to transfer the data collected into decision relevant information, available to a large and diverse group of distributed stakeholders and actors in increasingly participatory decision making processes. Issues of distributed databases, access and ownership, institutional structures, data quality, timeliness, and costs, as well as the transfer of data to information that is directly relevant to water resources management decisions, are discussed. The role of data for modeling, and the role of models, GIS, and expert systems in decision support are analyzed. Using examples from Mexico (Lerma River) and Malaysia (Kelantan River), some possible solutions for integrated basin-wide water resources information and decision support systems are presented and illustrated with practical applications.

1. Introduction

Water is a key resource for sustainable development and an increasingly precious one in many regions of the world. Increasing pressures ranging from the combined effects of population growth and economic development, with subsequent increases in per capita demands, to the potential impacts of climate change make water, together with energy, the strategic key resource of any sustainable development.

At the level of the European Union [1], water stress (defined as demand/supply imbalance due to quantitative or qualitative constraints) may generally be constant or may even decrease, but there are considerable regional and temporal differences:

- over-exploitation of aquifers and consequent salinization through salt water intrusion in the coastal zone, regression of wetlands, and land subsidies;
- reduction of river flow due to over exploitation, with subsequent water quality and environmental problems; at the same time, frequency and severity of floods are on the increase due to, *inter alia*, land use change in the catchments;
- while the pollution of major rivers is often declining due to better wastewater treatment, non-point source pollution primarily from agricultural sources remains high, leading to high nutrient concentrations and eventually to coastal eutrophication;
- nitrate concentrations in excess of the WHO standards or the EU Drinking Water Directive (50 mg/l) are often found in shallow aquifers, in particular in areas of intensive agriculture and livestock breeding [2].

All these problems, usually on a more dramatic scale, can be found in many other countries and regions as well.

Water resources management has two main components, which are closely linked: a physical one, which is basically a mass budget problem of distribution in time and space with an inherently stochastic element driven by climate, and a socio-economic one, which defines the level of investment in water related infrastructure and the allocation of water to different, often conflicting uses and users. To quote from COM (2000) 477 [3]: "Factors explaining this situation include barriers to the adoption of more efficient technologies, limited incentives to reduce water use, inadequate institutional framework, gaps in the integration of environmental concerns into sector policies, and lacking or poorly implemented environmental policies".

2. Data and Information

Data management is most often considered from a *supply* point of view, with emphasis on the technicalities of data storage and retrieval, metadata, data quality and integrity. From the *demand* side, the main question is not how to manage the data, but what information requirements they are supposed to address. A truly scientific view of water resources management would formulate policy options as hypotheses and then look for the experiment to test the hypotheses. Since a large system like a river basin is not easily subject to experimentation, one can only use observations (of current and historical events) and modeling to address these questions. Thus, one has to derive the information required to answer these questions from the data available and the analytical tools used to process them.

The kind of data structures one may encounter in any environmental analysis, including water resources management, can be derived from a small set of basic types:

1. symbols and scalars (which may refer to generic or spatially referenced objects); a generic object would be a pollutant chemical with its attributes like solubility or a decay rate; a geo-referenced object within the context of a water resources study might be a reservoir with an obviously larger set of attributes, or a monitoring station. In addition to more or less static properties, these objects will usually have associated data in the form of:
2. time series (of point observations, which can be aggregated/analyzed into graphs or fields), basically the same as (1) with the added attribute of a time stamp and a location again (Fig. 1);
3. graphs (basically reducible to the previous two classes with the added attribute of a topology);
4. fields (matrices/tensors) which again can be derived from (1) and (2) above, with the possibility of using implicit geometry;
5. maps of different kinds, which, however, are really the geo-referenced data type of (1), where the reference may be to structures other than points (lines, polygons, graphs, etc). These data constitute what is customarily shown as the legend of a map, associated with its geometrical and topological structures.

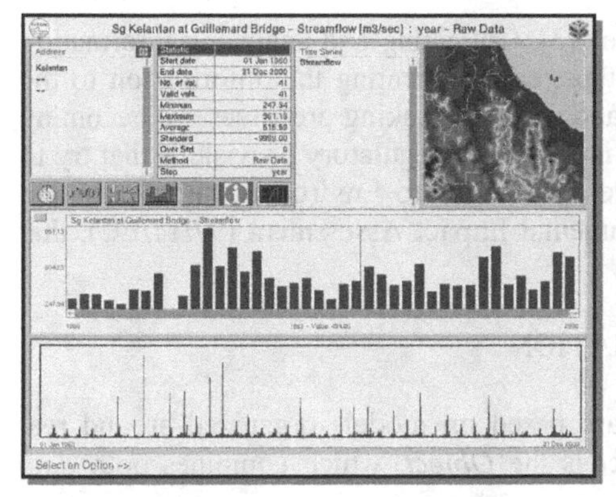

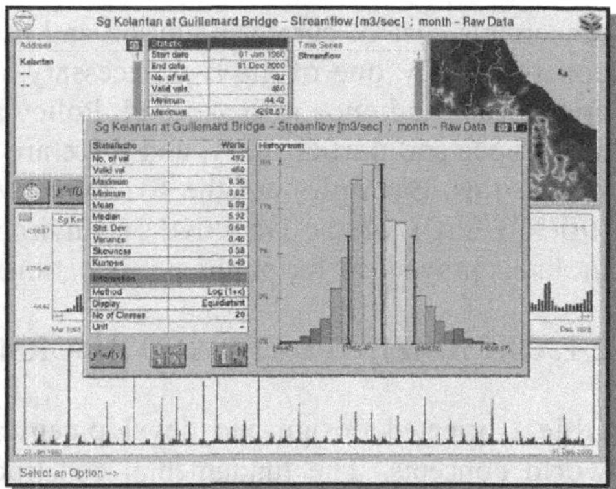

Figure 1. Monitoring data display and analysis: 40 years of streamflow data from the Kelantan River, Malaysia, display of daily and annual average flow, frequency distribution and basic statistics.

There is another block of data and information that is structurally different but equally relevant: information describing the relationship between the dynamic behavior of objects, i.e., the procedural information that may be expressed as algorithms or rules. Darcy's law and the kinematic wave equation are examples, but simple (and more operational) relationships like the release policy of a reservoir (relating storage to outflow) also fall in this class. Parameters for the algorithms or rules, e.g., slopes and conductivity, are then again data types in the above sense, their spatial and temporal resolution corresponding to the dimensionality of the model (from a lumped parameter rule to a 3D partial differential simulation model).

At the highest level of conceptual aggregation, we find rules for assessment against regulations (such as a water quality guideline) or even more general (and thus in need of interpretation) policies like those expressed in the European Unions Water Framework Directive [2]. Here, the information may no longer be qualitative and algorithmic, but semi quantitative (expressed as inequalities) or fully symbolic, qualitative.

For illustration, consider the Water Framework Directive. Two principles are to be taken into account for economic aspects of the River Basin Plans: firstly, the full recovery of the costs of water services, and secondly, the polluter pays principle. More specifically, member states are expected to ensure that "water policies provide adequate incentives for users to use water resources efficiently" and that there is an "adequate contribution from different water uses industry, households and agriculture to the recovery of the costs of water services" by 2010. An underlying concept of the EU Water Framework Directive (WFD) is the following: "water is not a commercial product like any other but, rather, a heritage which must be protected, defended and treated as such". The WFD, in Article 9, requires member states to "take account of the costs of water services" and describes, in Annex III, the economic analysis that should be conducted for this purpose. The goals of the Directive focus on water quality, and water-pricing strategies are seen as instruments to achieve "good ecological status" of water.

This leads to specific data requirements: “It is important to identify, test and make operational data collection methodologies (i.e., which data collection technology, at which spatial scale, with which temporal frequency) that provide a useful estimate at reasonable cost of current pollution and use "[3]. Collecting and analyzing information constitute only one of the two necessary steps; communicating this information to the appropriate audience is the second. Policy and decision making processes are becoming more open and participatory, and there are increasingly regulatory provisions that try to ensure this. Examples are the EU Directives on Access to Environmental Information (90/313/EEC), or the Directive on Environmental Impact Assessment (97/11/EC), that includes provisions for public participation.

2.1. ORGANIZING DATA AND INFORMATION

Object oriented design and development are based on models organized around real world concepts. The fundamental construct is the *Object,* which combines both *data structure* and *behavior* in a single entity [4].

Definition and language use varies. Rumbaugh [5] refers to classification as objects with the same data structures (attributes) and behavior (operations) grouped into a class. Shlaer-Mellor [6] uses the term "object" to be the abstraction of like things and refers to individual objects as instances. However, OOD can conveniently be described as including, or using, the following concepts:

- *Abstraction:* denotes the essential characteristics of an object, that distinguish it from all other kinds of objects.
- *Encapsulation:* is the process of hiding all the details of an object, that do not contribute to its characteristics.
- *Modularity:* is the property of a system that can be decomposed into a set of strongly cohesive and loosely coupled modules.
- *Hierarchy:* is the ranking or ordering of abstractions. Two kinds of hierarchical relationships are important here:
- *Aggregation*, i.e., *is part of* ... and
- *Inheritance*, i.e., *is a kind of* ....
- *Typing*: is the enforcement of the class of an object such that objects of different types may not be interchanged.
- *Concurrency*: is the property that distinguishes an active object, which has its own thread of control, from an inactive one, which does not.
- *Persistence*: is the property of an object through which its existence transcends time (i.e., the object continues to exist after its creator ceases to exist) and/or space (the object's location moves from the one in which it was created).

To bring these concepts to life in a concrete context, consider a river basin. One can describe any river basin, and any water resources management problem, by sets of interacting objects. In a very simple example, the river basin (or water resources management problem) consists of the river system and a number of water users. The river system, in turn, is a set of nodes and arcs (river reaches), with nodes being elements like sub-catchments, reservoirs, and water users. The users represent entities like municipalities, irrigation districts, or industrial enterprises. The behavior of the objects (summarized in a scenario) describes how they affect the water, e.g., how they generate it (a sub-catchment), how they store it (a reservoir), how they route it (a river reach), or how they consume it (a city).

An example of an object oriented approach is described by Fedra and Jamieson [7]. River basins and associated water resources management problems can be structured in terms of three types of objects. The first type, called RIVER BASIN OBJECTS, represent real world entities, such as reservoirs, sub-catchments, cities, or treatment plants. The second type, NETWORK OBJECTS, represent a different layer of abstraction, such as models of a river system or network. The final type are SCENARIOS, directly representing decision problems; they represent model oriented collections of instantiations of NETWORK OBJECTS that are partially derived from RIVER BASIN OBJECTS. All objects are spatially referenced; that is, they are known by location (map display and selection), as a single point (observation station), as a reference point designating a larger object (lake, city), as a rectangle including one or several points or polygons (irrigation district), or a as polygon (sub catchment).

Objects have two functions:

- they can obtain or update their current state (load, compute, infer, etc.) in a given context, referring to SOURCES (which may be other objects);
- they can report their current state or parts of the their state to CLIENTS (the screen, to each other, to models, a hardcopy device, the Internet, etc.).

For example, sub-catchment (Fig. 2, left) objects use a rainfall-runoff model to obtain the runoff from the catchment under a set of land use, internal water use, and meteorological conditions (the latter are obtained as time series from a climate stations object). This runoff, in turn, is used by the water resources model WRM as input for a start node (sub-catchment node). In the same way, demand nodes in WRM are linked to various river basin objects (settlement, industries, irrigation districts (Fig. 2, right) and obtain their detailed behavior over time (e.g., water demand, consumptive use coefficients, losses, etc.) from these objects. Through the location of objects, the linkage to the GIS layers is established so that spatial concepts (such as catchment, river reach, or the neighborhood of a point location) can be used for calculations (methods) by the objects.

The objects are grouped into CLASSES, which may include:

- monitoring stations such as Climate stations, Flow measurements, Water quality;
- demand nodes, including Abstractions, Settlements, Water works, Industries, Animal farms and feedlots, Irrigation districts;
- input nodes, such as Sub-catchments, Well fields, Interbasin transfers;
- flow control structures, including Dams and Reservoirs, Lakes, Weirs and falls, Gates and sluices;
- river reaches and cross sections;
- water quality related objects such as wastewater treatment plants;
- auxiliary nodes (geometry, control, scenic sites and reference points).

Each of these classes may have any number of elements. Each object class has a set of specific attributes, organized in a set of data structures, and associated METHODS that include any or all of the data types described above, i.e., symbols, scalars, time series, and more complex spatial data structures.

Objects may be linked to other objects; for example, a treatment plant may lead on to a flow and a water quality observation station and its data. Objects have hypertext files that provide further explanation, metadata, and context. Objects can also be hierarchically structured; for example, an irrigation district may contain any number (of again hierarchically grouped) sub-districts, which are automatically aggregated and kept consistent.

Objects have methods available, which allow them to obtain or update some of their dynamic or derived properties in a specific context. Many object properties are static and can be stored in their respective databases and files. Other object properties, such as the outflow from a sub-catchment or the monthly water requirements of an irrigation district, may describe current or historical situation, in which case the data may be coming from an on-line monitoring or data acquisition system. They may also refer to hypothetical or future situations in the context of a planning scenario or a forecast, and then depend on numerous controlling variables or plans, decisions, and assumptions. Models such as a rainfall-runoff model or an irrigation water demand estimation model can be triggered by the respective objects (i.e., sub-catchments, irrigation districts) to estimate some of their attributes. They can, in turn, be fed to a sub-catchment start node or an irrigation demand node in the river network and provide input to the corresponding simulation models.

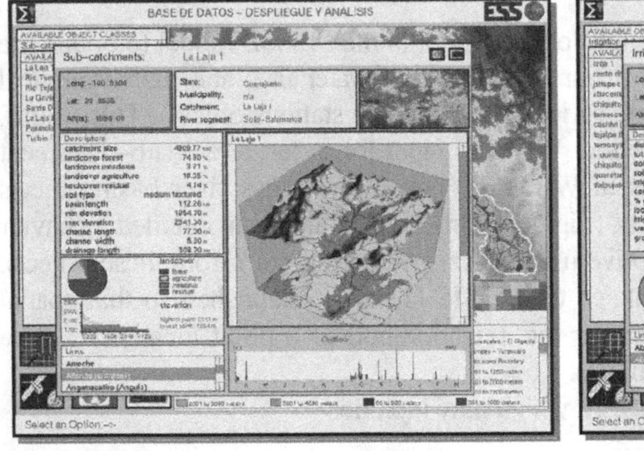

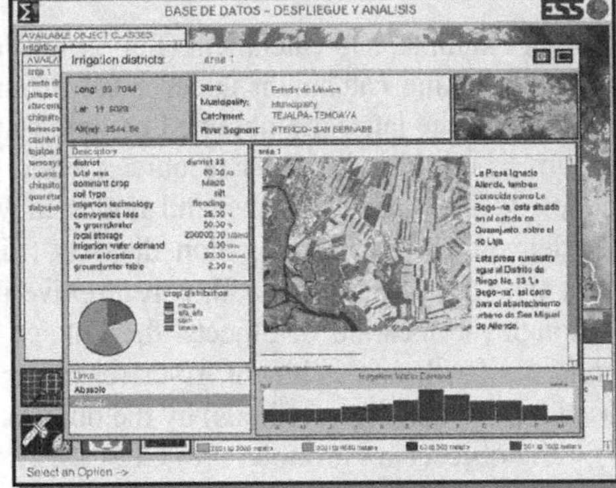

Figure 2. River basin objects: sub-catchment (left) and irrigation district (right).

The context for such an estimation is a time period, e.g., a default (reference) year, and all observations and data pertaining to it. Alternatively, the context can be defined by a model specific scenario (including, for example, the selection of a specific year or period and its hydrometeorological characteristics; or a hypothetical year of specific characteristics, such as very dry or very wet, and thus represented by a synthetic time series with the appropriate statistical characteristics derived from the historical time series of observations) and, within this constraint, by a set of user specified assumptions.

3. Decision Support

The objective of a computer based decision support system for water resources management is to improve planning and operational decision making processes by providing useful and scientifically sound information to the actors involved in these processes, including public officials, planners and scientists, various interests groups such as farmer as major water users, and possibly the general public. This information must be:

- timely in relation to the dynamics of decision problem;
- accurate in relation to the information requirements;
- directly understandable and usable;
- easily obtainable, i.e., cheap in relation to the implied costs of the problems (see, for example, [8]).

The ultimate objective is to ensure sufficient and sustainable water resources, thus contributing to the maximization of some (rather hypothetical) social welfare function.

Decision support is a very broad concept and involves both rather descriptive information systems, which just demonstrate alternatives, and more formal normative, prescriptive optimization approaches that design them. Any decision problem can be understood as revolving around a *choice between alternatives*.

These alternatives are analyzed and ultimately ranked according to a number of criteria by which they can be compared. These criteria are checked against the objectives and constraints (our expectations), involving possible trade-offs between conflicting objectives. An alternative that meets the constraints and scores the highest on the objectives is then chosen. If no such alternative exists in the choice set, the constraints have to be relaxed, criteria have to be deleted (or possibly added), and the trade-offs redefined.

However, the key to an optimal choice is in having a set of options to choose from, that does indeed contain an optimal solution. Thus, the generation or design of alternatives is an important, if not the most important step. In a modeling framework, this means that the generation of scenarios must be easy so that a sufficient repertoire of choices can be drawn upon.

The selection process is then based on a comparative analysis of the ranking and elimination of (infeasible) alternatives from this set. For spatially distributed and usually dynamic models (natural resource management problems most commonly fall into this category), this process is further complicated since the number of dimensions (or criteria) that can be used to describe each alternative is potentially very large. Since only a relatively small number of criteria can usefully be compared at any one time (due to the limits of the human brain rather than computers), it seems important to be able to choose almost any subset of criteria out of this potentially very large set of criteria for further analysis and modify this selection if required.

Approaches to decision support span a wide range of conceptual levels but they typically include Information systems, scenario analysis, comparative evaluation of scenarios and optimization, including discrete multi-criteria optimization.

3.1. INFORMATION SYSTEMS

To simply provide information about the state of a system, compare it to expectations, and possibly provide simple forecasts based on observed trends constitute an essential step in any decision support. A decision becomes necessary if and only if the current or expected future state of the system deviates from the expectations. Monitoring for compliance with targets and objectives and early warning are therefore important initial steps.

3.2. SCENARIO ANALYSIS

In a DSS framework, Scenario Analysis supports the exploration of a number of WHAT IF questions. The scenario is the set of initial conditions and driving variables (including any explicit decision variables) that completely characterize the system behavior, which is expressed as a set of output or performance variables (Fig. 3). Scenario variables that cannot be influenced by the decision maker, but have to be taken into account, include aspects like hydrometeorology (a dry or a wet year) or behavioral aspects (price elasticities, i.e., consumer reactions to changes in water prices or wastewater taxes).

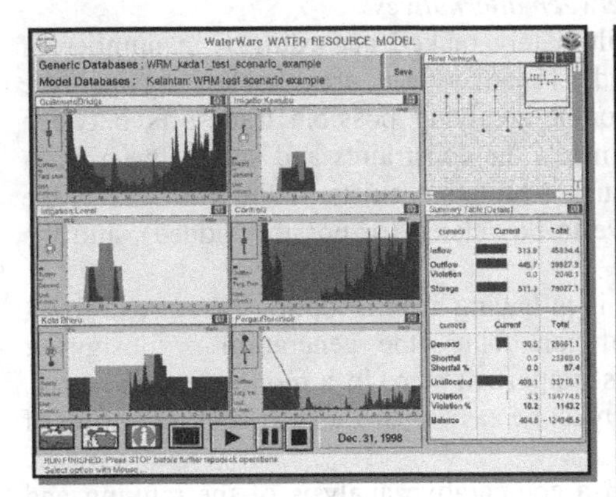

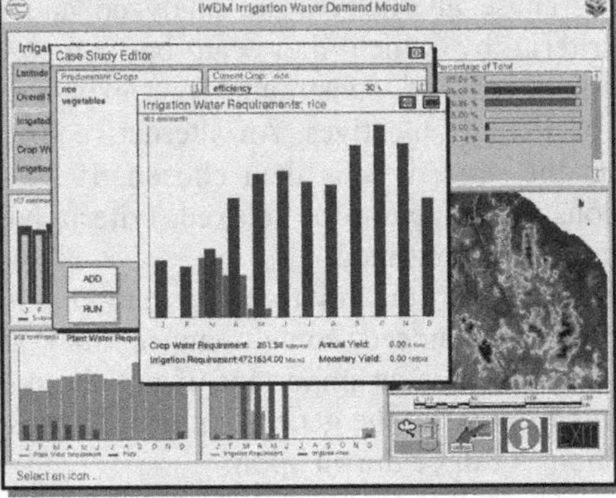

Figure 3. Simulation of a dynamic water allocation scenario, using the detailed simulation of irrigation water demand for the corresponding river basin nodes.

3.2.1. *Decision Variables*

The decision variables that define a water resources management scenario include, for example:

- water allocation patterns or water pricing;
- infrastructure development (a new reservoir);
- operational policies (like reservoir release rules);
- setting of thresholds and standards like water quality guidelines.

3.2.2. *Performance Variables*

The performance variables measure the overall behavior of the system (in terms of a set of criteria and objectives defined as direct or derived targets for these criteria) in an aggregate form. This is clearly necessary for simple reasons of cognitive limitations: a scenario run of one year at daily output intervals, for a given network with 100 links and nodes will produce up to 500,000 data items, depending on the number of attributes recorded at each node or reach. For comprehension (as an elementary step toward comparative evaluation), they must be summarized in a few performance variables. These could include:

- overall supply/demand balance, shortfall and unallocated surplus;
- average, maximum, and several spatial/temporal integrals of the various flow, demand, and supply or water quality values;
- reliability of supply, fraction of time (days) the targets are not met;
- net benefit (or cost) of the overall water supply/use system;
- compliance with environmental objectives and standards.

These criteria can be derived for the overall system, or any subsystem down to the individual nodes (e.g., city or irrigation district), the overall analysis period (typically a water year), or any seasonal sub-period down to individual days.

In summary, a simple scenario analysis results in a single (set of) result(s), that is (implicitly or explicitly) compared against a set of (absolute) objectives and criteria (expectations) and constraints such as environmental standards or some minimal requirements for flow.

3.3. COMPARATIVE EVALUATION

Comparative evaluation requires that the performance variables of more than one scenario (minimally two for direct pairwise comparison) are communicated (displayed) to the user simultaneously. For the spatially distributed network or domain-grid specific data, this is accomplished by displaying equivalent data sets in parallel display for direct visual comparison. For the performance variables, this is accomplished by the parallel display, tabular and graphical, of the respective numerical values or their symbolic classification. In both cases, graphical and numerical, the side-by-side display can be augmented by the calculation and display of relative and absolute differences (deltas) of the respective performance variables, for example, as a map of differential (increases and decreases) of groundwater tables from two model results representing two separate pumping scenarios.

In summary, comparative scenario analysis results in direct comparison of two (or a set of) result(s), that are explicitly compared against each other and interpreted in terms of improvement or deterioration of performance variables *vis a vis* the objectives and constraints. From the decision making point of view, this should lead to the selection of the preferred solution in this (possibly repeated) pairwise comparison.

3.4. OPTIMIZATION

Since each scenario is described by more than one performance variable or criterion, the direct comparison does not necessarily result in a clear ranking structure: improvements in some criteria may be offset by deterioration in others. This can only be resolved (and result in an eventual ranking and selection) through the introduction of a preference structure that defines the trade-offs between objectives. For a description of the underlying mathematics of iterative, multi-criteria decision analysis, see [9]. As an overall decision support tool, one can now attempt to optimize systems performance. This is basically inverting the logic of scenario analysis, asking HOW TO instead of WHAT IF: the target is given (at least in the sense of a direction, e.g., maximize net benefit while meeting all the constraints), and the set of decision variables required to get there is sought.

As an alternative, more appropriately for complex, dynamic, nonlinear and spatially distributed systems which are difficult, if not impossible, to optimize directly, one can use a discrete multi-criteria approach to find an efficient strategy (scenario) that satisfies all the actors and stakeholders involved in the water resources and environmental management decision processes. Discrete optimization selects from a set of (discrete) alternatives rather than trying to adjust a set of continuous parameters to an optimal setting. The preferences of decision makers can be conveniently defined in terms of a reference point, that indicates one (arbitrary, but preferred) location in the solution space, i.e., any hypothetical solution that seems desirable (but not necessarily achievable). Normalizing the solution space in terms of achievement or degree of satisfying each of the criteria between nadir (worst possible) and utopia (best possible) allows one to find the nearest available Pareto solution efficiently by a simple distance calculation.

4. Data Communication

An important aspect in all of the above is data communication. This is related both to the linkage of analytical tools and the data or information resources, as well as to the communication of information to the end users in the decision making process. One can distinguish three related phases: data acquisition, data management, and the communication with the end user or user interface.

Data acquisition covers a broad range of activities, from the slow and infrequent input of analytical laboratory data or major field surveys to the regular and automatic input of telemetry data from meteorological, hydrographical, or environmental monitoring stations.

One important aspect of these data streams is that they are distributed in space and usually also in terms of institutional ownership. That makes bringing them together into one consistent and integrated database difficult, as data collection strategies from sensors to data transmission protocols may be different. In the object oriented approach, these differences are addressed by different methods or filters that link the data streams to the monitoring station objects. For the basic layer of physical transmission, the use of GSM and the Internet helps to standardize protocols, i.e., using the http (hypertext transfer) protocol on top of relatively cheap carrier, i.e., public telephone.

For the data management layer, one finds a similar structure in that data resources (e.g., specific databases and collections) may reside in different locations with different institutions. Again these can be integrated into a virtual database through the object oriented architecture, a set of appropriate retrieval and filter methods, and the Internet protocols based on TCP/IP. One major advantage of keeping data distributed is that it also keeps the responsibility of data maintenance distributed and together with the ownership of the data. It does require, however, a high degree of reliable availability, as well as scrupulous adherence to the metadata agreed upon to ensure a reliable interpretation.

The Internet and its general-purpose user interface, the HTML browser, also provide a most convenient and easy way to use remote access for distributed users of information systems [10, 11]. While the restricted bandwidth limits client side interaction in HTML and its extensions, local flexibility can be achieved by Java applets at the price of higher data volumes to be transferred through the net.

5. Application Examples

5.1. WATERWARE SYSTEM

The application examples presented below are based on the *WaterWare* river basin management information system [7, 11, 12]. The system and its applications are also described on line: http://www.ess.co.at/WATERWARE.

WaterWare is an information and decision support system for water resources management, compliant with 2000/60/EC, that was originally developed within the framework of the EUREKA EUROENVIRON project EU 478. The system is designed to support river basin scale planning and operational management, monitoring, water allocation, pollution control, and environmental impact assessment tasks. *WaterWare* is designed to help compile and manage the information on river basins and basin management plans, as described by the Water Framework Directive 2000/60/EC [2].

The primary application domain is water resources management, including planning and operational applications; the system supports the compilation and integration of a consistent object oriented data basis in a dynamic, multi-media (hypermedia) style, that integrates data bases, text, GIS, CAD, graphics and imagery, and a dynamic expert system as well as interactive simulation models (rainfall-runoff, water allocation, water demand estimation, surface and groundwater quality, etc.). This forms the basis for decision support for water resources development and allocation, using a basin-wide dynamic water budget model coupled to a set of dynamic simulation and forecasting models of individual river basin components and functions for WHAT-IF type scenario analysis. A related function is the environmental impact assessment for water resources development projects.

WaterWare uses an object database to represent a broad range of river basin objects and scenarios of water resources development plans or management options. The primary simulation tool is a network-based water resources model that describes the daily supply-demand budget within the system. The individual nodes of the overall system represented by that model can be represented by specific models, which, in turn, describe the behavior of each node type in detail. Examples are a rainfall-runoff model for (ungaged) sub-catchments, or an irrigation water demand estimation model for irrigation districts.

5.2. LERMA-CHAPALA, MEXICO

The Lerma-Chapala catchment is situated in central Mexico. It covers a total area of more than 54,000 km2, with a difference in altitude of 1100 m. The main branch of the

Lerma rivers has a total length of more than 700 km. Five states have a share in the catchment, which includes 40 major subcatchments and tributaries.

With the by far dominant water use of above 70% in agriculture, extreme low flow events during the dry season, together with insufficient wastewater treatment, lead to severe water quality problems. The steadily growing water demand in the basin has led to a continuing drop in the level of Lake Chapala. This, together with the water quality problems of the lake, is a major concern.

5.3. KELANTAN RIVER, MALAYSIA

The Kelantan river basin largely coincides with the Kelantan province in northeast Malaysia. The upper catchment of the river (flow monitoring station at Guillemard bridge) covers an area of about 11,900 km2. Annual average streamflow over the last 40 years of observations varies from about 250 to almost 1000 m3/s, with a long term annual average of 515 m3/s (Fig. 1).

The application is specific for water resources management in a paddy growing area in North East of Penisular Malaysia, i.e., an area of around 30,000 ha under the Kelantan State Development Authority (KADA). This area is a coastal belt of Sg. Kelantan River.

The main problem in this area is water shortage during the dry season (March to May) and flooding in the monsoon season (November to January). Monthly flow, corresponding to the variability of the Monsoon driven climate, varies over two order of magnitude from a monthly low of 44 m3/s to a maximum of 4269 m3/s. The water management should be able to help in providing a sustainable water resources development with minimal conflicts between the needs of government, the local community, land developers and water users such as KADA, Water Supply Department, and the Drainage and Irrigation Department (DID).

The main objectives of the application are:

- to develop and implement a hydro information system for optimal, economically efficient allocation of water resources for improved paddy production;
- to determine optimum water resources management strategies during the dry period and droughts and the peak irrigation season;
- to identify environmental and ecological impacts of watershed patterns;
- to evaluate the water pollution and its impact on irrigation, industry, domestic water supply and also river fisheries.

6. Discussion

What holds for water resources management in particular equally applies to environmental planning and management in general. Environmental information and decision support systems (EDSS) have emerged over the last decades as important tools for environmental planning and management. Environmental problems, from urban and industrial pollution to natural and technological hazards, and certainly problems related to water resources, keep growing, driven by local and global population growth and ever growing consumption of energy and materials. However, especially in the industrialized countries, the most simple decisions with large pay-offs have already been taken: what remains in most cases is the fine-tuning of the relationship between technology, economy, and the environment.

These problems are complex in the physical domain and usually controversial in the socio-economic domain. Environmental systems are complex, dynamic, spatially distributed, and highly nonlinear. Their coupled processes operate on a multitude of interdependent scales in time and space. In addition, many of the governing processes are not directly observable and therefore not easily understood. On the socio-economic side, all decisions related to environmental planning and management are characterized by multiple and usually conflicting objectives and multiple criteria. One also faces the problem of uncertainty, data versus perceptions, beliefs, and fears, hidden agenda and plural rationalities, which are a necessary consequence of the increasingly wide public participation in the decision making processes. Environmental awareness has grown sharply over the last decades, and environmental legislation is introducing and tightening standards in many fields. Public participation and the right to know are mandated by law in many cases. This makes environmental information an important element of the policy making process in a civic society.

Society's response to the perceived (and some very obvious) environmental problems is the introduction of laws and regulations. Regulatory instruments or control policies take a variety of forms. These include a range of monetary instruments from taxes to subsidies; laws and regulations including planning requirements and process oriented controls, such as the requirements of best available technology (BAT); and more recently, mechanisms of self-regulation, the reliance on voluntary compliance and control, basically achieved through social and eventually market mechanisms and pressures rather than through any central government administered enforcement policy. Another important aspect of environmental legislation, however, comprises the provisions for free access to environmental information and public participation in decision and policy making processes, as they directly affect the design and implementation of environmental decision support systems.

Having established that there is a regulatory framework, and thus a real need for environmental information (in the broadest sense), the main issue is how to obtain - and communicate - policy and decision relevant information in a reliable, timely and cost effective way to the different audiences and participants in the decision making processes, ranging from the technical specialist to the general public. At the same time, information is becoming a commodity and a service in an explosively developing information society. Computing power is abundant and cheap, and access to broad-band communication is ubiquitous. These trends define the framework for the potential and the role of environmental information and decision support systems in the future.

Environmental information and decision support systems address environmental problems. Their development must equally consider the nature of planning and decision making processes and, thus, the users and audiences of such systems. A systems architecture, based on distributed client-server system and an object oriented design, provides the framework, multi-media formats and the user interface that can reach the diverse audiences involved in environmental decision making processes. This architecture implements and, thus, reflects the main features of a generic DSS approach: multi-layered (and distributed), open and flexible (adaptive), and interactive with a multi-media user interface. The implementation must also support the integration of a variety of tools including models, GIS, expert systems, and their shared databases. The system architecture proposed for environmental information and decision support systems can be understood at two levels. On the conceptual level, it describes the logical relationship of the elements (objects) used to represent, and manage, an environmental problem in a decision support system. This is based on an object oriented design. On the physical implementation level, it provides the actual operational hardware and software environment for running the decision support system. This is based on a client-server architecture that integrates various distributed information sources and supports different types of (display) clients. For an interactive decision support system, an intuitive problem representation and a graphical, symbolic user interface are important aspects; effective communication with a diverse audience is one of the basic requirements for a decision support system and the interactive user interface.

In the core of this approach are environmental simulation models that translate human interventions (e.g., emissions from techno-economic processes or the consumptive use of a natural resource) into environmental impacts and their subsequent socio-economic assessment. The basic principles behind these models are the conservation laws, resulting in systems of difference or partial differential equations describing the flow of environmental media (air, water), which, in turn, transport and diffuse pollutants. The control of these basic processes, responsible for the availability and quality of resources in space and time, including the transformation of emissions into ambient concentrations (emissions) is at the core of a large class of environmental problems [13].

Going beyond the classical numerical environmental models, which are primarily process oriented, geographic information systems provide the tools for spatial analysis. They can be linked to dynamic and spatially distributed simulation models to extend their basically static analytical capabilities.

Another class of tools is the rule-based expert systems; they use symbolic logic for deductive inference and can express qualitative concepts and conditional relationships that are difficult to formulate in purely numerical terms. Expert systems are empirical and heuristic systems, based on a more or less explicit, and usually qualitative, understanding of how things work. A perfect example of an ideal application area is law or, in the context of water resources, water rights and allocation problems. However, it is important to realize that expert systems are certainly no substitute for many time-tested quantitative and physically based methods and models, but they should be seen as complementary techniques which can improve many of these models. Obvious applications related to numerical models are in data pre-processing, parameter estimation, the control of the user interface, and the interpretation of results. There are certainly enough arts and crafts components in numerical modeling, that open attractive opportunities for AI techniques. Within the multi-tiered approach to decision support presented, expert systems can play an important part for many relationships that are difficult to quantify.

The integration of these different tools, i.e., models, GIS, and expert systems, provides the basis for a new generation of powerful information and decision support systems. Decision support systems (DSS) are based on information management and model-based decision support. They envision experts as the primary users, as well as decision and policy makers; and, in fact, the computer is seen as a mediator and a translator between the expert and the decision maker, and between science and policy. The computer is, thus, not only a vehicle for analysis, but also and even more importantly, a vehicle for communication, learning, and experimentation.

The three basic, interwoven elements of a DSS are:

- to supply factual information, based on existing data, statistics, and scientific evidence;
- to assist in designing alternatives and to assess the likely consequences of such new plans or policy options; and,
- to assist in a systematic multi-criteria evaluation and comparison of the alternatives generated and studied.

Practical systems are characterized by methodological pluralism. The individual components of the system are based on quite different concepts, levels of aggregation, and methods of analysis, namely, numerical simulation, mathematical programming, symbolic simulation, interactive database access, and rule- and inference-based information retrieval, all of which are integrated into one coherent system.

Rather complex environmental decision support systems are becoming feasible in part due to the explosive development of information technology. Broadband communication and the Internet are further changing the role and the nature of environmental information.

Not only do environmental problems continue to pose challenges, but they also get more difficult as the technological and economic constraints, as well as the regulatory framework, get tighter. At the same time, the nature of policy and decision making changes into a more participatory style. Different audiences and the direct public use of scientifically based information in increasingly open decision making processes call for new presentation formats, a new style of argumentation, and also a new role for the scientist.

The main change in attitude and approach is the integration of scientifically based analysis with public policy and decision making processes. This forces scientists to leave the fabled ivory tower and present their findings to a lay audience. This requires effective communication beyond a peer group, and this is increasingly achieved worldwide over the Internet. This changes the rules of the traditional academic discourse radically. The detached objective analysis, subject only to peer review and judged on a fine balance of innovation and a solid foundation in tradition as well as elegance, becomes embroiled in daily politics and is judged by public acceptance, usefulness, efficiency, and effectiveness. Concepts of correctness, precision, observability, experimental verification, and l4east-square correspondence with data, completeness, convergence, formal proof, or optimality are confronted with concepts of political feasibility, cost efficiency, expediency, acceptability (that is the ability to sell it to a majority), or simply good enough.

In the context of multi-criteria decision support, it is easy to demonstrate that, given a set of feasible alternatives, the efficient solution depends on the choice of criteria. This choice is ultimately a political one. What a decision support system then contributes is not so much an efficient mechanism to find an optimal solution, given any set of (more or less) debatable preferences, but a mechanism to make the entire process more accessible, open and transparent.

7. References

1. EEA (1999) Environment in the European Union at the turn of the century, European Environment Agency, 44 pp, Luxembourg Office for Official Publications of the European Communities.

2. CEC (2000) Directive 2000/60/EC of the European Parliament and the Council of 23 October 2000 establishing a framework for action in the field of Community water policy, Official Journal L327, 22/12/2000, p001-0073.

3. CEC (2000) Communication from the Commission to the Council, the European Parliament and the Economic and Social Committee: Pricing policies for enhancing the sustainability of water resources, COM(2000) 477 final, Brussels.

4. Booch, G. (1991) *Object Oriented Design with Applications*, Benjamin/Cummings, California, USA, ISBN 0-8053-0091-0.

5. Rumbaugh *et al.* (1991) *Object Oriented Modelling and Design*, Prentice Hall, NJ, USA. ISBN 0-13-629841-9.

6. Shlaer, S, and Mellor, S. (1988) *Object Oriented Systems Analysis, Modelling the World in Data*, Yourdon Press, NJ, USA. ISBN 0-13-622940-7.

7. Fedra, K. and Jamieson, D.G. (1996) An object oriented approach to model integration: a river basin information system example, in K. Kovar and H.P. Nachtnebel (eds.), IAHS Publ. no 235, pp. 669-676.

8. Fedra, K. (1997) Integrated Environmental Information Systems: from data to information, in N.B. Harmancioglu, M.N. Alpaslan, S.D.Ozkul, and V.P. Singh (eds.), *Integrated Approach to Environmental Data Management Systems*, Kluwer, Dordrecht, pp. 367-378.

9. Fedra, K. (2000) Environmental Decision Support Systems: A conceptual framework and application examples, Thése prèsentèe á la Facultè des sciences, de l'Universitè de Genéve pour obtenir le grade de Docteur és sciences, mention interdisciplinaire, 368 pp., Imprimerie de l'Universitè de Genéve, 2000.

10. Fedra, K. (2000) Environmental information and decision support systems, *Informatik/Informatique* **4/2000**, 14-20.

11. Fedra, K. (1996) Multi-media environmental information systems: Wide-area networks, GIS, and expert systems, *GIS: Geo-Informations-Systeme* **9/3**, 3-10.

12. Fedra, K. and Jamieson, D.G. (1996a) The WaterWare decision-support system for river basin planning: II. Planning Capability, *Journal of Hydrology* **177 (1996)**, 177-198.

13. Loucks, D.P., Kindler, J., and Fedra, K. (1985) Interactive water resources modeling and model use: An overview, *Water Resources Research* **21/2**, 95-102.

URBAN DRAINAGE, DEVELOPMENT PLANNING AND CATCHMENT FLOOD MANAGEMENT – GIS CONTRASTS IN THE U.K.

J. C. PACKMAN
Natural Environment Research Council
Centre for Ecology and Hydrology, Wallingford, U.K.

Abstract. GIS developments are compared for three sectors involved in urban flood management. *Private Water Companies* are responsible for water supply and drainage, holding databases of service areas, pipes and manholes. GIS allows model results to be linked to maintenance and failure data (blockages, overflows, flooding, etc.). *Local Government* is responsible for planning and regulating urban growth, involving wide-ranging consultations. They hold GIS databases of property boundaries, land use, population, facilities, environment, etc. They aim to include environmental guidance, legal frameworks, and links to models. The Environment Agency is responsible for flood defense and regulating water abstractions and discharges. They hold databases of rainfall and runoff, topography, land use, climate, etc. New "Flood Management Plans", linking GIS to model of flood discharge, depth and damage, will inform *Local Government* planning but will be separate from the other GIS.

1. Introduction

Urban areas are central to most water management problems, forming concentrations of population and capital investment that need both water supplies and protection from flooding. In the past, municipal governments were responsible for local water management, but the need to consider issues of water quality, resources and flooding on a wider regional and river basin scales has resulted in more complex institutional arrangements. In most of the UK (Scotland and N. Ireland differ in various respects), responsibility for water issues is split at least three ways. Local Authorities (**LAs**)[1] remain responsible for the planning and regulation of urban growth and for ensuring the effectiveness of *ordinary* (minor) *watercourses*. However, the Environment Agency (**EA**), a public body linked to the Department for Environment, Food and Rural Affairs (**DEFRA**), is responsible for licensing river and groundwater abstractions, for regulating foul, treated and surface water discharges to the environment, and for providing defenses against *main river* flooding. Regional Water Service Companies (**WSCs**), operating under the financial jurisdiction of the government's Office for Water Services (**OFWAT**) and the technical jurisdiction of the EA, are responsible for developing and delivering water supplies and urban drainage. Each of the organizations (LAs, EA, WSCs) operates a monitoring network of some sort, and each has, or is developing, GIS based procedures to help derive and display information relevant to their role. As each organization also uses consultants for some aspects of its work, GIS provides a way of retaining in-house knowledge and expertise, and of easily presenting and sharing relevant information. Concentrating on the issues of drainage and flooding,

[1] All acronyms and abbreviations are shown in bold where first defined.

this paper describes further the roles and responsibilities of each organization, the monitoring networks they operate, and developments they are making in the use of GIS.

2. Urban Drainage, Watercourses and River Channels: U.K. Roles and Responsibilities

In the UK, the responsibility for protecting land from flooding and for managing drainage so as not adversely to affect neighboring land lies principally with the landowner. However, various organizations have powers and responsibilities depending on the legal status of the drainage system, ranging from *private drains* and *public sewers* (essentially piped drainage), to *ordinary watercourses* and *main rivers.* These responsibilities define each organization's needs for monitoring networks and information production.

Private drains normally relate to individual properties (e.g., housing plots) and are the sole responsibility of the property owners. However, they include *highway drains* serving the carriageways and pavements of public roads (owned by Highway Authorities but often managed under contract by LAs). They also include drains in many industrial and business parks, where property is leased and infrastructure managed centrally. *Public sewers* generally form the main urban drainage system collecting flow from *private drains* and *sewers.* Owned by the WSCs and defined in asset inventories, they include surface water drains, foul (wastewater) sewers, and combined (surface & wastewater) sewers. Foul and combined sewers drain to treatment works, but surface water drains and storm overflows on combined sewers (**CSOs**) discharge in wet weather directly to rivers and watercourses. The WSCs invest about £600M of capital each year to reduce flooding and pollution caused by urban drainage. Charges that can be levied from their customers are set by OFWAT, based on various performance criteria, including service efficiency, numbers of properties at risk of flooding, environmental impact of sewage overflows and discharges, and prosecutions for non-compliance with discharge consents. The 2000-1 OFWAT report [1] shows an increase in unsatisfactory CSOs, and an increase in sewer flooding in some areas. Nationally, over 20,000 properties are at risk of sewer flooding once in ten years.

Ordinary watercourses are normally small open channels, ditches and streams draining areas of less than a few square kilometers. However, in urban areas, they may include surface water drains, culverted watercourses and hidden rivers draining areas of 10 km^2 or more. *Main rivers* are normally larger channels, specifically identified on maps held by DEFRA. However, they can include some quite small channels draining areas of under 5 km^2. Riparian owners alongside *ordinary watercourses* and *main rivers* have rights and responsibilities on flooding and drainage, but the LAs and EA respectively have 'permissive powers' (and thus responsibilities, but not duties) to carry out works for the common good (such as flood defense). LA activities seldom go beyond maintenance or trying to resolve persistent minor problems (e.g., at roadside ditches and culverts). In practice, the EA is responsible for most flood defense expenditure, with an annual investment of about £250M drawn from departmental grants and flood defense levies (collected by the LAs on its behalf). EA flood defense activities are overseen by local area committees that include nominees from the LAs. While the EA's responsibilities on flooding relate to *main rivers*, its responsibilities for water quality cover all rivers. The EA will thus identify and seeks to enforce pollutant limits on all unsatisfactory outfalls (industrial, surface water, and CSOs) to any watercourse.

Despite their involvement in *ordinary watercourses*, the LAs' main involvement in flood issues comes from their strategic planning and development control duties. Urban development in flood plains is at risk from flooding, and it reduces the floodable areas that hold back floodwater from downstream. Urban surfaces and drainage systems may also increase the volume and speed of runoff thus raising flood flows downstream. Updated guidance [2] on development and flooding has recently been issued by the Department for Transport, Local Government and the Regions (**DTLR**). This requires LA to assess flood issues in their strategic planning and to prioritize future development sites on the basis of flood risk. In assessing development applications, LAs should ensure that flood issues are properly addressed, that full flood-risk assessments are made where appropriate, and that the costs of flood defense works and long-term maintenance are fully covered. The LAs should consult with the EA on all flooding and drainage issues, but the final approval rests with the LA, and minor developments may be handled under a blanket agreement. The EA will usually oppose development in flood plain areas and recommend that runoff is controlled by the use of Sustainable Urban Drainage Systems (**SUDS**) - involving soakaways and local flood storage. Recent severe flooding has highlighted these issues, and the EA now supplies *indicative flood plain maps* for England and Wales. It has also initiated a national programme of Catchment Flood Management Plans (**CFMPs**) to assess land use impacts and flood defense strategies on a catchment-wide basis. Developers will usually discuss flood defense measures directly with the EA. They will also agree water supply and drainage details with the WSC, who, on completion of the development, will (usually) take ownership of the new public water infrastructure.

In addition to these responsibilities, LAs are also responsible for emergency services, including flood fighting, and the EA is responsible for providing flood warnings.

3. U.K. Urban Environment Monitoring Networks and Their Purpose

The above discussion has shown how the WSCs are responsible for urban runoff systems and for meeting discharge criteria to receiving watercourses; the LAs' for planning control; and the EA for river flood defense and for setting and monitoring criteria to which the WSCs and LAs should work. Most urban environment monitoring networks relate directly to these responsibilities.

The Environment Agency (EA) is the prime body concerned with environmental monitoring and data archiving in the UK, operating a permanent network of approximately 1300 river flow gauging stations throughout England and Wales. The Scottish Environmental Protection Agency (**SEPA**) operates a further 300 stations, and the Northern Irish Department of the Environment (**DoENI**) operates about 40 stations. Equivalent numbers of 'level only' stations are also operated, along with many temporary and short period stations. Historic data may have been digitized from chart records, but, since the late 1980s, most gauges are logged electronically at 15-minute intervals. More recently, many of the stations have been telemetered, and data may be downloaded at regular intervals (e.g., daily), or more frequently during incidents or when triggered by alarm conditions of flood or low-flow.

The EA/SEPA/DoENI flow monitoring programs are managed by up to 40 separate area offices throughout the UK, and the data are currently stored on separate database servers (though to a common database format within each organization). SEPA has recently established a unified flow archive, and EA is proposing a similar system (HARP). Currently, data are supplied to external users as computer files, but access over the Internet is being considered, allowing easier interpretation by the users' software. As described by Littlewood [3], the only integrated database of river flow is the National River Flow Archive (**NRFA**) held at CEH Wallingford. This comprises mean daily flows and monthly maxima for the 1300 principal EA, SEPA and DoENI flow gauges. However, a new archive of high quality flood data has recently been proposed by the EA, with access provided via the Internet.

The river flow monitoring network is not primarily operated to assess catchment flood behavior, but to assess water resources, allowable abstractions and discharges, and low-flow conditions when reduced abstraction rates are imposed. Telemetered flow gauges, with or without telemetered rain gauges, are also used for flood forecasting and warning purposes. The monitoring thus forms a *decision support* network, as described by Harmancioglu et al [4], with gauge locations chosen for operational needs rather than to represent catchment processes. However, data from 1000 gauges over the UK were considered suitable (longer records, stable rating curves, etc.) for developing statistical flood frequency procedures as described in the UK Flood Estimation Handbook (FEH) [5]. Data from 200 stations, where adequate recording rain gauge data were also available, were used to develop the FEH rainfall-runoff method of flood estimation. These subsets of gauges thus form pseudo *academic-curiosity* networks, though the site locations were not chosen to understand overall catchment response or to test specific hypotheses. In general, the data are better suited to determining local parameter values for a generalized model form than for testing new model structures. This is the approach adopted in the GIS based procedures for Catchment Flood Management Plans (**CFMPs**) described later in this paper.

Besides flow monitoring networks, the Environment Agencies also maintain groundwater observation wells, daily and recording rain gauges, and water quality sampling and monitoring sites. Water quality data, covering 65,500 locations throughout England and Wales (with regular sampling/monitoring at about 15,000 locations) is held on eight regional servers of a common Water Information Management System WIMS. Most of the data relate to weekly or monthly sampling, involving a few basic determinands (e.g., Dissolved Oxygen, Ammonium, BOD), or specific suites of determinands concerned with European directives (Bathing, Nitrates, Urban Waste Water Treatment, etc). As with flow, the data can be supplied in computer files, and access over the Internet is being considered. However, corresponding flow data for most locations is not available, and thus, neither instantaneous nor annual pollutant loads can be derived. For about 170 sites, mainly near the tidal limits of major rivers (including Scotland and Northern Ireland), data for a wide range of pollutants are abstracted and combined with appropriate flow data to produce the Harmonized Monitoring database specific to estimating loads to the sea (see Littlewood [3]).

For comparison with water flow and quality monitoring, air quality data are collected mainly by Local Authorities, but integrated into a National Air Quality Information Archive (covering 140 automatic hourly monitors and over 1350 weekly storage tubes) maintained by DEFRA. Details of monitor locations and data for a range of gases and particulates can be accessed freely over the Internet. Such access to data may reflect greater public awareness and concern for air quality than water, but it may also reflect the additional errors and uncertainty in water data, and the ultimate need to convert measured river levels and pollutant concentrations to flow and flux rates.

Although these environmental monitoring networks are quite extensive, there is generally a dearth of monitoring at sufficient spatial and temporal discrimination to define processes within the urban environment or to investigate strategies to mitigate urban impacts on flooding and water quality. Urban runoff response is flashy, derives from extremely varied land use, and is affected by rapidly changing flow characteristics in piped drainage systems (e.g., at overflows, due to backing up and surcharging in manholes, from flow dependent changes between sub and supercritical velocity, etc.). With additional problems over maintenance, representative urban flow and quality monitoring networks are rare, both in sewer systems and in downstream river networks.

3.1. URBAN DRAINAGE FLOW SURVEYS

The need to improve urban drainage methods has led to limited historical *academic-curiosity* monitoring in the UK, covering periods of 1-3 years at about 20 sites. From these data, individual flood events have been extracted [6] and generalized rainfall-runoff models developed, including the current HydroWorks [7] urban drainage model. However, a new monitoring strategy has developed around the WSCs' use of models to assess flood and water quality for current and future conditions. Intense monitoring networks, involving up to 20 flow monitors/km^2, are operated for short periods of typically 6-12 weeks (covering at least 3 storm events) in order to verify the model performance. The model is then run, using selected events from a longer observed or synthetic rainfall sequence to predict flood runoff and water quality impacts and to assess new management options. This ***flow survey*** is not for *decision support* or *academic-curiosity*, but for *model verification*. Its aim is not to develop modeling procedures but to allow minimum adjustments to generalized model parameters.

Given the character of urban discharges, *flow survey* monitoring is usually based on a special 'mouse', temporarily fixed to the drain invert, comprising a pressure transducer measuring flow depth, and an ultrasonic Doppler sensor measuring flow velocity in the vicinity of the mouse. Velocity profiling and field gauging are used to convert monitored depth and velocity to flow rate. Automatic water quality samplers may be triggered by depth or flow. Equipment in sewers must be *intrinsically safe* (cannot give sparks to ignite explosive gasses), and the installation and management require specialist equipment, work teams and safety procedures. Thus, *flow surveys* are normally carried out for the WSCs by specialist *flow survey contractors*. Model building, verification, and use in the assessment of system performance are also highly specialized and are usually carried out by a consultant working closely with the *flow survey contractor*. Studies are 'one-offs', and monitoring of long-term impacts or changing seasonal conditions is rare. As the data are project specific and seldom of sufficient extent for developing modeling processes, they have rarely been made available to external users. Older surveys are often poorly documented, and the data may be unusable or even lost.

However, the *flow survey* and modeling approach has proved effective for the WSCs, particularly in assessing renovation and rehabilitation strategies to reduce flooding within the sewer system (so called Drainage Area Studies, **DAS**). There is little incentive for WSCs to extend monitoring periods. However, studies to assess discharges and water quality impacts in the receiving watercourses (Urban Pollution Management, **UPM** [8]) are less convincing. UPM studies are mainly intended to address unsatisfactory CSOs, as identified by the EA through visual assessment (sewage products, erosion, sedimentation, etc.), and then included by OFWAT in the WSC work programmes (Asset Management Plans, **AMPs**). The EA is involved in supervising UPM studies and assessing proposals, but resources for independent assessment and post project appraisal are seldom available: *Flow survey* monitoring will have ceased, and any longer term EA water quality monitoring in the receiving watercourses will normally be at monthly intervals and exclude flow, making the identification of CSO impacts in wet weather impossible to identify.

4. GIS Procedures for Producing Information

4.1. WATER SERVICE COMPANIES (WSCs)

The use of urban drainage models (**UDMs**), as described above, requires detailed information on the drainage system, including:

(a) pipe lengths, slopes, diameters, roughness, and sediment depth;
(b) details of CSOs, weirs, tanks, sluices and other structures;
(c) manhole depths and diameters;
(d) paved, roof and pervious areas draining to pipe lengths; and
(e) for water quality studies, sediment and pollutant data for different land uses, industrial discharges, and sewer deposits.

UDMs are built on a manhole-to-manhole unit, and urban areas in the UK typically have a (public sewer) manhole density of about 400-600 per km^2. However, modeling each manhole has been found unnecessary, and it is usual to consider only those manholes where there is some significant change (e.g., of contributing area, diameter, or slope). Even so, a typical UDM will still include about 200 manholes/km^2 and may cover an area of 5 km^2. Collecting and managing such quantities of drainage system data is a sizeable challenge, often involving detailed site surveys of pipe connectivity and contributing areas.

However, digital databases of sewer assets were developing rapidly in advance of the 1989 privatization of the WSCs. Adopting the STC25 [9] manhole record card and combining information from field survey and sewer maps, PC-based databases were compiled, including manhole dimensions, locations (to the nearest meter), and incoming and outgoing pipes. As computer power and digital mapping developed, software packages could display the data graphically, allowing drainage boundaries and contributing paved, roof and pervious areas between manholes to be defined. These STC25 based packages were not true GIS, but they could be used to query manhole details, trace flow paths, and output system data in the format required by a UDM.

STC25 based packages are still in use, but limitations on the number of manholes they can hold, progressive centralization of WSC database functions, and the need for systems that are more task related have led to the development of newer GIS based procedures. Much of the STC25 data are 'public access', as previously supplied by LAs following requests from prospective developers and house buyers for details of the nearest sewer. The WSCs usually continue to provide such parts of the database for LAs and EA use. Similar information (but for every manhole) is needed for UDM studies, together with details of contributing areas and sewer controls (pumping stations, tanks, overflows, sluices, etc.). Asset Management tasks require more details of the type and frequency of maintenance, and OFWAT requires records to be kept of operational incidents such as flooding, its cause and extent. Although operational records do not provide conventional flow and level monitoring data, they can provide a useful overview of drainage system performance to corroborate model studies.

Consultants carrying out DAP and UPM studies (see previous section) are increasingly using GIS procedures, with maps, pipe layouts, and control structures on separate layers. Macros can export drainage system data files for external UDMs, and modeled flows and levels can be saved in a form accessible for display back in the GIS, either as color-banded spatial snapshots and sequences, or as time series plots at selected points (manholes). *Flow survey* data can be saved in the same way, allowing observed and modeled flows to be compared in the same spatial and point formats. Summary model outputs, such as event peak flows or the frequency of flooding can be also be displayed.

Figure 1 [10] shows one such GIS view, where the basic street and drainage layout is overlain by:

(a) modeled estimates of flood frequency (1-X years);
(b) those pipes requiring regular flushing; and
(c) the properties that are frequently flooded.

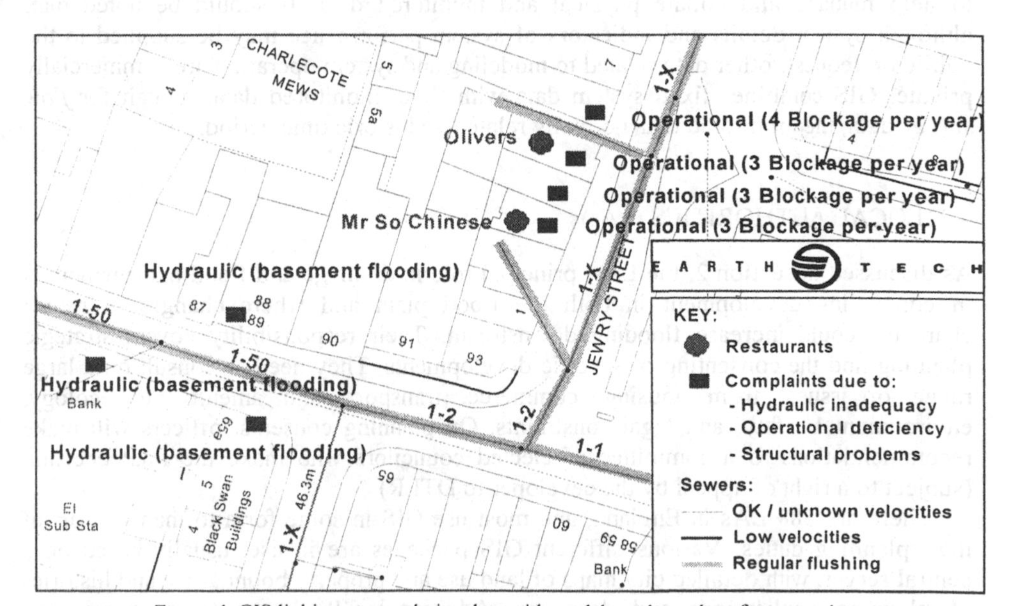

Figure 1. GIS linking map and pipe data with model results and performance data.

The difference between modeled and observed flooding was assessed as largely due to the accumulation of fats discharged from local restaurants. All the data were held together in one GIS project file on a single PC and could be supplied to the client on CDROM. Thus, the long-term stewardship of the data and model results are better assured.

The modeled results shown in Fig. 1 were based on the HydroWorks UDM, which has now been built into an integrated GIS based system called InfoWorks [11]. Figure 2 [12] gives another example of how GIS can help manage monitored data, showing the GeoPlan (pipe layout), Node (rainfall-runoff) and Results (*flow survey*) windows produced by InfoWorks. For clarity, the GeoPlan window has the street and building backgrounds switched off, but pipe width and shade are banded to indicate modeled instantaneous flow depth and surcharging during an observed event. Arrows on the pipes indicate flow velocity, and concentric circles round the nodes indicate levels of surface flooding. Buttons on the Toolbar allow the response to rainfall to be played and paused at any time, and clicking on a node gives the numerical results at that point. The node window shows the modeled inflow time-series at a selected point, and it also shows the time at which the GeoPlan window is stopped. The results window shows modeled and *flow survey* time-series of depth and flow at a selected node. Windows can be moved and resized and new windows opened, allowing a detailed view of system performance during an event.

InfoWorks is a PC Windows based package, and GIS and database files can be imported and exported in a number of standard formats. Basic spatial and time-series data are usually held on a central server, with users bringing working copies into their own filespace. Drainage system data will often need editing from that held by the WSCs, particularly if large areas of *private drains* need to be modeled. Extensive audit trail and case differencing procedures are included to improve stewardship of the data.

The above discussion is meant to indicate how GIS can be used in the WSC context to help manage and collate physical and monitored data. It should be noted that, although system details and indicators of system performance may be supplied to the public on request, other data related to modeling and system operation are commercially private. GIS combines fixed system data with 'live' monitored data, though for *flow survey* data, the monitored data normally relate to a historic time period.

4.2. LOCAL AUTHORITIES (LAs)

As discussed in section 2, the LAs' principal role in drainage and flood management is in controlling development in both the flood plain and where changes in runoff character could increase flooding downstream. Their responsibility covers strategic planning and the consenting of specific developments. They need to consult on a large range of issues, from housing, commerce, transport, and amenity, to ecology, environmental safety, and legal constraints. On planning consents, officers will make recommendations to a committee of elected councilors who make the final decision (subject to a right of appeal by the developer to DTLR).

There are 388 LAs in England, and most use GIS in some form to manage part of their planning duties. Various different GIS packages are in use, usually based on a central server, with detailed city maps of land use and property boundaries, and histories of planning applications and decisions. Additional GIS layers (often more than

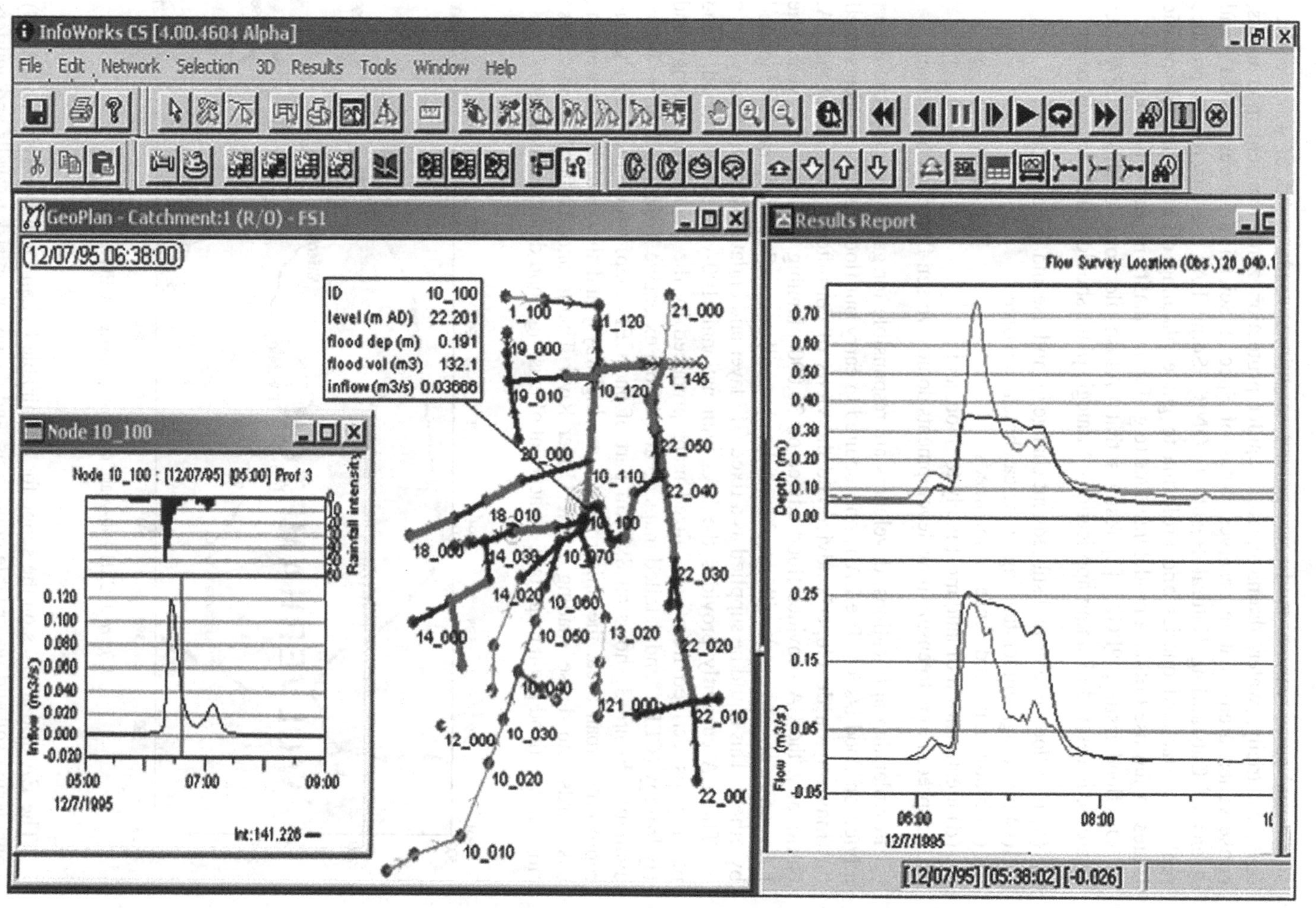

Figure 2. GIS based drainage modeling, displaying results and test data.

50) would cover various land uses, ground conditions, roads, rights of way, services (gas, electricity, water, drainage, etc), aquifer protection zones, ancient monuments, conservation areas, nature reserves, Sites of Special Scientific Interest (**SSSI**), and Areas of Outstanding Natural Beauty (**AONB**). Some LAs use macros to produce replies to standard queries and proforma to guide planning officers through specific issues. A few use GIS as a model interface (e.g., for air quality impacts of incinerators).

On flood issues, PPG25 [2] restates that LAs should consult with the EA over strategic issues and planning applications. Strategic plans should reflect EA advice on:

(a) past flood events and subsequent change in land use and flood defenses;
(b) hydrologic understanding of the catchments involved;
(c) biodiversity and conservation needs;
(d) identification of land most likely to flood; and
(e) precautions to ensure new developments do not worsen flooding downstream.

For planning applications, developers are responsible for gaining appropriate expert advice on flood issues. The LA is not required to carry out flood risk assessments itself but may rely on developer's advice and the views of consultees, particularly the EA. Nationally, the EA provides flood advice on 30,000 planning applications a year. The LAs rarely need to work with monitored data but will assess flood information supplied by others. This could be supplied as a fixed GIS layer and updated when necessary.

The EA currently provides the LAs with national, 100-year return-period, flood plain maps, derived from generalized maps produced by the Centre for Ecology and Hydrology (**CEH**), and detailed maps obtained from extensive hydraulic analysis where available. Figure 3 shows an 8x3 km part of one CEH map, where the gray shading represents urban area, the uniform lines are highways, and the variable line wending eastwards into dense shading is the Upper River Thame and its flood plain. These 'indicative' flood plain maps take no account of existing flood defense measures.

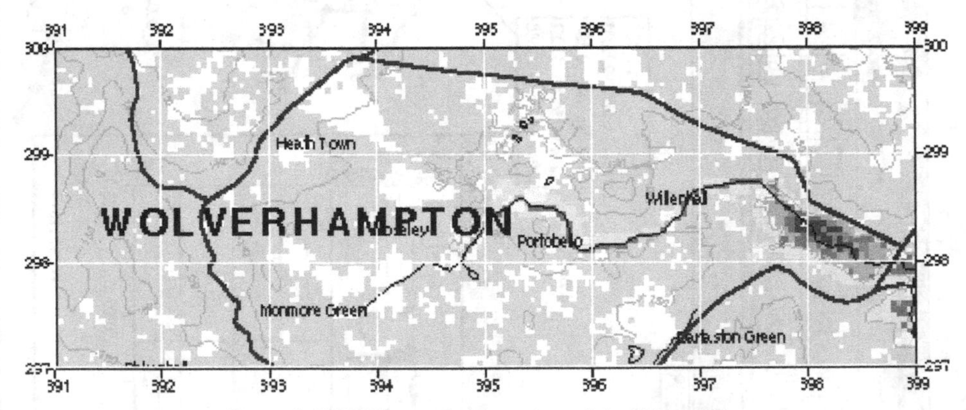

Figure 3. CEH 100-year flood map (copyright CEH & OS).

The EA also advises on the scope for Sustainable Urban Drainage Systems (**SUDS**), considering soil and water table conditions, and the pollution risk to wells and aquifers. Maps showing the suitability for SUDS have been prepared for some areas.

There is considerable scope to extend and standardize GIS approaches within the LAs. DTRL is supporting a three-year project under the Natural Environment Research Council's urban environment programme **(URGENT)** to develop a new GIS based Environmental Information System [13]. This will provide an interface to URGENT research data, models and scientific expertise, providing thematic maps, guidance, and proforma on specific environmental issues. These include: mining subsidence; unstable ground; contaminated land; water pollution; air quality; flood risk; conservation; and biodiversity. CEH has the lead role in water pollution and flood risk.

4.3. THE ENVIRONMENT AGENCY (EA)

As discussed previously, the EA maintains extensive monitoring networks and databases of rainfall, runoff, and water quality. They also hold various national gridded data sets (e.g., maps, topography, land use, climate), and historic flood event outlines. Using in-house teams and external consultants, they run groundwater, catchment and river models to assess abstractions, discharges and floods, to design flood defenses and to determine flood plain limits. They also assess modeling studies by WSCs and developers. They already use GIS or GIS-like procedures in some of these areas.

Recent flood incidents have highlighted a need to consider the impacts of changing land use and climate change on a catchment-wide basis. The EA is now commissioning a national coverage of Catchment Flood Management Plans (**CFMPs**) to be developed within a PC based GIS framework [14], providing a consistency of structure and approach. Figure 4 shows the steps involved in the CFMP process (on the left) where the dotted box forms the scope of the GIS framework (on the right). The framework has an open structure, comprising a core database (extracted from the EA's databases), linked to modeling tools. Although the models for hydrology (flood runoff) and river flow routing are external to the GIS, the assessment of flood depths over the flood plain and the conversion to damage costs are made within the GIS itself.

CFMPs are intended for strategic planning in catchments up to 10,000 km^2, and for:

(a) understanding flood generation processes;
(b) defining flood hydrographs over a range of return periods at key locations and sub-catchments (nodes);
(c) assessing flood management policies under current and future land use and climate scenarios.

The hydrology and flow routing tools must give appropriate detail and be readily applicable over the whole catchment. Thus, the FEH [5] statistical and rainfall-runoff methods are used to generate subcatchment flows, and the ISIS [15] variable diffusion model is used for river flow (allowing schematic flood plains and using rating curves to convert flow to depth). These models are calibrated as far as possible on monitored flood peak data. The **Case manager** provides the vital means of combining, saving, and querying the scenario and modeling options that lead to specific flood estimates.

The use of monitored data is fundamental to the FEH hydrology methods [5]. In the statistical method, from annual maximum flood data on 1000 catchments, median (*i.e.*, 2-year) annual flood ***(QMED)*** and ***growth factors*** to rarer floods were derived. A regression equation was found, relating *QMED* to catchment descriptors (*area, soil type, annual rainfall, etc.*) determined using a digital terrain model and gridded data. At an ungauged site, *QMED* is estimated by the regression equation and adjusted by the mean *estimation-error-factor* found at similar gauged catchments nearby. *Growth factors* are then found by 'pooling' factors from gauged sites most similar in *area, soil* and *rainfall*. The method is dynamic in that new-gauged data can contribute directly to the estimate.

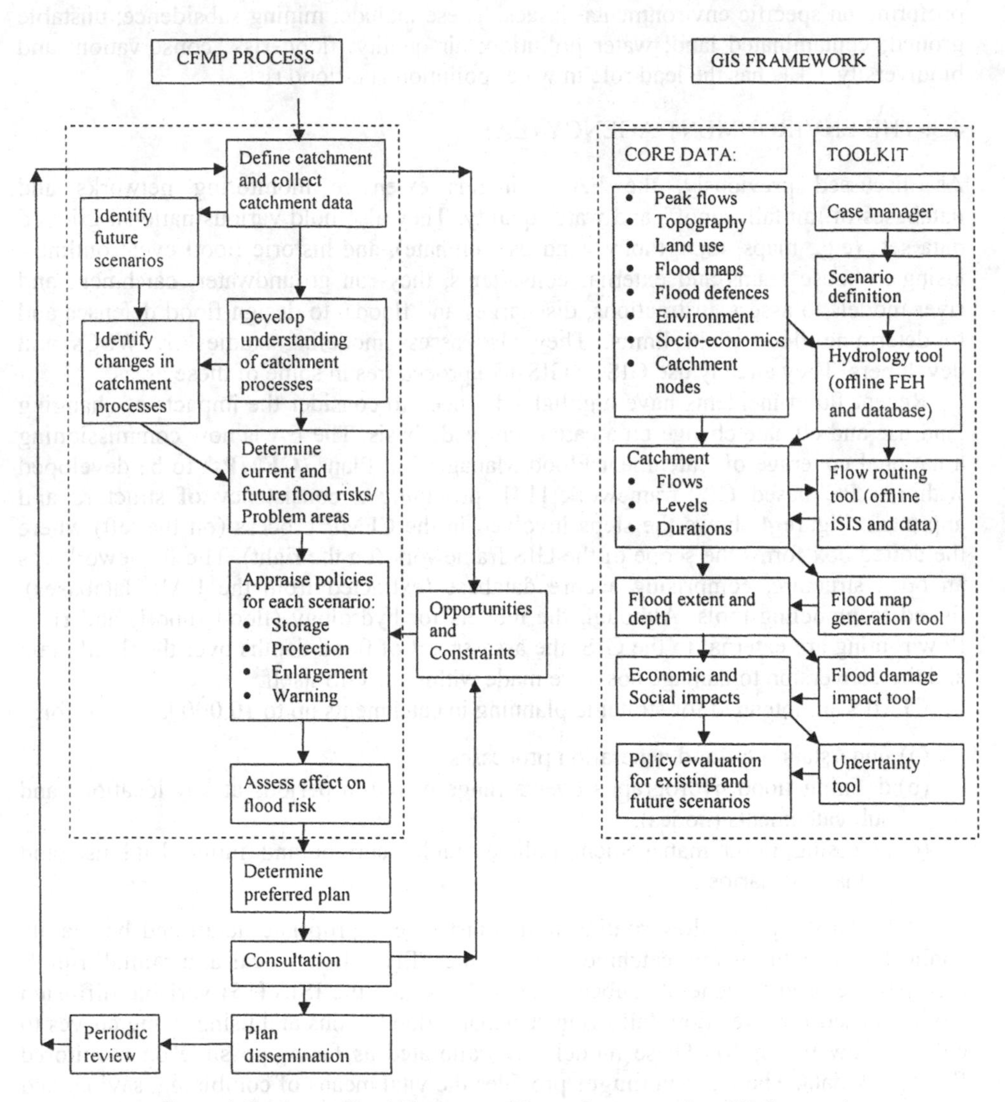

Figure 4. Catchment Flood Management Plan and corresponding GIS framework.

The FEH is accompanied by three CD-ROMS giving: catchment descriptors for every UK catchment of greater than $0.5km^2$; annual maximum data from the original model development; and the WINFAP software to derive pooled *growth curves* using additional data. Although the user should choose 'similar' catchments with care, a fully automated procedure has been applied to every catchment greater than $0.5km^2$. Thus GIS layers of 2,5,10, 25, 100 and 200 year *peak flows* (based on the original data) have been derived and will be incorporated in the CORE DATA of the GIS, allowing an immediate catchment-wide view of flood risk. These *peak flow* layers will be used to inform rather than define the CFMP, and the user should compare results with those obtained at key sites by manual WINFAP analysis, using updated annual flood data.

The FEH statistical method is based on direct analysis of flood data and provides the best estimate of T-year flood peak. However, it provides little information on flood generation, or the effects of confluences, flood plains, land use or climate change. The GIS framework thus uses the FEH rainfall-runoff model for its main scenario analysis. Local rainfall and runoff data to calibrate the model are comparatively rare though the FEH does present fitted model parameters for 200 catchments, and also *peak flow* data can be used to develop generalized parameter adjustments. Further details are beyond the scope of this paper, but it should be noted that the model is used both for individual subcatchments and as a component in a multiple subcatchment analysis involving channels and reservoirs. Different storms will become critical to the catchment as the analysis moves downstream, providing an additional challenge for the **Case manager**.

While the main processing of monitored data occurs in the offline Hydrology and Flow routing tools, the GIS framework provides the structure:

(a) to manage and display the model results;
(b) to extend flow depths across the flood plain;
(c) to derive damage costs; and
(d) to assess a range of flood defense strategies.

Consultation on the eventual preferred flood management plan will include the EA flood defense committees and the LAs that would have to implement any necessary development control. Initial indications are that the consultees believe recent flooding has been more affected by changing urban and rural land use practices than current monitoring and modeling would suggest. The CFMP process is thus generating renewed interest in such *academic-curiosity* monitoring and modeling.

5. Conclusions

These three GIS based examples of deriving information from environmental monitoring show how approaches differ with organizational needs. The LAs use GIS in a fairly conventional way, with a large number of fixed (but updateable) data layers available to many individual users on a central server. They make little direct use of monitored flow data but do use GIS layers derived from monitored data by the EA. The EA have a national data center, giving access to a range of GIS data, and monitoring archives are being centralized in a GIS like form. However, for the Catchment Flood Management Plans being developed by various consultants, a stand-alone GIS framework will be used to aid consistency of approach and manage the modeling and appraisal of a large number of scenarios. Monitored data are fundamental but will mainly be processed outside the GIS itself. The WSCs are probably the most advanced in using GIS to integrate monitored data and model results, but the short-term nature of *flow survey* data makes this easier. The benefits of catchment-wide urban modeling and calibration have long been recognized, but using a centralized GIS to link and maintain datasets provides new gains. Overall, these examples highlight the difficulties of seeking an integrated approach to urban water planning. Yet, although the various GIS cannot be linked directly, data layers can be easily transferred between them.

6. References

1. OFWAT (2001) *Levels of Service for the Water Industry in England and Wales 2000-2001 Report*, http://www.ofwat.gov.uk/pdffiles/los2001.pdf.
2. DTLR (2001) *Development and Flood Risk*, Planning Policy Guidance 25, PPG25, http://www.dtlr.gov.uk/ppg25/index.htm.
3. Littlewood, I.G. (2001) Integrated application of United Kingdom national river flow and water quality databases for estimating river mass loads, *This book*, Part VI, pp. 229-240.
4. Harmancioglu, N.B., Alpaslan, M.N. and Ozkul, S.D. (1996) Conclusions and recommendations, in N.B. Harmancioglu, M.N. Alpaslan, S.D. Ozkul and V.P. Singh (eds.), *Integrated Approach to Environmental Data Management Systems*, Series 2 Environment, Vol. 31, Kluwer Academic Publishers, Dordrecht, pp.423-436.
5. Institute of Hydrology (1999) *Flood Estimation Handbook*, 5 volumes, IH, Wallingford, U.K. http://www.nwl.ac.uk/ih/feh/.
6. Packman, J.C. (1992) *Wallingford Urban Runoff Database, User Manual*, Institute of Hydrology customer report to WRc.
7. Wallingford Software (1999) *HydroWorks, Version 5*, Wallingford, UK, http://www.wallingfordsoftware.com/products/hydroworks.asp.
8. Foundation for Water Research (1998) *Urban Pollution Management Manual, second edition,* FWR, Medmenham, U.K.
9. DoE/NWC (1980) *Sewer and Watermain Records*, STC Report No. 25, London, U.K.
10. Osborne M P (2001) *Personal communication,* Earth-Tech, Abingdon, U.K.
11. Wallingford Software (1999) *InfoWorks,* Wallingford, UK, http://www.wallingfordsoftware.com/products/infoworks.asp.
12. Andrews, A.J. (2001) *Personal communication,* Wallingford Software, Wallingford, U.K.
13. Leeks, G., Bridge, D., and Duffy, T. (2001) Decision support systems for planners, Draft report to DTLR, CEH Wallingford, U.K.
14. HR Wallingford, Halcrow, CEH Wallingford and Flood Hazard Research Centre (2001) *Development of a Modelling and Decision Support Framework (MDSF) for Catchment Flood Management Planning,* Inception Report to EA/DEFRA.
15. Wallingford Software and Halcrow (2001) *iSIS Version 2.0*, Wallingford, UK, http://www.wallingfordsoftware.com/products/isis.asp.

METADATA AS TOOLS FOR INTEGRATION OF ENVIRONMENTAL DATA AND INFORMATION PRODUCTION

E. VYAZILOV, N. MIKHAILOV, V. IBRAGIMOVA and N. PUZOVA
National Oceanographic Data Centre
All Russia Research Institute of Hydrometeorological
Information World Data Centre
6, Korolyov St., Obninsk
Kaluga reg., 249035 Russia

Abstract. The metadata structure required for information on a condition of the environment is analyzed. Metadata are organized as several systems: the integrated database "Oceanography", as data storehouse; the CD-ROM on "information resources and production" for distribution; and as metadata on Web sites. Metadata are used for the development of integrated technologies for environmental monitoring and information production. The directory on "information resources and productions of Russia" is presented.

1. Introduction

Within the last decades, thousands of initial, inverted and processed global, regional and local data sets are created. Information about ten thousands of expeditions is assembled; only in Russia, data on more than 33,000 expeditions are collected. Hundreds of various softwares are developed, on the basis of which it is possible to obtain thousands of computational characteristics. Information on the state of the environment is presented on the Web. Various sorts of referral information about a state of the natural environment are defined by the distributed character of the medium studied and by the irregular allocation of observation sites in the territory. This requires knowledge on the characteristics of the monitoring network, including a variety of monitoring platforms, parameters, methods and their retrieval [1, 2, 3].

The metadata are created at the moment of observation, and further they are supplemented by new attributes. Metadata describe properties of the data, their structure, allowable meanings, format, interrelation with other data, and other characteristics of the data, which help to interpret and use the data correctly. Metadata are information on data to be referred to during data processing.

With respect to information flow, it is necessary for the user to be guided very quickly [4, 5]. For a successful search of high-quality data, various metadata (information on data sets, formats of their exchange, software for their processing, organizations assembling and storing data, and others) are needed.

Metadata allow the integration of data, information production, and softwares. If one knows the address of an environmental database, information production or softwares on the Web, it is possible to retrieve them easily, using metadata.

2. Types of Metadata

Before considering problems in the development of metadatabases, it is necessary to specify their role in a data processing system. Table 1 presents the contents metadata and their particular areas of use. The analysis of the Table indicates that metadata arise at the moment of observation, and further they are generalized and supplemented by new attributes. Figure 1 presents the metadata levels for environmental data. The basic metadata levels are:

- *Sources of data*: observing network, platforms - Research Vessels (RV), coastal stations, satellites, buoys, projects, programmes, organizations, experts, equipment, cruise descriptions;
- *Information sources*: descriptions of data sets, information on maps, information on bibliographic, socioeconomic databases, methodical documents;
- *Information for data management and processing*: formats, codes, processing software, models, etc.;
- *Information on production*: online data, climatic data, atlases, electronic handbooks, and prognoses.

TABLE 1. Levels and types of metadata.

Technological Level	Types of metadata
Observation	Information on networks and methods of observation, definition of methods for hydrochemical and polluting parameters, information on measuring means and places of control equipment
Platforms for observations	Research vessels (RVs), coastal stations, satellites, planes, buoys
Data acquisition	Information on technologies for data acquisition, formats of data transfer, descriptions of transmitted data sets, standards of representation, and data and metadata transfer on the Internet
Accumulation of data	Description of data sets, organizations, data suppliers, owners, users, formats of the acquisition, storage and exchange of data, monitoring projects, cruises of RV, content of observations, information on parameter codes, information on technologies, methods of data quality check
Interdepartmental and international data exchange	Information on technologies and formats, description of data sets, information on monitoring projects and programs
Storage and protection of data	Information on technologies, storage areas
Processing	Information on methods of processing and analyzing data, software, algorithms for parameter calculation, data content
Modeling	Information on models, methods, formats of output data
Data distribution	Information on data distribution (analyses, bulletins, monthly journals, year-books, climatic directories, forecasts), forms of distribution (tables, diagrams, maps, text, sound, editions), spatial / temporal scales of data representation, software, editions (handbook, atlases, including electronic Web resources)
Protection of nature	Information about impacts of the natural environment and extreme situations on economic conditions

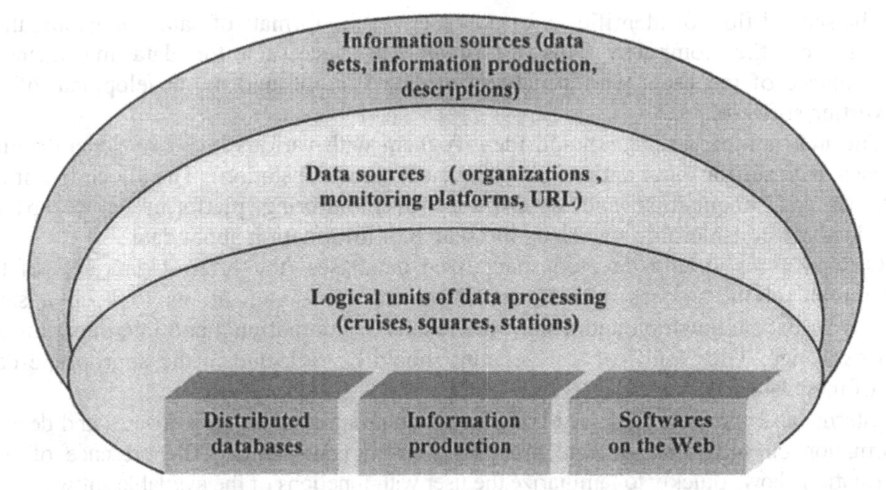

Figure 1. Metadata for integration of environmental data, software, productions.

A characteristic feature of metadata is their low frequency of change since the majority of their contents change very little in time. The process of developing metadata includes the following operations: the choice of metadata characteristics, development of the database structure, preparation of documents, and entering the data on the computer.

3. Metadata in a Data Processing System

Information is needed about data sets, observation methods and equipment, methods of determining hydrochemical and polluting parameters, available softwares, and geographical attributes. Information on oceanographic stations, buoys, marine meteorological observations, formats of data storage and their structure is the most detailed type of identifier information. The identifier information serves to simplify data use and data processing. The information on stations includes spatial-temporal coordinates of the observation site. The identification of stations is carried out in time (year, month, number, time) and in space (latitude, longitude). The absence of any of the listed attributes puts the expediency of storage of such a station under doubt. Besides the information common to all stations (observation on sections, polygons, etc.), it is necessary to mark the additional characteristics inherent in each form of observation. The additional information about time series, including those on oceanographic weather stations, cover attributes such as the start and end date of observations, amount of data and terms of observation. For sea sections, these attributes include the name and the number of a section, numbers and constant coordinates of points of the section.

On the basis of information on spatial/temporal coordinates of oceanographic stations, it is possible to calculate the characteristics of data coverage in space and time (the amount of observations in squares with various spatial scales from 0.5^0 up to 10^0 and in time from one month to a climatic year). The data coverage is obtained by statistical processing of spatial / temporal coordinates of stations with irregular observations and is intended for a preliminary assessment of the completeness of data.

The second flow of identifier information cover the formats of data storage and their structure on the computer. This information is necessary for data management, acquaintance of the users with structures of data storage, and the development of the converting software, etc.

The information on data should identify them with various levels of generalization (cruises, data set for the month of observation at a coastal station). The allocation of the units of data acquisition and descriptions of monitoring platforms in a separate metadatabase considerably speeds up the search of information about data.

Organizations should develop specialized databases (the referral data sets of the information about sources). It is very important for organizations to keep a strict account of data acquisition and also of the results of international and interdepartmental data exchange. The results of this account should be reflected in the appropriate data sets of metadata.

Internal and external users need both the common information on softwares and detailed information on softwares for data processing and documentation. The presence of such information allows quickly to familiarize the user with functions of the available software.

A new metadata type is information on products, representing organizations on Web sites. If such information is available, one may search very quickly the needed links (URL address or ftp - files). Information on products is published in different sources, such as [6]. The structure of data for an information product (climatic, forecast, other) is presented in Table 2.

For example, in RIHMI-WDC, large numbers of metadata are used. They are:

- database for cruises (RVs) on historical observations for a long period (more than 33,500 records);
- data set of coordinates of oceanographic stations (2 500 000 records);
- data set of information about observation over currents (7500 buoy stations);
- database of information about coastal and estuary stations and posts (450 stations);
- information on monitoring platforms (more than 800 former USSR RVs);
- format descriptions of data storage (15 formats);
- information about marine organizations of the former USSR (70 organizations);
- information on data sets and databases on technical carriers in various organizations (more than 250 descriptions);
- bases of the dictionaries - codes (vessels, countries, organizations, geographical areas, parameters, equipment etc.) with more than 10 thousand parameters.

4. Organization of Metadata

Global metadata should be stored at World Data Centres, and national metadata, at sites of the appropriate national organizations. Metadata allow the retrieval of initial data on several levels of organizations, depending on the qualification of the users and their rights [7].

The top level (the most accessible) metadata over the Web are designed for the wide user. It covers common information on data sets and databases available in various national and international organizations; information on marine organizations; monitoring projects; the experts; and sources on environment, available in the Internet. There are full lists of data sets, softwares, and information products. For each type of metadata, dynamic access to the

TABLE 2. Structure of data for a information product (climatic, forecast, other).

Description of fields	Remarks
Name of information product	
Country - developer of an information product	
Organization	
Department	
Author (s)	
Discipline	Ice, oceanography, etc.
Spatial scale for information production	Fixed point, square - 1^0,2^0
Production type	On-line, off -line
Production form	Analysis, forecast, map, text, table, graphic, database
Geographical region	
Address of information product	URL, directory of files
Periodicity of presenting product	Hourly, daily, weekly, monthly, early
Time of presenting product	
Period of product generalization	Day, week, month, years
In advance forecasts	
Language	
Units	
Limitations on use	
Parameters	
Name and issue for periodical publishing of the product	
Methods of presentation	e-mail, ftp, http, xml
Methods of development	
Product description	
Used database	
Areas of use	
Monitoring network	
Monitoring platforms	
Equipment	
Method of observation	
Software	
Bibliography	

information is provided on the basis of languages such as Extensible Markup Language (XML) and Resource Description Framework (RDF). Thus, there is an opportunity to standardize the representation and the search for metadata by the use of Web technologies.

The second level metadata are designed for experts or hydrometeorologists, who want to estimate the amount, quality, and the completeness of the initial data. Therefore, metadatabases, representing information on monitoring platforms (RVs, coastal stations, satellites, etc.), methods of observation, equipment, and the detailed data coverage of a region are used. This level of metadata is created in a relational DataBase Management System by the opportunities of data distribution over the Internet as dynamic pages in XML and RDF languages.

The third level metadata are intended for search of the initial information on databases and data distribution in a form convenient for the professional users. These metadata basically are intended for a database manager. Such metadatabases allow the search for physical addresses of data storage or the procedure of data processing. To distribute the initial data to the external user, codes are used. At this level, the primary factor is the classification of data. The basic inputs to the system of search of the initial data cover the name of a parameter, method of reception, spatial/temporal scale of representation, type of data (text, factographic), spatial data and conditions of search (geographical area, period of observation) to receive physical addresses of data storage (name of the table in database) or the name of the application software for calculation of parameters.

All metadata are stored in a DataBase Management System, where their urgency is supported, and they are used periodically (at the top level) for updating the dynamic information on a Web site. Further, after development of the metadata structure in XML and RDF languages, the reorganization of the base metadata is possible.

For the organization and transfer of metadata, it is common to use the XML language and object model DOM (Document Object Model), being the standards of W3C (World Wide Web Consortium, http://www.w3.org). The use of XML tables allows not only the transformation of formats, but also the manipulation of data after the reception of metadata on the screen of the user. For example, the user may sort, make search, or add the information directly from browser. DOM allows transforming metadata from a format of a database in XML. The XML language allows standardizing not only the structure of attributes of various types of metadata, but also their names and types of data.

5. Development of a Metadata Directory

The first attempt to develop a system with a set of metadata is made in the Russian National Oceanographic Data Centre. The directory of "Information sources and product of Russia " is created [8]. The directory is prepared to represent information sources on a sea environment. The directory is intended for search of information about data, formats, organizations etc. and includes:

1) *Sources of the information* (monitoring networks and equipment, expeditions, projects and programs, marine organizations, information about RVs, including current ones, information on coastal stations and posts over the seas of Russia, satellites, experts).
2) *Information sources:*
   - The descriptions of data sets and databases with information about formats, description of data structure, maps, data coverage for various kinds of observations (deep-water, operative data transmitted on channels of global telecommunication systems, currents, coastal, hydrometeorological, satellite, ice, synoptic, disasters, international and national projects, foreign sources of the information available in Russia, spatial data);
   - Scientific and technical information;
   - Legal information;
   - Methodical documentation;
   - Socioeconomic information.
3) *Management facilities and data processing:*
   - Codes used in various organizations;
   - Formats of the acquisition, storage and exchange of data (interdepartmental and international);
   - Software for the acquisition and processing of data.

4) *Information products*:
   - Online data (SINOP, BATHY, TESAK, SHIP, BUY, TRACOB, GRID, STORM, others);
   - Climatic information (maps, tables, databases);
   - Forecasts (maps, tables, text);
   - Recommendations.

In Fig. 2, the page for viewing the Directory on "Information sources and product of Russia" is shown. The directory includes 13 types of metadata, about 30 000 units of descriptions (cruises, data sets, projects, organizations etc.). In Table 3, information on the quantitative characteristics of metadata in the directory is given.

Information products

- Current information
  - Actual weather
    - Weather reports
    - Synoptic chart
    - Natural phenomena data
    - Satellite data
  - Hydrometeorological conditions of sea
  - Ice conditions
  - Pollution of environment
- Prognostic and diagnostic information
- Climatic information
  - Hydrometeorology of marine ports
  - Hydrometeorology of sea
    - Water temperature fields
    - Salinity fields
  - Ice conditions
  - Catalogue of natural phenomena
  - Complex presentaion of climatic information
- Information for decision making
  - List of products
  - Products on the map
- General information on ESIMO

Sort by organizations | Sort by kinds of products | Sort by geographical regions

ESIMO informatiom products (examples)

Arctic and Antarctic Research Institute

Type	Kind of products	Geographical region	Name
Operational data	Analysis	Gulf of Finland	**Ice information for January 21, 2000**
Operational data	Forecast	Arctic	**Pressure field forecast**
Operational data	Forecast	Arctic	**Wave forecast**
Operational data	Forecast	Arctic	**Ice drift forecast**
Operational data	Forecast	Arctic	**Forecast of currents**
Operational data	Observations	World Ocean	**Reception of satellite information**
Delayed mode data	Data	Arctic	**Arctic ice data in SIGRID format**
Delayed mode data	Map	Antarctic	**Map of ice situation around Antarctic**
Delayed mode data	Map	Antarctic	**Monthly maps of ice situation around Antarctic**
Delayed mode data	Map	Arctic	**10-day ice maps of Arctic**
Delayed mode data	Map	Arctic	**7-day ice maps of Arctic**
Delayed mode data	Map	Arctic	**Climate ice characteristics**

Figure 2. The page for viewing the Directory on "Information sources and product of Russia".

The search for information about data sets and databases is realized as HTML pages on the basis of navigation for three directions of search (kinds of observation, geographical areas and organizations/owners or storage of data). For search of the information in sections (organization, cruises etc.), XML language is used. At present, metadata from this Directory are stored on Web site (http://www.oceaninfo.ru) and Web sites of organizations RIHMI-WDC (http://www.meteo.ru), AARI (http://www.aari.nw.ru), and FEHRI (http://www.hydromet.com.ru).

TABLE 3. Information on metadata.

Metadata objects	Number	
	Records	Attributes
Information on data sets and databases	230	35
Format descriptions	15	20
Information on organizations	60	12
Information on RVs	850	13
Information on the action of RVs	122	5
Information on coastal stations	450	65
Information on satellites	2	10
Information on monitoring networks	10	17
Information on experts	110	12
Information on monitoring projects	70	15
Information on RVs cruises	32500	21
Information on models, software	10	13
Information on equipment	150	17
Information on marine maps	100	19

6. Monitoring of Information Sources

It is possible to carry out analytical inquiries and to receive the aggregated characteristics, using metadata, i.e., to carry out the analysis of data from various organizations. For example, it is possible to receive:

- amount of data sets on organizations, regions;
- amount of RV cruises;
- amount of oceanographic stations on squares, periods, parameters, etc.

For effective data management, it is necessary to know the condition of information sources and products, which can be reflected:

- by the condition of monitoring networks;
- by the estimation of information flows;
- by the distribution of information on the computer;
- by the generalized characteristics of databases;
- by the amount of the executed inquiries by the users to reflect their information needs.

The characteristics of data flows are the amount of data sources (for example, cruises) and the data volumes (number of observations for one year, on the average). In Table 4, the characteristics of the entrance information and volumes of data from various monitoring platforms are given. The characteristics of data flows in processing of ocean data are:

- *the phase of observation* – the amount of information sources (RVs, coastal stations, buoys; Table 5), volumes of the received information from one source (daily, monthly, annual);
- *level of data in time* - volumes in information processing at the center (daily, monthly, annual) in bytes;
- *level of data in space* - station, cruise, territory, region;
- *stage of data processing* - volume of processed information, time of processing;

- *stage of information distribution* - volume of output information, periodicity of representation (day, week, decade, month, year); spatial data integration (region, sea); sorting of data (in time, space and space/time).

TABLE 4. Characteristics input information in NODC of Russia.

Observation type	Cruise number	Size	Average increase per year	Period of observations	Increase in 2000
Deep sea water	33500	2300000 profiles	250 cruises	1885-2000	168 cruises
BT	4812	3200000 profiles	-	1949-1991	-
CTD	520	305000 profiles	-	1974-2000	-
Marine hydrometeorological	-	36000000 observations	500000	1870-2000	1200000
Coastal observations	-	400 stations	-	1900-2000	-
Air craft observations	-	1024 maps	15 maps	1970-1987	-
Buoy	-	3500000 observations	1500000	1989-2000	1300000
Current	1762	7500 stations	100	1936-1996	1
Data BATHY, TESAK	-	472000 observations	14400	1981-2000	13000
Chemical pollution	2520	220000 profiles	50 cruises	1957-1995	-
Marine meteorological data of RVs	1745	450000 observations	35 cruises	1911-2000	-
Observations of solar energy in sea	793	190000 observations	16 cruises	1934-1994	-
Marine aerology	714	80000 observations	14 cruises	1938-1987	-
Ozone	126	7000 observations	4 cruises	1970-1987	-
Radiochemical	497	-	10 cruises	1970-1990	-
Data of meteorological rockets	89	-	3 cruises	1970-1990	-

TABLE 5. Number of monitoring platforms.

Characteristics	Platforms					
	Coastal stations		RVs		Satellites	
	1985	2000	1985	2000	1985	2000
Number of platforms	450	186	360	120	5	2
Observations for deep sea water	150	170	500	500	500	2500

The information on equipment used and measuring systems reflects the quality of a monitoring network; therefore, collecting information about the equipments of a network is an important factor in processing of metadatabases. The distribution of data over the basic regions of World Ocean (Table 6) reflects regional coverage of data; therefore, such a table is useful to the potential user. The distribution of the information on computerized carriers (Table 7) shows readiness of data for processing.

7. Conclusion

Metadata are organized in several systems: the integrated database "Oceanography" as data storehouse; the CD-ROM on "information sources and products" for distribution; and on sites of organizations and participants of the program "World Ocean".

The development of metadata (creating different metadatabases and their integration, dynamic presentation of metadata, connection of metadatabases in a database management system and Web technologies, establishment of distributed metadatabases, etc.) use integrated technologies for environmental monitoring and information production.

TABLE 6. Number of oceanographic data for regions of World Ocean.

Oceans and Russian Seas	Number cruises	Number stations
Azov Sea	593	33600
Aral Sea	51	1500
Baltic Sea	4067	269800
Barents Sea	2491	120700
White Sea	919	47300
Bering Sea	320	16500
Kaspy Sea	4035	50400
Laptev Sea	419	12600
Ohkotsk Sea	800	68500
Black Sea	1659	94500
Japan Sea	2244	106000
Pacific ocean	6467	444000
Atlantic ocean	5484	647600
Indian Ocean	815	86300

TABLE 7. The distribution of the information on computerized carriers, in %.

Observation type	Hardcopy	Magnetic tape	CD-ROM	Web sources
Deep see water	5	95	95	5
BT	10	90	70	1
Current	40	60	60	0
Marine hydrometeorological observations	10	90	25	25
Air craft observations	100	0	0	0

8. References

1. Vyazilov E.D. (2001) Organization and usage of metadata, *CD-ROM, TIEMS 2001*, June 19-22, Holmenkollen Park Hotel, Oslo, Norway. 9 pp.
2. Vyazilov E.D., Lebedev I.V., and Puzova N.V. (2001) Development of metadata bases in field oceanography, - *4-th Russian Conference "Modern state and problems of navigation and oceanography" (HO-2001)*, Collection reports, Vol. 2, June 6-9 2001, Sankt- Petersburg, Navy, p. 240-245.
3. Vyazilov E.D. (2001) Metadata as basis of global environmental data management, *Electronic Journal «NewsLetter ESIMO»* 7, 24 pp. (http://www.oceaninfo.ru/news/newsl7.htm).
4. ISPRS (2000) Tateishi R. and D. Hastings (eds.) *Global Environmental Databases - Present Situation; Future Directions*, International Society for Photogrammetry and Remote Sensing, Working Group IV/6 (1996-2000), 233 pp.
5. Vyazilov E.D. (2001) *Information Resources on Environment Data*, Moscow, URRS, 312pp.
6. IOC, WMO (1986) Integrated Global Ocean Services System: Oceanographic Products issued by National Centres, *Information Service Bulletin*, March, No. 7, 112 pp.
7. Besprozvannykh A.V., Mikhailov N.N., and Vyazilov E.D. (2000) Web-technology for data access and exchange GETADE-8: Document 20, IOC/IODE-TADE-VIII/20, Paris, 3 March 2000, (http://ioc.unesco.org/iode/expertise/getade/getade8/document 20 - web.doc).
8. Vyazilov E.D. Mikhailov N.N., and Puzova N.V. (2001) Electronic dictionary on CD-ROM "Information resources for World Ocean", - *4-th Russian Conference "Modern state and problems of navigation and oceanography" (HO-2001)*, Collection reports, Vol. 2, June 6-9 2001, Sankt-Petersburg, Navy, p. 254-257.

PERSPECTIVE DECISIONS AND EXAMPLES ON THE ACCESS AND EXCHANGE OF DATA AND INFORMATION PRODUCTS USING WEB AND XML APPLICATIONS

V. M. SHAIMARDANOV, N. N. MIKHAILOV and
A. A. VORONTSOV
All-Russian Institute of Hydrometeorological
Information-World Data Centre
6, Korolev St., Obninsk, Kaluga region
Russian Federation, 249020

Abstract. It is very important to facilitate the access and the integration of data and information for the acquisition, accumulation, modeling and transformation of environmental data into information required for decision making and planning of the environment. Examples on the use of new WEB-technology for access and exchange of marine data are given. The use of XML-applications for retrieval of new information is presented.

1. Introduction

In recent years, there have been significant changes in the field of information technologies. In response to these changes, new ideas and decisions are required for marine data management to ensure efficient information support for various marine activities.

The present paper considers preliminary design decisions and the conceptual approach used in the establishment of *a decentralized marine data system* within the framework of a generalized scheme for integrated data management. The term "scheme" has a specific meaning here. It defines: (1) what the future data management technologies will have under control; (2) how data, data management systems, software applications, and networks fit within the integrated data management process; (3) who performs data management and where.

The task of integrating marine data and information products within the unified information space in the form of a virtual oceanographic data center is a key issue in the conceptual scheme of the decentralized marine data system.

This paper discusses the approaches and decisions employed in construction of the unified system of information on the world ocean state within the framework of the "World Ocean" Federal target program in the Russian Federation.

2. The Demand for Data Management

The international and national marine information systems were established a long time ago. Since then, National, Regional and World Centers of these systems have developed a variety of extensive data sets and databases on numerous aspects of the marine environment, which are widely applied in scientific and practical activities. The marine information sources are increasing in volume every year. The achievements are impressive, but there are still some problems in the development of marine data management systems. The question is whether it is necessary to radically modernize the currently operating information systems for data exchange and dissemination. If so, what objectives should be pursued and what key directions of its modernization should be selected?

There are two main reasons which make it necessary to take immediate actions on marine data management: (i) radical changes in user requirements for data and information products on the marine environment; (ii) fast development of new information technologies.

2.1. EXPECTED INFORMATION DEMANDS

The demand grows every year for marine information sources in various fields of marine activities involving science, economy, social, and other spheres. The expected demand for marine data are the most clearly formulated as part of the GOOS/GCOS/WCRP climatic module, where the set of tasks for climate studies and appropriate physical oceanic systems is determined and where recommendations concerning the requirements for ocean data are worked out. These requirements basically determine the necessity to prepare long series of marine meteorological and oceanographic data at fixed points of the ocean on the basis of historical and current data.

In recent years, the GOOS coastal component became much more active with respect to marine data requirements. This component faces problems which, along with the basic data on physical states of the sea and the ocean, require data on a great number of non-physical (bio-chemical, litodinamical, and other) aspects of the marine environment so that the behavior of coastal zones and their response to natural and anthropogenic impacts can be determined.

Reliable and diversified data on the marine environment together with effective data management can serve as a significant factor in obtaining new information (in fact, in increasing our knowledge) on marine phenomena and processes. Therefore, the critical point in meeting user requirements is the availability of necessary data and information products to the user and the integration, joint processing and the analysis of available data and information products in an interactive mode.

It is possible to formulate the specific requirements for development of marine data management systems. Marine information systems should ensure preparation of coordinated data and information products on physical, biological and chemical parameters of the ocean. These data should reflect both the variability of specific marine phenomena and the interaction of these phenomena, representing various spheres of the marine environment such as water, air, and coastal zones in a wide spectrum of conditions such as dynamics, pollution, dangerous phenomena, etc.

Another important issue in the development of data management tools is the design of a fully complete and integrated (" end-to-end ") process of data management: from data collection to dissemination of complex and reliable information products. In this context, it is desirable to ensure information exchange and access to information sources at each stage of the technology.

In view of the current and the expected information demands, marine information systems should comprise a system of coordinated data and information products on physical, biological and chemical parameters of the ocean and the underlying water surface, which ensures "end-to-end" data management in real time (Fig.1).

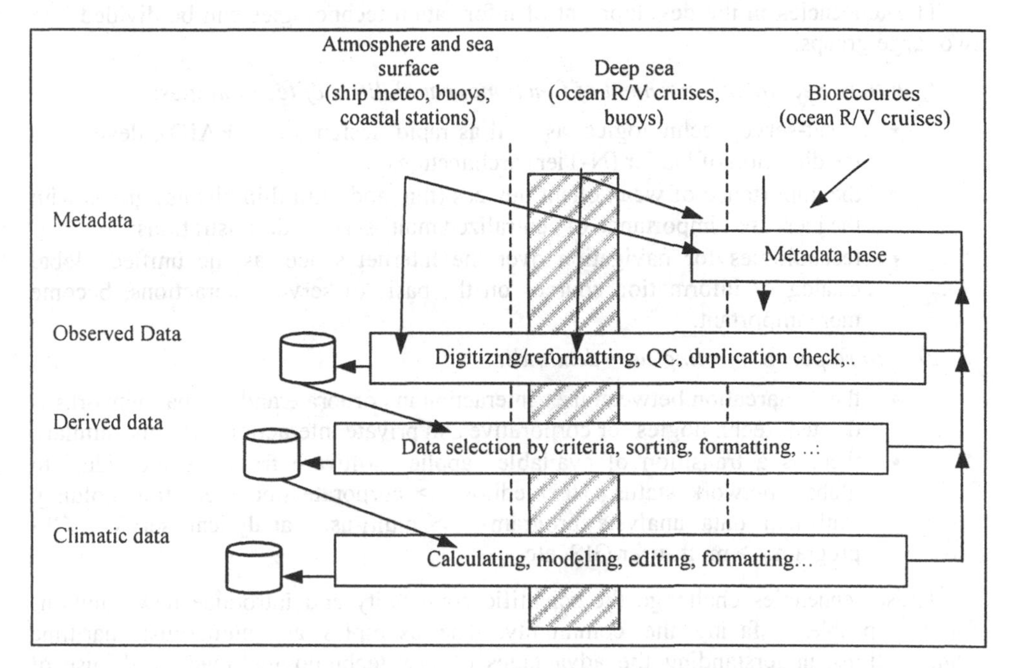

Figure 1. Structure of an " end-to-end " marine data management process.

2.2. NEW INFORMATION TECHNOLOGIES

The present-day market of instrumental information technologies is overfilled with various advanced products, such as database management systems (Oracle, Informix, Sybase, IBM DB/2, MS SQL and others) and Geoinformation systems (ArcInfo, ArcView, MapInfo with their various applications). In recent years, a rapid progress is observed in the development of object oriented technologies used to construct distributed information systems and data storage structures such as Web technologies (W3C standard protocols - TCP/IP, HTTP, SOAP, etc.; languages such as HTML and XML) and middleware used to construct client-server technologies such as CORBA, Microsoft COM, and languages of visual programming and developments of remote applications such as Java, Delphi, Visual Basic, Visual C and others.

Technologies oriented at flexible and coordinated dissemination of large-scale information sources and applications among individual scientists and institutes are also fast progressing. These technologies are designated by the term "Grid" and are often associated with virtual organizations. The GRID protocols and standards, integrated in GlobusToolkit, are similar to Web technologies in many aspects. However, there are certain distinctions between GRID and Web technologies, which will be overcome at the Internet 3 level. GRID technology, including Globus ToolKit, is used in large-scale projects for various subject domains such as: Grid Physics Network project, European Data Grid, Particle Physics Data Grid; www.ppdg.net and Network for Earthquake Engineering Simulation Grid.

The tendencies in the development of information technologies can be divided into two large groups:

1) with respect to enhancement of functional capabilities of technologies:

- client-server technologies, as well as rapid design tools (RAID), develop in the direction of ladder (N-Tier) architectures;
- the importance of weak client devices (thin and ultra thin clients) grows with the increased importance of centralized multi-server administrations;
- technologies for navigation over the Internet space, as the unified global catalog of information sources on the basis of server interactions, become more important.

2) with respect to meeting user demands:

- the demarcation between user interaction in corporate and global networks is deleted; technologies for corporative and private interactions become similar;
- there is a transition of available applied software from the individual to global- network status (text editor - > corporate documentation holding; statistical data analysis program - > multi-user analytical service; GIS program - > multi-user GIS, etc.).

These tendencies challenge the scientific community and introduce new solutions for the problems facing the community. The examples are numerous; maritime organizations, understanding the advantages of new technologies, made wide use of DBMS, GIS and Web technologies for marine data processing.

The fact is that new information technologies are no more than powerful toolkits to build applied information systems in the appropriate subject domain. Their employment does not exclude (and probably even intensifies) the necessity to solve traditional problems such as data exchange between organizations, interfaces between various blocks of technology, building specific marine applications into technologies, and others. Even if new instrumental tools are easy to use, it should be decided what is necessary for the new generation information systems to make them flexible and effective in the future.

3. Model of the Decentralized Marine Data Management System

3.1. CONCEPTUAL APPROACH AND ARCHITECTURE

At present, the volume of electronically accessible marine data and information products (hereafter referred to as "data" if no special explanations are required) grows rapidly. Most of the available and newly collected marine data are held in the national and world centers of international information systems such as IOC IODE, World

Meteorological Organization (WMO), Global Ocean Observing System (GOOS), and others, which use various data storage and management systems, computers and networks based on various platforms. Here, data and appropriate hardware-software and network facilities of the centers of marine information systems are defined as *data sources* and form *the information space of the systems.*

The marine data are held in sequential files of flat and hierarchical structures and in relational/object oriented databases. Some of the data held in various centers are similar in content. Data (each separate file or a family of files, a collection of data, a separate database) usually have a relatively strict structure, organized in the form of a preassigned database scheme or data format; but types and forms of representation of these structures (database schemes, formats, codes of parameters, etc.) are varied. For this reason, it is possible to state that the existing marine data are not homogeneous in structure. In most cases, interaction with data of the centers is possible only through the interface which is either provided by the appropriate center or is specially designed to access the data of this center.

In other words, the system of the world marine data represents an inhomogeneous information space, including a great number of data sources, which are distributed geographically and have a high level of independence as far as data structure, contents, storage, processing and management are concerned. In essence, the aim in marine data management technologies is to change from the inhomogeneous information space to the unified information space on the environment by the integration of marine data, based on telecommunication interactions. In this case, users will be able to perceive the data as a whole, and the data will acquire the status of a corporate information source.

To create a unified information space on marine environment, an appropriate methodology is required for integration of data and for providing access to data. In this connection, a significant restriction should be noted, which stems from the independence of marine data sources, i.e., the transition to the unified information space should not cause essential reconstruction of data structures and data management techniques available in current and future data sources.

Generally, two approaches may be used for data integration: centralized (Centralized Data Warehouse, CDW) and virtual (decentralized) (Virtual Data Warehouse, VDW) data holdings. In the context of the model for marine data management, the term " Data Warehouse " is used for two reasons: first, this model is based to a certain degree on the concept of "Data Warehouse" as an element of the classical distributed information systems; and second, the use of this term stems from the desire to more clearly define a range of problems to be solved, i.e., exchange and dissemination of the completed marine data and information products.

When the first approach is adopted, data from various sources arrive at the data warehouse, and then all user requests are met using these data. For example, the CDW approach is applied in the IOC GODAR (Global Ocean Archeology and Rescue Project), where data from 15-20 sources around the world are accumulated for 3-4 years at the World Data Centre – a centralized data warehouse. After completion of this 3-4 year period, the data are made accessible to users through a series of laser disks (WA'94 and WA'98) or the appropriate Web site. The CDW approach does not guarantee the urgency of data (in volume, set of parameters, and forms of representation) since a new version of data warehouse is created in the course of a sufficiently long period of time without any coordination with data sources and user requirements. The advantage of this approach is that the tools required to create a data warehouse of such type and meeting of user requests are relatively simple, and the time required to meet user requests is not too long.

When the virtual approach is assumed, the data are retained in data sources, and a user request addressed to VDW is decomposed at a stage of execution into a series of requests addressed to individual data sources. In this case, data are not replicated, and their urgency at the stage of processing a request is guaranteed. On the other hand, since data sources are independent, their integration requires much more sophisticated methods and tools of navigation through decentralized data sources and optimization of requests and of their execution. The virtual approach is more suitable for construction of systems where the number of data sources is large. Such systems are distinguished by inhomogeneous structures; the data are frequently changed, and data structure, content and management control within data sources is weak. Therefore, the decentralized virtual approach is taken as a model of the IODE data system. In recent times, methods and tools of this approach have been actively developed for various subject domains.

To consider the virtual approach in more detail, it should be noted that most of the methods used in the context of this approach are also suited for centralized data warehouses when slightly modified.

Considering the organization of a virtual data warehouse, it is possible to identify two groups of components: *logic and physical.* The VDW logic component includes an internal data model and other means to unify data from various sources and to provide access to these data; including the dictionary of parameters, the language of requests addressed to VDW, the language for interaction with data sources, and other means. The VDW logic component is used for representation of data taken from all sources. Therefore, the unified access to all data, which are integrated is made available for users. One of the key requirements imposed on the model is the transparency of access to external data sources, i.e., a user perceives the external data held in distributed data sources as local data in the context of the VDW data model and takes no care of data source access control.

Before considering the physical architecture of VDW, it should be noted that fixed standards are not yet available in this field. The proposed architecture is based on the classical concept of distributed information systems and basic decisions on GRID construction. The VDW architecture includes components at four levels:

Data Sources: The data sources present the lower level (basis) VDW and include data storage (management) systems, data files and (or) databases, computers, and other facilities available at the IODE centers.

Integrators: The integrators ensure data-pipeline between data sources and VDW and provide the following services: access to a source data on requests from the Navigator; conversion of data from a data source to the internal data model; interaction between the data source and the model to meet user requests. Simply stated, the Integrator is a wrapper to bind a data source so that it becomes an element of the VDW information space.

Navigator: The navigator represents the next level of the VDW architecture and provides a number of services at the collective level of the coordinated management of data from multiple data sources, namely: decoding of user requests, their formalization in the language of requests and decomposition for data sources, transmission of requests to data sources, and monitoring of their execution.

User requests: The user requests represent the final level of the VDW architecture and include user applications through which the access to VDW is provided and the use of the VDW data is initiated. It is necessary to emphasize that the labeled "application" level is meant to designate a point of access to VDW, i.e., "application" is more a logical than a physical component. In practice, "application" can be the program which is started from a user computer. It provides the user with forms and other visual means required to formulate a request for the VDW data. "Application" can also be a fragment of the IODE Web-portal, which represents a unified point of user access to VDW. In the latter case, a user uses the usual Web-browser to access VDW.

In Fig. 2, the basic components of the VDW architecture and a preliminary list of services under each component are shown.

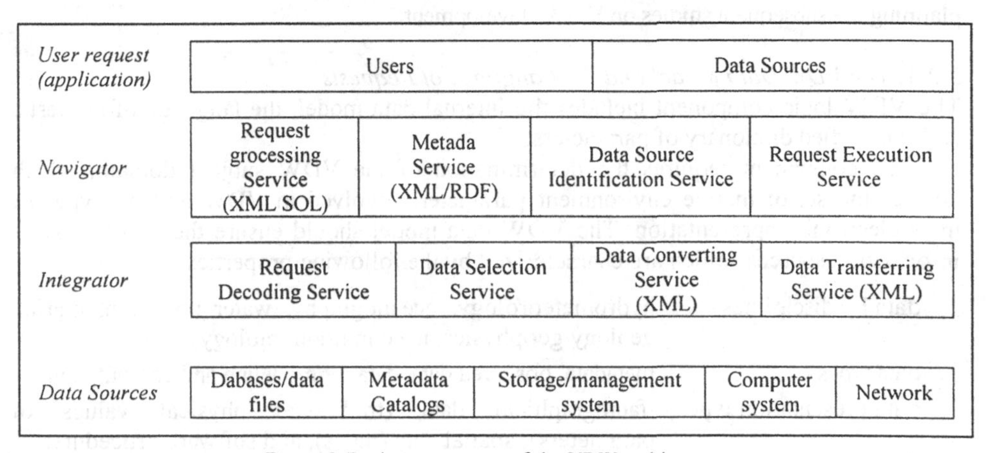

Figure 2. Basic components of the VDW architecture.

As mentioned above, VDW does not directly interact with local data storage/ management systems of data sources. Instead, to obtain data, the *Navigator* interacts with numerous *Integrators*, whose basic task is to organize the conversion of required data to the data warehouse format and to initiate data transmission from a source to the designated user address.

It is important to note that the *Integrator* does not accumulate data in the VDW format and can interact with both the Navigator and the other Integrators. Hence, it is possible to construct a complex network of Integrators interacting with each other. This network will allow the integration of data from various sources to meet the needs of various applications interacting with the virtual data warehouse. In this case, the user interacts with the marine data system only at the "application" level with a possibility of access to the data as the unified information space on the marine environment.

The second important dissimilarity between VDW and the traditional systems is that the user does not formulate requests in direct terms of a database scheme or a data file format which is used by the data source for data storage. One of the main VDW objectives is to save the user the trouble of knowing specific features of data sources and having to interact with each of them. Instead, the user formulates requests in terms of a natural language or by using the appropriate forms of request entry at the 'application' level of the VDW architecture. Further, they are converted to the request language terms compatible with terms of the so-called intermediate or internal VDW data model.

3.2. PROBLEMS RELATED TO VDW IMPLEMENTATION TECHNOLOGIES AND TOOLS

If the proposed architecture is used, the construction of VDW is reduced to the development of its logical and physical components: the Navigator, Integrators, and Application. Design and development of the VDW components are exclusively complicated, and the relevant specific technical decisions require a careful and long-term study. Therefore, the presented paper considers no more than the general design proposals, which can form the basis for planning the subsequent studies on VDW development.

3.2.1. *The VDW Data Model and the Language of Requests*

The VDW logic component includes the internal data model, the language of requests, and the unified dictionary of parameters.

This component requires the determination of the VDW subject domain, which defines the set of marine environment parameters involved in VDW and the types of their electronic representation. The VDW data model should ensure the description of information sources, which are characterized by the following properties:

- data by disciplines - hydrometeorology, oceanography, water pollution, marine geology/geophysics, ice condition, biology;
- data types- metadata, observed data, data for modeling and computations;
- data presentation types - factographical data (traditional physical values of parameters), spatial data (maps), and software procedures.

Data Model. A great variety of data structures used by data sources does not allow the development of data transmission formats covering the whole range of requirements without using the logical environment for data unification. Development of the data description language automatically recognized by physical components of the system, such as Navigator, Integrators and Application, can be used as an alternative approach.

When the HTTP transport protocol is used for interaction within VDW, the XML language providing a platform to share data in a common syntax seems to be the best decision. The important factor is that this language is the W3C (World Wide Web Consortium) standard for the HTTP protocol. Obviously, the XML extension, or the marine XML, will be required, which will ensure data integration at syntactic and semantic levels. Along with mere language tools, the marine XML should include a library of utilities to support basic data management operations in the marine XML framework, such as extraction/loading of XML file fragments by semantic and syntactic criteria, support of multiple XML sources, translation among DTD's, obtaining the standard XML file description, and others.

The development of the marine XML is an independent and complex problem, and it should be considered under a separate cover. The marine XML should ensure the unified syntactic and semantic description of the marine environment data structures and types which are to be placed in the system. The data types are listed above. To fulfill its task, the marine XML should include at least a language model, tools (schemes, structures and rules of description) for factographical, spatial and semi-structured data, and a unified vocabulary of parameters to support data semantics.

Language for Description of Requests. To perform successful interactions between the VDW physical components, it is important to choose a unified language of requests. The language of requests should be based on the marine XML for the same reasons as the data description language. At present, there is no standard for the language of requests; however, the W3C group spends efforts to solve this problem.

The Unified Dictionary of Parameters. The currently operating IODE data sources use a wide range of vocabularies for codes of parameters, and they do not allow a full-scale data exchange or joint data processing. The dictionary of parameters should play a major role in the conversion of data from a data source local structure to the internal VDW data model. The dictionary should include at least a definition of the parameter, the name of the parameter, parameter code, brief name, unit of measurement, and other attributes.

3.2.2. *Environment for Interaction of Components*

From the developer's point of view, VDW is a set of information resources and program applications interacting with each other in a certain environment (middleware). The interaction environment includes models and protocols of interfaces, Web and application servers. In the VDW context, it should ensure transparent information and program interfaces between components of the system - Application, Navigator, Integrators, and Data Sources. It should be noted that the interaction environment should be located in each data source (IODE Center) which is involved in VDW.

The choice of the interaction environment is one of the most complicated steps in the development of distributed information systems. It is an independent problem, and careful studies are required to solve it.

As it is mentioned in section 2.2, the construction of the interaction environment can be: (i) based on Web-oriented object information technologies; (ii) based on the GLOBUS system software, where many elements of interaction are implemented.

Figure 3 shows a general scheme of an interaction environment based on Web-technology and J2EE standard. Such a scheme is currently operating in RIHMI-WDC.

3.2.3. *Navigator*

The basic function of the Navigator is to analyze a user request in the context of the VDW structure and contents, to define the list of data sources able to meet the request, and to ensure interaction with these data sources.

Figure 4 illustrates a scenario where Application (end-user) sends a request to VDW, and this request is executed by the Navigator services to identify the location of data sources and to organize the transmission of the required data to the user.

The scenario is implemented in the following way:

(i) upon receiving a user request in the natural form, the Navigator addresses the Request Processing Service for decoding the request and its conversion to the list of request attributes (data type, geographical area, time, and others) (1) and then transmits the request attributes to the Metadata Service catalogs;

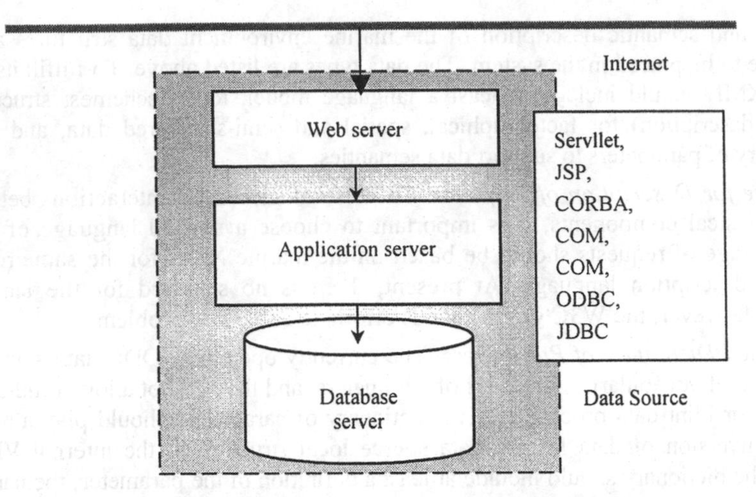

Figure 3. A general scheme for data sources interaction environment.

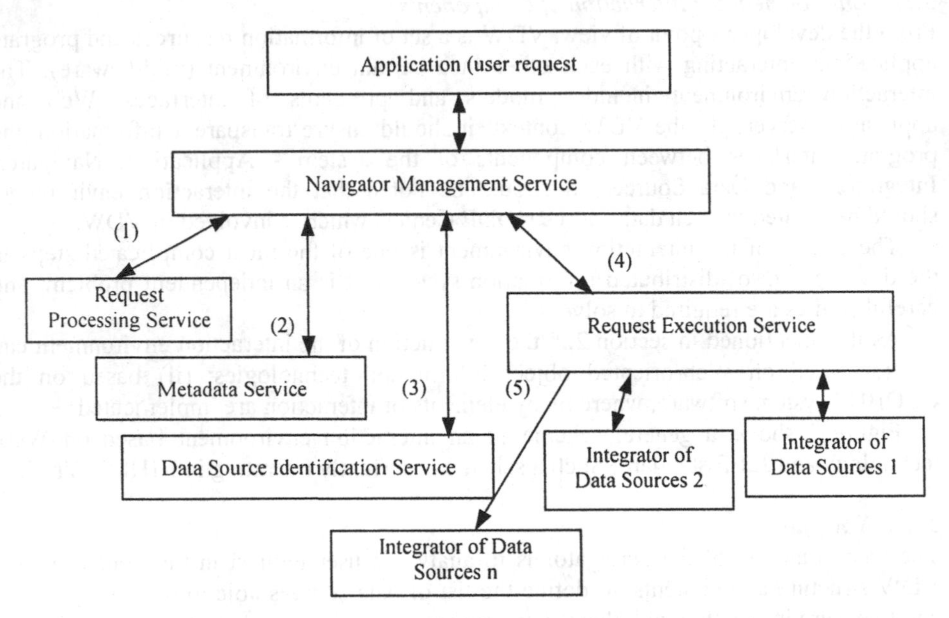

Figure 4. The data management scheme at the Navigator level when a scenario of access to the VDW data is implemented.

(ii) the Metadata Service (2) requests attribute-based indexes and produces indexes of data sources, which contain data with specified characteristics;

(iii) the Navigator transmits the list of these indexes to the Data Sources Identification Service, which sends back the physical locations (URL and other) of the needed data (3);

(iv) the Request Execution Service (4) addresses the Integrators of appropriate data sources and receives from them data in the form of marine XML-based VDW data model (5). It also ensures monitoring of the execution of requests and compilation of responses, based on data provided by separate data sources.

It is evident that the synchronization of updating, the integrity of metadata, and addresses of physical locations at the Data Sources level and at the Navigator level is one of the most important factors to make this scheme work. The development of a mechanism to detect and delete duplicated data obtained by request from various sources is another important factor.

3.2.4. *Integrators*

The function of Integrators is to execute specific elementary requests for data which arrive at Data Sources Service from the Navigator. To illustrate the data management scheme at this VDW level, let's consider the scenario of Fig.5:

(i) an elementary request comes to specific Data Source Integrator and is transmitted to the Request Decoding Service to be converted to semantics of a request inquiry to the Data Source local database;
(ii) the Data Selection Service addresses the data source catalogs, connects with the data source management /storage system, and transfers data management to this system so that data sample can be prepared (7);
(iii) the selected data pass through the Data Converting Service, where data are converted from the data source local structure (format) to the VDW data model (to the form of the XML document) (8);
(iv) the XML document is sent to Navigator to be incorporated into the integrated XML document meeting user request specifications (9).

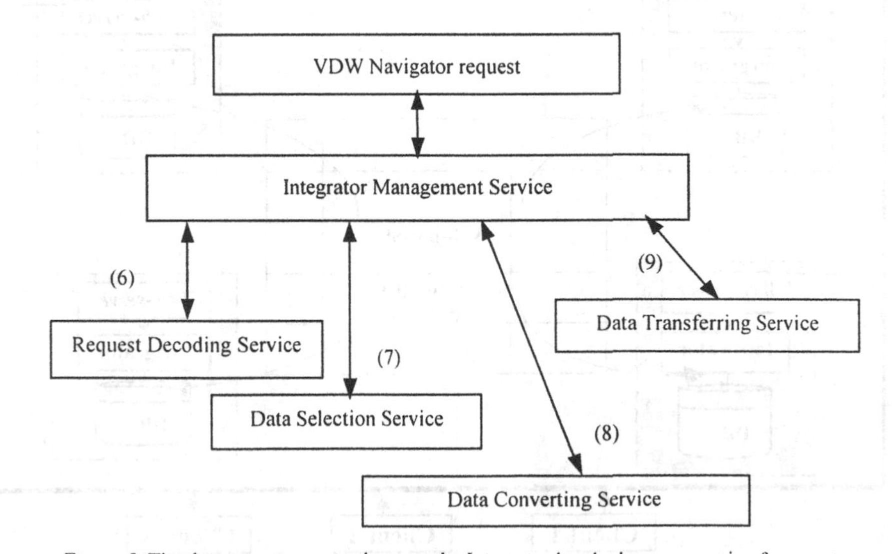

Figure 5. The data management scheme at the Integrator level when a scenario of access to the VDW data is realized.

Conversion of data from a local database/file format to the VDW data model is the important problem at this stage of data management in VDW. Here, the quality of the unified vocabulary parameters is of special importance, as well as the mechanism and tools of coordination and setting of conformity between the unified vocabulary of data and the vocabulary (vocabularies) of data used in the data source base (bases).

3.3. VIRTUAL OCEANOGRAPHIC DATA CENTRE

The concept of the decentralized IODE data system was considered above on the basis of the approach named "virtual data warehouse". VDW, by definition, is the technology of the world ocean unified information space, and its development, support and use are somewhat more complicated.

The actual problem, which has to be solved by the decentralized IODE data system, is the coordinated sharing and supporting of oceanographic information sources. The sharing, as mentioned, is not just data exchange. It should be under strict control with due regard to the rights and responsibilities of data providers and end-users. The following should be clearly defined: what is shared, who is allowed to share, and what are the conditions under which sharing may take place. A group of data IODE data centers, who cooperate with each other, with data providers and data users on the basis of the VDW technology and in accordance with the rules and regulations of sharing mentioned above, together with ocean information sources, made up what is called the "Virtual Oceanographic Data Center (VODC)" of the marine information system IODE. Figure 6 presents the scenario of VODC operation in terms of providing access to the marine information sources.

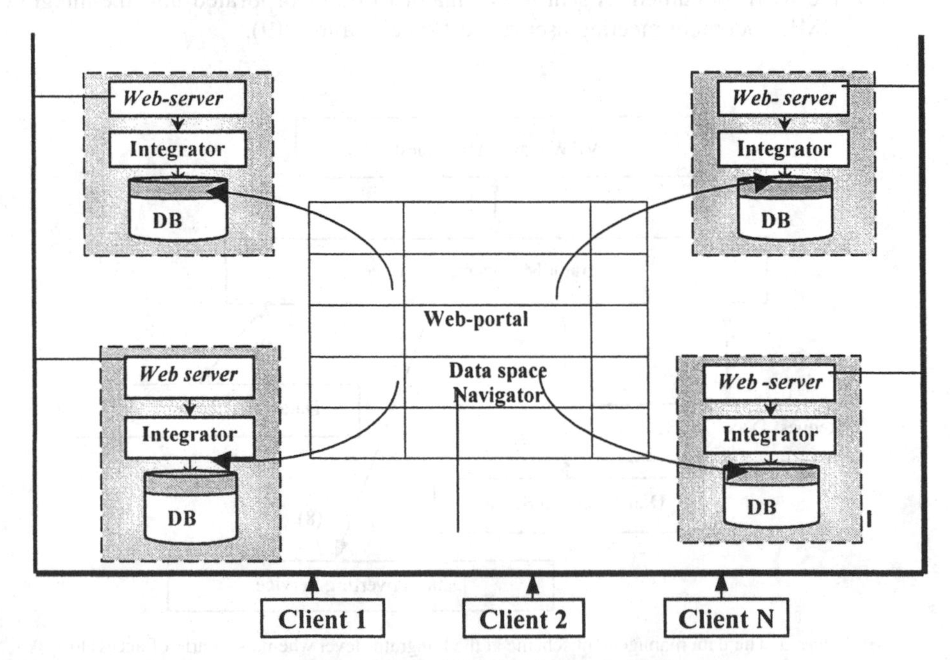

Figure 6. The scheme of distributed access to the VODC.

The centers of international and national marine information systems prepare metadata, data and information products in accordance with their obligations, specified in the rules regulating the VODC use and support. The information sources generated by these data centers are placed on their local servers in the form of data sets and bases, connected to the VDW Integrator through special interface.

World and regional data centers, on the basis of metadata and other information coming from national data centers, prepare a special VDW Navigator database, which is placed in the Web-portal. The external user or any other center addresses the Navigator via the Internet, which, in turn, addresses the appropriate Integrators whose data sources contain the data specified in request. The integrators execute elementary requests and transmit the results to the Navigator, which compiles the complete response to the request and transmits data to the user.

VODC can be used for the construction of "virtual end-to-end" data management technologies. If one considers IODE system in this case, specialized centers of projects related to the implementation of separate sequential data management stages will act as the IODE Centers.

The stages are as follows:

- Collection and processing of primary data, such as GTSPP data, coming via GTS;
- Formation of global data sets on the basis of current and historical data;
- Data processing and retrieval of climatic products.

The specialized centers of such projects place databases, which are prepared in the framework of projects, in VDW according to the considered scheme and ensure their on-line accessibility.

4. Conclusion

Considerations described above are basically of a conceptual character. The main and the only aim here is to show what "virtual data warehouse" and "virtual oceanographic data center" are and to give an idea of how currently operating systems, blocks, databases, applications and network media may be "placed" in the unified marine information space.

The final conceptual decentralized marine data system is explained by several definitions:

1. The decentralized marine data system represents the centers of international and national information systems, interacting on the basis of VDW technologies and performing their functions on a regular support of databases in accordance with the assumed obligations and agreed rules. For the decentralized marine data system user, only the "virtual oceanographic data center" will be visible.
2. The VDW technologies, models, and protocols will cover the complete life cycle of marine data, starting with data collection and extending to the retrieval of final information products and their delivery to users. Hence, VDW can be considered as the tool for end-to-end data management.
3. Information and technological compatibility of the decentralized marine data system elements should be achieved through the development and the use of the

relevant VDW standards and components on the basis of modern information technology.

The ideas proposed here are open to discussion. Thus, the presented paper deliberately does not consider any practical activity concerning VDW/VODC construction because, before any attempt, one should have a clear idea of what he wants to construct and how it will look like. In other words, the immediate task is to develop a clear, well-balanced and understandable concept of the decentralized marine data system. Therefore, any remarks, proposals and comments are encouraged.

First of all, it should be noted that the development of the decentralized marine data system should envisage long-term design decisions. It means that all developers should use the agreed concept (model) and the set of design documentation, which regulates the construction of VDW/VODC basic components. In this connection, it is reasonable to suggest that GETADE should prepare technical design decisions concerning VDW/VODC, and the prepared document should be used as a manual on the development and sequential establishment of the decentralized IODE data system. Obviously, in the course of time, the VDW/VODC design decisions may be adjusted and upgraded with regard to the experience gained, changes in user requirements, technical progress, and other circumstances.

The marine XML, the common data model, and the language of requests based on marine XML, which creates the VDW/VODC logical environment, are key elements of this technology. Thus, the initial steps should be devoted to the development of these key elements.

Part X

CONCLUSIONS AND RECOMMENDATIONS

CONCLUSIONS AND RECOMMENDATIONS

N. B. HARMANCIOGLU[1], P.GEERDERS[2], and O. FISTIKOGLU[1]
[1]*Dokuz Eylul University, Faculty of Engineering*
Tinaztepe Campus, Buca 35160 Izmir, Turkey

[2]*P. Geerders Consultancy*
Kobaltpad 16-3402 JL Ijsselstein, The Netherlands

1. Integration in Environmental Data Management Systems

As noted earlier in Part I of this volume, the keyword in the designated objectives of the 1996 and 2001 NATO Advanced Research Workshops (ARW) is integration. The major diagnosis made at the recent ARW is that advances in global environmental and water resources management are not primarily limited by a lack of data and information but by a lack of proper data and information management. At present, there exist huge amounts of different types of environmental data, which are not merged on a routine basis for the effective production of information. Modern data management offers various ways and tools, such as geographic information systems, to reduce, condense, integrate and analyze such data. Furthermore, modern data types are not limited to routine ground-based observations; they include new data types such as remote sensing data from satellites and airborne platforms and data from real-time sensors and systems, producing high volumes of data. Moreover, numerical models provide another powerful source of data, especially for forecasts and simulation. The availability of such data and the advances in data collection technologies has increased the need for “integration” in EDM systems (see Parts I and V).

It follows from the above that the essence of the problem lies basically in inadequate data and information management rather than in a lack of data. There are further impacts of poor data management, including (see Part I):

- ineffective exchange of knowledge;
- potential loss of valuable historical data;
- significant amounts of redundant work involved in information production;
- lack of efficiency in assembling the relevant information required for the solution of a given environmental problem;
- increased budget required for data organization in particular projects or programmes.

Integrated data management is not solely required for scientific and technical purposes. It is the basis for environmental decision making where community participation has become a significant component. Thus, production of sound information on environmental problems should also serve to inform the public in order to broaden the basis for the decision making process.

2. General Results

Several subgroups were formed at the 2001 ARW in order to discuss specific topics of environmental data management, to reach related conclusions and to make specific recommendations. The results of their discussions show that integrated environmental data management still mainly belongs to the academic context and that practical implementation of the related principles by environmental projects, including integrated environment management, is rare. This is largely due to the fact that the community of potential users is hardly aware of the considerations and recommendations made by meetings and groups such as the recent NATO ARW.

As a consequence, environment related projects still often fail to establish sound data and information systems, as a basis for their activities. This specifically relates to aspects such as the standardization of measurements and observations, to the usage of a common terminology, to a definition of the form and format of data and information products, and to the management of the data and information archives, including their long-term accessibility. Lacking awareness on existing, recommended and proven methodologies, references and systems often leads to "reinventing the wheel", causing an unnecessary expenditure of manpower and resources, as well as leading to a lack of interoperability between data and information from different sources.

Nevertheless, the demand for integrated management of the environment and natural resources leads to the requirement for integrated, multidisciplinary investigations, including that for specific measurements and observations. An integrated approach to data and information management forms the road towards a scientifically responsible merging of data and information into dedicated, application-oriented results for science and for decision making.

The participants of the 2001 ARW concluded that, at this moment, time has come to actively pursue and promote the application of the considerations and recommendations in the context of existing and planned environmental programmes and projects. The participants agreed that this goal could best be reached through the following CONCRETE ACTIONS:

- promote the conclusions of the 2001 and the preceding 1996 workshops through a ***website*** dedicated to integrated environmental data and information management topics;
- include on this website ***links to sources of reference information*** for potential users on existing, recommended and proven methodologies, references and systems for integrated management of environmental data and information;
- ***inform*** relevant regional and global organizations, projects and programmes (including EU, EEA, UNESCO/IOC and UNESCO/IHP, WMO, FAO, GBIF, ESA, CEOS, UNEP, ...) on the conclusions and recommendations of this and preceding ARW, and ***invite their cooperation*** in reaching a global consensus on integrated management of environmental data and information;
- ***promote and facilitate*** concrete applications of integrated EDM in the context of regional (e.g., Mediterranean) or international (e.g., EU, UNESCO/IOC, WMO, etc.) programmes and projects;

- establish related to the website, an ***e-mail forum*** or similar mechanism for exchange and promotion of views and ideas on various aspects of integrated environmental data and information management;
- establish an ***international society or association*** on integrated environmental data and information management, as a platform for the coordination of the relevant developments, for the implementation of awareness raising efforts and for ensuring the continuing attention for this specific and highly important subject.

Specific efforts and issues related to these actions, should be considered and implemented in the context of a follow-up meeting to be held as soon as possible to further elaborate on CONCRETE ACTIONS.

3. Conclusions and Recommendations on Monitoring and Sampling

One of the group discussions at the recent 2001 ARW was devoted to monitoring and sampling, including physics and statistics of sampling, instrumental aspects, design of monitoring networks, data quality, remote sensing, and environmental modeling. The conclusions and recommendations derived as a result of this group work are summarized in the following.

For rational management of the environment, it is necessary to understand environmental processes. Thus, it is believed that monitoring systems need to reflect the relationship between data collection, models, and science This implies a rethinking of environmental monitoring and a move to connecting the data collection process to network design and to process understanding. To achieve this general objective, it is necessary to define specific goals for a new monitoring and sampling framework, namely that there are feedback and feed-forward connections between these traditionally independent elements (Fig. 1). Monitoring networks thus coupled will better support decision making, from basic engineering to sustainable development based upon scientific understanding.

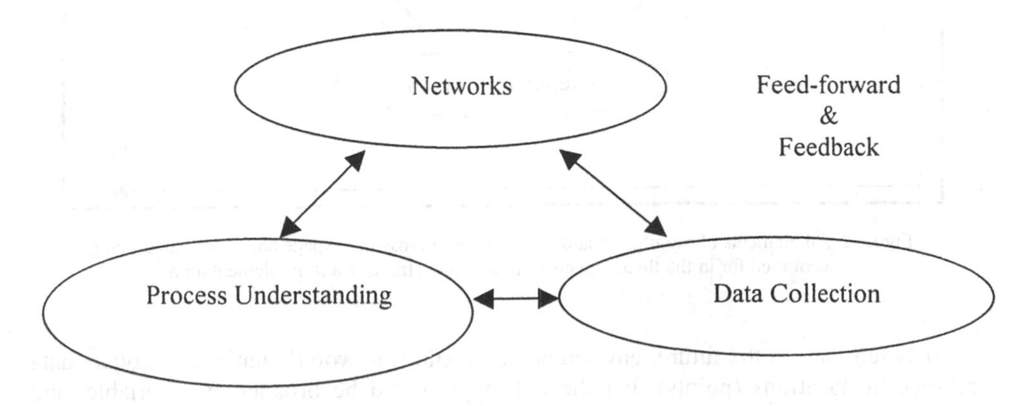

Figure 1. Structure of Coupled Network.

Such networks should be developed on an interdisciplinary basis and engage different communities – particularly the social sciences, economics and environmental sciences. The interchange between these broad communities is essential to ensure that communication and understanding is shared. Successful environmental management cannot be based upon a situation where one community builds the agenda and excludes the others. It must also be stressed that being interdisciplinary does not mean generalist: communication and sharing among specialists is essential.

The view of monitoring, particularly environmental monitoring continues to evolve. Frequently, monitoring was established as making repeated measures on an ongoing basis. Later in time, there was a shift to an understanding that the data monitoring provided should be designed to meet a goal or objective. Such objectives might include [1] developing an academic science understanding of process or causal mechanism; [2] supporting environmental, social and economic decision making, and [3] providing some form of contingency against unforeseen change. Of particular interest is that, at the 1996 ARW, these three alternatives were viewed as "or"; but in 2001 and forwards, they are seen as "and" (i.e. – not mutually exclusive). Another change in perspective is the development of a broader understanding of scales – both time and space. There is a need to appreciate the scope of an environmental process and how it may, or may not be, scalable in space and/or in time (Fig. 2).

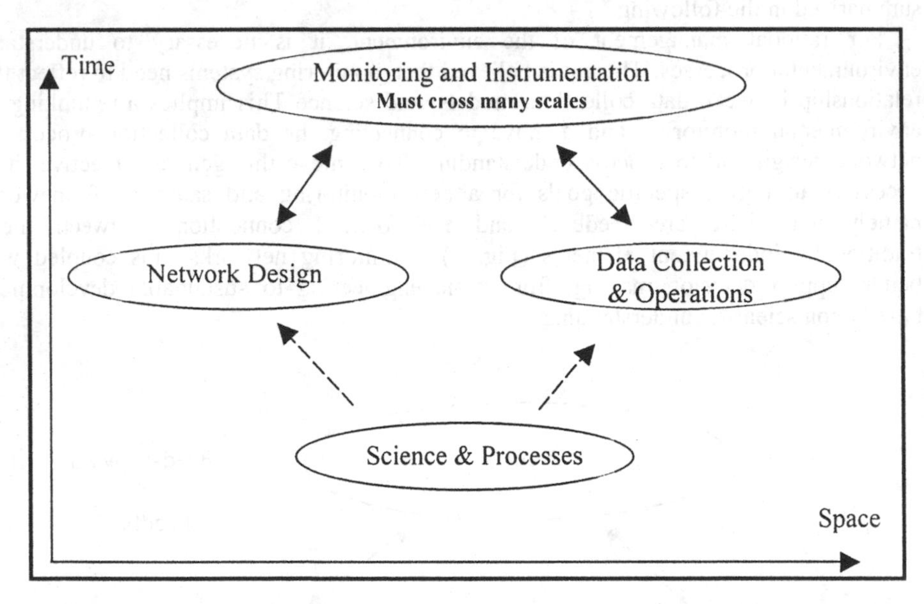

Figure 2. All elements of monitoring and sampling must cross both space and time scales, being accounted for in the three aspects of monitoring structure and implementation.

It is felt that, in the future, environmental monitoring would continue to collect data at specific locations (points); but the data type would be broader multivariable, and variables will be monitored, using a sufficient number of robust instruments covering the spatial and temporal scales required to fulfill the objectives at local, regional, and global levels. It is also expected that monitoring systems will include (automatic) tools for quality assurance and quality control. At present, there is often a considerable delay between the observation itself and its availability from operational databases. There is also a lack of worldwide standards for data acquisition, quality control and data handling. We are frequently faced with difficulties of understanding since we have not made a practice of relating process-related measurements to environmental monitoring. Examples are: micro scale and macro pore measurements, measurements of pipe flows and of floodplain velocities. Solutions to these difficulties would include: establishing better links between researchers, manufacturers and users; developing an improved synthesis of monitoring at alternate scales; and an improved balance between quality and quantity observations, especially for time and space dimensions of water quality, e.g., river mass loads. It is also recognized that implementation of these solutions requires improving understanding and awareness at political levels.

The future design of environmental monitoring networks will be based upon scientifically based guidelines and will involve social and economic considerations. Largely due to their scientific background and origin, present networks are ad-hoc and lack exchangeability. At the same time, many countries are reducing their investment in monitoring networks. This will make it more difficult to manage and solve environmental problems, to assess situations, and to simulate outcomes of measures and actions. The solution to this dilemma is to move towards a design that will be based upon scientifically based principles, guidelines, and that will include social and economic aspects. The group recommended to adopt international guidelines where possible and develop new (but compatible) ones where needed; to commit to maintain existing key monitoring stations; and to support the establishment of robust monitoring networks in developing countries.

Data quality continues to be an underlying difficulty in many environmental monitoring programs. It is suggested that, in the future, we will see quality assurance as an integral part of the monitoring process. As a result, the data will include documentation, and will adhere to established standards and guidelines, thus improving transferability and compatibility of the data. One part of the data quality problem is the lack of a common dictionary of definitions. Presently, environmental monitoring networks are often deficient in space and time scales, and data from one location is incompatible with other locations, even within the same jurisdiction. The path to a solution is to make the data generated by networks to be of a known quality.

Remote Sensing technology has a strong potential to provide a linkage between environmental monitoring programs. However, space agencies should develop a broader understanding of the requirements of environmental applications, which should result in specific, dedicated space sensors. Presently, the sensors that are deployed are often inappropriate for environmental applications, and the group felt that, often, the choice of sensors was based upon strong lobby groups rather than upon scientific considerations. Environmental applications demand an improved coverage in time (more frequent), space (global), spectral bands (hyper-spectral) and resolution (more detail). The user community would be able to benefit more from space data, if the data would be made available more widely. The environmental monitoring community should seek a more appropriate cost recovery basis for users.

Efforts need to be made to ensure that, in the future, modeling should become an integral and interactive portion of network design and consistent with the monitoring and sampling scales. Presently, there is a lack of consistency between model scales and sampling scales. Models are often developed to meet operational needs and then cannot be calibrated to accommodate other environmental processes. The solution to this problem could be to ensure that models are built to fit a specific purpose. This aspect needs to be integrated into the development of international guidelines.

We need to make better use of the basic sciences that environmental science is built upon: physics, chemistry and ecology. We should better integrate science into the network design in order to better interpret the data obtained and to feedback to the design. Frequently, networks are designed without consideration of these basic sciences. This takes place in countries of every level of development, and the resulting data are often not interpretable. Principles and practices of network design such as those presented and discussed at this workshop may form the basis for generic guidelines.

4. Conclusions on User Aspects

4.1. MAJOR ISSUES

At the 2001 ARW, another group discussion was held on user aspects. The main topic addressing the user perspective was divided in a number of individual subtopics, namely:

- Awareness on data availability
- Data availability
- Data quality
- Access to data
- Data products
- GIS
- Special Interest group
- Actions

For each sub-topic, the group tried to address the status, future trends, problems and solutions as well as suggestions for specific actions where appropriate.

4.2. AWARENESS (ON DATA AVAILABILITY)

The following comments were made, regarding awareness on data availability:

- STATUS: while there is a vast and growing volume of environmental and water resources data available in principle, there is limited awareness on what is available where and under which conditions.
- FUTURE: with more and more data becoming available, there is no indication that awareness by potential users is improving.
- PROBLEMS: the main problems identified are missing metadata, language problems, and the lack of dedicated information sources. There is neither an established market place nor a brokerage service for this type of data.

- SOLUTION: the solution is obviously to improve awareness by dedicated information products, better use of search engines, but also a better presentation and publishing of data holdings. An ideal solution would be to institute a Single Point of Contact (SPOC), combining e.g., classical approaches like an international journal with a web portal, news group, mailing lists etc. This raises the question of funding and long-term economic viability of such an institution or service.

4.3. DATA AVAILABILITY

Regarding data availability, the group reported the following:

- STATUS: data availability is characterized between the extremes of "too much" and "not enough", for any specific requirement.
- FUTURE: this is not likely to get better, as there is no clear relationship between data acquisition on the one hand and the requirements for data on the other.
- PROBLEMS: the mismatch of data versus *specific* problems is a principle problem. Any data collection procedure is based on specific needs, and needs is changing as new problems arise. Specific needs are unpredictable and depend very much on users and their specific problems. A complete generic coverage in terms of parameters (spatial and temporal coverage) is simply infeasible with current technology (and funding imaginable). At the 2001 ARW, the issue of cost was discussed (see below), but no agreement found; the participants found themselves split between proponents of free access to environmental information and those favoring a market approach, each with its advantages and disadvantages.
- SOLUTION: a satisfactory solution is difficult at a generic level. A minimal step in the right direction is to ensure that nothing gets lost. Another solution may be found in flexible and adaptive collection strategies (nanotechnology, autonomous microprobes, sensors on every airplane), which, however, cannot solve the problem of historical and baseline data requirements.
- ACTION: a specific recommendation includes the improvement of archiving. Environmental data should be considered potentially invaluable assets, and a world heritage just like landscape and architecture (which would put them under the auspices of organizations such as UNESCO, but raises the issues of institutional responsibility: WDC, UN/GRID, or the national level).

4.4. DATA QUALITY

The following conclusions are drawn, regarding data quality:

- STATUS: data quality is at best variable, mostly unknown, not an integrated part of many data sets, which is true in particular for many data sources on the Internet and older historical data.
- FUTURE: data quality is slowly improving (in terms of awareness, data quality standards, sensor technology, spatial and temporal coverage), but that the main issue is institutional if not personal rather than technical.

- PROBLEMS: while most valuable for any long-term and comparative studies, historical data often lacks explicit quality standards; in all domains, we find a lack of consistent metadata.
- SOLUTION: standardization is an obvious approach, which raises the question whether there is an appropriate standardization body (like ISO) for environmental data.
- ACTION: there is ample room for activities by professional societies and national groups to work toward standardization and creation of a better awareness on data quality issues, and on existing procedures and guidelines for quality control and assessment.

4.5. ACCESS TO DATA

Comments regarding access to data are:

- STATUS: again this varies widely from public domain to proprietary and to classified; from real-time to delayed publication; from electronic to paper formats.
- FUTURE: with more data becoming available, access should be generally improving; at the same time, reduced public funding and the need for institutional cost recovery introduces new economic barriers in many fields.
- PROBLEMS: the main access limitations are still institutional. The aspect of costs, publicly funded versus market prices, i.e., "who should pay", was discussed; but, at the 2001 ARW, no satisfactory solution found.

4.6. DATA PRODUCTS

Conclusions regarding data products are the following:

- STATUS: here, the assessment depends again mainly on the specific problem. Data products require a market (or regulatory mandate) to be viable in the long-term, so the issue is primarily one of costs.
- STATUS: good (for presentation and tools, in part thanks to the Internet); but, again, user and problem dependent.
- FUTURE: the future may see even more improvements with increasing demand (a larger market) and, thus, more potential economic incentives.
- PROBLEMS: are obviously user and problem dependent, as affordable products imply a larger group of users with common needs.

4.7. GEOGRAPHIC INFORMATION SYSTEMS

Geographic Information Systems are just one of the data management and analysis technologies available, but provide a powerful paradigm of the thematic map.

- STATUS: the status of current GIS is good; it is a generally available enabling technology.

- FUTURE: improving based on a dynamic industry.
- PROBLEMS: problems are mainly related to some basic limitations with regard to the temporal dimension (4D GIS) and the generally limited analytical capabilities for more complex analysis and modeling tasks.
- SOLUTION: an obvious and simple solution is to wait for the industry to develop missing functionality. The alternative is a do-it-yourself approach (using, for example, open source software tools, e.g., GRASS is available in source code for individual development and adaptations).

4.8. IMPLEMENTATION OF PROPOSED ACTIONS

In paragraph 2 of section 1, several concrete follow-up actions were proposed. These could form a new approach to a better coordination of demand for and supply of environmental data. Some of these actions, such as a website and a portal, could be implemented between the participants of the workshop, possibly as additional elements of their already existing facilities. However the establishment and professional management of an international society or association on integrated environmental data management would require additional resources and funding.

The current EU Framework 6 Programme provides a mechanism to set up and maintain, at least for several years, an international "Network of Excellence" on a specific subject. Due to new regulations, besides the EU countries, the Mediterranean countries and countries from Eastern Europe can participate fully in this network. Such a network could be used to further develop the issue of integrated environmental data management towards a set of concrete, international guidelines, possibly accompanied by several pilots or demonstrations. The network could, in due time, establish a solid team of partners for the submission of large and long-term Research Projects for FP6.

SUBJECT INDEX

— D —

— E —

—H—

—I—

— J —

— K —

—L—

—M—

LIST OF ABBREVIATIONS AND ACRONYMS

ADB	Archived Data Bank
AMSR	Advanced Microwave Scanning Radiometer
ANN	Artificial Neural Networks
ANNGIE	Artificial Neural Networks Graphical Integrated Environment
APWL	Accumulated Potential Water Loss
AQUASTAT	Global information system of water and agriculture developed by the Land and Water Development Division of FAO
ASAR	Advanced Synthetic Aperture Radar
ASTER	Advanced Space-borne Thermal Emission & Reflection Radiometer
AVHRR	Advanced Very High Resolution Radiometer
AVN	AViatioN model
AWIPS	Advanced Weather Interactive Processing System
CDW	Centralized Data Warehouse
CEH	Centre for Ecology and Hydrology
CEOS	Committee on Earth Observing Satellites
CFMPs	Catchment Flood Management Plans
CMIS	Conical MW Imager Sounder
COADS	Comprehensive Ocean-Atmosphere Data Set
CSDA	Confirmatory or exploratory Spatial Data Analysis
CZCS	Coastal Zone Color Scanner
DAAC	Distributed Active Archive Center of Goddard Space Flight Center
DATAGRID	MM5 modeling system
DAQ	Data AcQuisition software
DBMS	DataBase Management Systems
DCMI	Dublin Core Metadata Initiative
DEM	Digital Elevation Models
DETR	UK Department of Environment, Transport and Regions
DHTML	Dynamic HyperText Markup Language
DMSP	Military satellite of USA
DoENI	Northern Irish Department of the Environment
DOM	Document Object Model
DSS	Decision Support Systems
DTLR	UK Department for Transport, Local Government and the Regions
DTM	Digital Terrain Models
EA	Environment Agency
ECMWF	European Centre for Medium-Range Weather Forecast
EDAS	Eta Data Assimilation System
EDM	Environmental Data Management
EDSS	Environmental Information and Decision Support Systems

EEA	European Environment Agency
EIA	Environmental Impact Assessment

TENTS

HEC-HMS	Hydrologic Modeling System designed to simulate the precipitation-runoff processes of dendritic watershed systems (Hydrologic Engineering Center)
HMS	Harmonized Monitoring Scheme
HRU	Hydrological Response Units
HSU	Hydrological Similar Units
HTML	HyperText Markup Language
HWRP	Hydrology and Water Resources Programme (WMO)
HYCOSs	Regional Hydrological Cycle Observing Systems
ICES	International Council for the Exploration of the Seas
IDB	Integrated Data Bank
IGBP	International Geosphere-Biosphere Programme
IGRAC	International Groundwater Resources Assessment Centre
INAG	Portuguese Water Institute
IIT	Integrated Information Technology
IKONOS	Remote sensing satellite
INTERREG IIC	North Sea Region Programme of EU
IOC	Intergovernmental Oceanographic Commission of UNESCO
IOCARIBE	Caribbean and adjacent regions (IOC)
IOCEA	Central Eastern Atlantic - 'Western Africa' region (IOC)
IOCINCWIO	North and Central Western Indian Ocean - 'Eastern Africa' region (IOC)
IOCINDIO	Central Indian Ocean region (IOC)
IODE	International Oceanographic Data and Information Exchange program (IOC)
IOWDM	Watershed Data Management program of US Geological Survey
ISO	International Standard Organization
ISP	Internet Service Providers
IST	Information System Team
IWRA	International Water Resources Association
KADA	Kelantan State Development Authority
KDD	Knowledge Discovery in Databases
LAN	Local Area Network
Landsat	remote sensing satellite
LAs	Local Authorities
LDA	Linear Diffusion Analogy Model
LOICZ	Land-Ocean Interactions in the Coastal Zone
LOIS	UK Land-Ocean Interaction Study
LRF	Linear Rapid Flow Model
LVQ	Learning Vector Quantization
MEDI	Metadata system of IODE
MERIS	Medium Resolution Imaging Spectrometer
METEOSAT	geo-stationary satellite
MHICrimea	Marine Hydrophysical Institute of Ukraine
MLM	Maximum Likelihood Method
MME	Maximum Entropy

MMS	Modular Modeling System
MODIS	Moderate Resolution Imaging Radio Spectrometer
MOM	Method Of Moments
NARI	Network Analysis for Regional Information
NCAR	National Center for Atmospheric Research
NCEP	National Center for Environmental Prediction
NDVI	Normalized Difference Vegetation Index
NERSC	Nansen Environmental and Remote Sensing Center
NIERSC	Nansen International Environmental and Remote Sensing Center
NNs	Neural Networks
NOAA	National Ocean and Atmospheric Administration, USA
NODC	National Oceanographic Data Centre of Russia
NRFA	National River Flow Archive
ODIN	Ocean Data and Information Network capacity
OECD	Organization for Economical Cooperation and Development
OFWAT	UK Government's Office for Water Services
OLE 2.0	Microsoft's Object Linking and Embedding
OLS	Ordinary Least Squares
OSPAR	Oslo and Paris Commission
PCA	Principal Component Analysis
PMW	Passive Microwave
POA	Problem-Oriented Applications
PWM	Probability Weighted Moments
QPF	Quantitative Precipitation Forecast
RMSE	Root Mean Square Errors
RV	Research Vessels
SAR	Synthetic Aperture Radar
SEPA	Scottish Environment Protection Agency
SIMN	Servizio Idrografico e Mareografico Nazionale
SIRIUS	Saarbrücken Information Retrieval and Interchange Utility Set
SOM	Kohonen's Self-Organizing Maps
SPI	Standardized Precipitation Index
SPOT	Remote sensing satellite
SRM	Snowmelt Runoff Model
SSM/I	passive microwave sensor of DMSP, USA
SWE	Snow Water Equivalent
TCP/IP	W3C standard protocol
TLRN	Time-Lagged Recurrent Network
TOPEX/POSEIDON	satellite
TOPMODEL	rainfall-runoff model
UDMs	Urban Drainage Models
UIR	Unit Impulse Response Function
UN-ACC/SCWR	UN Administrative Committee on Coordination Subcommittee on Water Resources
UNCED	The United Nations Conference on Environment and Development

UNEP	United Nations Environment Programme
UNESCO	United Nations Educational, Scientific and Cultural Organization
UNESCO/IHP	International Hydrological Programme of UNESCO
UNESCO/IOC	Intergovernmental Oceanographic Commission of UNESCO
UPM	Urban Pollution Management
URGENT	UK Natural Environment Research Council's urban environment programme
USACE	United States Army Corps of Engineers
USGS	US Geological Survey
VCL	Vegetation Canopy Lidar
VDW	Virtual Data Warehouse
VRML	Virtual Reality Markup Language
W3C	World Wide Web Consortium
WDC	World Data Centre
WESTPAC	Western Pacific region (IOC)
WFD	EU Water Framework Directive
WHO	World Health Organization
WHYCOS	World Hydrological Cycle Observing System
WIS	Water Information System
WLS	Weighted Least Squares
WMO	World Meteorological Organization
WRCP	World Climate Programme
WSCs	Regional Water Service Companies
WWAP	World Water Assessment Programme
WWDR	World Water Development Report
WWW	World Weather Watch
WWW	World Wide Web
XML	Extensible Markup Language